Proton Transfer Reaction Mass Spectrometry

Proton Transfer Reaction Mass Spectrometry

Principles and Applications

ANDREW M. ELLIS
Department of Chemistry, University of Leicester, UK

CHRISTOPHER A. MAYHEW
School of Physics & Astronomy, University of Birmingham, UK

WILEY

This edition first published 2014

Registered office
John Wiley & Sons Ltd, The Atrium, Southern Gate, Chichester, West Sussex, PO19 8SQ, United Kingdom

For details of our global editorial offices, for customer services and for information about how to apply for permission to reuse the copyright material in this book please see our website at www.wiley.com.

Library of Congress Cataloging-in-Publication Data

Ellis, Andrew M. (Andrew Michael), 1963–
Proton transfer reaction mass spectrometry : principles and applications / Andrew M. Ellis, Christopher A. Mayhew.
pages cm.
Includes index.
ISBN 978-1-4051-7668-2 (cloth)
1. Proton transfer reactions. 2. Mass spectrometry. I. Mayhew, Christopher A. II. Title.
QD501.E545 2014
539′.60287–dc23

2013028864

A catalogue record for this book is available from the British Library.

Set in 10/12pt Times by Aptara Inc., New Delhi, India.
Printed and bound in Malaysia by Vivar Printing Sdn Bhd

1 2014

Contents

Preface **xiii**

SECTION 1 PRINCIPLES

1 Background **3**
1.1 Volatile Organic Compounds in the Earth's Atmosphere 3
1.2 Volatile Organic Compounds in Other Environments 5
1.3 Techniques for VOC Measurements 6
1.3.1 Gas Chromatography 6
1.3.2 Ion Mobility Spectrometry 9
1.3.3 The Flowing Afterglow Technique 11
1.3.4 The Selected Ion Flow Tube 14
1.4 Emergence of Proton Transfer Reaction Mass Spectrometry 15
1.4.1 Historical Background 15
1.4.2 Compound Identification Using PTR-MS 17
1.4.3 An Introduction to Quantitative Aspects of PTR-MS 21
1.4.4 A Comparison between PTR-MS and SIFT-MS 22
References 23

2 Chemical Ionization: Chemistry, Thermodynamics and Kinetics **25**
2.1 Introduction 25
2.2 Proton Transfer 27
2.2.1 Energy Units 27
2.2.2 Thermodynamics of Proton Transfer 27
2.2.3 Kinetics of Proton Transfer 31
2.2.3.1 Background 31
2.2.3.2 Theoretical Prediction of Proton Transfer Rate Coefficients 33
2.2.3.3 Illustrative Calculations of Proton Transfer Rate Coefficients and Comparison with Experiment 37
2.2.4 Reagents and Mechanisms 38
2.2.4.1 Chemistry of H_3O^+ Reactions 38
2.2.4.2 Reactions of Hydrated Hydronium Clusters 42
2.2.4.3 Alternative Proton Donors 43
2.3 Other Chemical Ionization Processes 44
References 45

3 Experimental: Components and Principles **49**
3.1 Introduction 49
3.2 Ion Extraction and Ion Optics 50
3.2.1 Ion Acceleration 51
3.2.2 Ion Steering 53
3.2.3 Ion Lenses 54
3.2.4 Simulation of Ion Trajectories 56
3.3 Ion Sources 57
3.3.1 Hollow Cathode Discharge Ion Source 57
3.3.2 Ion–Molecule Chemistry Leading to H_3O^+ Production 59
3.3.3 Alternative Ion Sources 61
3.3.4 Generating Reagent Ions Other Than H_3O^+ 63
3.4 Drift Tubes 64
3.4.1 Practical Aspects 64
3.4.2 Ion Mobility and Transit Times 69
3.4.3 Ion–Molecule Collision Energies 71
3.4.4 Ion Cluster Distributions 73
3.5 Mass Spectrometry 76
3.5.1 Some Important Definitions 77
3.5.1.1 Ion Mass and Mass-to-Charge Ratio 77
3.5.1.2 Mass Resolution 78
3.5.1.3 Transmission and Dynamic Range 79
3.5.2 Quadrupole Mass Spectrometry 81
3.5.2.1 Basic Principles of the Quadrupole Mass Spectrometer 81
3.5.2.2 Practical Issues 83
3.5.3 Quadrupole Ion Trap Mass Spectrometry 85
3.5.3.1 Basic Principles 85
3.5.3.2 Collision-Induced Dissociation 87
3.5.3.3 Three-Dimensional Quadrupole Ion Traps in PTR-MS 88
3.5.3.4 The Linear Ion Trap in PTR-MS 90
3.5.4 Time-of-flight Mass Spectrometry 90
3.5.4.1 Basic Principles of TOF-MS 90
3.5.4.2 Improving the Resolution: Spatial Focusing 92
3.5.4.3 Reflectron TOF-MS 93
3.5.4.4 Mass Calibration in TOF-MS 94
3.5.4.5 Advantages and Limitations of TOF-MS 95
3.5.4.6 TOF-MS Analysers in PTR-MS 96
3.6 Ion Detectors 97
3.6.1 Discrete Dynode Detector 98
3.6.2 Channel Electron Multiplier 100
3.6.3 Microchannel Plate Detector 101
3.7 Analogue versus Digital Signal Processing 103
References 106

4 Quantitative Analysis **111**
4.1 Introduction 111
4.2 Extracting the Concentration of a Trace Gas from PTR-MS 111

4.3 Normalized Counts per Second 113
4.4 Why Calibrate? 113
4.5 Calibration Techniques 116
4.5.1 Static Gas Calibration 116
4.5.2 Dynamic Methods 117
4.5.3 Alternative Dynamic Calibration Procedures 119
4.6 Effect of Humidity 120
4.7 Accuracy, Precision and Limit of Detection 122
4.8 Validation of PTR-MS 125
References 126

SECTION 2 APPLICATIONS

5 PTR-MS in the Environmental Sciences 131
5.1 Background 131
5.2 Use of Reagent Ions Other Than H_3O^+ 138
5.3 Biogenic VOCs 141
5.3.1 General Details 141
5.3.2 Forest Emissions 142
5.3.2.1 Tropical Rainforests 142
5.3.2.2 Coniferous Forests 144
5.3.2.3 Deciduous Forests 146
5.3.2.4 Eddy Covariance Measuring Methodologies 147
5.3.2.5 Forest VOCs and *m/z* Assignments 150
5.3.3 Plantations 150
5.3.4 Various Land Emissions 151
5.3.4.1 Woodland and Grassland Savannahs 151
5.3.4.2 Shrubland 152
5.3.4.3 Alfalfa and Grass Fields 152
5.3.5 Oceans and Seas 154
5.3.5.1 Norwegian Fjord 154
5.3.5.2 Coastal Regions 155
5.3.5.3 Indian Ocean 156
5.3.5.4 Tropical Atlantic Ocean 156
5.4 Anthropogenic VOCs 157
5.4.1 Background 157
5.4.2 VOCs in Urban and Rural Sites 157
5.4.2.1 Innsbruck 157
5.4.2.2 Caracas 158
5.4.2.3 Houston 158
5.4.2.4 Tokyo 159
5.4.2.5 Barcelona 159
5.4.2.6 Manchester and London 160
5.4.2.7 Mexico City 160
5.4.2.8 Toronto and Environs 162
5.4.2.9 Paris 163
5.4.2.10 Boston, New York and Los Angeles 163

5.4.3 Diesel Engine Emissions 164
5.4.4 Aircraft Emissions 164
5.4.5 VOC Emissions Associated with Farming 164
5.4.5.1 Cattle 165
5.4.5.2 Pigs and Sheep 165
5.4.6 Other Studies of Anthropogenic VOCs 166
5.4.6.1 Air Quality 166
5.4.6.2 Firework Emissions 166
5.5 Biomass Burning 166
5.6 Applications of PTR-MS to Laboratory Studies of Atmospheric Chemistry 169
5.6.1 Laboratory Studies of Biomass Burning 171
5.6.2 Reaction Products and Reactive Species 173
5.6.3 Simulation Chamber and Container Measurements 176
5.7 Plant Studies 181
5.7.1 Isoprene Emissions 182
5.7.2 Acetaldehyde Emissions 185
5.7.3 Pollination 187
5.7.4 Roots and Soil 188
5.7.5 Other Plant Studies 189
5.7.5.1 Root-secreted VOCs 189
5.7.5.2 Methanol Release and Bacterial Growth: Plant-Methylobacterium Association 190
5.7.5.3 Comparison of VOC Emissions from Young and Mature Leaves 190
5.7.6 Stress-Related Emissions 190
5.7.7 VOC Emissions Linked to Plant Damage 193
5.7.7.1 Mechanical Wounding 193
5.7.7.2 Weather Damage 194
5.7.7.3 Harvesting and Mowing 195
5.7.7.4 Biofuel Crops 196
5.7.7.5 Herbivore Attack by Small Predators 197
5.7.7.6 Large Herbivore Attack 202
5.7.8 VOC Uptake by Plants 202
5.8 Outlook for Atmospheric and Environmental Applications of PTR-MS 203
References 203

6 PTR-MS in the Food Sciences 221
6.1 Background 221
6.2 Combined GC–MS and PTR-MS Studies for Food Analysis 223
6.3 Mass Spectral Fingerprinting 226
6.4 Flavour Release and Perception 227
6.4.1 Drinks 228
6.4.1.1 Coffee 228
6.4.1.2 Tea 231

6.4.1.3 Carbonated Drinks 232
6.4.1.4 Fruit Juices 233
6.4.1.5 Wine 233
6.4.1.6 Vodka 234
6.4.1.7 Infant Formula 235
6.4.2 Food 235
6.4.2.1 Cheese 235
6.4.2.2 Bread 237
6.4.2.3 Onions 237
6.4.2.4 Wheys 237
6.4.2.5 Fruit 238
6.4.3 Flavour Release: Food Texture, Composition and Physiological Effects 240
6.5 Food Classification, Food Quality and Food Control 245
6.5.1 Geographical Location 245
6.5.1.1 White Truffles 245
6.5.1.2 Butter 246
6.5.1.3 Olive Oil 247
6.5.1.4 Roe 247
6.5.1.5 Dry-Cured Ham 247
6.5.1.6 Cumin Cheese 248
6.5.2 Food Classification and Quality 249
6.5.3 Food Freshness and Ripening 250
6.5.3.1 Meat Degradation 250
6.5.3.2 Fruit and Vegetables: Ripening, Storage and Monitoring 251
6.5.3.3 Ripening of Cheese 253
6.5.4 Process Monitoring and Biochemical Processing 253
6.6 Outlook for Food Science and Technology Applications 256
References 257

7 PTR-MS in the Medical Sciences **267**
7.1 Background 267
7.2 Breath Analysis 268
7.2.1 Smoking and Breath Volatiles 271
7.2.2 Isoprene in Breath 272
7.2.3 Acetone in Breath 275
7.2.4 Lung Studies: Cancer and Emphysema 276
7.2.5 Other PTR-MS Breath Studies 278
7.2.5.1 Crohn's Disease and Ulcerative Colitis 278
7.2.5.2 Carbohydrate Malabsorption 278
7.2.5.3 High Mass-Resolution PTR-TOF-MS Breath Studies 278
7.2.5.4 Kidney Function and PTR-MS 280
7.2.5.5 Liver Disease 280
7.2.6 Drug Monitoring and Pharmacokinetics Using Breath Analysis and PTR-MS 281

7.2.7 Breath VOC Levels Measured Using PTR-MS versus Blood Concentrations 284
7.2.8 Breath Sampling and PTR-MS 285
7.2.8.1 Offline Breath Sampling 286
7.2.8.2 Online Breath Sampling 287
7.2.9 PTR-MS and Breath Analysis: Requirements and Future Directions 287
7.3 Online PTR-MS Measurements of Volatile Emissions from Microbial Cultures 290
7.3.1 Bacteria 290
7.3.2 VOC Emissions from Fungi 296
7.3.3 Concluding Remarks on Microbial Emissions 297
7.4 Other Medical Applications 297
7.4.1 Urine Headspace Analysis 301
7.4.2 Skin Emissions 301
7.4.3 VOC Emissions from Human Cells 301
7.4.4 VOCs in Clinical Environments 302
References 302

8 Applications of PTR-MS to Homeland Security: The Detection of Threat Agents 311
8.1 Background 311
8.2 Explosives 312
8.2.1 Forensic Issues 312
8.2.1.1 The Unambiguous Detection of TNT 315
8.2.1.2 High Mass Resolution PTR-TOF-MS Measurements of TNT 318
8.2.1.3 Reagent Ion Switching and Explosives Detection 319
8.2.1.4 PTR-MS and the Detection of Traces of Explosives 320
8.2.2 Environmental Aspects and Explosives 320
8.3 Chemical Warfare Agents and Toxic Industrial Chemicals 321
8.4 Narcotics 322
8.5 Date Rape Drugs 325
8.6 Ion Mobility Mass Spectrometry and PTR-MS: A Brief Comparison for Homeland Security Applications 326
8.7 Future Directions 327
References 328

9 Liquid Analysis Using PTR-MS 331
9.1 Determination of Henry's Law Constants Using PTR-MS 331
9.2 Analysis of Liquids 333
References 336

***Index* 337**

Nothing tends so much to the advancement of knowledge as the application of a new instrument.

Sir Humphry Davy

Preface

Proton transfer reaction mass spectrometry (PTR-MS) is widely used for the detection of volatile organic compounds in air. The historical development of this technique can be traced back to measurements of the rates of ion–molecule reactions in the 1960s and 1970s using flow tube methods. The technology introduced for measuring the kinetics of these reactions morphed into the analytical technique we now know as PTR-MS in the mid-1990s. The seminal work in this area was performed by Werner Lindinger and his team at the University of Innsbruck.

Remarkably, the development of PTR-MS by the Innsbruck group very nearly did not happen! Early requests for funding of this work were rejected by various research agencies, as the acknowledgements in some of the early PTR-MS research papers forcibly testify. It was only through Lindinger's perseverance, support from Fa. Nestle (Nestec Ltd, Switzerland) and initial funding from the "*Gesellschaft für Strahlenforschung*" (GSF, Neuherberg, BRD), that made it possible for the developmental work on PTR-MS to be pursued. With limited resources, and together with Armin Hansel, Alfons Jordan and other colleagues, PTR-MS as we know it today began to emerge. This developmental work was followed up with pioneering applications in the environmental sciences, food science and medicine, all in collaboration with colleagues at Innsbruck and external establishments. By the time of his untimely death in 2001, Werner Lindinger had demonstrated the applicability of PTR-MS as an analytical tool for monitoring trace gases in several different environments and had established a spin-out company for the commercial exploitation of this novel technology.

The initial growth of PTR-MS research was so fast that by the beginning of the twenty-first century it warranted its own international conference. The first took place in 2003 in Innsbruck and it has now become a regular event taking place every two years, usually in Obergurgl, Austria. These conferences are a testament to the growth of PTR-MS as an analytical technique, with new applications being regularly reported and new researchers from many disciplines becoming involved in this exciting technology. The various books of abstracts for these conferences provide a unique summary of the activities pursued and the diverse range of applications, as well as an illustration of how fast the technology is evolving. For the interested reader, these books of abstracts are available as downloadable pdf files and they provide a valuable resource which is impossible to duplicate in this book.

PTR-MS has become a remarkably versatile tool, with applications in many areas of science and technology. A review of the research literature at the time of writing indicates that approximately 50% of all reported activities are based on studying VOCs (anthropogenic and biogenic) in the environment, and in particular for atmospheric science. Approximately 30% of publications are based in the area of food science/technology while another major field of application is to be found in the health sciences, representing approximately 15%

of PTR-MS publications to date. The remaining 5% of publications deal with a multitude of topics.

The diverse range of applications inevitably means that the users of PTR-MS come from a variety of backgrounds. While we suspect that the majority of users are likely to have received strong training in the chemical sciences, others may be drawn into the field from, for example, the environmental, biological or medical sciences. To learn about PTR-MS, one can turn to the research literature and consult the original publications that describe key developments in the technique or focus on a certain application. Equally, there are several excellent reviews on PTR-MS available in the published research literature. However, the research literature can be a rather terse and fragmented source of information, geared as it is towards specialists in the subject field. Furthermore, there have been several important new developments in PTR-MS in recent years, especially with regard to instrumentation. It would therefore seem timely to gather much of this information in one place.

Given the diverse user community and the increasing maturity of PTR-MS as a technique for gas monitoring and gas analysis, it is surprising that no book dedicated to PTR-MS has been published. Here we have attempted to fill this hole. Our primary aim has been to produce a book that is particularly targeted at those who are relatively new to PTR-MS, although more seasoned investigators may also benefit in some ways from its content. Broadly speaking, this book is divided into two parts. In the first part, which consists of Chapters 1–4, we describe the underlying principles of the PTR-MS technique, including the relevant ion–molecule chemistry, thermodynamics and reaction kinetics. Details are provided on practical aspects of PTR-MS, including a discussion of ion sources, drift tubes and mass spectrometers. As we proceed on this journey we give, where appropriate, some brief historical narrative.

The second part of the book, which spans Chapters 5–9, turns its attention to some of the many applications of PTR-MS. Here we want to demonstrate the scope and benefits, as well as the limitations, of PTR-MS. Our aim here has been to give a thorough but not exhaustive coverage of applied PTR-MS. We particularly want to try and show that PTR-MS, while already in widespread use, is rapidly finding new avenues where it can be applied. Four key areas of application will be described, namely environmental science, topics associated with food/drinks, medicine and homeland security, and each of these receives a dedicated chapter (Chapters 5–8). A short chapter (Chapter 9) will also deal with the applications of PTR-MS in liquid analysis. Chapters 5–9 build upon the material presented in Chapters 1–4 and are essentially self-contained reviews of the specific topics mentioned above. Consequently, the reader can dip into those that are of particular interest to him/her and if desired can safely ignore those of more peripheral interest.

Overall, we have tried to pitch the content of the book at a level which can be followed by an advanced undergraduate or early stage graduate student with a decent background in chemistry, and in particular physical chemistry. However, we also hope that people approaching PTR-MS from other disciplines who have a more modest knowledge of physical chemistry can follow much of the text. Our overarching aim has been to provide the reader, particularly one who is relatively new to PTR-MS, with a level of understanding of the technique which will then allow them to approach the research literature in this field with confidence. If we have come anywhere near this aim, then we will be delighted.

We need to thank several people for assisting with the preparation of the manuscript including Dr Peter Watts, Professor Paul Monks, Dr Franco Biasioli, Dr Philipp Sulzer,

Dr Valerie Gros, Dr Jens Herbig, Dr Margaret O'Hara, Dr Robert Blake, Dr Wilfried Szymczak, Shane Barber, Vicky Orson-Wright and Kathleen Hynes. We are also grateful to the publisher, and in particular our primary contact at Wiley, Sarah Tilley, for patience as each revised deadline came and went! Most of all, we are grateful for the enduring support and patience of our families while this book was being written.

Andrew M. Ellis
Christopher A. Mayhew
April 2013

Section 1
Principles

1

Background

1.1 Volatile Organic Compounds in the Earth's Atmosphere

Air usually contains many volatile organic compounds (VOCs).[1] These VOCs can derive from numerous sources, including emissions from plants and animals, as well as man-made sources such as motor vehicles and factories. The air we breathe consists mainly of inorganic gases: predominantly nitrogen and oxygen, but there are also other inorganic gases at lower concentrations, such as argon, water and carbon dioxide (see Table 1.1). By comparison the quantity of organic compounds in air is extremely small, of which the most abundant organic gas by far, methane, forms only around 1 part per million by volume (ppmv). At much lower concentrations still are other organic compounds such as methanol, acetone and isoprene. The exact quantities of these and other VOCs can vary depending on where any measurement is made. However, given that such compounds are usually present at extremely low levels (often referred to as *trace* levels) relative to the inorganic compounds in air, with even the more abundant VOCs often being in the low parts per billion by volume (ppbv) range,[2] it would seem safe to assume that VOCs are of negligible importance in almost any context. This assumption is incorrect.

There are many reasons why it is important to know the identities and the quantities of VOCs in the Earth's atmosphere. One of the major motivations is the desire to understand the impact of human activities on the natural atmosphere. Broadly speaking, the sources of VOCs in air can be divided into two groups: natural sources, also known as *biogenic* emissions, and man-made sources, known as *anthropogenic* emissions [2–4].

After methane the principal biogenic compounds are isoprene and monoterpenes. It is the release of monoterpenes that is responsible for the characteristic smell of forests, particularly

[1] The distinction is sometimes made between volatile and semi-volatile organic compounds but no such distinction is employed in this book.

[2] A billion here refers to a multiplier of 10^9, so 1 ppbv is one part in 10^9 by volume. For an ideal gas this implies one gas atom/molecule out of a total of 10^9 atoms/molecules.

Proton Transfer Reaction Mass Spectrometry: Principles and Applications, First Edition. Andrew M. Ellis and Christopher A. Mayhew.

Table 1.1 *The most abundant gaseous constituents of air and their typical mixing ratios*

Gas[a]	Mixing ratio[b] (ppmv)
Nitrogen	781 000
Oxygen	209 000
Argon	9 300
Carbon dioxide	387
Neon	18
Helium	5.2
Methane	1.7
Hydrogen	0.53

Source: Numerical data extracted from Reference 1.
[a]The amount of water vapour (not shown in the table) strongly varies depending on the humidity and can range from near-zero up to 40 000 ppmv (i.e. 4% of atmospheric gas).
[b]Mixing ratio (volume fraction) expressed in parts per million by volume (ppmv).

pine forests. Much of the emission of these compounds is linked to the photosynthesis cycle in plants, but there are other reasons for their emission, including as a defence mechanism against insect damage. However, many other compounds, including oxygenated VOCs such as alcohols, aldehydes and ketones, can be emitted from plants in response to various stimuli. In addition to their obvious biological interest, the emission of biogenic VOCs is extremely important in atmospheric chemistry. Biogenic emission is the dominant source of VOCs in the Earth's atmosphere and many of these compounds play an important role in atmospheric oxidation processes leading, among other things, to the formation of ozone.

Anthropogenic VOCs can disturb the natural atmospheric oxidation processes. Sources range from fossil fuel emissions to industrial solvent emissions and biomass burning. Significant compounds can include a variety of hydrocarbons, both aliphatic and aromatic, as well as oxygenated compounds. One of the well-known detrimental aspects of anthropogenic VOC emission is the production of photochemical smog, which forms when there are excessive levels of both nitrogen oxides and VOCs in combination with sunlight. The relatively high levels of ozone and particulate matter that this produces can make breathing difficult for many people and in worst-case scenarios can lead to serious injury or even death through respiratory diseases, such as chronic asthma. More generally, many anthropogenic VOCs may be highly toxic, carcinogenic and/or mutagenic, and emissions in the urban environment, in factories, or perhaps even in the home are of considerable concern.

Given the importance of VOCs in the environment, a variety of techniques have been developed for their measurement. The ideal technique does not currently exist, and thus one must assess the pros and cons of each available method in order to decide which is the most suitable for a given application. For example, the requirement might be to determine the amount of one or more compounds in the atmosphere on a large scale, perhaps over a region, a country or even a continent. In such circumstances, one must look towards techniques like satellite observation, which can explore large areas very rapidly, albeit only for a relatively small number of compounds. Aircraft measurements offer a compromise, where small-scale instruments can be carried over large ranges, but with limited coverage. On the more local scale, there is a greater variety of instrumentation that can be deployed, ranging from differential optical absorption spectroscopy (DOAS) to even more local techniques such as gas chromatography (GC), ion mobility spectrometry (IMS) and proton transfer reaction mass spectrometry (PTR-MS), all of which essentially sample air in the immediate vicinity of the instrument. However, in addition to spatial distribution, time-resolved VOC measurements may also be important to the atmospheric scientist. Are measurements every hour satisfactory, or might information be required every minute of every day? In some circumstances, it is important to be able to follow transitory events and thus a slow measurement technique may be of little or no use.

The focus in this book is on one particular technique, PTR-MS, which is widely used for measuring trace-level VOCs. Along with other applications, we aim to show that this is an important and versatile technique for atmospheric scientists. It provides a means of making local measurements with good sensitivity and at relatively high speeds. However, in this chapter we shall also describe complementary techniques, most notably gas chromatography–mass spectrometry (GC–MS), and competitor techniques such as selected ion flow tube mass spectrometry (SIFT-MS), to give some context. In each case, we will provide an assessment of the major strengths and weaknesses of each technique.

1.2 Volatile Organic Compounds in Other Environments

Instruments that can identify and quantify trace levels of VOCs have many potentially important applications beyond atmospheric science. For example, VOC emissions from foods and drinks are critical in our perception of their taste and smell. Food manufacturers are keen to understand the factors that contribute to flavour perception, and correlating VOC measurements with human trials of sensory perception is an important way of doing this. VOC emissions from foods and drinks can also be an indicator of quality. In particular, the onset of serious degradation may be characterized by the excessive emission of one or more VOCs, and these marker compounds therefore offer the key to an automated indication of food and drink decay. This is an area of investigation that is still very much in its infancy, but which may have very considerable implications for the food and drinks industries.

Another area where VOC measurements are still in their infancy is in the diagnosis of diseases through the measurement of the VOC composition of human breath. Breath is a mixture dominated by the common inorganic constituents of air, but it also contains a small fraction of VOCs in the ppbv to pptv (parts per trillion by volume) range. VOCs can be produced anywhere in the body and may be transported via the bloodstream to the lungs, where they are exhaled in breath. It has long been known by the medical community that a characteristic smell on the breath of a patient can sometimes indicate a specific medical condition. However, until recently no technique was available which could reliably and quickly assess the identities and quantities of the many VOCs present in human breath. That barrier is beginning to disappear and with it comes the prospect of using breath analysis as a serious means of non-invasive screening for specific diseases.

The above list of applications is by no means complete, and there are many other areas of science and technology that might benefit from rapid and sensitive VOC measurements, such as botany, forensic science and security screening. Some of these topics are considered in detail in the later chapters of this book (Chapters 5–9). In this short overview, we hope to have convinced the reader that trace VOC measurements are beneficial across a wide range of science and industry.

1.3 Techniques for VOC Measurements

It is not the aim of this book to provide an exhaustive account of analytical techniques that can be used to determine trace levels of VOCs in the gas phase. Our focus is, of course, on PTR-MS. However, like any measurement technique, PTR-MS has its strengths and weaknesses and it is important to put these into context by making a comparison with other types of instrumentation. As we will see, important criteria to be considered by any potential user include sensitivity, linearity of response, accuracy, specificity and speed of measurement, together with the cost and size of the instrumentation. Some of the major alternatives to PTR-MS for trace VOC measurements are described in this section.

We start with a description of what is currently the single most important and widely used technique for VOC gas analysis, *gas chromatography*. We will then turn our attention to techniques that typically show a much faster response, starting with *ion mobility spectrometry* and its variants. *Flowing afterglow* and *selected ion flow tube* studies of ion–molecule reactions are then considered, since these set the scene for the two main direct mass spectrometry techniques for quantifying gaseous VOCs, PTR-MS and SIFT-MS. The chapter closes with an introductory account of both PTR-MS and a comparison of PTR-MS with SIFT-MS.

1.3.1 Gas Chromatography

Gas chromatography (GC) is, in many respects, the 'gold standard' for trace gas analysis. The basic principle behind GC is that the constituents of a flowing gas mixture (the *mobile* phase) can be separated by passage over a suitable liquid or solid (the *stationary* phase). Partitioning between the mobile and stationary phases can lead to different *retention times* for different compounds due to differences in the way each compound in the gas interacts with the stationary phase. It is this difference in retention times that underlies GC.

In the early days of GC, the stationary phase was a particulate solid and would be packed into a tube known as a *column*. Nowadays much narrower diameter columns, known as *capillary columns*, are prevalent and the stationary phase is frequently a thin layer (typically 1 μm thick) of viscous liquid coated on the inside of the column wall. The columns are coiled and are normally very long, with lengths up to 60 m not being uncommon. As illustrated in Figure 1.1, the analyte mixture is injected into a flowing carrier gas, usually helium, upstream of the column and then proceeds through the column and onwards to a detector at the end of the instrument. The column is located within an oven to maintain a well-regulated temperature for the elution process. Whatever type of column is used, the aim is to ensure that the gas constituents are able to interact intimately and frequently with the stationary phase as they make their way through the column, with those interacting most strongly taking the longest time to reach the detector. Consequently, a plot of detector signal versus time should give a series of peaks at retention times that are characteristic for specific compounds. Such a plot is known as a gas chromatogram.

There are a number of important issues to consider in the practical application of GC for air analysis. First and foremost is the sensitivity, which is insufficient to measure the very

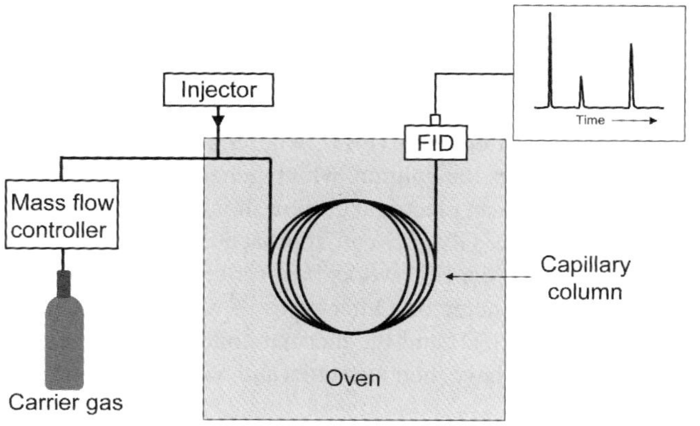

Figure 1.1 *Schematic diagram of a GC instrument with an FID. An analyte gas is injected into a flow of inert carrier gas at the injector. The gradual separation of compounds in the column leads to elution of distinct compounds at different (retention) times, as measured by the FID, leading to a chromatogram such as that shown in the upper right of the diagram.*

small quantities of VOCs in a single syringe injection of air. Instead a pre-concentration process is required, in which the desired analyte gas is accumulated over a period of time in an appropriate vessel before rapid release in a concentrated burst into the GC column. One way of doing this is to collect the analyte in a cryogenic trap. This will not condense the common constituents of air, such as O_2 and N_2, but will condense organic compounds. If this approach is adopted, a suitable collection container is required. One possibility is a stainless steel canister, but the inside surfaces must be coated with some relatively inert material, such as Teflon or silica, to minimize the possibility of surface-catalysed reactions. Once sufficient material is in the trap, it can be released by flash heating, which can be done electrically or by simply immersing the trap in hot water. Relatively simple cryogenic traps are rarely used for pre-concentration, partly because they are indiscriminate and thus collect all condensable materials, including water. Furthermore, it is relatively inconvenient to rely on a cryogenic approach, particularly if measurements are made in environments well away from standard laboratory facilities. As an alternative, traps consisting of suitable adsorbent materials, such as a carbon-based material (e.g. charcoal) or a polymer, can be employed. An example of the latter is a material known as Tenax, which works well for trapping compounds such as aromatics and terpenes and is widely used.

Another important issue is the column. The length of the column is one significant factor, since a longer column should improve the separation, but the choice of the stationary phase is critical. No single stationary phase is suitable for all types of compounds. For example, a stationary phase composed of a non-polar substance is best for separating hydrocarbons, particularly volatile hydrocarbons, whereas a polar stationary phase is essential for oxygenated VOCs such as alcohols, aldehydes and ketones. There are many types of stationary phases available and an important aspect of GC is selecting the right one for the particular application.

Water from the analyte is a major problem in GC work, not least because it interferes with the performance of the column. Consequently, efforts are made to remove as much water as possible before it reaches the column. Cold traps or traps packed with drying agents can be used for this purpose, but one must always take care to ensure that these traps do not also remove the organic constituents of the analyte gas.

There are several types of detectors that can be coupled to a gas chromatograph. One example is the flame ionization detector (FID), where, as the name implies, the effluent flows through a flame as it leaves the column, which generates ions as one of the products. These ions are then detected via an electrical current monitor. When coupled with suitable pre-concentration of the analyte, GC-FID can approach compound detection sensitivities as low as a few pptv. However, the FID works best for compounds such as hydrocarbons and is therefore not a universal detector. Other types of well-known GC detectors include the electron capture detector (ECD) and the thermal conductivity detector (TCD), and as with FID these alternatives also have their strengths and weaknesses but we will not discuss these in this book.

Unfortunately, compound resolution is not as easy as the discussion above implies. Many distinct compounds are difficult to separate fully on a column. Even if the compounds can be separated, the FID, ECD or TCD does not provide any compound identification: it merely registers the fact that a compound is leaving the column at a given moment in time. As implied earlier, compounds can sometimes be identified on the basis of the time taken to leave the column, that is, from their retention times. The retention time of a particular compound can be pre-determined by adding that specific compound to the GC instrument in a calibration procedure. Of course, there is the possibility, particularly with complex mixtures, that two or more compounds may have very similar retention times and therefore cannot be distinguished in this way. Furthermore, a calibration approach will only work if you already have a pretty good idea of what might be present in the analyte.

To get around these limitations, the FID, ECD or TCD can be replaced with a mass spectrometer equipped with an electron impact ionization source. In most cases, the mass spectrometer is of the quadrupole variety (see Chapter 3 for further details). The instrument works by recording a whole series of mass spectra, one after the other, as the analyte elutes through the column. In this way, important analytical information is obtained from both the retention time and the mass spectrum. The mass spectrum recorded for a particular GC peak can be compared with those stored in a library on the control computer, which usually allows compound identification. GC–MS is an extremely valuable analytical tool, but it suffers from a lower detection sensitivity than GC-FID because of the scanning time of the mass spectrometer, and so for air analysis it is often used more for compound identification than for quantification.

From the description above, some of the limitations with GC and its variants become clear. It is not a universal technique, since the choice of trap and column will affect the sensitivity and accuracy towards certain classes of compounds. In particular, oxygenated VOCs are more of a problem than hydrocarbons for GC analysis. The principal problem with oxygenated VOCs in GC stems from their polarity, which requires polar or semipolar column materials to achieve sufficient compound separation. Unfortunately, these columns also easily degrade when they come into contact with water, and therefore stringent efforts must be made to avoid any moisture. A particularly serious matter in GC analysis is the speed of measurement which, because of the need for sample collection and also some degree of pre-concentration, is often limited to a single measurement every few minutes, at best. Consequently, if rapidly evolving gas systems are being explored, GC techniques are not suitable.

Despite these limitations, GC remains the analysis technique of choice for many applications. It frequently achieves good compound separation, has good sensitivity, is reasonably cheap, is very well established and is reliable. Furthermore, developments such as two-dimensional GC techniques offer new scope [5]. The two-dimensional GC technique employs two columns in series but the two columns have different separation characteristics. This allows resolution of a much wider range of compounds than standard GC and has the potential to allow the exploration of very complex VOC environments in great detail.

1.3.2 Ion Mobility Spectrometry

Ion mobility spectrometry (IMS) has become a widely used analytical technique in the past two decades. It has found particular popularity in military and security circles because relatively cheap, simple, robust and compact IMS devices can be constructed which are capable of rapidly detecting a wide range of VOCs with high sensitivity. Figure 1.2 shows the basic structure of an IMS instrument. The analyte gas is injected and subjected to ionization at one end of the instrument and an ion detector is located at the other end. The tube in-between contains a series of electrodes which generate an electric field along its axis, drawing ions towards the detector. A tube equipped with these electrodes is called a *drift tube* and the underlying principle of IMS is the separation of ions according to their *mobilities* through a gas, which is usually air at atmospheric pressure.

We can draw a very loose analogy with GC in that the ions in IMS act like the mobile phase and the neutral gas as the stationary phase. The mobility quantifies the ease with

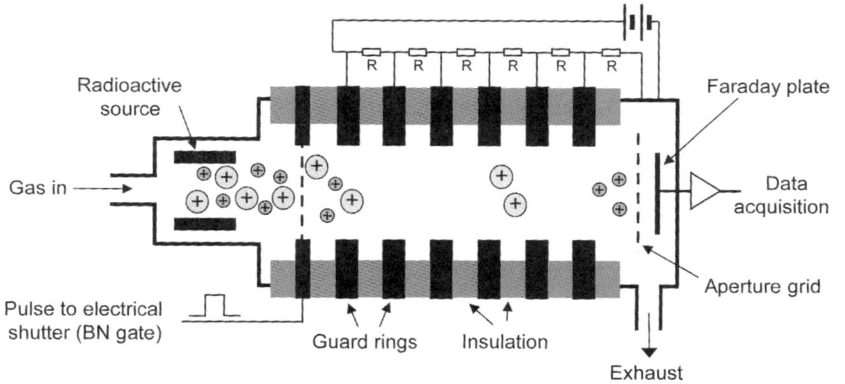

Figure 1.2 *Typical arrangement for a conventional ion mobility spectrometer. Ions are produced in the upstream region (left-hand side of the figure), in this case via a radioactive source, and are then drawn from left to right by an electric field applied through a series of electrodes (the 'guard rings'). Ions are injected in pulses using an electrical shutter (a Bradbury–Nielson (BN) gate) and the time taken to reach the detector is then determined. The ion detector in the figure is a simple Faraday plate (see Section 3.6).*

which a particular ion can move through a specified buffer gas at a given temperature and pressure when drawn along by an electric field of known magnitude. Light ions will tend to have higher mobilities than heavy ions, and thus the mobility through a gas serves to discriminate ions according to their masses. However, the size and structures of the ions are also important, since they will affect the collision cross section between neutral gas molecules and the ions. Consequently, large ions will tend to have smaller mobilities than small ions, but there is clearly room for some subtleties here because the collision cross section is also influenced by intermolecular forces, which in turn can depend on factors such as the charge distribution in the ion. Thus, while the mobilities of ions through the neutral gas may not always be easily predictable, they do offer a means of separating different types of ions on the basis of the time they take to pass through the buffer gas.

A radioactive strip is commonly employed in IMS to create ions, although other sources are also in use including photoionization and corona discharge sources, with the latter becoming increasingly popular for commercial instruments. In positive ion mode, the presence of water vapour in the background gas leads to the formation of hydrated hydronium ions, $H_3O^+(H_2O)_n$, where $n = 0, 1, 2$, etc.[3] These ions are the initiators of ionization in the analyte gas molecules and they do this by transferring charge to organic gases (M) by reactions of the type

$$H_3O^+(H_2O)_n + M \rightarrow (H_2O)_m MH^+ + (n + 1 - m)H_2O \quad (1.1)$$

To separate the different types of ions, a short burst of ions must be injected into the drift tube. This is achieved via a fast-acting electronic shutter, which is actually an ion deflection device known as a Bradbury–Nielsen gate. Injection of this pulse of ions has the effect of starting a clock and the arrival of ions at the detector is then measured relative to this starting time. Although there have been a number of different designs, the basic constituents of most drift tubes are a series of metal electrodes of circular cross section (sometimes called guard rings) interspersed by insulators. The electrical potential along the tube is chosen so as to draw the reagent and product ions towards the detector (see Figure 1.2). If the drift tube was operated under high vacuum conditions, such that ion–molecule collisions were essentially eliminated, then the speed of the ions would increase continuously as they travelled from one end of the drift tube to the other. However, IMS devices are not operated under high vacuum conditions and therefore collisions with the background gas results in the ions quickly reaching a constant, terminal velocity, the so-called drift velocity, which depends on the factors mentioned earlier and is directly proportional to the electric field strength.

The ion signal is measured as a function of time and should therefore consist of a series of peaks corresponding to ions with different mobilities arising from different chemical compounds in the analyte. Drift tube transit times depend on the length of the tube but are typically on the order of several tens of milliseconds. By comparison, the injection time for ions is <1 ms, and until all of the ions in the injected bunch reach the end of the drift tube, a second bunch of ions cannot be injected. Consequently, the *duty cycle*, which is a measure of the fraction of ions that reach the detector out of the total number of ions that could reach the detector if the experiment was not pulsed, is rather low and is typically 1%. This is an important factor in limiting the sensitivity of IMS. Nevertheless, detection of compounds

[3] The mechanism by which hydrated hydronium ions form in a radioactive or electrical discharge source is described in Chapter 3.

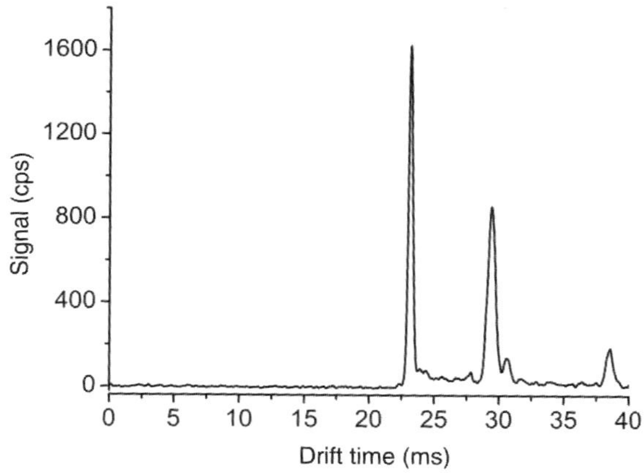

Figure 1.3 *An example of an ion mobility spectrum. In this example, the IMS was operated in positive ion mode using air as the buffer gas. Di(propylene glycol) methyl ether (DPGME), which is a commercial solvent, was introduced into the reaction region of the drift tube. Moving from left to right the first peak corresponds to the reagent ions $H_3O^+(H_2O)_n$, while the second and third peaks correspond to $DPGME(H_2O)_nH^+$ and $(DPGME)_2(H_2O)_nH^+$, respectively. On the vertical axis cps = ion counts per second.*

present in quantities as low as 1 ppbv or so can be reached, which is often adequate for many applications such as security and military uses. An illustrative ion mobility spectrum is shown in Figure 1.3.

The main drawback with IMS is its inherently poor selectivity. As with GC, many compounds cannot be fully separated by IMS, and even if they are separated it may not be easy to establish their identities. To try and rectify this, IMS has been coupled with mass spectrometry [6], but this comes at the expense of increased cost, complexity and size of the instrument. A notable development is Hadamard transform IMS [7, 8], which promises to resolve the problem of the low duty cycle of conventional IMS and should therefore result eventually in a significantly improved sensitivity, although again this delivers a more complex instrument.

1.3.3 The Flowing Afterglow Technique

PTR-MS has its origins in the development of the flowing afterglow (FA) method for the study of ion–molecule reaction kinetics. This so-called ion-swarm technique was introduced in the 1960s by Ferguson and co-workers and it revolutionized the study of ion–molecule reaction kinetics and thermodynamics [9, 10].

Figure 1.4 shows a schematic of the apparatus. The main feature is a flow tube,[4] which provides the means of extracting kinetic information from relatively fast reactions. The idea here is that reactions between some atomic or molecular ion, X^+, and neutral molecules, which we designate as Y, are explored. The production of reagent ions, X^+, is a two-stage

[4] A flow tube is distinct from a drift tube in that the transport mechanism in the former is gas flow driven by a pressure difference between the two ends of the tube, that is, no electric field is involved in transporting ions.

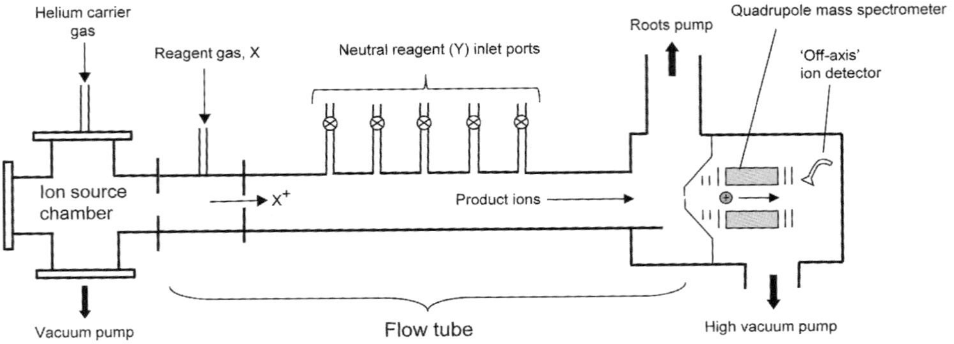

Figure 1.4 *Schematic of a flowing afterglow apparatus. With the apparatus shown reagent Y can be added at any one of the inlet ports positioned along the flow tube. An off-axis ion detector, as shown in the figure, is often used in quadrupole mass spectrometers to minimize noise from radiation and excited neutral molecules generated upstream in the ion source and in the flow tube. Ions are deflected towards the detector by electrodes (not shown).*

process. First, ions are created in an inert buffer gas, which is usually helium. This is normally achieved either by an electron impact using a heated filament electron source or via an electrical discharge. These ions are then carried along a flow tube by the flowing buffer gas (again usually helium) and further downstream they are mixed with gas X. Gas X can be ionized, typically by charge transfer from He^+, but also by Penning ionization from metastable electronic excited states of helium that are also produced in the initial ionization region. The advantage of keeping the gas X away from the initial point of ionization is that it helps to minimize the formation of unwanted fragment ions, which would be more of a problem if gas X was exposed directly to a highly energetic electrical discharge or direct electron bombardment.

A bright glow extends along the upstream part of the flow tube due to the light emitted by electronically excited constituents of the discharged gas. The extensive glow of this plasma is the origin of the name *flowing afterglow*. An essential feature of the FA technique is that it attempts to explore ion–molecule reactions at thermal energies. This is achieved by using a large excess of buffer gas and ensuring that the reactant ions are formed well upstream of the point where the neutral reagent gas, Y, is added. The flow tube is usually fairly long, a typical length being 1 m, and with gas flow speeds in the region of 100 m s^{-1}. This means that the ions have a period of several milliseconds to thermalize prior to reaction, although cases are still known where thermalization is incomplete due to slow collisional energy transfer from electronically excited and vibrationally excited states [11].

The neutral reactant gas, Y, is added at a specific entrance port along the flow tube. The pressure in the flow tube is maintained at approximately 1 mbar. After mixing with the buffer gas and reactant ions, a portion of the flow tube effluent passes through a small orifice at the end of the flow tube and enters a mass spectrometer, which is usually a quadrupole mass spectrometer (see Chapter 3). Reaction essentially stops as the mixture passes through the orifice since this mass spectrometer chamber must be operated at a low pressure ($\leq 10^{-4}$ mbar), and therefore further reactive collisions virtually cease. One way of determining the reaction kinetics is to vary the flow rate of the incoming neutral gas while monitoring the ion current of the reactant ions with the mass spectrometer. As the flow rate of the neutral gas increases, the observed count rate of the reactant ions, X^+, should increase since the reaction time, t, which is determined by the distance from the injection point of Y to the pinhole at the entrance to the mass spectrometer, will decrease. Alternative approaches to extracting kinetic information involve varying the length of the reaction zone, as shown in Figure 1.4 where Y can be injected at different positions along the flow tube, or by varying the flow rate of the reactant Y while fixing the overall flow rate of gas through the flow tube.

If X^+ represents the reactant ions and Y represents the neutral reagent gas, then the rate of a bimolecular reaction between these two species is given by

$$-\frac{d[X^+]}{dt} = k[X^+][Y] \tag{1.2}$$

where k is the rate coefficient[5] for the $X^+ + Y$ reaction and the square brackets around each reagent denote the concentration. Conditions are chosen such that $[Y] \gg [X^+]$, which means that [Y] is effectively constant and therefore Equation 1.2 can be solved to yield

$$[X^+]_t = [X]_0 e^{-k[Y]t} \tag{1.3}$$

where $[X^+]_0$ and $[X^+]_t$ refer to the concentration of X^+ at the point of mixing and at some later time, respectively. The reaction time, t, can be determined provided the rate of gas flow through the flow tube is known. Thus the distance from the point where X^+ and Y are mixed through to the aperture at the end of the flow tube (where reaction is assumed to stop) corresponds to a fixed and determinable reaction time for a given gas flow rate. Consequently, the quantity of X^+ detected by the mass spectrometer is a function of [Y] and it is then a simple matter to extract the rate coefficient from the logarithmic version of Equation 1.3, namely

$$\ln\left(\frac{[X^+]_{[Y]}}{[X]_{[Y]=0}}\right) = \ln\left(\frac{i(X^+)_{[Y]}}{i(X^+)_{[Y]=0}}\right) = -kt[Y] \tag{1.4}$$

Here $i(X^+)_{[Y]}$ is the measured ion current at the mass of X^+ for a known Y concentration and $i(X^+)_{[Y]=0}$ is the value when no Y is added. The rate coefficient can then be extracted from a simple straight-line graph using Equation 1.4. A more detailed account of the kinetic analysis procedure can be found elsewhere [11].

Once established, the FA technique was used to determine the kinetics of numerous ion–molecule reactions. It also has the potential to be used as a technique for quantitative analysis of gas Y, providing the rate coefficient for the $X^+ + Y$ reaction is known. However, while a major step forward in the study of ion–molecule reactions, there are some weaknesses of this technique. On the experimental side, it requires a relatively large piece of apparatus. In particular, the flow rate of gas through the flow tube is substantial and yet gas pressures in the region of 1 mbar are required inside the tube. Consequently, a relatively large pumping system, usually based on a Roots pump, is required to handle the gas throughput.

A particular difficulty with the FA technique is avoiding interferences from reactions of unwanted ions. If the ion source gas, X, is a diatomic gas then the number of possible ions

[5] Throughout this book we will use the term *rate coefficient* rather than *rate constant* but the two are interchangeable.

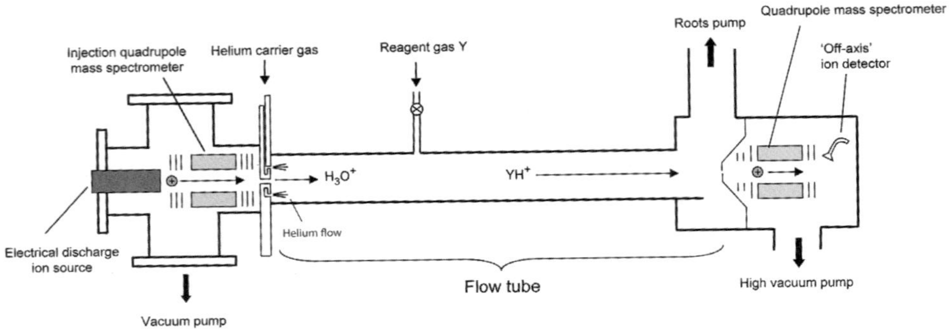

Figure 1.5 *Schematic of a SIFT instrument. In contrast to a flowing afterglow experiment the reagent ion is mass-selected using a quadrupole mass filter from the many possible ions produced in the discharge source and in this particular figure the ions selected are shown to be H_3O^+. Proton transfer creates product ions YH^+.*

that might be formed when X encounters the afterglow plasma is relatively small, with the likelihood that one particular reagent ion dominates the observed chemistry. However, this is a much more serious problem for polyatomic molecules, where the range of possible fragmentation products may be large. This weakness of the FA technique was the catalyst for the development of a new flow tube technique, which is described in the next section.

1.3.4 The Selected Ion Flow Tube

The *selected ion flow tube* (SIFT) solved the key limitation of the FA technique, the tendency for several distinct primary ions to form when polyatomic gases are used in the discharge, by employing a mass filter to allow ions of only one specific m/z into the flow tube. This breakthrough originated from work by Adams and Smith and led to a technique which is still in use today [12].

The basic components of a SIFT instrument are illustrated in Figure 1.5. Since much of the technology is similar to the FA experiment, we focus here only on the key differences and the general performance characteristics of SIFT. The main difference in SIFT is that a quadrupole mass filter[6] is fitted upstream of the flow tube to deliver reagent ion selection. A potential problem arises from the different pressure requirements in the apparatus, since the quadrupole filters at both ends of the instrument require a pressure of $\leq 10^{-4}$ mbar for effective operation while the flow tube is operated at pressures near 1 mbar. To minimize backflow of the inert carrier gas (typically helium, as in the FA) into the injection quadrupole filter the carrier gas is input through a Venturi inlet. This consists of a series of small holes concentrically arranged around the ion entrance aperture which are shaped to direct the gas flow away from the exit aperture of the injection system, as illustrated in Figure 1.5. This, and the use of a high-speed pump at the downstream end of the flow tube, allows satisfactory pressure regimes to be obtained throughout the apparatus.

A major difference between SIFT and FA is that the ion selection process results in an ion number density in the flow tube that is several orders of magnitude lower for SIFT than

[6] The operating principles of quadrupole mass filters are described in Section 3.5.2.

for FA. This results in a lower detection sensitivity for SIFT. Nevertheless, the ability to select specific reagent ions has led to a wealth of fundamental kinetic data on ion–molecule reactions of importance in many areas, including atmospheric and interstellar chemistry.

1.4 Emergence of Proton Transfer Reaction Mass Spectrometry

1.4.1 Historical Background

The development of PTR-MS owes much to its precursors, the FA and SIFT techniques. However, PTR-MS is intended as an analytical technique rather than a means of measuring kinetics and it achieves this by essentially inverting the original aims of the FA and SIFT techniques. Instead of measuring the rate coefficient for a specific reaction, it is now presumed known from earlier kinetic studies or it can be deduced from calculations, as described in Section 2.2.3.2, or it may be omitted completely if gas calibration is employed (as described in Chapter 4). This information can then be used to determine the concentration of a compound, as shown later in Section 1.4.3.

A key milestone in the development of PTR-MS was the application of a selected ion flow drift tube (SIFDT) technique as an analytical tool by Werner Lindinger and co-workers in 1994 [13]. This employed a specific mass-selected reagent, H_3O^+, for subsequent reaction with an analyte gas. The choice of this ion is highly significant. Proton donation from H_3O^+ to many organic molecules is exothermic because the so-called *proton affinities* of these molecules exceed those of H_2O (see Section 2.2.2 for more details). Only a few organic molecules, notably small alkanes and ethene, have proton affinities below that of H_2O. Exothermic proton transfer reactions are generally very fast, proceeding typically at the collision-limited rate. However, the common constituents of air, inorganic species such as N_2, O_2 and CO_2, have lower proton affinities than H_2O. Consequently, proton transfer from H_3O^+ is almost entirely selective to the organic constituents of air, which makes H_3O^+ an outstanding proton source for trace VOC studies.

Another benefit that accrues from using H_3O^+ is that proton transfer tends to be a *soft* ionization process; that is, there is little or no fragmentation for many organic gases (although some exceptions are known and these will be discussed in Chapter 2). Thus, by locating a mass spectrometer downstream of the flow drift tube, the mass spectrum can be measured and in this ideal limit will derive from protonated organic molecules whose mass can be readily deduced. This is important when dealing with complex gas mixtures, since in the best-case scenario interference from fragment ions could lead to false compound identification and, in the worst-case scenario, could lead to a hopelessly congested spectrum with little or no analytical value.

A crucial difference between the SIFDT technique and the previously discussed SIFT method is that the former employs an electric field along the flow tube axis to increase the velocities of the ions. The electric field is incorporated through a series of electrodes added along the flow tube, in a similar manner to those employed in IMS (see Figure 1.2). However, the SIFDT technique still employed a conventional flow tube design in which a neutral reagent gas is diluted in helium and then fed into a relatively long flow tube, albeit one that now also contains electrodes. The original purpose of the electric field was to be able to deliver extra energy to ion–molecule collisions by raising the mean velocity of the ions, thereby allowing the effect of kinetic energy on ion–molecule reactivity to be measured [14]. However, in the application of SIFDT as an analytical technique the electric field served the purpose of minimizing cluster ions in the flow drift tube. In that early study, Lindinger and co-workers were interested in measuring concentrations of gases such as acetone, methanol and ethanol in human breath, but breath is generally saturated with water vapour [13]. Even though the mass filter is capable of highly selective H_3O^+ injection into the flow tube, the abundant water molecules in the humid analyte gas will react with the H_3O^+ resulting in the formation of hydrated hydronium clusters, that is, $H_3O^+(H_2O)_n$. We discuss the significance of these clusters in the context of PTR-MS in some detail in later chapters, but the bottom line is that proton transfer from $H_3O^+(H_2O)_n$ clusters may proceed at different rates from H_3O^+, or in some cases entirely different reaction processes may occur (e.g. Reaction 1.1), thus complicating the data analysis and making it difficult to extract reliable VOC concentrations. The flow drift tube employed by Lindinger and co-workers resolved this difficulty by dissociating the $H_3O^+(H_2O)_n$ clusters in energetic collisions with the neutral gas (primarily the helium buffer gas), thus regenerating H_3O^+ as the reagent ions. Notice that the aim here was to deliver enough additional kinetic energy to achieve efficient fragmentation of $H_3O^+(H_2O)_n$ clusters so that the predominant reagent ion in the flow drift tube was H_3O^+,[7] while at the same time avoiding (as much as possible) an energy that would cause significant fragmentation of protonated organic ions formed by the reaction between H_3O^+ and VOCs in the analyte gas. The best compromise seems to be a mean ion–neutral collision energy in the region of 0.15–0.25 eV, which is an order of magnitude higher than the average kinetic energy possessed by the molecules in a room temperature gas.

PTR-MS appeared on the scene shortly thereafter from work also carried out by Lindinger and his team [15]. Two major changes transformed the SIFDT approach into what we know today as PTR-MS. The first change was the use of a short drift tube in place of the flow drift tube. Lindinger and co-workers avoided the inert carrier gas employed in SIFT and SIFDT and simply injected the analyte gas, usually an air sample of one form or another, into the drift tube. With H_3O^+ as the proton source, the major constituents of air, N_2 and O_2, serve as the buffer gas in place of helium, thus enabling collision-induced dissociation to minimize cluster formation. In this way, the quantity of analyte gas delivered into the instrument can be some two orders of magnitude higher than if using a SIFT approach, thus yielding a corresponding increase in detection sensitivity for VOC detection. Furthermore, because a long flow tube with a large flow of helium was replaced with a short drift tube (typical length 10–20 cm) with a much lower incoming flow of gas, the instrument became much shorter and the pumping system originally used in the SIFDT approach could be drastically reduced in size.

A second innovation introduced by Lindinger and co-workers yielded a simpler and even more compact instrument. This was achieved by replacing the mass-selected H_3O^+ source with a hollow cathode discharge, which with a suitable design (see Chapter 3) was found to deliver an almost perfectly pure (>99.5%) source of H_3O^+ without any need for a mass

[7] $H_3O^+(H_2O)_n$ clusters form by termolecular reactions of the type $H_3O^+(H_2O)_{n-1} + H_2O + X \rightarrow H_3O^+(H_2O)_n + X$, where X is some third body. In addition to increased fragmentation of $H_3O^+(H_2O)_n$ clusters at an elevated kinetic energy, the rate of formation of these cluster ions is reduced because the rate coefficient has a strongly negative energy dependence.

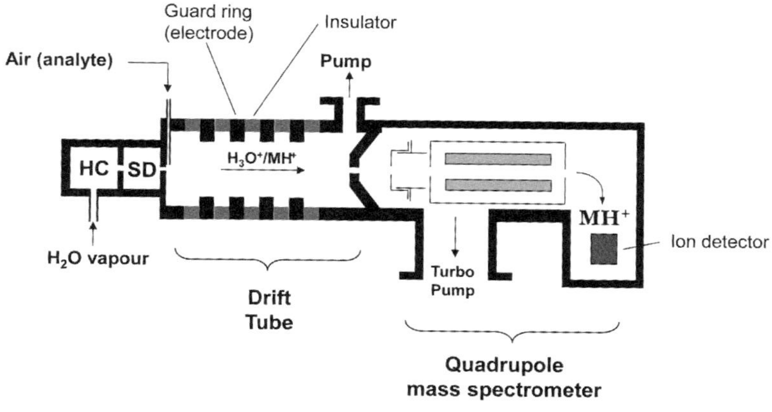

Figure 1.6 *Schematic of a PTR-MS instrument, where in this case the mass spectrometer is a quadrupole device. HC signifies a hollow cathode discharge ion source and SD is known as the source drift region. These, along with other components of this instrument, are discussed in more detail in Chapter 3.*

filter. The resulting overall instrument is illustrated schematically in Figure 1.6. Chapter 3 provides a more detailed account of the various components within a PTR-MS instrument.

The years following the seminal work by Lindinger and his team have seen a variety of new developments. For example, time-of-flight and ion-trap mass spectrometers have been employed in PTR-MS instead of quadrupole mass spectrometers. Furthermore, other types of ion sources have been introduced, including radioactive sources. Developments such as these will be covered in detail in Chapter 3.

1.4.2 Compound Identification Using PTR-MS

PTR-MS allows the monitoring of compounds and in some cases the identification of specific compounds in a gas. The key to doing this is mass spectrometry, and in this section we give a brief overview of mass spectra derived from PTR-MS.

As mentioned in the previous section, the principal source of organic ions in PTR-MS is proton transfer from H_3O^+, as illustrated in the following reaction

$$H_3O^+ + M \rightarrow MH^+ + H_2O \tag{1.5}$$

which is a special case of Reaction 1.1 where the formation of hydrated hydronium ions, $H_3O^+(H_2O)_n$, has been minimized and is hopefully negligible.

Having generated ions, it is now possible to record a mass spectrum. A mass spectrum is a plot of measured ion abundance (y-axis) as a function of m/z values of the product ions (x-axis). Here m refers to the mass of the ion and z refers to the charge of the ion as a multiple of the fundamental unit of charge, e ($=1.60210 \times 10^{-19}$ C). In PTR-MS, we always deal with singly charged ions and thus $z = 1$ in all cases, and so we could justifiably discuss the mass and ignore the charge. However, to be consistent with other types of mass spectrometry we will tend to employ the notation *m/z* in this book, although alternative possibilities are discussed in Section 3.5.1. Essentially, *m/z* is treated as a quantity without units and is derived from either the *nominal* integer ion mass (e.g. $^{1}H = 1$ and $^{16}O = 16$ for atomic ions) or, for high-resolution mass spectra, as the *monoisotopic* ion mass (the exact mass of a molecule calculated from the most abundant isotope of each constituent atom, that is, $^{1}H = 1.0078$ and $^{16}O = 15.9949$). For example, the nominal and monoisotopic *m/z* values for H_3O^+ are 19 and 19.0184, respectively.

Although not always achieved, the intention in PTR-MS is to use ionization conditions where fragmentation of the protonated organic molecules is kept to a minimum, as discussed in the preceding section. In mass spectrometry parlance, the non-fragmented ion is referred to simply as the *parent ion*, or sometimes as the *molecular ion*. In the case of proton transfer, we will refer to MH^+ as the protonated parent molecule.

Illustrative mass spectra from PTR-MS measurements are shown in Figures 1.7 and 1.8. Both of these spectra were recorded in the laboratories of the authors. Figure 1.7 shows part of a mass spectrum recorded on a PTR-MS instrument equipped with a quadrupole mass analyser, whereas Figure 1.8 was recorded using a time-of-flight mass analyser. It will be clear from these spectra, and in particular the expanded view of the spectrum shown in Figure 1.8(b), that the mass resolution is much higher for the time-of-flight mass analyser than for the quadrupole analyser, something that is discussed in more detail in Chapter 3. Some of the main inorganic species seen in the mass spectrum are highlighted in Figure 1.7, notably NO^+, $O_2{}^+$ and $H_3O^+(H_2O)$. Note that the spectrum in Figure 1.7 starts at $m/z = 20$, which is intentional because the very large signal for H_3O^+ at *m/z* 19 would otherwise saturate and possibly damage the ion detector. Nevertheless, a signal for H_3O^+ can be seen at *m/z* 21, which arises from the ^{18}O isotope of oxygen, which has a low natural abundance. Since the naturally occurring ^{16}O:^{18}O abundance ratio is close to 500:1, the *m/z* 21 signal

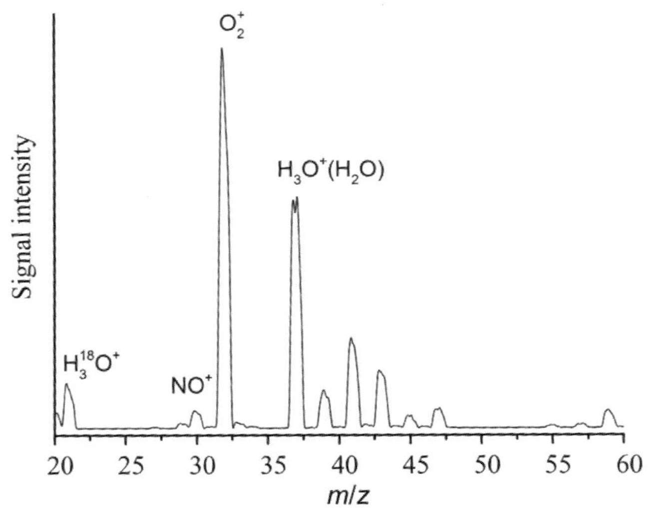

Figure 1.7 *Mass spectrum derived from a PTR-MS instrument containing a quadrupole mass analyser. The analyte was laboratory air.*

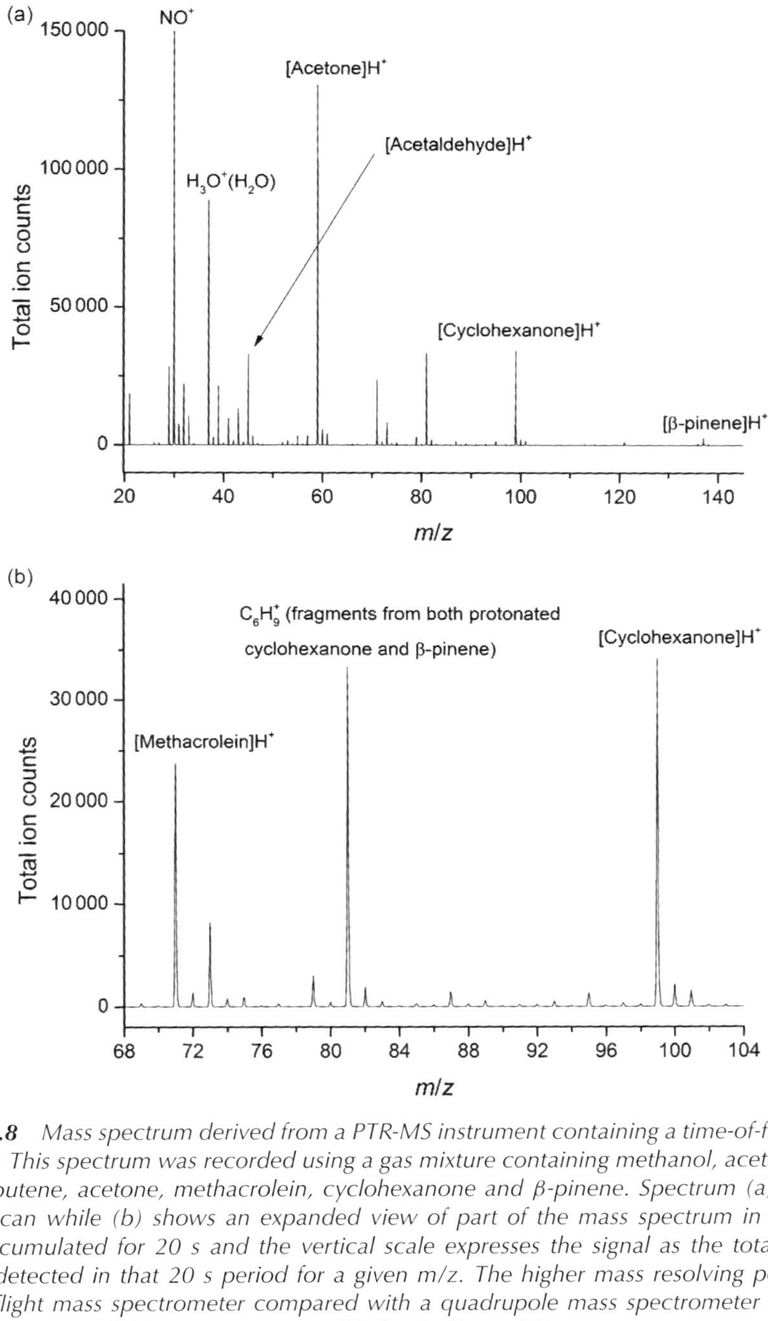

Figure 1.8 *Mass spectrum derived from a PTR-MS instrument containing a time-of-flight mass analyser. This spectrum was recorded using a gas mixture containing methanol, acetaldehyde, trans-2-butene, acetone, methacrolein, cyclohexanone and β-pinene. Spectrum (a) shows a survey scan while (b) shows an expanded view of part of the mass spectrum in (a). Data were accumulated for 20 s and the vertical scale expresses the signal as the total number of ions detected in that 20 s period for a given m/z. The higher mass resolving power of a time-of-flight mass spectrometer compared with a quadrupole mass spectrometer is evident from comparison of this mass spectrum with that in Figure 1.7.*

can be multiplied by 500 to deduce the abundance of H_3O^+ ions in the drift tube. In particular, the ratio of the ion abundances for H_3O^+ versus MH^+ is essential information for the quantitative determination of VOCs using PTR-MS, as discussed in the next section. We can also deduce from Figure 1.7 that the quantity of H_3O^+ locked up in the form of $H_3O^+(H_2O)$ clusters ions is about 1% of the amount of free H_3O^+ in the drift tube.[8]

For the spectrum shown in Figure 1.8(a), a gas mixture containing several VOCs of known composition was deliberately added into the drift tube of the PTR-MS instrument. Prior to entering the drift tube, the gas mixture was diluted yielding mixing ratios on the order of tens of ppbv for each compound. Similar to the spectrum in Figure 1.7, the spectrum in Figure 1.8(a) begins at $m/z = 20$. Among the other peaks seen on the low m/z side of the mass spectrum are those from $H_3O^+(H_2O)$ and NO^+. From the relative intensities of the $H_3O^+(H_2O)$ signal at $m/z = 37$ and the $H_3{}^{18}O^+$ signal at $m/z = 21$, only about 1% of the hydronium ions are in the form of $H_3O^+(H_2O)$, which is similar to that found from Figure 1.7. Contaminant ions in the drift tube, such as NO^+ and $O_2{}^+$, can form when air from the analyte gas back-diffuses into the ion source region of the PTR-MS instrument (see Chapter 3). For the instrument used to record the spectrum in Figure 1.7, $O_2{}^+$ is much more prominent than NO^+, whereas the opposite is true for the instrument used to record the spectrum in Figure 1.8(a). However, in both cases the NO^+ and $O_2{}^+$ levels are very low relative to the abundance of H_3O^+ and therefore the presence of these contaminant ions will usually have a negligible effect on the VOC analysis.

Some of the peaks from protonated parent organic ions are labelled in Figure 1.8(a). Several peaks are very prominent and can be assigned unambiguously, such as those from protonated acetone (m/z 59) and acetaldehyde (m/z 45). However, compound identification with PTR-MS is not without its difficulties, as illustrated in an expanded view of a section of the mass spectrum in Figure 1.8(b). The prominent peak at m/z 81 is not from a protonated parent ion but is instead a peak which has contributions from fragment ions with the molecular formula $C_6H_9{}^+$. These fragment ions derive from protonated cyclohexanone and β-pinene, showing that the protonated parent molecules are not the exclusive products for these particular molecules. Potential contributions to the mass spectrum from fragment ions are something that all PTR-MS users should be aware of.

Other peaks can be seen in Figures 1.8(a) and 1.8(b), including many weak peaks. Some of these will be from fragment ions of the various known compounds in the gas mixture, but there is also likely to be a background of various trace organic compounds which may derive from a number of sources, including trace contaminants on the walls of the gas inlet line and inside the PTR-MS instrument. These contaminant contributions can be partly removed by recording a background spectrum in the absence of the analyte and then subtracting this from the analyte spectrum (see Section 4.4). However, in Figures 1.7 and 1.8, we deliberately show raw mass spectra where no background subtraction has been applied.

As mentioned above, time-of-flight mass spectrometry usually has very good mass resolution and so, as can be seen in Figure 1.8(b), peaks differing in nominal m/z by one unit are easily resolved. In fact even higher mass resolution is possible than that shown in

[8] This assumes that the mass spectrometer transmits and detects H_3O^+ and $H_3O^+(H_2O)$ with the same efficiency. In practice for an accurate assessment of the relative abundances of these ions, one must take into account the mass-dependent differences in ion transmission by the mass spectrometer, as discussed in more detail in Chapters 3 and 4.

Figure 1.8 and this can sometimes be used to distinguish between compounds which have the same nominal mass but different accurate masses, as will be discussed in Section 3.5.4.6.

1.4.3 An Introduction to Quantitative Aspects of PTR-MS

The measurement of ion signals by the mass spectrometer in PTR-MS provides a means for determining the absolute concentration of a specific constituent of a gas mixture, as will be shown below. With H_3O^+ as the proton source and assuming reaction with only a single organic gas, designated M, the proton transfer reaction is as shown in Reaction 1.5. This reaction is a second-order elementary reaction and will satisfy the rate equation

$$-\frac{d[H_3O^+]}{dt} = k[H_3O^+][M] \tag{1.6}$$

Assuming that $[M] \gg [H_3O^+]$, which is reasonable since M is a neutral gas (even if present at trace levels), then [M] is constant (the reaction is pseudo first order) and Equation 1.6 can be integrated to yield

$$[H_3O^+]_t = [H_3O^+]_0 e^{-k[M]t} \tag{1.7}$$

The reaction time t is the time it takes for the reagent ion, in this case H_3O^+, to travel from the point where it is first mixed with the analyte to the end of the drift tube (beyond which reaction essentially stops). So far the kinetic analysis is directly analogous to that already provided for the FA technique in Section 1.3.3. The concentration of H_3O^+ can be related to that of MH^+ by

$$[MH^+]_t = [H_3O^+]_0 - [H_3O^+]_t \tag{1.8}$$

Rearranging Equation 1.8 so that $[H_3O^+]_t$ is the argument and substituting into Equation 1.7 gives

$$[MH^+]_t = [H_3O^+]_0[1 - e^{-k[M]t}] \tag{1.9}$$

Two final assumptions yield a particularly simple expression. First, we assume that conditions are chosen such that only a small proportion of H_3O^+ is consumed by the reaction, that is, $[H_3O^+]_0 \approx [H_3O^+]_t$, or equivalently $[MH^+] \ll [H_3O^+]$. A second assumption is that $k[M]t \ll 1$, which is usually true providing M is present at well below the parts per million level. A Taylor expansion of the exponential term in Equation 1.9 and retaining only the first term in the expansion then leads to the following:

$$\frac{[MH^+]}{[H_3O^+]} = \frac{i(MH^+)}{i(H_3O^+)} = k[M]t \tag{1.10}$$

The concentrations of the ions on the left-hand side of Equation 1.10 refer to those at the downstream end of the drift tube and thus for convenience the subscripts have now been omitted. The ratio of currents, i, for MH^+ and H_3O^+ at the end of the drift tube is assumed to be equal to the ratio of the measured count rates for these two ions at the detector of the mass spectrometer. Equation 1.10 then reveals that it is possible to calculate the concentration of gas M in the drift tube having measured the ratio $i(MH^+)/i(H_3O^+)$. In principle, the only other information required is the rate coefficient, which is assumed known, and the reaction time t, which can be estimated from a calculation (see Section 3.4.2) or can be measured by pulsing the ion source [15].

In practice, there are many approximations involved in deriving Equation 1.10, some of which have been mentioned above, along with others that have not yet been discussed, but which will be encountered in later chapters. To extract accurate concentrations for specific compounds by PTR-MS, calibration of the instrument is essential, as described in Chapter 4. We also note here that accurate concentrations can usually only be obtained if a single compound contributes to the signal observed at a given m/z.

It is worth emphasizing that the concentration of M in the drift tube determined from application of Equation 1.10 is not the same as that in the original analyte gas, since the concentration of analyte is reduced by entering a low pressure tube. The process of calculating the external concentration is relatively simple and is illustrated in Section 4.2. We simply note here that, as already met earlier in the chapter, we will tend to express the abundances of VOCs as *mixing ratios* in this book, rather than absolute concentrations. The mixing ratio expresses the quantity of a material as a proportion of the total amount of material present (chemists will know that this is similar to a mole fraction). Assuming ideal gas behaviour, the volume occupied by a pure gas is proportional to the number of molecules present and thus the mixing ratio is sometimes known as the volume mixing ratio. Thus one part in 10^6 is known as a part per million (1 ppmv = 1 part per million by volume), one part in 10^9 as a part per billion (1 ppbv) and one part in 10^{12} as one part per trillion (1 pptv). Use of mixing ratios is convenient because they are independent of temperature and pressure. It is also straightforward to convert these quantities into an absolute concentration if the total pressure of the analyte gas is known.

1.4.4 A Comparison between PTR-MS and SIFT-MS

The conversion of SIFT into an analytical technique was pioneered by Smith and Španěl and took place at around the same time that PTR-MS was developed [16]. The technique is known as SIFT-MS and is analogous to PTR-MS in that the basic idea is to employ the kinetics taking place in a flow tube to determine the concentration of one or more constituents of a gas sample. However, there are important differences between SIFT-MS and PTR-MS. A clear advantage of SIFT-MS is that no electric field is employed and it is therefore a realistic prospect to carry out ion–molecule reactions under thermal conditions where the kinetic behaviour is well known. In contrast, PTR-MS operates at higher effective temperatures and while this temperature can be estimated (see Section 3.4.4) the kinetic behaviour at this higher temperature is often not known. Another advantage of SIFT-MS stems from the mass-selectivity of the ion source, which makes it straightforward to switch from one ion reagent to another without any concerns about impurities. For example, switching from H_3O^+ to NO^+ is possible. Clearly NO^+ cannot supply protons but it can undergo charge exchange reactions with some gases (see Section 2.3) and this difference in chemistry can sometimes be useful in distinguishing gas molecules with the same nominal mass and which are difficult to discriminate between using SIFT-MS based on H_3O^+ chemistry. However, despite not having the degree of selectivity achieved in SIFT-MS, sources of NO^+ and other ionization reagents which are highly pure have also been developed for use on PTR-MS instruments (see Sections 2.2.4.3 and 2.3).

The principal weaknesses of SIFT-MS when compared with PTR-MS are its lower sensitivity and the larger size of the instrument. In terms of sensitivity, it rarely achieves better than a few ppbv, although there has been a report of the detection of phosphine down to 190 pptv [17]. A sensitivity of a few ppbv is insufficient to detect all but the most abundant VOCs in air and this, coupled with the bulkiness and cost of SIFT-MS instruments, has curtailed the number of users.

References

1. Wayne, R. P. (1991) *Chemistry of Atmospheres*, 2nd edn, Oxford University Press.
2. Hewitt, C. N. (1999) *Reactive Hydrocarbons in the Atmosphere*, Academic Press, San Diego, CA.
3. Hester, R. (1995) *Volatile Organic Compounds in the Atmosphere*, Royal Society of Chemistry.
4. Koppmann, R. (Ed.) (2007) *Volatile Organic Compounds in the Atmosphere*, Blackwell Publishing.
5. Ramos, L. (Ed.) (2009) *Comprehensive Two Dimensional Gas Chromatography*, Elsevier Science.
6. Kanu, A. B., Dwivedi, P., Tam, M. *et al.* (2008) Ion mobility–mass spectrometry. *J. Mass Spectrom.* **43**, 1.
7. Clowers, B. H., Siems, W. F., Hill, H. H., Massick, S. M. (2006) Hadamard transform ion mobility spectrometry. *Anal. Chem.* **78**, 44.
8. Szumlas, A. W., Ray, S. J., Hieftje, G. M. (2006) Hadamard transform ion mobility spectrometry *Anal. Chem.* **78**, 4474.
9. Fehsenfeld, F. C., Ferguson, E. E., Schmeltkopf, A. L. (1966) Thermal-energy ion-neutral reaction rates. III. The measured rate constant for the reaction $O^+(^4S) + CO_2(^1\Sigma) \rightarrow O_2{}^+(^2\Pi) + CO(^1\Sigma)$. *J. Chem. Phys.* **44**, 3022.
10. Ferguson, E. E., Fehsenfeld, F. C., Schmeltkopf, A. L. (1969) Flowing afterglow measurements of ion-neutral reactions. *Adv. At. Mol. Phys.* **5**, 1.
11. Adams, N. G., Smith, D. (1988) Flowing afterglow and SIFT. In: *Techniques for the Study of Ion–Molecule Reactions* (Eds J. M. Farrar, W. H. Saunders), John Wiley & Sons, Inc., New York, pp. 165–220.
12. Adams, N. G., Smith, D. (1976) The selected ion flow tube (SIFT): a technique for studying ion-neutral reactions. *Int. J. Mass Spectrom. Ion Phys.* **21**, 349.
13. Lagg, A., Taucher, J., Hansel, A., Lindinger, W. (1994) Applications of proton transfer reactions to gas analysis. *Int. J. Mass Spectrom. Ion Processes* **134**, 55.
14. Howorka, F., Fehsenfeld, F. C., Albritton, D. L. (1979) H^+ and D^+ ions in He: observations of a runaway mobility. *J. Phys. B: Atom. Molec. Phys.* **12**, 4189.
15. Hansel, A., Jordan, A., Holzinger, R. *et al.* (1995) Proton transfer reaction mass spectrometry: on-line trace gas analysis at the ppb level. *Int. J. Mass Spectrom. Ion Processes* **149–150**, 906.
16. Smith, D., Španěl, P. (2005) Selected ion flow tube mass spectrometry (SIFT-MS) for on-line trace gas analysis. *Mass Spectrom. Rev.* **24**, 661.
17. Milligan, D. B., Francis, G. J., Prince, B. J., McEwan, M. J. (2007) Demonstration of selected ion flow tube MS detection in the parts per trillion range. *Anal. Chem.* **79**, 2537.

2

Chemical Ionization: Chemistry, Thermodynamics and Kinetics

2.1 Introduction

Chemical ionization mass spectrometry (CI-MS) is a mature technique for chemical analysis that was first developed in the mid-1960s [1,2]. Since the ionization process in PTR-MS is a form of CI, it is useful to first consider some general aspects of CI-MS before considering the thermodynamics and kinetics of proton transfer.

An attractive reason for introducing CI was that it provided much softer ionization than the dominant ionization technique available at that time, electron ionization (EI).

$$e^- + M \rightarrow M^+ + 2e^- \tag{2.1}$$

Typical EI energies employed in mass spectrometry are in the region of 70 eV, well above the first ionization energies of all chemical compounds. Some of the excess energy available can find its way into the cation, which often leads to copious fragmentation, that is, the molecular ion M^+ dissociates. Useful information can be extracted from the detection of fragment ions, such as the identification of specific functional groups present in the original molecule. However, if fragmentation is excessive it can be difficult or impossible to find a peak in the mass spectrum that can be attributed to the molecular (parent) ion, which in turn makes it difficult to identify the compound(s) responsible for the mass spectrum. This is a particular problem if the material under investigation, the analyte, is a complex mixture of compounds. While the fragmentation process can be reduced by using electron impact energies close to the ionization threshold, there is also a major loss of sensitivity because EI cross sections are much lower at these energies than near 70 eV.

In CI, the ionization of the desired molecule or molecules is achieved through reaction with an atomic or molecular ion. Various processes are possible, ranging from simple charge (electron) transfer through to chemical reactions (formation and/or the breaking

Proton Transfer Reaction Mass Spectrometry: Principles and Applications, First Edition. Andrew M. Ellis and Christopher A. Mayhew.

of chemical bonds) such as proton transfer, hydride ion transfer, or ion–molecule adduct formation. These particular CI processes are shown below.

$$X^+ + M \rightarrow M^+ + X \qquad \text{Charge transfer} \tag{2.2}$$

$$XH^+ + M \rightarrow MH^+ + X \qquad \text{Proton transfer} \tag{2.3}$$

$$X^+ + MH \rightarrow M^+ + HX \qquad \text{Hydride } (H^-) \text{ transfer} \tag{2.4}$$

$$X^+ + M + Z \rightarrow MX^+ + Z \qquad \text{Adduct formation} \tag{2.5}$$

In Reaction 2.5, a third body, specified as Z, is needed to remove some energy, otherwise the adduct MX^+ will dissociate rapidly.

Bimolecular ion–molecule reactions such as those in Reactions 2.2–2.4 are often very fast, by which we mean that they can often occur on every collision, which makes them useful for efficient ionization of target molecules in mass spectrometry. Furthermore, the energy available to deposit into the product cation is typically low, and may be well under 1 eV if an appropriate CI reagent is chosen. This low excess energy means that CI-MS is considered to be a soft ionization technique since it tends to leave parent ions intact. However, it is also wise to bear in mind that some ion–molecule reactions can occur via concerted mechanisms which lead to fragmentation even with much smaller excess energies, so soft ionization is not necessarily fragment-free.

CI-MS can be operated in positive or negative ion modes. The most commonly used CI reagents in positive ion CI-MS have been Brønsted acids, such as the XH^+ reagent in Reaction 2.3. As will be seen in the sections that follow, such reagents are effective when the proton affinity (see Section 2.2.2) of the acceptor molecule, M, exceeds that of the substituent X, which makes the proton transfer thermodynamically allowed. In low-pressure CI, the main CI reagents have tended to be ions such as CH_5^+ and NH_4^+, which can be formed by electron bombardment or via an electrical discharge of the precursor gases CH_4 and NH_3, respectively, followed by ion–molecule reactions in the ionization source which result in the terminal reagent ion. CH_5^+ has been a particular favourite of the CI-MS community because CH_4 has a proton affinity substantially below that of many non-polar organic compounds. The result is that many such compounds undergo proton transfer reactions with CH_5^+ which yield the protonated parent molecule, as well as some fragments, and the combination of these two pieces of information assists compound identification (providing the compound of interest is not part of a complex mixture, otherwise the fragmentation may be disadvantageous, as discussed earlier).

An alternative use of CI is in *atmospheric pressure chemical ionization*, or APCI for short. APCI is frequently combined with liquid chromatography, since it interfaces well with liquid samples. Liquid chromatography provides a means of separating the constituents of a solution and this provides simplification *prior* to the mass spectral analysis. In APCI, the key CI reagent is H_3O^+ and its hydrated analogues, $H_3O^+(H_2O)_n$. These protonated water clusters are secondary ions formed in abundance in atmospheric air from the primary H_3O^+ ions generated in an electrical (corona) discharge source. The proton affinity of water is higher than that of methane and thus use of H_3O^+ tends to result in less fragmentation than with CH_5^+.

H_3O^+ is by far the most widely used reagent ion in both PTR-MS and SIFT-MS analytical techniques. The aim of this chapter is to introduce some basic aspects of proton transfer

Table 2.1 Conversion factors for commonly used energy units

Unit	eV	kcal mol^{-1}	kJ mol^{-1}
eV	1	23.06	96.48
kcal mol^{-1}	0.0434	1	4.184
kJ mol^{-1}	0.0104	0.239	1

chemistry, including thermodynamics and kinetics, as well as presenting an overview of H_3O^+ reactions. In addition, some brief consideration will also be given to alternative proton donors, as well as to other CI reagents that have seen use in suitably adapted PTR-MS instruments.

2.2 Proton Transfer

Gas-phase proton transfer to some molecule M is defined by Reaction 2.3, in which all reactants and products are in the gas phase. Reaction 2.3 represents non-dissociative proton transfer, that is, the proton simply hops from one molecule to another with no further changes. However, we will see later that this proton transfer process sometimes results in fragmentation of the recipient molecule, which is referred to as dissociative proton transfer.

The discussion below deals first with the thermodynamics and kinetics of proton transfer reactions in the gas phase. This is followed in Section 2.2.4 by a general account of proton transfer chemistry, including possible proton transfer reagents.

2.2.1 Energy Units

Before discussing the thermodynamics of proton transfer, some brief comments on energy units are appropriate. Generally, the unit used for energy in thermochemistry is kJ mol^{-1}, and we will predominantly use this unit for specifying the energetics of reactions in this book. However, it is not uncommon in the PTR-MS community to find other units being used owing to its interdisciplinary nature. These include the electron volt (eV) and kcal mol^{-1}, and Table 2.1 provides the conversions necessary to transform one of these units into another.

2.2.2 Thermodynamics of Proton Transfer

Thermodynamics can be used to predict whether or not a reaction can occur under given conditions. The key quantity required is the Gibbs energy change for a reaction, ΔG, where G represents the Gibbs energy and Δ refers to the difference in values between products and reactants. Reaction 2.3 will be thermodynamically spontaneous if the Gibbs energy change for the process is negative.[1] A helpful way of interpreting this quantity for a proton transfer

[1] Thermodynamics reveals the direction and extent of change for a chemical or physical process, but not the speed at which it occurs. In thermodynamics, the word *spontaneous* merely identifies the direction in which change will occur and should not be taken to imply that it is necessarily fast.

reaction is to recognize that it derives from a difference in the Gibbs energy changes for the following two reactions.

$$M(g) + H^+(g) \rightarrow MH^+(g) \tag{2.6}$$

$$X(g) + H^+(g) \rightarrow XH^+(g) \tag{2.7}$$

Since Gibbs energy is a function of state, Hess's law applies and so the Gibbs energy change for Reaction 2.3 can be expressed as

$$\Delta G_T^0(2.3) = \Delta G_T^0(2.6) - \Delta G_T^0(2.7) \tag{2.8}$$

where T is the temperature and the 0 superscript refers to the standard state.

The Gibbs energy change is linked to the equilibrium constant, K_{eq}, for Reaction 2.3 through the following expressions:

$$\Delta G_T^0 = -RT \ln K_{eq} \tag{2.9}$$

$$K_{eq} = \frac{[MH^+][X]}{[M][XH^+]} \tag{2.10}$$

In the above equations T is the absolute temperature, R is the universal gas constant, and the square brackets around the chemical compounds refer to their concentrations in the gas phase.

By accepting protons, reactants M and X in Reactions 2.6 and 2.7 are acting as *bases*. The willingness of a substance to accept a proton is conveniently represented by its *basicity*, which is defined as the negative of the Gibbs energy change for the proton acceptance process. Imagine mixing roughly equal quantities of M and XH^+ in a closed vessel and obtaining roughly equal amounts of MH^+ and XH^+ when equilibrium is reached. This corresponds to an equilibrium constant with a value close to unity, since the numerator and denominator in Equation 2.10 will be almost equal, and in turn this implies that $\Delta G_T^0 \approx 0$. If compound M has a basicity above that of compound X, then ΔG_T^0 is negative for Reaction 2.3 and the equilibrium position is tilted towards the products. This can be quantified using Equation 2.9. For example, if compound M has a basicity some 10 kJ mol^{-1} above that of compound X, substitution into Equation 2.9 yields $K_{eq} \approx 60$ at a temperature of 298 K. Clearly a relatively small difference in basicities will result in a large bias of the reaction towards products or reactants.

Comprehensive tables of gas-phase basicities for a wide range of compounds have been compiled by Hunter and Lias [3] and a list of basicities for some illustrative classes of compounds is provided in Table 2.2. These gas-phase basicities can be used to ascertain the spontaneity of a particular proton transfer reaction, but it is more common to make use of *proton affinity* instead. The proton affinity of a compound is defined as the negative of the *enthalpy* change for the proton acceptance reaction (e.g. Reaction 2.6 or 2.7). Thus if proton acceptance is an exothermic process then the proton affinity is a positive quantity. The gas phase basicity (B) and proton affinity (PA) are related by

$$PA = B - T\Delta S_T^0 \tag{2.11}$$

which follows from the well-known expression

$$\Delta G_T^0 = \Delta H_T^0 - T\Delta S_T^0 \tag{2.12}$$

Table 2.2 *Some illustrative gas phase basicities and proton affinities*

Classification	Molecule	Basicity (kJ mol^{-1})	Proton affinity (kJ mol^{-1})
Inorganic gases	O_2	396	421
	N_2	465	494
	CO_2	516	541
	O_3	626	626
	H_2O	660	691
	NH_3	819	854
Alkanes	Methane	521	544
	Ethane	570	596
	Propane	608	626
	i-Butane	671	678
	Cyclopropane	722	750
Alkenes	Ethene	652	681
	Propene	723	752
Alkynes	Acetylene	617	641
	Propyne	723	748
Aromatic hydrocarbons	Benzene	725	750
	Toluene	756	784
	o-xylene	768	796
	p-xylene	767	794
	Naphthalene	779	803
Alcohols	Methanol	725	754
	Ethanol	746	776
	Propan-1-ol	756	787
	Propan-2-ol	763	793
	Phenol	786	817
Ethers	Dimethyl ether	765	792
	Methyl ethyl ether	781	809
	Diethyl ether	801	828
Aldehydes	Formaldehyde	683	713
	Acetaldehyde	737	769
	Propanal	754	786
Ketones	Acetone	782	812
	Butanone	796	827
	Pentan-2-one	801	833
	Pentan-3-one	807	837
Carboxylic acids	Formic acid	710	742
	Acetic acid	753	784
	Propanoic acid	766	797
Esters	Methyl formate	752	783
	Methyl acetate	791	822
	Ethyl formate	768	799
	Ethyl acetate	805	836
Nitriles	Acetonitrile	748	779
	Benzonitrile	781	812

(continued)

Table 2.2 (Continued)

Classification	Molecule	Basicity (kJ mol^{-1})	Proton affinity (kJ mol^{-1})
Amines	Methylamine	865	899
	Dimethylamine	897	930
	Ethylamine	878	912
	Aniline	851	883
Organosulfur compounds	Dimethyl sulfide	801	831
	Dimethyl disulfide	783	815
	Methane thiol	742	773
	Ethane thiol	758	790
Halocarbons	Fluoromethane	572	599
	Chloromethane	621	647
	Chloroethane	667	693
	Fluorobenzene	727	756

Source: Taken from a compilation by Hunter and Lias [3].
The values for proton affinity and basicity listed are for a temperature of 298 K. For some specific molecules, such as H_2O, CO_2 and ethane, there is a considerable array of experimental and theoretical data to justify an estimated error margin of $\pm$1–3 kJ mol^{-1}. For most other molecules the error margins are somewhat larger, with Hunter and Lias estimating a maximum of error of $\pm$8 kJ mol^{-1}.

where H is the enthalpy and S is the entropy. The assumption behind the use of an enthalpy change (as in the proton affinity) instead of the Gibbs energy change (as in basicity) to judge reaction spontaneity is that the entropy change, ΔS_T^0, in Equation 2.11 is negligibly small. In most instances, this is likely to be a good approximation since a proton is a simple entity whose transfer from one molecule to another is unlikely to alter the net system entropy in any substantial way. Thus, in almost all practical cases, the proton transfer process represented in Reaction 2.3 will be thermodynamically spontaneous if the proton affinity of M is greater than that of X.

Table 2.2 also lists proton affinities. The most commonly used proton donor in PTR-MS is H_3O^+, and therefore the relevant proton affinity is that of its conjugate base, H_2O. The best available value for the proton affinity of H_2O is 691 $\pm$ 3 kJ mol^{-1}, which is lower than those of all the organic compounds listed in Table 2.1, with the exception of some of the aliphatic hydrocarbons and haloalkanes. Consequently, from a thermodynamic point of view, H_3O^+ will be a willing proton donor to a whole swathe of organic compounds and this fact, which is central to PTR-MS, has already been commented on in the previous chapter. In contrast, and once again as commented on earlier, it can be seen from Table 2.2 that the reactions of H_3O^+ with commonly encountered inorganic compounds in air, such as nitrogen and oxygen, are endothermic and are therefore thermodynamically unfavourable at room temperature.

The justification for employing proton affinities can be seen from inspection of Table 2.2. For example, ethanol has a substantially higher proton affinity than water, the difference being 85 kJ mol^{-1}. In terms of gas-phase basicities, the difference is 86 kJ mol^{-1}, which is almost identical to the proton affinity difference and well within the expected error margins for the experimental determinations of these quantities. Similar conclusions can be reached for other pairings in Table 2.2, demonstrating that it is indeed valid to employ proton affinities in place of basicities for determining the direction of spontaneous change in most proton transfer reactions.

Before we proceed to move to the kinetics of proton transfer, a comment on the effect of temperature is needed. It is usual for the drift tubes in PTR-MS instruments to be operated at elevated temperatures and temperatures up to 100°C are commonly used. This variation in applied drift tube temperature is not particularly important as far as the thermodynamics is concerned, since ΔS_T^0 is expected to be near to zero for proton transfer and ΔH_T^0 will show only a small variation over a temperature range of a few tens of degrees. However, as will be seen in Chapter 3, the effective temperature for ion–molecule reactions inside the drift tube in PTR-MS is expected to be much higher than the nominal drift tube temperature, and indeed will be above 1000 K, because of increased collision energies caused by the applied electric field. At this substantially elevated temperature, entropic effects, as represented by the $-T\Delta S_T^0$ term in Equation 2.11, might be significant, as might the temperature dependence of ΔH_T^0. Thus the use of thermodynamic parameters specified at 298 K to predict which processes may be thermodynamically spontaneous in PTR-MS needs to be applied with some caution.

2.2.3 Kinetics of Proton Transfer

2.2.3.1 Background

Employing thermodynamics to determine whether or not a reaction is likely to take place is, of course, only half the story. The other important factor is the rate of the reaction, and it is possible for a reaction that is thermodynamically allowed to be very slow in practice. A slow reaction will result if there is an excessive energy barrier (activation energy) for the reactants to surmount in order to transform themselves into products. Assuming that entropic effects are small, an endothermic reaction has, by definition, a minimum energy barrier equal to the enthalpy of the reaction. An exothermic reaction has no such *minimum* energy barrier but it is still possible that a large barrier might have to be surmounted for reaction to occur. A reaction may also be very slow if it requires a very specific orientation of reactants in order for the reaction to proceed (passing through a so-called *tight transition state*).

Extensive early work by Bohme and co-workers, using the flowing afterglow method (see Chapter 1), found that *exothermic* proton transfer reactions are fast [4]. More generally, this is typical of bimolecular *ion–molecule reactions*, a class of reaction to which proton transfer reactions belong. The reasons for this are twofold. Firstly, because such reactions usually involve the transfer of a charged entity from one place to another without any rupture of 'normal' chemical bonds, there is little or no activation energy. Secondly, long-range attractive forces between the ion and neutral molecule can serve to increase the collisional cross section beyond that which would be expected on the basis of the nominal sizes of the individual molecules.[2] In other words, an ion and molecule on a trajectory that would

[2] The collisional *cross section*, which is normally given the symbol σ, is a measure of the effective size of the colliding particles, expressed as an area. The link between the rate coefficient and σ is $k = <\sigma v>$, where v refers to the relative velocity of the colliding molecules and the brackets represent a Boltzmann-weighted average of the cross section (which depends on the collision energy) and the relative collision velocity.

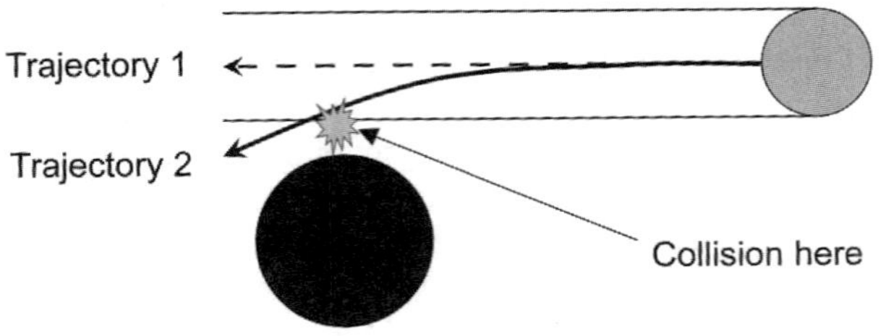

Figure 2.1 Trajectories for two hypothetical spherical molecules. Two trajectories are shown by arrows. Trajectory 1 (dashed line) is the trajectory that would be taken in the absence of any attractive intermolecular forces, which would see the two molecules miss each other. This trajectory is modified by long range attractive forces (trajectory 2, solid line) which create an effective collision cross section that is larger than the hard-sphere limit. In this particular case, the modification leads to a successful collision. The effective increase in collision cross section over the hard sphere limit ($\pi(r_1 + r_2)^2$ where r_1 and r_2 are the radii of the spherical particles) is very significant if one of the molecules is charged, since this can give rise to substantial charge-induced dipole attractive forces.

otherwise result in a near miss can be drawn together and made to collide with the help of the long-range forces, as illustrated in Figure 2.1. Of course long-range (van der Waals) forces are also in operation in neutral–neutral reactions but these are much weaker and therefore less effective in influencing the reaction cross section, whereas in an ion–molecule system there are much stronger electrostatic forces present such as charge–dipole and/or charge-induced dipole forces [5]. Consequently, in ion–molecule reactions the pre-exponential term in the Arrhenius equation, $k = A\ \exp(-E_a/RT)$, is enlarged when compared with neutral–neutral reactions. This should result in ion–molecule reactions being substantially faster than barrierless neutral–neutral reactions and this is borne out in numerous kinetic studies. Fast bimolecular ion–molecule reactions typically show rate coefficients $\geq 10^{-9}$ cm^3 s^{-1}, whereas in neutral–neutral reactions they are at least an order of magnitude smaller.[3] A compilation of the early kinetic data for proton transfer reactions can be found in the book by Ikezoe and co-workers [6] while a more up-to-date version is available online [7].

A comprehensive set of thermal rate data (300 K) for proton transfer reactions involving H_3O^+ is available, thanks largely to SIFT-MS studies by Smith and Španěl (see for example a recent summary in Reference 8), along with contributions from other groups such as Viggiano and co-workers [9–12] and Arijs and co-workers [13–15] using either SIFT-MS or flowing afterglow measurements. Rate coefficients obtained for some selected H_3O^+ reactions with organic molecules are presented in the final column of Table 2.3. Rate coefficients for proton transfer from H_3O^+ to many different classes of organic molecules are now available but the conclusion remains the same as from the earlier flowing afterglow work, namely that the exothermic proton transfer reactions are invariably fast, and in most cases agree very closely with theoretical rate coefficient predictions based on

[3] A second order rate coefficient is usually expressed either as a molar quantity, with units of dm^3 mol^{-1} s^{-1}, or in cm^3 s^{-1}. The latter is sometimes written as cm^3 molecule^{-1} s^{-1} to specify that it is 'per molecule' but since 'molecule' is not actually a unit it can be omitted and this is the practice adopted in this book.

Table 2.3 Comparison of calculated and experimental rate coefficients (expressed in 10^{-9} cm^3 s^{-1} and for a temperature of 300 K) for proton transfer reactions involving H_3O^+ and some illustrative VOCs (the dipole moments and polarizabilities of the neutral reactant are also listed). Definitions of k_L, k_{ADO} and k_{cap} are provided in section 2.2.3.2.

VOC	μ_D (Debye)[a]	$10^{30} \times \alpha$ (m^3)[a]	k_L	k_{ADO}	k_{cap}	k_{expt}
H_2S	0.97	3.95	1.33	1.69	1.92	1.9 ± 0.4[b]
NH_3	1.47	2.10	1.13	2.08	2.67	2.4 ± 0.5[b]
Propene	0.37	6.26	1.62	1.64	1.74	1.7[c]
Propyne	0.78	6.18	1.62	1.81	1.98	1.8[c]
Benzene	0	10.3	1.92	–	–	1.8[d]
Toluene	0.38	12.3	2.06	2.07	2.16	2.3[d]
Methanol	1.70	3.29	1.23	2.16	2.70	2.7 ± 0.8[b]
Ethanol	1.69	5.84	1.54	2.29	2.74	2.8 ± 0.7[b]
Formaldehyde	2.33	2.45	1.07	2.58	3.55	3.4 ± 0.9[b]
Acetaldehyde	2.75	4.6	1.38	2.97	3.99	3.6 ± 0.9[b]
Acetone	2.88	6.4	1.56	3.09	4.06	3.9 ± 1.0[b]
Formic acid	1.41	3.4	1.18	1.83	2.22	2.7 ± 0.8[b]
Acetic acid	1.70	5.1	1.39	2.14	2.60	3.0 ± 0.9[b]

[a]Reference 16.
[b]Reference 17.
[c]Reference 18.
[d]Reference 19.

barrierless ion–molecule capture processes (see next section). Consequently, the absence of an experimental value for the rate coefficient of a particular proton transfer reaction of importance in a PTR-MS study is not necessarily an impediment since a theoretical value can be derived that is likely to possess a margin of error comparable to that in any experimental determination. In the next section, we describe some of the theoretical models developed for predicting the rate coefficients of ion–molecule reactions.

2.2.3.2 *Theoretical Prediction of Proton Transfer Rate Coefficients*

There are a number of theoretical prescriptions for determining rate coefficients of exothermic ion–molecule reactions. The simplest originates from concepts developed more than a century ago by Langevin and applies to a reaction between a point ion and a spherical (non-polar) molecule [20]. The dominant long-range attractive force in this case is the ion-induced dipole interaction, which takes the form

$$V(r) = -\frac{1}{4\pi\varepsilon_0}\frac{\alpha q^2}{2r^4} \tag{2.13}$$

where q is the charge on the ion, α is the polarizability of the neutral reactant molecule and ε_0 is the permittivity of free space. The above expression gives the potential energy as a function of the ion–molecule separation, r. The polarizability is a measure of the ease with which an electric dipole moment can be induced in a molecule by an applied electric field. An approaching ion provides such an electric field and will distort the charge distribution in the molecule such that a region of opposing charge will be drawn towards the approaching ion, creating a dipole moment in the neutral molecule. This ion-induced dipole interaction is attractive and the more polarizable the neutral molecule, the stronger the attraction will be for a given ion–molecule separation, as illustrated in Equation 2.13.

Incorporation of this attractive potential into a simple collision model, which we shall not present here, allows the prediction of the reaction cross section and, ultimately, the reaction rate coefficient. The Langevin rate coefficient, as it is often known [21], is given by

$$k_L = \sqrt{\frac{\pi \alpha q^2}{\mu \varepsilon_0}} \tag{2.14}$$

where μ is the reduced mass of the colliding partners ($\mu = m_1 m_2/(m_1 + m_2)$ where m_1 and m_2 are the masses of the ion and the neutral molecule, respectively). This and other rate coefficient expressions presented below are also commonly referred to as *capture* rate coefficients because the long-range attractive force 'captures' the approaching reactants and can bring them into a favourable trajectory for reaction. Equation 2.14 predicts the collision rate coefficient for many reactions between ions and non-polar molecules quite accurately. It is also noteworthy that Equation 2.14 predicts that the rate coefficient is unaffected by temperature.

Equation 2.14 also provides a good starting point for calculating rate coefficients of ion–molecule reactions involving polar molecules (i.e. molecules with a permanent electric dipole moment). However, in the case of polar molecules it is necessary not only to account for the ion-induced dipole interaction but also to consider the interaction between the ion and the permanent electric dipole moment (μ_D) of the neutral molecule, which takes a very different form from the potential energy expression in Equation 2.13. The potential energy for a charge–dipole interaction is given by

$$V(R, \theta) = -\frac{1}{4\pi\varepsilon_0}\frac{q\mu_D}{2r^2}\cos\theta \tag{2.15}$$

Notice that in this case the potential energy is a function of both the atom–molecule separation, r, and the angle, θ, between a line joining the ion to the centre of charge in the neutral molecule and the direction of the permanent dipole moment vector. Notice that the dependence on r is now r^{-2} rather than r^{-4}, so on this basis alone if the Langevin model was used it will underestimate the rate coefficient for a reaction between a point ion and a polar molecule.

A good theoretical prediction of the rate coefficient for a reaction between a point ion and a polar molecule should account for both the charge-induced dipole interaction and the ion-permanent dipole interaction. However, the latter interaction is non-trivial to deal with because θ is not a fixed quantity (the molecule can rotate) and therefore some means must be found to account for its variation in real colliding systems. An early solution was to ignore any such variation and to assume instead that $\theta = 0$, since this is the lowest energy orientation when a positive ion approaches a polar molecule and one might assume that this would be statistically preferred. This is known as the locked-dipole model, but it is unsatisfactory because it tends to overestimate rate coefficients by unrealistically discounting a range of thermally accessible dipole orientations that would give weaker charge–dipole interactions.

A more realistic description was provided by Su and Bowers and is known as the average dipole orientation (ADO) theory [22, 23]. The ADO rate coefficient is given by

$$k_{\text{ADO}} = \sqrt{\frac{\pi \alpha q^2}{\mu \varepsilon_0}} + \frac{C \mu_{\text{D}} q}{\varepsilon_0} \sqrt{\frac{1}{2\pi \mu k_{\text{B}} T}} \tag{2.16}$$

where k_{B} is the Boltzmann constant. The first term on the right hand side of Equation 2.16 is just the Langevin contribution seen already in Equation 2.14. The ion-permanent dipole interaction is accounted for in the second term. C is a dimensionless '*locking*' parameter ($0 \leq C \leq 1$) that accounts for the average orientation of the permanent dipole moment of the neutral molecule. When $C = 1$, Equation 2.16 is equivalent to the locked-dipole theory. In practice C turns out to be a function of the dipole moment and polarizability, μ_{D} and α, respectively, of the neutral molecule, and specifically the ratio $\mu_{\text{D}}/\alpha^{1/2}$. It is also a function of temperature. The locking parameter has been parameterized as a function of $\mu_{\text{D}}/\alpha^{1/2}$ and thus, provided μ_{D} and α are known for the neutral molecule of interest, it is then a simple matter to extract the value of C [24]. Values of the locking parameter, expressed as a function of $\mu_{\text{D}}/\alpha^{1/2}$, are shown in Figure 2.2.

In the next section, some example calculations of rate coefficients will be given but we note here that a comparison of the rate coefficients predicted from the ADO procedure with experimental values shows that the ADO values tend to underestimate the rate coefficients,

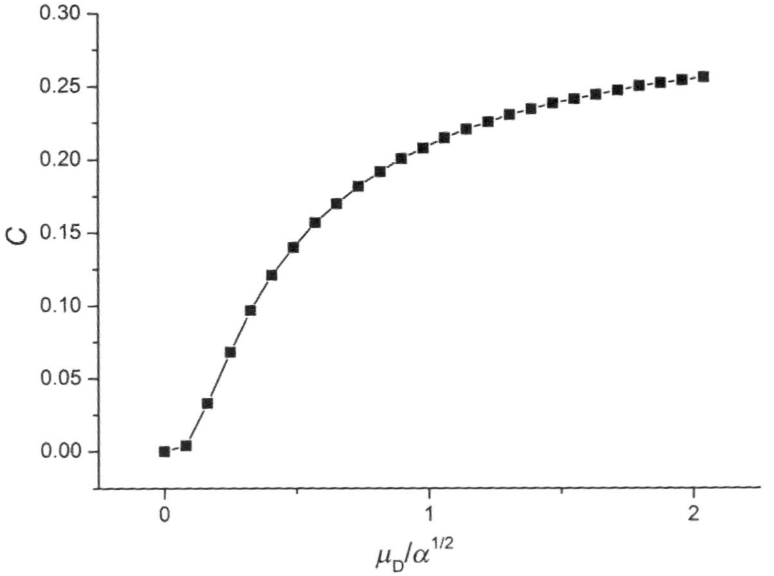

Figure 2.2 *Variation of locking parameter C with the ratio $\mu_{\text{D}}/\alpha^{1/2}$. These data were taken from Reference 24 and apply for a temperature of 300 K. The effect of temperature is small over a range of a few tens of degrees and so can usually be neglected. Note that μ_{D} has been expressed in Debyes and α in Å^3 to obtain the values of $\mu_{\text{D}}/\alpha^{1/2}$ shown on the horizontal axis. Reprinted with permission from [23]. Copyright © 1973, American Chemical Society.*

typically by 10–20% [25, 26]. This is attributable in large part to the neglect of the dipole moment of the charged reagent, which is regarded as a point charge in the standard ADO treatment. Nevertheless, the ADO theory provides a means of obtaining decent estimates of rate coefficients for proton transfer reactions where the experimental values are unknown.

Instead of seeking an analytical expression for the rate coefficient on the basis of classical expressions for the ion–molecule interaction potential, an alternative approach is to model the reaction process through a series of classical trajectory calculations. The details are somewhat complicated and so the interested reader should consult the original references. Importantly, the results of such trajectory calculations have been parameterized by Su and Chesnavich to give expressions which allow calculation of the rate coefficient [27]. The resulting rate coefficient, termed the capture rate coefficient, k_{cap}, is given by

$$k_{cap}(T) = k_L \times K_{cap}(T_R, I^*) \tag{2.17}$$

where k_L is the Langevin rate coefficient. The quantity $K_{cap}(T_R, I^*)$ is a parameterized quantity determined by Su and Chesnavich via several thousand trajectory calculations, giving a multiplier of k_L that incorporates dipole locking effects. Both T_R and I^* are quantities that depend on α and μ_D of the neutral molecule according to

$$T_R = 4\pi\varepsilon_0 \frac{2\alpha k_B T}{\mu_D^2} \tag{2.18}$$

$$I^* = \frac{\mu_D I}{\alpha q \mu} \tag{2.19}$$

where I is the moment of inertia of the neutral molecule and all other symbols are as defined earlier. It turns out that K_{cap} is insensitive to the value of I^* when

$$I^* < \frac{0.7 + x^2}{2 + 0.6x} \tag{2.20}$$

where

$$x = \sqrt{\frac{1}{T_R}} \tag{2.21}$$

For reacting systems of the type we encounter in PTR-MS the values of α, μ_D and I are such that K_{cap} is in the insensitive region and thus inequality Equation 2.20 applies. In such circumstances, the parameterization process shows that

$$K_{cap} = \begin{cases} \dfrac{(x + 0.5090)^2}{10.526} + 0.9754; & x \leq 2 \\ 0.4767x + 0.6200; & 2 \leq x \leq 3 \\ 0.5781x + 0.3165; & 3 \leq x \leq 35 \\ 0.6201x - 1.153; & 35 \leq x \leq 60 \\ 0.6347x - 2.029; & x \geq 60 \end{cases} \tag{2.22}$$

Consequently, for most reacting systems K_{cap} depends only on T_R and is simple to calculate using a combination of Equations 2.18, 2.21 and 2.22. An application of the above will be given in the next section.

In the past 25 years, various improvements have been made to the capture theory, including parameterization over a wide temperature range [28] and the extension to reactions between ions and quadrupolar linear molecules [29, 30].

If experimental values of the polarizability and/or the dipole moment of the neutral molecule are unavailable, as is the case for many molecules, then these quantities can be calculated from *ab initio* or density functional theory (DFT) quantum chemical calculations. Zhao and Zhang have provided a comprehensive evaluation of this approach by carrying out calculations on 78 hydrocarbons and 58 non-hydrocarbons [31]. A quite modest level of theory (DFT with the B3LYP functional and a 6–31G(d,p) Gaussian basis set) gave polarizabilities and dipole moments which yielded good agreement with experiment (where such values were available). The calculated dipole moments and polarizabilities were then employed in ADO calculations to obtain thermal rate coefficients. For most molecules, the predicted proton transfer rate coefficient was within $\pm 25\%$ of the value determined from SIFT measurements.

2.2.3.3 *Illustrative Calculations of Proton Transfer Rate Coefficients and Comparison with Experiment*

To show how the theoretical prescriptions described in the previous section can be applied, we shall work through the calculation of rate coefficients for an example reaction, namely the protonation of acetone by H_3O^+.

$$H_3O^+ + CH_3COCH_3 \rightarrow H_2O + [CH_3COCH_3]H^+ \tag{2.23}$$

In order to attempt the calculation, the polarizability and permanent dipole moment of acetone are required. Literature values of these quantities are $\alpha = 6.4 \times 10^{-30}$ m^3 and $\mu_D =$ 2.88 D [16].[4] The reduced mass for this system is 2.38 $\times$ 10^{-26} kg and other quantities required are $q = e = 1.602 \times 10^{-19}$ C and $\varepsilon_0 = 8.854 \times 10^{-12}$ J^{-1} C^2 m^{-1}. There is now sufficient information to calculate the Langevin rate coefficient, which yields

$$k_L = 1.57 \times 10^{-15}\ \text{m}^3\ \text{s}^{-1} = 1.57 \times 10^{-9}\ \text{cm}^3\ \text{s}^{-1}$$

Next up is the ADO rate coefficient, which takes into account the permanent dipole moment of acetone. The locking parameter, *C*, is obtainable from Figure 2.2 or from Reference 24, but this requires the correct units for $\mu_D/\alpha^{1/2}$. In units of D/Å$^{3/2}$, $\mu_D/\alpha^{1/2} =$ 1.138, which from Figure 2.2 leads to $C = 0.220$. Finally, before calculating k_{ADO} we must convert the electric dipole moment of acetone into SI units, which gives $\mu_D = 9.61 \times 10^{-30}$ C m and hence

$$k_{ADO} = 3.10 \times 10^{-15}\ \text{m}^3\ \text{s}^{-1} = 3.10 \times 10^{-9}\ \text{cm}^3\ \text{s}^{-1}$$

Finally, we turn to the capture rate coefficient, and begin by obtaining a value for T_R.

$$T_R = 4\pi\varepsilon_0 \frac{2\alpha k_B T}{\mu_D^2} = 4 \times \pi \times 8.854 \times 10^{-12} \times \frac{2 \times 6.4 \times 10^{-30} \times 1.381 \times 10^{-23} \times 300}{(9.61 \times 10^{-30})^2}$$
$$= 0.064$$

[4] The Debye, symbol D, is a non-SI unit commonly used to indicate the magnitude of an electric dipole moment. 1 D = 3.33564 $\times$ 10^{-30} C m.

Thus $x = 3.95$ (see Equation 2.21) and we can therefore use the expression $K_{cap} = 0.5781x + 0.3165$ to obtain $K_{cap} = 2.60$. Consequently, this gives

$$k_{cap} = 2.60 \times k_L = 4.08 \times 10^{-9}\ \text{cm}^3\ \text{s}^{-1}$$

The experimental rate coefficient for the reaction of H_3O^+ with acetone has been determined to be $(3.9 \pm 1.0) \times 10^{-9}$ cm^3 s^{-1} at 300 K. Clearly the Langevin rate coefficient lies significantly below the experimental rate coefficient, which is expected given the neglect of the ion-permanent dipole interaction. Both the ADO and the trajectory theories do significantly better, with the latter in particular yielding excellent agreement with experiment.

Table 2.3 shows a comparison of the calculated and experimental rate coefficients for several other compounds. The comments made for the H_3O^+ reaction with acetone apply also to this larger group of reactions. Where supplied, the error margins for the experimental rate coefficients are about 20–30% of the mean value. In every reaction listed in Table 2.3, the capture rate coefficient lies within the stated error margins and in most cases is close to the quoted experimental mean value. From this comparison it seems fairly safe to assume that the calculated k_{cap} rate coefficient is likely to be every bit as reliable as any experimentally determined value and thus, as mentioned earlier, in the absence of an experimental rate coefficient, trajectory methods provide the best option for determining a reaction rate coefficient for subsequent use in analysing PTR-MS data.

2.2.4 Reagents and Mechanisms

2.2.4.1 Chemistry of H_3O^+ Reactions

We have seen that exothermic proton transfer reactions in the gas phase tend to be fast, usually occurring at the collisional rate. Attention now turns to the chemical outcome of these reactions. In particular, we wish to know whether or not proton addition to the recipient molecule results in a single cationic product (the protonated parent molecule via non-dissociative proton transfer) or whether fragmentation occurs (via dissociative proton transfer). We draw here mainly upon extensive information available from SIFT-MS studies. Broadly speaking, the proton transfer chemistry in SIFT-MS and PTR-MS should be similar, although the additional collision energy brought about by the use of an electric field in the drift tube in PTR-MS (see Section 3.4.2) can enhance the degree of fragmentation, so this should always be borne in mind when drawing upon the information below. Our aim here is not to provide an extensive account of the proton transfer chemistry, since that can be found in the original research publications. Instead, the purpose of this section is to try to identify some general trends for certain classes of VOCs and to show that simple non-dissociative proton transfer is by no means the only possible reaction.

Proton transfer is generally seen as a soft form of ionization and this is claimed to be one of the strengths of PTR-MS, since it should result in only one product ion for each VOC present in the analyte gas, thus avoiding the added complexity of ion fragmentation. However, soft ionization does not necessarily mean that no fragmentation of ions can occur and such fragmentation is possible in the ionization of some molecules even without the extra collision energy provided by the drift tube in PTR-MS. A good example is proton transfer from H_3O^+ to propan-2-ol. On the basis of the proton affinities (see Table 2.2) proton transfer from H_3O^+ to propan-2-ol is exothermic by 102 kJ mol^{-1}. However, the dehydration reaction

$$H_3O^+ + C_3H_7OH \rightarrow 2H_2O + C_3H_7{}^+ \tag{2.24}$$

can be calculated to be almost thermoneutral, using known enthalpies of formation for the reactants and products, and the resulting endothermicity is only around 4 kJ mol^{-1}. Consequently, the excess energy from the initial proton transfer is more than sufficient to initiate subsequent dehydration and indeed this is the dominant process seen.

Table 2.4 summarizes the main product channels for proton transfer reactions involving H_3O^+ with VOCs for several different functional groups. This table is intended to be illustrative rather than exhaustive and some other categories of VOCs have been studied in addition to those featured in the table. Some of the main observations are highlighted in the discussion which follows.

Light alkanes do not undergo exothermic proton transfer reactions with H_3O^+. However, as the molecular mass increases, the proton affinity of the alkane increases and therefore for larger alkanes the proton transfer process eventually becomes exothermic. Arnold and co-workers have estimated that the endothermic/exothermic crossover point occurs at hexane, such that all heavier alkanes should have clear exothermic proton transfer reactions with H_3O^+ [10]. For heptane and higher alkanes, fast reaction with H_3O^+ is seen but the rate coefficient falls short of that predicted from the collision-limiting models presented in Section 2.2.3.2. Furthermore, reaction is dominated not by proton transfer but instead by association, that is

$$H_3O^+ + C_nH_{2n+2} + X \rightarrow H_3O^+(C_nH_{2n+2}) + X \tag{2.25}$$

where X is a third body.

In contrast to non-cyclic alkanes, reaction of H_3O^+ with cyclic alkanes is exothermic, even for cyclopropane (proton affinity = 750 kJ mol^{-1}). Reaction is by simple non-dissociative proton transfer in the case of cyclopropane, but larger cyclic alkanes can undergo dissociative channels. For example methylcyclohexane reaction is dominated by H_2 ejection (90% of the products are produced in this reaction channel) [18].

Reactions between short chain alkenes and H_3O^+ are more favourable thermodynamically but the reaction with ethene is still (marginally) endothermic and relatively slow [32]. However, propene and higher alkenes undergo proton transfer at the collisional rate and the small alkenes show 100% production of the protonated molecule. However, dissociative channels are important for 1-heptene and higher alkenes [33].

The proton transfer reaction between H_3O^+ and acetylene (C_2H_2) is substantially endothermic and therefore cannot occur [32, 33]. However, propyne (C_3H_4), which has a proton affinity greater than H_2O, undergoes fast non-dissociative proton reaction with H_3O^+.

In aromatic hydrocarbons with saturated sidechains the protonation takes place at the aromatic ring. This leads to very simple behaviour in their reaction with H_3O^+, showing fast reaction (collision limited) with 100% production of the protonated parent species [33].

All alcohols react with H_3O^+ at the collisional rate. Methanol and ethanol show almost 100% non-dissociative proton addition at 300 K but it is worth emphasizing again that in PTR-MS the proton transfer reactions take place under conditions which cannot be described as anywhere near 300 K. In fact, the reaction outcome can be strongly dependent

Table 2.4 *Main product channels for reaction of H_3O^+ with volatile organic compounds at 300 K*

Functional group	Reaction products	References
Alkanes	No reaction for C_5 and lighter acyclic alkanes. C_6 and higher species give only association products (H_3O^+.M) but reaction is much slower than the collisional limit. Cycloalkanes react at the collisional limit	[10, 18, 33]
Alkenes	Fast reaction for all alkenes except ethane. 100% MH^+ production for small alkenes but fragmentation important for C_7 and higher alkenes	[18, 32, 33]
Alkynes	Slow (endothermic) reaction for acetylene but fast reaction for larger alkynes dominated by MH^+ production	[18, 32, 33]
Aromatic hydrocarbons	Yield almost exclusively MH^+	[33]
Alcohols	C_3 and higher alcohols show an increasing tendency to undergo dehydration on protonation. Dehydration is the *only* product channel for tertiary alcohols	[34, 35]
Ethers	Main product MH^+, although dissociative channels grow in importance as the complexity of the ether increases	[36]
Aldehydes	100% production of MH^+ for C_3 and lower aldehydes but increasing tendency to eject H_2O for C_4 and higher species	[37]
Ketones	Almost 100% production of MH^+ regardless of chain length	[37]
Carboxylic acids	Dominated by MH^+ formation, but accompanied by small amount of dehydration product	[38]
Esters	MH^+ formation is main channel for small esters but increasing propensity for major ion fragmentation, particularly once the alcohol conjugate is a propyl or higher unit	[38, 39]
Nitriles	100% production of MH^+	[40]
Amines	MH^+ main product, with minor channels involving alkene or H_2 loss also found for some amines	[40]
Organosulfur compounds	100% production of MH^+	[41, 42]
Organohalides	Most halomethanes undergo slow reaction. Heavier halomethanes undergo a variety of reactions, some of which include termolecular association with H_3O^+. On the other hand, aromatic halides show 100% MH^+ production in fast reactions	[43, 44]

on the reduced electric field conditions employed in the drift tube, as represented by the ratio of the electric field strength to the molecular number density in the drift tube, *E/N* (see Section 3.4.1). This was demonstrated in a PTR-MS study of the effect of *E/N* for various alcohols [34]. For example, at a fairly typical *E/N* of 115 Td the dehydration product for reaction between H_3O^+ and ethanol, $C_2H_5^+$, is already significant and becomes comparable in abundance to the protonated ethanol at 140 Td. Dehydration reactions dominate for all C_3 and higher alcohols, as mentioned earlier for propan-2-ol [35]. These reactions switch on with increasing size even near room temperature because the cationic hydrocarbon fragment increases in stability with size (as reflected in a lowering of the ionization energy of the corresponding neutral fragment) and thus water ejection becomes thermodynamically favourable. Under typical drift tube conditions employed in PTR-MS, the protonated parent molecule for C_3 and higher alcohols is not seen at all [34].

For small ethers, non-dissociative proton transfer is the major channel and indeed for the lightest ethers, such as diethyl ether, it is the sole product channel [36]. When dissociative channels occur, they involve either alkene ejection or alcohol ejection.

Aldehydes and ketones respond differently on reaction with H_3O^+ [37]. Ketones show only non-dissociative proton addition, whereas in aldehydes there is the possibility of dehydration products. Formaldehyde, acetaldehyde and propanal all show 100% non-dissociative reaction but the dehydration channel switches on at butanal for aldehydes with saturated carbon chains, although the non-dissociative channel still dominates for butanal and pentanal.

Carboxylic acids show strongly exothermic reactions with H_3O^+ and should therefore react at the collisional limit. As with alcohols and aldehydes, dehydration reactions have been identified for all but the smallest acids, formic and acetic [38]. However, for the small number of carboxylic acids whose reactions have been characterized, the non-dissociative channel dominates ($\geq$90%) in almost every case. Dehydration is also an issue for esters, although non-dissociative proton transfer has been established as the dominant reaction channel for many relatively simple esters [38]. The formates, methyl and ethyl formate, are somewhat unusual in that their dissociative channels, both accounting for $<$5% of product yield at 300 K, involve CO ejection rather than dehydration. With esters derived from propyl alcohol and higher alcohols, fragmentation is ubiquitous and can involve a number of reaction channels [39].

Proton transfer from H_3O^+ to amines is strongly exothermic and reactions are fast. Non-dissociative proton transfer dominates and for some reactions, such as for dimethylamine and aniline, this is the sole product channel [40]. Nitriles are similarly sturdy, with acetonitrile and benzonitrile yielding 100% protonated parent species when reacting with H_3O^+ [40].

Simple thiols and organic sulfides, such as methylthiol, ethanethiol, dimethyl sulfide and dimethyl disulfide, give 100% production of the protonated parent [41, 42], with the reactions taking place at the collisional rate limit.

Many haloalkanes are unfavourable in their reactions with H_3O^+. The halomethanes tend to react slowly on account of their reaction being endothermic or nearly thermoneutral [43]. Slightly heavier compounds, such as chloroethanes, show much faster reaction near the collisional limit but there is a tendency for some of these compounds to form adducts with H_3O^+ through termolecular reaction, rather than undergo simple proton addition. Halobenzenes show much simpler chemistry paralleling that of aromatic hydrocarbons, with 100% production of the protonated species without dissociation [44].

2.2.4.2 *Reactions of Hydrated Hydronium Clusters*

Ideally, if H_3O^+ is the chosen proton donor, then that should be the only ion reacting with the analyte gas in the drift tube of a PTR-MS instrument. In practice, the presence of unreacted water vapour from the ion source, from degassing of the walls of the drift tube, and from the analyte gas, inevitably leads to some formation of cluster ions of the type $H_3O^+(H_2O)_n$ via the general process

$$H_3O^+(H_2O)_{n-1} + H_2O + X \rightarrow H_3O^+(H_2O)_n + X \tag{2.26}$$

where X is a third body. In PTR-MS, attempts are made to minimize the proportion of $H_3O^+(H_2O)_n$ ions ($n \geq 1$) relative to H_3O^+ in the drift tube through the use of collision-induced dissociation. Nevertheless, despite these efforts, hydrated hydronium ions are still frequently observed in mass spectra from PTR-MS, particularly when investigating humid samples and/or when operating under low *E/N* conditions, and so it is important to be aware of the impact they may have on the ion chemistry.

A key point to note is that water clusters possess higher proton affinities than the bare water molecule (691 $\pm$ 3 kJ mol^{-1}) [3]. For example, the water dimer, $(H_2O)_2$, has a proton affinity of 808 $\pm$ 6 kJ mol^{-1} [45]. The higher proton affinity is the result of the added stability of the H_3O^+ brought about by sharing the positive charge with an additional water molecule. In other words, the H_3O^+ cation is stabilized through solvation by a water molecule. As more water molecules are added, the proton affinity of the water cluster increases but the incremental effect declines in magnitude as the cluster grows.

Important consequences result from hydrated hydronium formation. First of all, the higher proton affinity means that some reactions that occur with H_3O^+ do not occur with $H_3O^+(H_2O)_n$. A good example is acetaldehyde, whose proton affinity (see Table 2.2) lies between those of H_2O and $(H_2O)_2$ and will therefore accept a proton from H_3O^+ but not from $H_3O^+(H_2O)$. This brings us to the second important point about water cluster ions, which is that proton transfer is not the only possible reaction channel. In the case of acetaldehyde, it is known that reaction with H_3O^+ does indeed proceed by proton transfer and occurs at the collision-limiting rate. However, as shown by flowing afterglow [17] and SIFT [11] studies, reaction of $H_3O^+(H_2O)$ with acetaldehyde also occurs at the collision-limiting rate, but in this case proceeds via so-called ligand switching:

$$H_3O^+(H_2O) + CH_3CHO \rightarrow H_3O^+(CH_3CHO) + H_2O \tag{2.27}$$

The rate coefficient for the reaction between hydrated hydronium ($H_3O^+(H_2O)_n$) and acetaldehyde decreases as the number of water molecules, n, increases, as illustrated in Table 2.5. One factor that contributes to this decrease is an increase in the reduced mass of

Table 2.5 *Rate coefficients (10^{-9} cm^3 s^{-1}) for reaction of $H_3O^+(H_2O)_n$ (n = 0, 1, 2 and 3) with acetaldehyde*

H_3O^+	$H_3O^+(H_2O)$	$H_3O^+(H_2O)_2$	$H_3O^+(H_2O)_3$
3.6 $\pm$ 0.9	3.1 $\pm$ 0.8	2.6 $\pm$ 0.7	2.3 $\pm$ 0.6

Source: Taken from Reference 17.

the colliding partners (see Equation 2.14). Thus ligand switching both complicates the mass spectrometry, since it generates product ions beyond the simple proton transfer product, and it introduces chemistry where the rate of reaction depends on the degree of hydration. As this example shows, wherever possible it is good practice to try and minimize the proportion of $H_3O^+(H_2O)_n$ ions in the drift tube, but when exploring a humid sample this may be challenging. The consequences this can have on quantitative analysis are discussed again in Section 4.6.

Toluene is another molecule that possesses a lower proton affinity than the water dimer and therefore if any reaction with $H_3O^+(H_2O)$ occurs then it would be expected to be dominated by a ligand-switching process analogous to Reaction 2.27 above. In fact, SIFT studies show that proton transfer is the dominant process and occurs at roughly half the expected capture rate at room temperature, even though the reaction is endothermic by 25 kJ mol^{-1} [12]. The reason why this reaction occurs, even though it is predicted to be thermodynamically forbidden on the basis of proton affinities, is unclear but possible explanations have been discussed in Reference 12. The simple take-home message is that some caution is warranted in using proton affinities to determine whether or not proton transfer occurs.

2.2.4.3 *Alternative Proton Donors*

H_3O^+ has been the overwhelming proton donor of choice in PTR-MS. The reasons for this were given earlier and include the proton affinity of H_2O, which lies conveniently between those of the common inorganic constituents of air and those of most VOCs, making the latter but not the former group susceptible to proton transfer from H_3O^+.

There has been far less interest in using other proton sources in PTR-MS. However, relatively early on in the development of the technique, an application using $NH_4{}^+$ in place of H_3O^+ was reported [46]. In contrast to H_2O, it can be seen from Table 2.2 that most organic compounds have proton affinities below that of NH_3, so simple proton donation from $NH_4{}^+$ is not an option for most classes of organic compounds. The exceptions are amines and a number of other types of nitrogen-containing compounds. In a rather specific application, Lindinger and co-workers showed that for a mixture of pinene ($C_{10}H_{16}$) and 2-ethyl-3,5-dimethylpyrazine ($C_8H_{12}N_2$), both of which have the same nominal molecular masses, the introduction of $NH_4{}^+$ in place of H_3O^+ provided a means of distinguishing between these two compounds [46]. Both compounds willingly accept a proton from H_3O^+ but only 2-ethyl-3,5-dimethylpyrazine, an aromatic heterocyclic compound, has a higher proton affinity than NH_3.

Since proton transfer from $NH_4{}^+$ is out of the question for most organic compounds on energetic grounds, the chemistry is dominated by association reactions. The association reactions lead to adducts of the type $M.NH_4{}^+$. The factors affecting the reaction outcome have been explored in detail by Keough and DeStefano [47]. Studies of the reactions of $NH_4{}^+$ with a variety of organic reagents in both SIFT-MS and PTR-MS have also been reported [48, 49].

While the scope for applications of $NH_4{}^+$ in SIFT-MS and PTR-MS is limited, other proton donors derived from conjugate bases with proton affinities between those of H_2O and NH_3 might have more value. In the case of PTR-MS, the production of such 'intermediate' proton donors, which might be one of any number of protonated organic molecules, directly in the ion source (e.g. in a hollow cathode discharge) is likely to be a problem because of extensive fragmentation and subsequent ion–molecule reactions. However, a rather ingenious way of generating a clean organic proton donor has been demonstrated by Inomata and Tanimoto [50]. This is achieved simply by adding a sufficient excess of the appropriate conjugate base in the upstream part of the drift tube such that all the H_3O^+ ions transfer their available protons to the recipient organic molecule. This protonated organic molecule then becomes the proton source for reaction with an analyte added further downstream in the drift tube. A specific application that has already been demonstrated with this approach is the ability to distinguish between isomeric aldehydes and ketones [51]. This is possible because aldehydes generally possess proton affinities that are substantially below the corresponding ketone(s). Consequently, by choosing a proton donor derived from a conjugate base with a proton affinity lying between the aldehyde and ketone, it is possible to switch off any contribution from the aldehyde. It is important to recognize, though, that a change of proton donor will also bring about a change in the proton transfer rate coefficient, which should be taken into account in any attempt to use this approach to determine the concentration of a specific compound.

2.3 Other Chemical Ionization Processes

In principle, other types of CI reagents which do not employ proton transfer could be used to detect and quantify VOCs. Many such CI reagents have been investigated in SIFT-MS, but in PTR-MS the options are more limited since no mass selection of CI reagents is employed and therefore a naturally clean source of these ions is required. Strictly speaking, when proton transfer is no longer used, we should not refer to the technique as PTR-MS, but such semantics need not concern us here. Motivated by previous SIFT-MS work using NO^+ and $O_2{}^+$, these ions have seen use in a PTR-MS context, that is, as CI reagents in a PTR-MS instrument with little or no instrument modification [49, 52]. The chemistry and kinetics of these two reagents have been extensively investigated in SIFT-MS studies [32, 33, 35–40, 42–44].

The basic ion–molecule chemistry of NO^+ is well-known from a number of CI-MS investigations and has been neatly summarized in the book by Harrison [2]. Several modes of reaction are possible and a key quantity in determining the reaction outcome is the first ionization energy of NO, which is 9.26 eV [53]. Molecules with ionization energies below 9.26 eV may undergo rapid charge exchange reactions (as per Reaction 2.2) with NO^+. However, for molecules with ionization energies above this value simple charge transfer is not possible and other modes of reaction are found instead. The two dominant processes are association (Reaction 2.5) and hydride exchange (Reaction 2.4). However, the branching ratios of all reaction channels depend on the reaction conditions. For example, benzene and toluene show predominantly charge-exchange reactions with NO^+ in a SIFT-MS environment, since both of these aromatic molecules have lower ionization energies than NO. However, there is also a minor channel involving adduct formation. By way of contrast in a PTR-MS instrument, no adduct formation was seen [49].

Reactions of $O_2{}^+$ with organic molecules are dominated by charge exchange, since O_2 has a relatively high first ionization energy of 12.1 eV [53]. However, given that most organic molecules have first ionization energies in the range 8–11 eV, there is a substantial amount of excess energy on charge transfer that will be deposited in the product ion. This gives rise to the strong tendency for ion dissociation seen in the reactions of O_2^+ [49, 52]. For this reason, potential applications of O_2^+ in both SIFT-MS and in a PTR-MS context may be limited. Nevertheless, application of O_2^+ for the detection of NH_3 has recently been demonstrated by Norman and co-workers [54]. The much higher proton affinity of NH_3 compared to H_2O was discussed in the previous section, which suggests that H_3O^+ could be used to protonate NH_3 in a classic PTR-MS type of experiment. Unfortunately, there is a small but significant level of NH_4^+ emanating from the ion source. This NH_4^+ is assumed to be a minor product from reactions involving N_2 and H_2O and is difficult to reduce below a certain level. To tackle this problem, Norman and co-workers switched from water to oxygen as the ion source gas, giving a clean source of O_2^+ [54]. This undergoes almost exclusively charge transfer reaction, yielding NH_3^+ at *m/z* 17, where any background interference is negligible.

References

1. Field, F. H., Munson, M. S. B., Becker, D. A. (1966) Chemical ionization mass spectrometry. III. *Adv. Chem. Ser.* **58**, 167.
2. Harrison, A. G. (1992) *Chemical Ionization Mass Spectrometry*, 2nd edn, CRC Press, Boca Raton.
3. Hunter, E. P. L., Lias, S. G. (1998) Evaluated gas phase basicities and proton affinities of molecules: an update. *J. Phys. Chem. Ref. Data* **27**, 413.
4. Schiff, H. I., Bohme, D. K. (1975) Flowing afterglow studies at York University. *Int. J. Mass Spectrom. Ion Phys.* **16**, 167.
5. Maitland, G. C., Rigby, M., Smith, E. B., Wakeham, W. A. (1987) *Intermolecular Forces: Their Origin and Determination*, Clarendon Press.
6. Ikezoe, Y., Matsoka, S., Takebe, M., Viggiano, A. (1987) *Gas Phase Reaction Rate Constants through 1986*, Maurzen Company Ltd, Tokyo.
7. Anicich, V. G. (2003) An index of the literature for bimolecular gas phase cation–molecule reaction kinetics. JPL publication 03–19, http://trs-new.jpl.nasa.gov/dspace/bitstream/2014/7981/1/03-2964.pdf (accessed 8 August 2013).
8. Smith, D., Španěl, P. (2005) Selected ion flow tube mass spectrometry (SIFT-MS) for on-line trace gas analysis. *Mass Spectrom. Rev.* **24**, 661.
9. Arnold, S. T., Thomas, J. M., Viggiano, A. A., Morris, R. A. (1998) Reactions of $H_3O^+(H_2O)_n$ and $H^+(H_2O)_n(CH_3COCH_3)_m$ with CH_3SCH_3. *Int. J. Mass Spectrom.* **179/180**, 243.
10. Arnold, S. T., Viggiano, A. A., Morris, R. A. (1998) Rate constants and product branching fractions for the reactions of H_3O^+ and NO^+ with C_2-C_{12} alkanes. *J. Phys. Chem. A* **102**, 8881.
11. Midey, A. J., Arnold, S. T., Viggiano, A. A. (2000) Reactions of $H_3O^+(H_2O)_n$ with formaldehyde and acetaldehyde. *J. Phys. Chem. A* **104**, 2706.
12. Midey, A. J., Williams, S., Arnold, S. T., Viggiano, A. A. (2002) Reactions of $H_3O^+(H_2O)_{0,1}$ with alkylbenzenes from 298 to 1200 K. *J. Phys. Chem. A* **106**, 11726.

13. Schoon, N., Amelynck, C., Vereecken, L. *et al.* (2004) A selected ion flow tube study of the reactions of H_3O^+, NO^+ and $O_2{}^+$ with some monoterpene oxidation products. *Int. J. Mass Spectrom.* **239**, 7.
14. Michel, E., Schoon, N., Amelynck, C. *et al.* (2005) A selected ion flow tube study of the reactions of H_3O^+, NO^+ and $O_2{}^+$ with methyl vinyl ketone and some atmospherically important aldehydes. *Int. J. Mass Spectrom.* **244**, 50.
15. Amelynck, C., Schoon, N., Kuppens, T. *et al.* (2005) A selected ion flow tube study of the reactions of H_3O^+, NO^+ and $O_2{}^+$ with some oxygenated biogenic volatile organic compounds. *Int. J. Mass Spectrom.* **247**, 1.
16. Lide, D. R. (Ed.) (1992) *Handbook of Chemistry and Physics*, 72nd edn, CRC Press.
17. Bohme, D. K., Mackay, G. I., Tanner, S. D. (1979) An experimental study of the gas-phase kinetics of reactions with hydrated hydronium (1 +) ions (n = 1–3) at 298 K. *J. Am. Chem. Soc.* **101**, 3724.
18. Milligan, D. B., Wilson, P. F., Freeman, C. G. *et al.* (2002) Dissociative proton transfer reactions of $H_3{}^+$, N_2H^+, and H_3O^+ with acyclic, cyclic, and aromatic hydrocarbons and nitrogen compounds, and astrochemical implications. *J. Phys. Chem. A* **106**, 9745.
19. Španěl, P., Pavlik, M., Smith, D. (1995) Reactions of H_3O^+ and OH^- ions with some organic molecules; applications to trace gas analysis in air. *Int. J. Mass Spectrom. Ion Processes* **145**, 177.
20. Langevin, M. (1905) Une formule fondamentale de théorie cinétique [A fundamental formula of kinetic theory]. *Ann. Chim. Phys.* **5**, 245.
21. Gioumousis, G., Stevenson, D. P. (1958) Reactions of gaseous molecule ions with gaseous molecules. V. Theory. *J. Chem. Phys.* **29**, 294.
22. Su, T., Bowers, M. T. (1973) Theory of ion–polar molecule collisions. Comparison with experimental charge transfer reactions of rare gas ions to geometric isomers of difluorobenzene and dichloroethylene. *J. Chem. Phys.* **58**, 3027.
23. Su, T., Bowers, M. T. (1973) Ion–polar molecule collisions. Proton transfer reactions of $H_3{}^+$ and $CH_5{}^+$ to the geometric isomers of difluoroethylene, dichloroethylene, and difluorobenzene. *J. Am. Chem. Soc.* **95**, 1370.
24. Bass, L., Su, T., Chesnavich, W. J., Bowers, M. T. (1975) Ion–polar molecule collisions. A modification of the average dipole orientation theory: the model. *Chem. Phys. Lett.* **34**, 119.
25. Su, T., Bowers, M. T. (1973) Ion–polar molecule collisions: the effect of ion size on ion–polar molecule rate constants; the parameterization of the average-dipole-orientation theory. *Int. J. Mass Spectrom. Ion Phys.* **12**, 347.
26. Mackay, G. I., Betowski, L. D., Payzant, J. D. *et al.* (1976) Rate constants at 297 K for proton-transfer reactions with hydrocyanic acid and acetonitrile. Comparisons with classical theories and exothermicity. *J. Phys. Chem.* **80**, 2919.
27. Su, T., Chesnavich, W. J. (1982) Parametrization of the ion–polar molecule collision rate constant by trajectory calculations. *J. Chem. Phys.* **76**, 5183.
28. Su, T. (1994) Parametrization of kinetic energy dependences of ion–polar molecule collision rate constants by trajectory calculations. *J. Chem. Phys.* **100**, 4703.
29. Bhowmik, K., Su, T. (1986) Trajectory calculations of ion–quadrupolar molecule collision rate constants. *J. Chem. Phys.* **84**, 1432.
30. Turulski, J., Su, T., Niedzielski, J., Pezler, B. (2002) Classical theories of the ion/linear quadrupole capture. *Int. J. Mass Spectrom.* **216**, 115.

31. Zhao, J., Zhang, R. (2004) Proton transfer reaction rate constants between hydronium ion (H_3O^+) and volatile organic compounds. *Atmos. Environ.* **38**, 2177.
32. Diskin, A. M., Wang, T., Smith, D., Spanel, P. (2002) A selected ion flow tube (SIFT) study of the reactions of H_3O^+, NO^+ and $O_2{}^+$ ions with a series of alkenes; in support of SIFT-MS. *Int. J. Mass Spectrom.* **218**, 87.
33. Španěl, P., Smith, D. (1998) Selected ion flow tube studies of the reactions of H_3O^+, NO^+, and $O_2{}^+$ with several aromatic and aliphatic hydrocarbons. *Int. J. Mass Spectrom.* **181**, 1.
34. Brown, P., Watts, P., Märk, T. D., Mayhew, C. A. (2010) Proton transfer reaction mass spectrometry investigations on the effects of reduced electric field and reagent ion internal energy on product ion branching ratios for a series of saturated alcohols. *Int. J. Mass Spectrom.* **294**, 103.
35. Španěl, P., Smith, D. (1997) SIFT studies of the reactions of H_3O^+, NO^+ and $O_2{}^+$ with a series of alcohols. *Int. J. Mass Spectrom.* **167/168**, 375.
36. Španěl, P., Smith, D. (1998) SIFT studies of the reactions of H_3O^+, NO^+ and $O_2{}^+$ with several ethers. *Int. J. Mass Spectrom. Ion Processes* **172**, 239.
37. Španěl, P., Ji, Y., Smith, D. (1997) SIFT studies of the reactions of H_3O^+, NO^+ and $O_2{}^+$ with a series of aldehydes and ketones. *Int. J. Mass Spectrom. Ion Processes* **165/166**, 25.
38. Španěl, P., Smith, D. (1998) SIFT studies of the reactions of H_3O^+, NO^+ and $O_2{}^+$ with a series of volatile carboxylic acids and esters. *Int. J. Mass Spectrom. Ion Processes* **172**, 137.
39. Francis, G. J., Milligan, D. B., McEwan, M. J. (2007) Gas-phase reactions and rearrangements of alkyl esters with H_3O^+, NO^+, and $O_2{}^+$: a selected ion flow tube study. *J. Phys. Chem. A* **111**, 9670.
40. Španěl, P., Smith, D. (1998) Selected ion flow tube studies of the reactions of H_3O^+, NO^+, and $O_2{}^+$ with several amines and some other nitrogen-containing molecules. *Int. J. Mass Spectrom.* **176**, 203.
41. Williams, T. L., Adams, N. G., Babcock, L. M. (1998) Selected ion flow tube studies of $H_3O^+(H_2O)_{0,1}$ reactions with sulfides and thiols. *Int. J. Mass Spectrom. Ion Processes* **172**, 149.
42. Španěl, P., Smith, D. (1998) Selected ion flow tube studies of the reactions of H_3O^+, NO^+, and $O_2{}^+$ with some organosulphur molecules. *Int. J. Mass Spectrom.* **176**, 167.
43. Španěl, P., Smith, D. (1999) Selected ion flow tube studies of the reactions of H_3O^+, NO^+, and $O_2{}^+$ with some chloroalkanes and chloroalkenes. *Int. J. Mass Spectrom.* **184**, 175.
44. Španěl, P., Smith, D. (1999) Selected ion flow tube studies of the reactions of H_3O^+, NO^+, and $O_2{}^+$ with several aromatic and aliphatic monosubstituted halocarbons. *Int. J. Mass Spectrom.* **189**, 213.
45. Goebbert, D., Wenthold, P. G. (2004) Water dimer proton affinity from the kinetic method: dissociation energy of the water dimer. *Eur. J. Mass Spectrom.* **10**, 837.
46. Lindinger, W., Hansel, A., Jordan, A. (1998) On-line monitoring of volatile organic compounds at pptv levels by means of proton-transfer-reaction mass spectrometry (PTR-MS): medical applications, food control and environmental research. *Int. J. Mass Spectrom. Ion Process.* **173**, 191.

47. Keogh, T., DeStefano, A. J. (1994) Factors affecting reactivity in ammonia chemical ionization mass spectrometry. *Org. Mass Spectrom.* **16**, 527.

48. Adams, N. G., Babcock, L. M., Mostefaoui, T. M., Kerns, M. S. (2003) Selected ion flow tube study of NH_4^+ association and product switching reactions with a series of organic molecules. *Int. J. Mass Spectrom.* **223**, 459.

49. Blake, R. S., Wyche, K. P., Ellis, A. M., Monks, P. S. (2006) Chemical ionization reaction time-of-flight mass spectrometry: multi-reagent analysis for determination of trace gas composition. *Int. J. Mass Spectrom.* **254**, 85.

50. Inomata, S., Tanimoto, H. (2008) Differentiation of isomeric compounds by two-stage proton transfer reaction time-of-flight mass spectrometry. *J. Am. Soc. Mass Spectrom.* **19**, 325.

51. Blake, R. S., Patel, M., Monks, P. S. *et al.* (2008) Aldehyde and ketone discrimination and quantification using two-stage proton transfer reaction mass spectrometry. *Int. J. Mass Spectrom.* **278**, 15.

52. Wyche, K. P., Blake, R. S., Willis, K. A. *et al.* (2005) Differentiation of isobaric compounds using chemical ionization reaction mass spectrometry. *Rapid Commun. Mass Spectrom.* **19**, 3356.

53. NIST Chemistry WebBook, NIST Standard Reference Database Number 69, Eds. P. J. Linstrom and W. G. Mallard.

54. Norman, M., Hansel, A., Wisthaler, A. (2007) O_2^+ as reagent ion in the PTR-MS instrument: detection of gas-phase ammonia. *Int. J. Mass Spectrom.* **265**, 382.

3

Experimental: Components and Principles

3.1 Introduction

The apparatus required to perform proton transfer reaction mass spectrometry (PTR-MS) is illustrated in a simplistic fashion in Figure 3.1 and includes the following:

1. *An ion source*. Since H_3O^+ is normally used as the proton transfer reagent in PTR-MS, a source which generates this ion in high yield with little contamination from other ions is essential.
2. *A drift tube*. This has two purposes: (i) it provides a reaction zone of fixed length, which is important for quantitative analysis; (ii) the electric field, E, along the drift tube raises the kinetic energies of the ions relative to the neutral gas molecules. The resulting *drift velocity* of the positive ions, which is dependent on the ratio of E to the gas number density, N, in the drift tube, helps to prevent the formation of cluster ions such as $H_3O^+(H_2O)$ and $MH^+(H_2O)$ through moderately energetic collisional dissociation with neutral molecules, where M represents an analyte molecule.
3. *A mass spectrometer*. This provides a means for separating ions according to their mass/charge ratio, m/z. Quadrupole, time-of-flight and ion trap mass spectrometers have been employed in PTR-MS instruments and therefore the basic principles of these types of mass spectrometers will be discussed in some detail later in this chapter.
4. *An ion detector*. A sensitive ion detector and associated data acquisition electronics is critical to achieve the requisite overall detection sensitivity in PTR-MS.
5. *An instrumental control system*. A computer is used to collect and display data and for overall control of the instrument.

The aim of this chapter is to describe the design and performance of items 1–4 listed above. Before doing this, an overview will be provided of ion extraction and ion focusing optics,

Proton Transfer Reaction Mass Spectrometry: Principles and Applications, First Edition. Andrew M. Ellis and Christopher A. Mayhew.

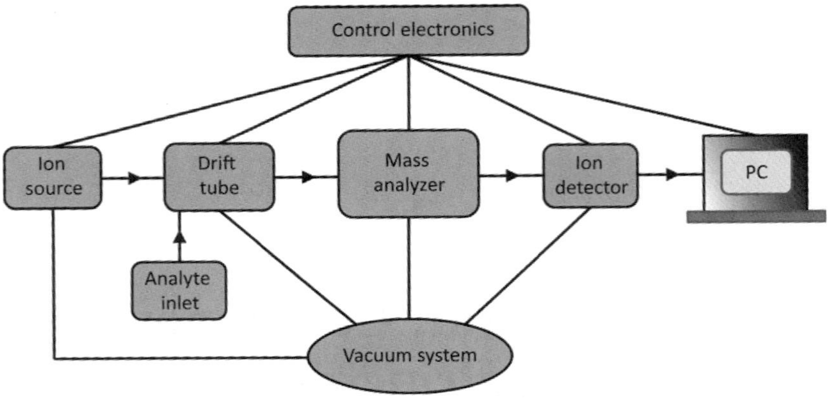

Figure 3.1 *Main constituents of a PTR-MS instrument.*

since these are important parts of any reasonably complex instrument that involves the manipulation and detection of ions and this includes PTR-MS.

Another important issue that underlies PTR-MS is vacuum technology. The ion source, drift tube, mass analyser and ion detector are all under some degree of vacuum. Different pressures are required in different parts of the instrument. For example, the operating pressure in the drift tube is several orders of magnitude higher than the pressure in the mass spectrometer. In the drift tube, multiple collisions between ions and molecules are essential for protonation of sufficient quantities of organic molecules in the analyte. On the other hand, a much higher vacuum (lower pressure) is required in the mass spectrometer because any collisions of ions with background gas molecules will degrade the performance of the mass spectrometer (an exception here is the ion trap mass spectrometer, where some collisions with background gas molecules are important and which is discussed later in this chapter). A good vacuum is also essential to protect the sensitive ion detectors used in most mass spectrometers, as discussed later in this chapter. It is important for the operator of a PTR-MS instrument to be aware of these issues, and a good working knowledge of vacuum technology is useful. However, a description of this technology is beyond the scope of this book and so for further details the reader is directed to a number of very good monographs on this topic, such as those by Chambers *et al.* [1], O' Hanlon [2] and Yoshimura [3].

3.2 Ion Extraction and Ion Optics

We start with a basic description of the fundamentals of ion beam manipulation, since this is relevant to the operation of the ion source and the drift tube, but more so the mass spectrometer and the transfer region between the drift tube and the mass spectrometer. For example, it may be necessary to accelerate or decelerate ions to specific kinetic energies. It is also frequently necessary to be able to deflect ions in a specific direction or to alter the shape of the incoming ion beam, that is, some degree of focusing might be necessary. The latter is the domain of charged particle optics, a sophisticated topic which we will not cover in detail. Instead, the aim here is to give any newcomer to this topic sufficient

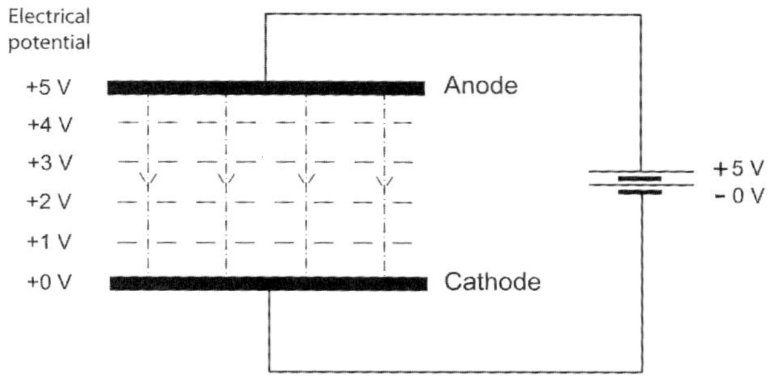

Figure 3.2 *Illustration of the relationship between electrical potential and electric field for parallel electrodes supplied by a 5 V DC power supply. Lines of equipotential are shown by the horizontal dashed lines and are perpendicular to the electric field, where the latter is shown by the vertically oriented lines (dot-dash). By convention the arrow on the electric field lines runs from higher to lower potential. This diagram is oversimplified because in practice the field lines do not run perpendicular to the inner surface of the electrodes near the left and right edges due to field fringing effects (see text for more details and also Figure 3.3).*

information to appreciate the basic elements required for the practical manipulation of ions. Only components that exploit electrostatics will be considered so ion optics based on magnetic forces will not be discussed, nor will components that exploit dynamic electric fields such as quadrupole and octopole ion guides. For more involved discussions of ion optics, the interested reader is directed to a number of books devoted to this topic, including those by Dahl [4], Szilagyi [5] and Liebl [6].

3.2.1 Ion Acceleration

The acceleration and deceleration of ions can be achieved through the use of an electric field. Suppose we have two plane-parallel electrodes, such as those shown in Figure 3.2, and apply different electrical potentials to the two electrodes, V_2 and V_1, such that we have a potential difference of $V = V_2 - V_1$ between the two plates. This can be achieved by connecting the electrodes to a direct current (DC) power supply, such as that shown on the right-hand side of the figure. This will generate an electric field between the electrodes, as illustrated by the field lines shown in the figure. These field lines are exactly perpendicular to the lines of equipotential, which are also shown in Figure 3.2 as dashed lines. The electric field between the two electrodes is notionally uniform and its strength is denoted by the symbol E, which is given quantitatively by

$$E = \frac{V}{d} \tag{3.1}$$

where d is the distance between the plates.[1] Since the SI unit of voltage is the volt, which is given the symbol V, the SI units of electric field are V m^{-1}. However, because the gaps

[1] Strictly speaking E is a vector quantity but in the case of a simple pair of parallel plates the direction of the field is restricted to a single dimension and an obvious direction, that is, perpendicular to the surfaces of the plates.

between electrodes are frequently of the order of a centimetre or so, it is common to see the electric field quoted in units of V cm^{-1}.

Any charged particle that is initially stationary between these plates will be driven into motion once the electric field is applied. In the absence of any background gas, that is, by operating under good vacuum conditions such that collisions of the charged particles with uncharged molecules are improbable, the charged particle will travel in a direction exactly along the field lines. Positively charged particles will travel towards the electrode with the more negative potential while negatively charged particles will move in the opposite direction. The electric field is a measure of the electrical force provided per unit charge, and so the greater the strength of the field, the greater the force acting on the charged particle. Our primary interest in this book is with positively charged ions, and so the discussion from here onwards will concentrate on these species.

If $V_2 > V_1$ and the ion starts its journey from the surface of the electrode of higher potential, then the ion will be exposed to the full range of the electrical potential difference across the electrode gap and will receive the maximum possible kinetic energy as a result of ion acceleration by the electric field. The kinetic energy, T, acquired by an ion of mass m (in kg) and charge q (in coulombs, C) as it reaches the other electrode is given by

$$T = q(V_2 - V_1) = \frac{1}{2}mv^2 \qquad (3.2)$$

where v is the final velocity of the ion. The simple relationship described in Equation 3.2 makes it possible to calculate the conditions necessary to accelerate or decelerate ions to any desired velocity in a perfect vacuum. Although the kinetic energy acquired can naturally be expressed in joules (J), it is quite common to see it expressed in *electron volts*, which is a non-SI unit encountered later in the chapter and elsewhere in this book (including Table 2.1). The electron volt, symbol eV, is the energy required to accelerate a particle with the fundamental electronic charge, e, through a potential difference of 1 V and is equal to 1.602×10^{-19} J to four significant figures. Since we are primarily interested in ions with a charge equal to e, expressing energies in eV often produces numbers of very convenient magnitude.

The electrode arrangement shown in Figure 3.2 would, of course, be impractical as a means of accelerating or decelerating ions since the ions will inevitably hit one or other electrode and will be unable to escape. To convert this into a practical means of accelerating ions for use elsewhere, the downstream electrode must be capable of transmitting ions. This can be achieved by using a wire mesh in place of a solid electrode. The best mesh electrodes possess small holes and thin wires. This combination ensures that the electric field between the electrodes remains close to uniform (see also below) while at the same time allowing good ion transmission ($\geq$90% is typical for a coarse grid, although finer grids will have a smaller transmission factor).

The kinetic energy attained by the ion as it arrives at the downstream mesh electrode will depend on its initial location when the voltage is switched on, as hinted at earlier.[2] For example, an ion midway between the two electrodes will initially be subjected to an electrical potential of $(V_2 - V_1)/2$. It will then follow the field lines and move towards

[2] This scenario could be achieved by initially having $V_2 = V_1$ and then suddenly switching the potential on one electrode to create a potential difference between the two electrodes that will set the ion in motion.

electrode 1 gaining an energy $q(V_2 - V_1)/2$. In other words, it will only gain half the kinetic energy that an ion starting at the surface of electrode 2 (the anode) would gain. In the specific case of Figure 3.2, an initially stationary positive ion starting at a position equidistant between the two electrodes would have a kinetic energy of 2.5 eV by the time it reached the cathode. This dependence of the ion kinetic energy on the starting position of an ion between two parallel electrodes is of considerable importance and will be met again later in this chapter when time-of-flight mass spectrometers (TOF-MS) are considered (Section 3.5.4).

3.2.2 Ion Steering

It is frequently necessary to alter the path of a beam of ions. This may be required in order to steer the beam through an aperture, perhaps because of the possibility of a slight misalignment of the optical elements, or towards an ion detector. Alternatively, it may be necessary to deflect the ion beam along a completely different path, for example, turning it almost 90° for injection into a flight tube (see Section 3.5.4).

Whatever the purpose, the most commonly used and simplest form of deflector is a pair of plane-parallel electrodes. With different potentials on the two electrodes, any positive ions that pass between the electrodes will be deflected towards the electrode with the least positive potential. If the velocity of an ion perpendicular to the surface of the plates is initially zero, then as soon as the ion enters between the plates and experiences the electric field it is deflected in a vertical direction, where the vertical direction is perpendicular to the surface of the electrode surface. For an ion with a charge e, the force experienced vertically is given by eE, and thus from Newton's second law the acceleration given to the ions is

$$a = \frac{F}{m} = \frac{eE}{m} \tag{3.3}$$

where m is the mass of the particle. The vertical deflection, y, at time t after entering the deflecting field follows from simple kinematics and is given by

$$y = \frac{at^2}{2} = \frac{eEt^2}{2m} \tag{3.4}$$

Instead of expressing the deflection in terms of time, we can express it in terms of horizontal distance travelled, x, where $x = vt$ and v is the horizontal velocity that is unchanged by a force acting in a perpendicular direction. Thus we end up with

$$y = \frac{eEx^2}{2mv^2} \tag{3.5}$$

which describes a parabolic deflection trajectory for the ion in a uniform electric field.

Figure 3.3 shows a contour plot of the calculated electrical potential of two plane-parallel electrodes and the trajectories of ions traversing between these two plates. Note that in reality there is a distortion of the electric field at the edges of the plates which was previously ignored in the simplified diagram in Figure 3.2, that is, the field does not simply disappear beyond the edge of the plates. The curvature of the field lines (which are perpendicular to the electrical potential contour lines) is an example of field *fringing*. Fringe fields are found near sharp edges of charged ion optics and arise because of the strong surface curvature of the optical element at these points. These fringe fields have an

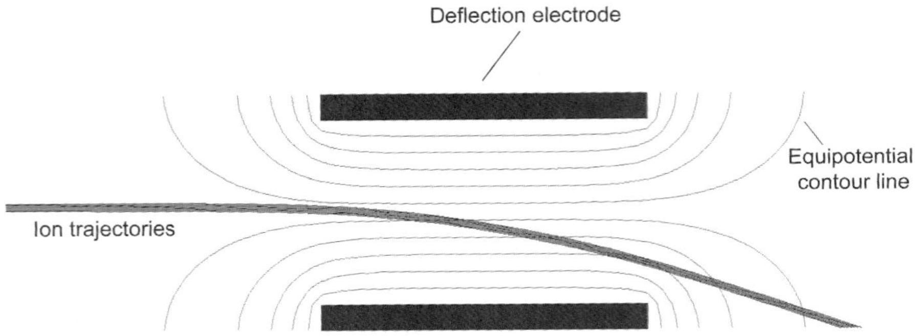

Figure 3.3 *Simulation of the electrostatic potential of a parallel-plate deflector and its effect on positively charged ions with m/z 50 and an incoming kinetic energy of 10 eV entering from the left of the diagram. This simulation was carried out using SIMION® and assumed plates of circular cross section of diameter 23 mm, thickness 2 mm, and separation 15 mm. The potentials on the upper and lower plates were +2 and 0 V, respectively.*

impact on the ion trajectory when compared to the idealized plane-parallel deflector but they do not change the basic function of the device, that is, it remains an ion deflector. However, fringe fields are extremely important elsewhere in ion optics and provide the underlying basis for ion focusing in electrostatic lenses, as will be seen in the next section.

3.2.3 Ion Lenses

Charged particle lenses are a key component in most mass spectrometers and are essential in many other areas of science and technology where ion and electron beams are employed. Broadly speaking, there are two important uses of ion lenses in mass spectrometry. First of all, a particular ion source may yield a strongly divergent beam of ions upstream of the inlet to a mass spectrometer inlet which, if uncorrected, would result in many ions following trajectories that either prevent them from entering the mass spectrometer in the first place or, if they do enter, will not make it all the way through to the detector. This potential loss of signal can be dramatic and may be the difference between a highly sensitive instrument and one which is so insensitive that it has little or no practical viability. An appropriately placed ion lens can help to correct this problem by gathering as many of the incoming ions as possible and converting them into a collimated beam for subsequent mass analysis.

A second application of an ion lens is to focus an ion beam to a relatively small point. This might be required, for example, when a small aperture is employed to separate two regions which must be held at very different pressures, that is, the aperture prevents too much gas escaping from the higher pressure region into the lower pressure region. If ions are required to pass through this aperture, perhaps to enter a mass spectrometer, then a substantial loss of signal will ensue if no ion lens is used, since most ions will simply hit the plate and therefore never make it into the mass spectrometer. Of course any ions that are focused through the aperture will form a divergent beam beyond it as they pass through the focal point and so one or more additional lenses may be required downstream of the aperture to collimate the beam.

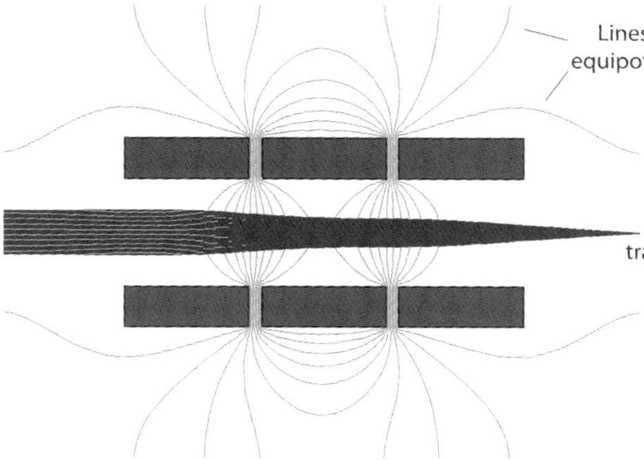

Figure 3.4 *Simulation of the electrostatic potential across a three-element cylinder lens (cross-sectional view). Also shown are lines of equipotential. Trajectories for a collection of 70 ions incoming from the left show the focusing properties of this lens. This simulation was carried out using SIMION® and assumed incoming ions of m/z = 50 and a kinetic energy of 10 eV. Potentials on the first, second and third electrodes (from left to right) were 10, −15 and 10 V, respectively. The lens elements are drawn to scale and for this simulation the internal diameter of each cylinder was chosen as 14 mm, the length of each cylinder was 15 mm, and the gap between adjacent cylinders was 2 mm.*

Two types of electrostatic ion lenses see common usage: cylinder lenses and aperture lenses. Figure 3.4 shows how the focusing effect occurs in the case of a three-element cylinder lens. The three cylinders are shown adjacent to each other with a small gap between each of them. Field fringing results in two sets of curved field lines with opposite curvatures in the inter-cylinder gaps and it is these fringe fields that enable ion focusing. For the example shown in Figure 3.4, the first and third cylinders are at the same electrical potential, while the middle cylinder is at a higher potential. Now consider what will happen to a positively charged ion travelling through cylinder 1 in a direction parallel to but off the central (optical) axis. As the ion nears the end of cylinder 1, it begins to experience the first fringing field. Remember, the force imparted by an electric field has a tendency to push ions in a direction perpendicular to the lines of equipotential. The extent to which this is successful is dependent on the kinetic energies of the ions, that is, slow ions of a given mass will be much more strongly affected than fast ions. As an ion enters the first fringing field, the curvature of the equipotential lines will steer the ion on a trajectory that moves it towards the incoming axis. As the ion passes through the gap and just inside the middle cylinder the fringing field here will cause the trajectory to change and the ions will now be deflected away from the optical axis. At first sight it might seem that the two effects will cancel, so nothing interesting will happen. However, this is incorrect because the ions move more quickly in the second fringing region. Consequently, there will be a net focusing effect because ion deflection is smaller when the ion speed is raised.

From the above, we can see that two cylinders will provide a converging lens. However, they generate outgoing ions at a different electrical potential to the incoming ions. For some

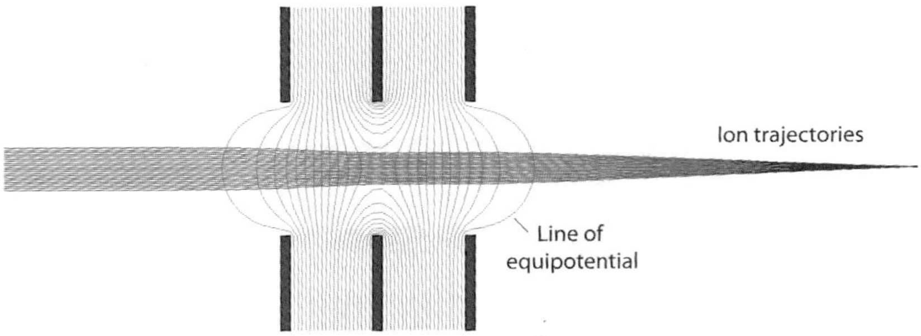

Figure 3.5 *Cross-sectional view of a three-element aperture lens. Also shown are lines of equipotential. Trajectories for a collection of 21 incoming ions show the focusing property of this lens. This simulation was carried out using SIMION® and assumed incoming ions of* m/z = *50 and a kinetic energy of 10 eV. Potentials on the first, second and third electrodes (from left to right) were 10, −7.5 and 10 V, respectively. Each lens element was given a 36 mm diameter, an aperture diameter of 16 mm, and a thickness of 1 mm. The spacing between the electrodes in the simulation was 9 mm.*

applications focusing is desirable without any change in ion energy and the three-element lens, as shown in Figure 3.4, with its equal potentials on the first and third electrodes, can deliver this. This type of lens is called an *Einzel* lens and it is very widely used.

Lens components can also be constructed out of disc electrodes with circular apertures. Here the key fringing fields fan out from the apertures and it is straightforward to extend the arguments above to explain why these *aperture lenses* exhibit a focusing effect. Figure 3.5 shows a simulation of the focusing response of a three-element aperture lens.

The decision as to whether to use an aperture or a cylinder lens depends very much on the practical requirements for a particular instrument. Clearly an aperture lens is more compact than a cylindrical lens, and therefore if space is paramount then the likelihood is that an aperture lens would be chosen. On the other hand a cylindrical lens will generally show higher ion transmission and therefore would be the optimum choice for maximum signal. Furthermore, cylinder lenses have fewer aberrations and because of their large apertures they are more tolerant of slight misalignment than would be the case with an aperture lens.

3.2.4 Simulation of Ion Trajectories

The properties of a wide variety of ion lens configurations have been systematically explored by many research teams and the underlying properties of these lenses are well understood and predictable in terms of the voltage ratios and dimensions of the lens components. However, it is common practice these days to design ion optics with the aid of computer simulations of the ion trajectories. Several programs have been developed for such applications, but the best known and most widely used is SIMION®. This is commercial software that has been through a number of revisions over many years and is available for operation on a desktop PC. The user specifies the electrode geometries, separations and applied potentials, from which SIMION can map out the inter-electrode potentials and the corresponding electric fields. Ion trajectory calculations can then be generated by the software. Other features include the incorporation of time-dependent electric fields (such as encountered in the quadrupole mass filter – see Section 3.5.2) and the ability to account for collisions between ions and background gas molecules. The simulations presented in Figures 3.3, 3.4 and 3.5 were generated using SIMION. Further details can be found from the SIMION web site (www.simion.com).

3.3 Ion Sources

3.3.1 Hollow Cathode Discharge Ion Source

The great majority of PTR-MS experiments carried out to date have employed a DC hollow cathode electrical discharge as the means of generating ions. Before considering the details of this ion source as it is applied to PTR-MS, some background information on electrical discharges and the hollow cathode effect will be given.

A simple electrode configuration consisting of two planar electrodes can be used to generate an electrical discharge in a gas. The electrode with the more positive potential is called the *anode* while the other electrode is known as the *cathode*. Gases are normally excellent electrical insulators, but can be converted into conducting media by application of a sufficiently high potential difference between the two electrodes. Stray ions and electrons between the electrodes lead to a self-sustaining avalanche of charge carriers once a voltage threshold, which is known as the breakdown voltage, is exceeded. It is important to recognize that electrical breakdown depends not only on the voltage applied across the electrodes, but also on the composition of the gas, the pressure of that gas and the separation and geometry of the electrodes (which determines the electric field). In particular if the pressure is too low (a high vacuum) then the density of charged particles becomes too low to sustain an electrical discharge.

There are a number of different types of electrical discharges which can be classified according to their voltage-current characteristics. At the very low end of the current scale ($\leq 1\ \mu A$) is the Townsend discharge, while at the other extreme is the arc discharge (current > 1 A). Between these two limits is the *glow discharge*, so-called because of a visible luminous glow produced within part of the discharge. It is the glow discharge that is employed in PTR-MS discharge sources.

It is worth taking a brief look at the structure of a glow discharge (more comprehensive but highly readable descriptions can be found in the book by Marcus [7] and a review by Pillow [8]). Figure 3.6 shows some of the main regions of a glow discharge, which is known to contain other regions not shown in Figure 3.6, including a so-called positive column and an Aston dark space. The length of the cathode dark space (sometimes alternatively called the Crookes dark space) and the negative glow region are determined by the gas identity, pressure and applied voltage. Once these are set, the length of the positive column can be reduced to almost zero by moving the anode towards the cathode. The positive column is not required for many applications of DC discharges and the conditions would usually be chosen which minimize its length.

For PTR-MS purposes the key region is the negative glow. This is a zone where there is a high density of positive ions and electrons, creating a plasma with a near-zero net charge. The glow from this region is the result of collisions of electrons with atoms or molecules,

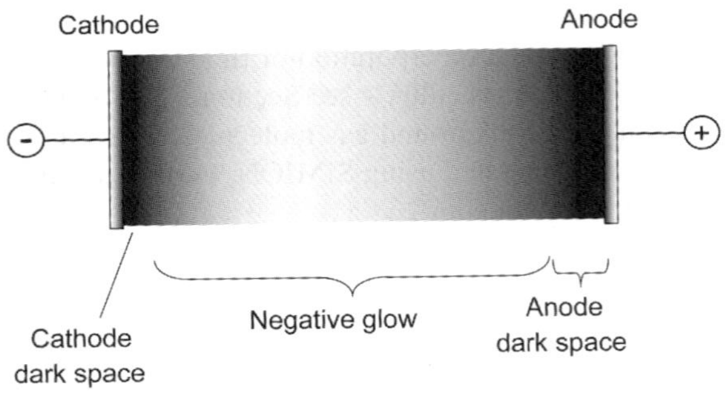

Figure 3.6 *Key regions in a hollow-cathode discharge. See text for further details.*

which produce electronically excited species that subsequently decay by light emission. Since the cathode surface is maintained at a high negative voltage and the negative glow has a near-zero voltage drop across its length, there must be a large drop in electrical potential across the cathode dark space. This has the effect of accelerating positive ions into the cathode surface, which can result in secondary electron emission. These electrons are then accelerated away from the cathode surface and gain energy of typically hundreds of eV as they accelerate through the cathode dark space. The high energies of these electrons are unfavourable for electronic excitation or ionization, which is why the cathode dark space appears dark. However, by the time the electrons reach the negative glow a significant proportion have undergone sufficient inelastic collisions with the gas to reach energies where electronic excitation and/or ionization is more favourable.

In the hollow cathode discharge (HCD), as the name implies, the cathode is no longer a simple planar electrode. It is typically either a cup-shaped electrode or a cylindrical electrode with both ends open, as shown in Figure 3.7. The shape of the anode in such a device is

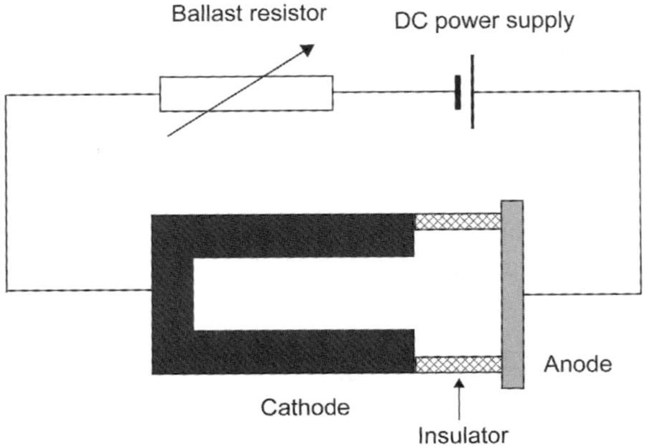

Figure 3.7 *A simple arrangement for a hollow cathode discharge. The variable resistance provided by the ballast resistor limits the amount of current drawn and improves the stability of the discharge.*

less important, but this is commonly a ring electrode or another cylindrical electrode. Conditions can be chosen in an HCD such that the negative glow fills virtually the whole of the cathode interior. The key feature of an HCD which makes it more useful than a planar electrode configuration is that far higher current densities are achievable. There are several reasons for this, all related to the cathode geometry. In a planar electrode configuration only those positive ions drifting in the direction of the cathode will escape the negative glow and be accelerated into the cathode surface, thereby generating secondary electrons. However, in a hollow cathode no matter which way these ions are moving the majority will eventually come into contact with the cathode surface, thus causing a huge increase in secondary electron production. The same argument also applies to ultraviolet photons that can also be produced in the negative glow and produce electrons from the cathode surface by photoemission. Finally, electrons ejected from the cathode surface that pass through the negative glow would not return in a planar electrode configuration. However, in a hollow cathode arrangement many of these electrons will encounter the strongly repulsive cathode fall-off potential on the other side of the cylinder and will be thus be accelerated back into the negative glow, whereby they can contribute to further ion production.

The hollow cathode electrode configuration in a PTR-MS ion source is rarely more complex than a simple hollow cathode adjacent to a planar anode, as illustrated in Figure 3.7. Ideally the discharge takes place in an atmosphere of water vapour, fed into the discharge region at a constant flow rate usually in the range of 5–10 STP $cm^3\ min^{-1}$.[3] The typical operating pressure in the HCD is ~1 mbar. Notice that in practice some air enters the discharge region via backstreaming from the drift tube, the consequences of which will be seen shortly. The voltage between the anode and cathode is normally a few hundred volts leading to an operating current of several mA. Such sources eventually become contaminated with surface deposits in a way that affects the discharge stability. The length of time required for this to become a problem depends on the operating conditions employed, but in the authors' experience HCD sources typically need cleaning every 2–3 months to restore their original performance stability.

3.3.2 Ion–Molecule Chemistry Leading to H_3O^+ Production

It is not immediately obvious that an electrical discharge in water vapour will lead to overwhelming production of H_3O^+. After all, electron impact ionization of H_2O can yield H_2O^+ or it can produce fragment ions such as H^+, $H_2{}^+$, OH^+ and O^+ ($H_2{}^+$ is a very minor product but we include its mention here and below for completeness) [9]. To make H_3O^+ with a relatively high purity, some secondary chemistry is needed. H_2O^+ is readily converted into H_3O^+ by the fast reaction

$$H_2O^+ + H_2O \rightarrow H_3O^+ + OH\ (k = 1.8 \times 10^{-9}\ cm^3\ s^{-1}) \tag{3.6}$$

where the rate coefficient is an experimental value at 300 K [10].

---

[3] Gas flow rates are often quoted in standard cubic centimetres per minute, which is written as STP $cm^3\ min^{-1}$ (and sometimes abbreviated to sccm). The term 'standard' refers to the conditions of standard temperature and pressure (STP) for gases, which are 1 atmosphere and 0°C (273.15 K), respectively. The advantage of working with volume flow rates expressed as STP quantities is that it allows flow rates recorded at different operating temperatures and pressures to be directly compared, as will be shown in a specific example later in this chapter.

However, what about the fragment ions in the discharge? It turns out that these too can be converted to H_3O^+, either through a fast, direct ion–molecule reaction or through formation of H_2O^+ and subsequent reaction via Reaction 3.6 [11]. The chemistry is summarized below and rate coefficients (at 300 K) are taken from the compilation by Ikezoe and co-workers [10]:

$$OH^+ + H_2O \rightarrow H_3O^+ + O \quad (k = 1.3 \times 10^{-9}\ \text{cm}^3\ \text{s}^{-1}) \tag{3.7}$$

$$\rightarrow H_2O^+ + OH \quad (k = 1.8 \times 10^{-9}\ \text{cm}^3\ \text{s}^{-1}) \tag{3.8}$$

$$O^+ + H_2O \rightarrow H_2O^+ + O \quad (k = 2.6 \times 10^{-9}\ \text{cm}^3\ \text{s}^{-1}) \tag{3.9}$$

$$H_2^+ + H_2O \rightarrow H_3O^+ + H \quad (k = 3.4 \times 10^{-9}\ \text{cm}^3\ \text{s}^{-1}) \tag{3.10}$$

$$\rightarrow H_2O^+ + H_2 \quad (k = 3.7 \times 10^{-9}\ \text{cm}^3\ \text{s}^{-1}) \tag{3.11}$$

$$H^+ + H_2O \rightarrow H_2O^+ + H \quad (k = 8.2 \times 10^{-9}\ \text{cm}^3\ \text{s}^{-1}) \tag{3.12}$$

The large rate coefficients shown above are all indicative of ion–molecule reactions that occur at essentially a collision-limiting rate. To allow these reactions to fully take place, so that a high yield of H_3O^+ is attained, a so-called *source drift region* may be added between the HCD and the drift tube.

As mentioned earlier, ingress of some air from the analyte gas into the discharge region of a PTR-MS instrument is inevitable. This can generate other types of ions such as $N_2{}^+$ and $O_2{}^+$. Fortunately, $N_2{}^+$ can be readily removed by a charge exchange reaction with H_2O [12],

$$N_2^+ + H_2O \rightarrow H_2O^+ + N_2\ (k = 2.8 \times 10^{-9}\ \text{cm}^3\ \text{s}^{-1}) \tag{3.13}$$

and so the presence of $N_2{}^+$ ultimately results in H_3O^+ formation via reaction of H_2O^+ with H_2O (Reaction 3.6).

Oxygen is more of a problem, since O_2 has a lower first ionization energy than H_2O and therefore $O_2{}^+$ cannot undergo a charge exchange reaction with H_2O. Consequently, once $O_2{}^+$ is formed there is no readily available route for the loss of this ion, and $O_2{}^+$ is frequently seen as a low-level contaminant ion emanating from the ion source in PTR-MS experiments. Another contaminant ion is NO^+ that can be formed by the reaction [12]

$$N^+ + O_2 \rightarrow NO^+ + O\ (k = 2.6 \times 10^{-10}\ \text{cm}^3\ \text{s}^{-1}) \tag{3.14}$$

As with $O_2{}^+$, once NO^+ is formed it is unable to undergo charge-exchange reactions with other constituents in the source gas mixture because of the low ionization energy of NO. Consequently, NO^+ is another well-established contaminant ion from discharge ion sources.

Although $O_2{}^+$ and NO^+ are persistent contaminants which can undergo their own reactions with many analyte molecules, it is usually possible, through a combination of careful experimental design and appropriate gas flow conditions, to reduce them to levels where their impact is negligible. Conditions are attainable where >99% of the ions leaving the source drift region are H_3O^+. Consequently, using a simple and compact HCD a relatively clean source of H_3O^+ ions can be achieved without requiring a mass filter, thus yielding a simplified experimental set-up when compared to SIFT-MS.

3.3.3 Alternative Ion Sources

An HCD is the ion source used in the majority of PTR-MS instruments. However, other ion sources have been explored and exploited in PTR-MS and we briefly illustrate some of these in this section. For full details the reader is referred to the original publications.

Instead of a hollow cathode electrical discharge, an alternative plane electrode DC discharge source has been reported by Inomata and co-workers and is shown in the upper parts of Figure 3.8 [13]. As in the hollow cathode source, there is both a primary discharge

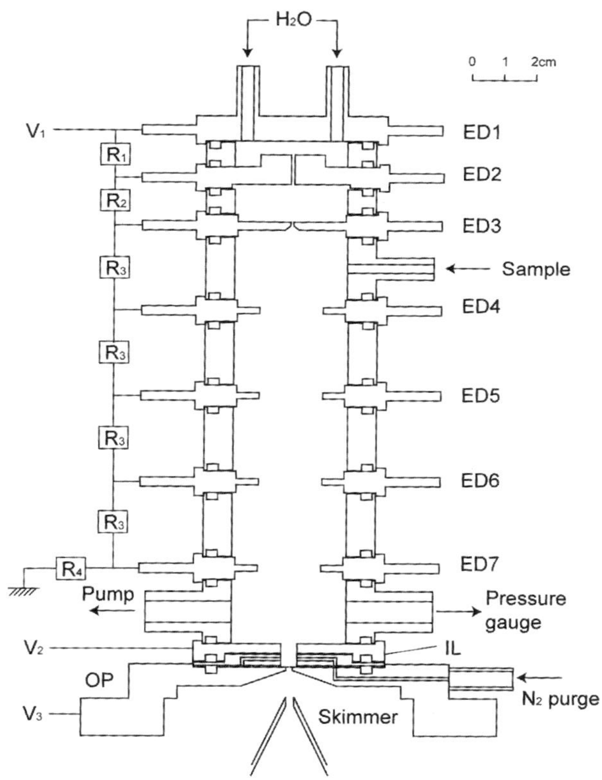

Figure 3.8 Discharge source and drift tube assembly used by Inomata and co-workers [13]. The primary discharge region consists of elements ED1–ED2 and is followed by a source drift region (ED2–ED3). The drift tube runs from ED3 through to the interface plate (shown as IL). Only ions that pass through the skimmer enter the mass spectrometer (not shown). Voltages are maintained along the electrode network via a potential divider using the resistors shown along the left side of the electrode chain. Reproduced with permission from [13]. Copyright © 2006 John Wiley & Sons, Ltd.

region (ED1-ED2) and a source drift region (ED2-ED3) prior to the drift tube. The discharge is initiated by the entrance of water vapour between anode and cathode plates located ~5 mm apart with a potential difference of 500 V. To help confine the discharge to this primary region, the primary and source drift regions are separated by a capillary of length 12 mm and diameter 1 mm in the cathode. One of the advantages claimed for this source over a more conventional hollow cathode source is its compatibility with a higher drift tube pressure, which is in the region of 7 mbar rather than the more usual 1–2 mbar (see Section 3.4.1). By operating at higher pressure, more analyte gas can be added to the drift tube and thus the ultimate detection sensitivity increases accordingly. This ion source has been tested on a PTR-MS instrument equipped with a TOF-MS (see Section 3.5.4 later) and provides a stable source of H_3O^+ with current levels similar to the standard HCDs. Contamination from NO^+ and $O_2{}^+$ was reported but was found to be small (~0.5%).

A plane electrode discharge source has also been designed by Hanson *et al.* [14]. This operates under sub-normal glow discharge conditions, which means that the discharge draws ≤100 μA of electrical current rather than the several mA drawn by a normal glow discharge. Furthermore, because the discharge takes place in an annular region away from the aperture leading into the drift tube, this source effectively acts like a combined discharge source and source drift region. This simplified design allows the drift tube to be operated at the relatively high pressure of 11 mbar, thus delivering more gas into the instrument and thereby increasing the sensitivity. Stable operation of this source for several months without cleaning has been reported.

A more radical departure from the HCD is to use a radioactive ion source, an idea first reported in PTR-MS work by Hanson *et al.* [15]. This exploits a low-level α particle emitter, ^{241}Am, to ionize water vapour and generate H_3O^+. A version of the Hanson radioactive source has been built in the laboratory of one of the authors and is illustrated in Figure 3.9. The α source is deposited on a metal strip located on the inside wall of a protective stainless steel cylinder, which replaces the discharge region in a hollow cathode source. The energetic (5 MeV) α particles cause multiple ionization events and a chain of subsequent ion–molecule reactions, similar to that described in the previous section, can be invoked to account for the efficient H_3O^+ production observed. The H_3O^+ current injected into the drift tube is in the region of 300 pA, or 2×10^9 ions per second.[4] With this type of ion source there is no external current driver and the long-term stability of the ion current is excellent. Furthermore, contamination from stray analyte gas entering the ion source region was found to be minimal, as judged by the low level of contaminant ions such as NO^+. This was attributed to the high operating flow of gas into the ion source (up to 20 STP cm^3 min^{-1} of an N_2/H_2O mixture), which minimizes backstreaming. The drift tube (see next section) could be operated at an even higher pressure, up to 13 mbar, than that in the planar discharge source described in the previous paragraph. Ion count sensitivities at

[4] It is possible to determine this quantity by measuring the current incident on the metal plate separating the source drift region from the drift tube. The current is small but measurable using a picoammeter. The plate has an aperture to allow ions into the drift tube but since the diameter of the plate (d_{plate}) and the diameter of the aperture ($d_{aperture}$) are known, then the ion current entering the drift tube (assumed to be dominated by H_3O^+) is equal to the measured current multiplied by $(d_{aperture}/d_{plate})^2$ Of course this assumes that the current is evenly distributed across the end plate, which may not necessarily be the case. An alternative approach is to measure the ion current incident on the plate at the end of the drift tube, but this will not fully reflect the rate of ions entering the drift tube because scattering losses and collisions with the walls through radial diffusion will inevitably reduce this current.

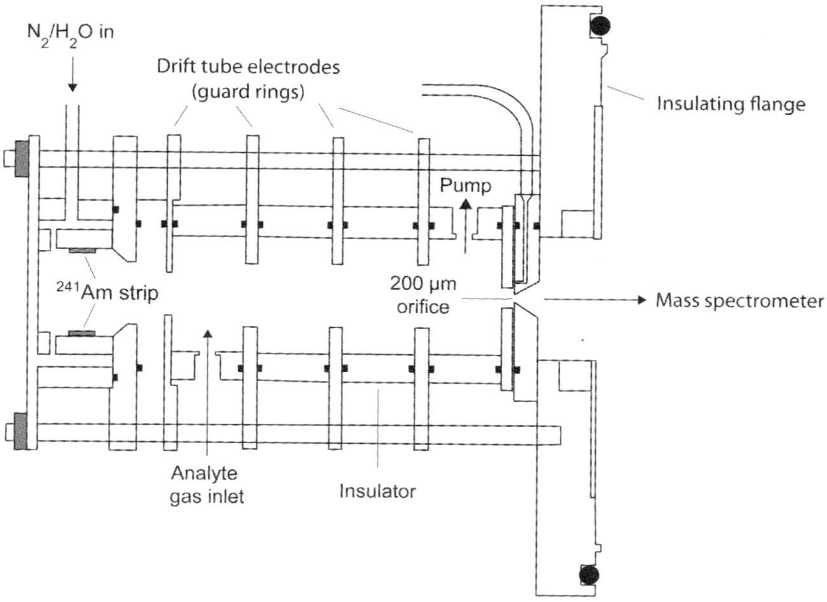

Figure 3.9 *Diagram of a radioactive ion source and drift tube. The radioactive source is a thin metallic strip impregnated with ^{241}Am which delivers α particles. The α particles interact with an incoming flow of nitrogen and water vapour: the nitrogen is used to ensure that a sufficient flow of gas is achieved to prevent backstreaming of the analyte gas into the ion source region. Adapted from Reference 15 with permission from Elsevier. More details about this source and the associated drift tube can be found in the original research paper. Adapted from [15] with permission from Elsevier.*

the mass spectrometer of several hundred Hz per ppbv were achieved for common VOCs such as acetone and isoprene, which delivers detection sensitivities of a few tens of pptv for individual VOCs in a ~1-s measurement time.

3.3.4 Generating Reagent Ions Other Than H_3O^+

The radioactive source shown in Figure 3.9 has also been used as a clean source of chemical ionization (CI) reagents other than H_3O^+ [16, 17]. In particular, both NO^+ and $O_2{}^+$ ion streams have been generated by this means and offer useful alternatives to H_3O^+ for certain applications, as already discussed in Section 2.3.

Discharge sources have also been used to generate alternatives to proton transfer reagents. A clean source of $O_2{}^+$ from a HCD in a conventional PTR-MS instrument was first demonstrated by Norman *et al.* [18]. This was employed as a means of detecting NH_3 which, as already discussed in Section 2.3, is difficult to quantify when using H_3O^+ as the CI reagent because of the residual $NH_4{}^+$ formed by reactions from N_2 leaking into the ion source. $O_2{}^+$ reacts with NH_3 (and almost all VOCs) by charge transfer, since O_2 possesses a significantly higher ionization energy than NH_3. More recently other workers have described discharge sources of NO^+ [19, 20] and $O_2{}^+$ [19].

The possibility of generating Kr^+ with quite a high purity (>93%) from a conventional HCD source in a PTR-MS instrument has also been demonstrated [21]. The motivation here is similar to the use of $O_2{}^+$, but with Kr^+ a much wider range of gases, including CH_4, CO and CO_2, can now be detected and quantified because Kr has a higher ionization energy than all of these molecules, and so Kr^+ can ionize each of them by charge transfer. However, there are complications, including the need to dilute the analyte gas in order to avoid excessive charge transfer from Kr^+ to O_2 in air samples (N_2 is not a problem since it has a higher first ionization energy than Kr). This dilution will accordingly reduce the sensitivity, but it is not necessarily a major problem if interest is focused on quantifying inorganic gases, since these are often present in the analyte at much higher concentrations than VOCs.

The use of reagents such as NO^+, $O_2{}^+$ and Kr^+ moves us into a realm beyond the strict definition of *proton transfer* reaction mass spectrometry, but this additional flexibility in the means of analyte ionization does have the potential to broaden the scope of the technique and, in the case of Kr^+, rather dramatically so. With the availability of several alternative ionizing reagents it would be more appropriate to refer to the technique as chemical ionization reaction mass spectrometry (CIR-MS) than PTR-MS. Indeed the former term has already been used in the research literature when referring to the availability of H_3O^+, NO^+ and $O_2{}^+$ reagent ions from a single source in a PTR-MS instrument [16]. However, the term PTR-MS is now well established in the literature and at present there seems to be no widespread desire to change to CIR-MS.

3.4 Drift Tubes

3.4.1 Practical Aspects

The drift tube is a critical part of a PTR-MS instrument. In its most common form it is essentially a series of electrodes, equally spaced and separated by insulating spacers to maintain the voltages applied to each electrode. The aim is to generate a uniform electric field that draws ions along the drift tube and delivers an increased migration velocity for the ions, as explained in Section 3.4.2. However, we first briefly consider possible structures for a drift tube.

The arrangement in Figure 3.8 is illustrative of the most common design for a drift tube. The drift tube is the region located downstream of electrode ED3. The tube is of circular cross section and contains several electrodes (sometimes called guard rings), labelled ED4–ED7 in Figure 3.8. The electrodes in most drift tubes employed in PTR-MS are constructed from stainless steel, since this is strong and relatively affordable while also being less susceptible to corrosion than some other metals. Separating the electrodes is an insulating material, and in the case of Figure 3.8 the material used was static dissipative Teflon. This form of Teflon is a good insulator but also has a sufficient conductivity to prevent the build-up of excess static charge, which would otherwise produce undesirable electric fields that can impede ion motion. The aperture of the electrodes ED4–ED7 is smaller than that of the cylindrical Teflon blocks, which means that the electrodes protrude into the drift tube.

Although there are variations, as already pointed out in the previous section, the typical pressure inside a PTR-MS drift tube is in the region of 1–2 mbar. Analyte gas is added to the drift tube at an upstream part of the drift tube where it then mixes with the ion stream emanating from the source drift region. This is normally achieved with a simple inlet tube, the assumption being that turbulence rapidly delivers good mixing of these gas streams and thus we obtain well defined reaction kinetics inside the drift tube. However, if desired the analyte gas can be added via a Venturi-type inlet in order to deliver a radially uniform distribution of gas into the drift tube.

To maintain the desired pressure inside the drift tube, continuous pumping is required. The drift tube must be air-tight, which means that separate components such as the Teflon spacers and the stainless steel electrodes must be properly sealed together to prevent external gas intrusion. This can be achieved by machining O-ring grooves in the electrodes. Once the O-rings (e.g. Viton) are put in place the electrodes and insulators are firmly clamped together and the system should be leak-tight.

There are two options for delivering a constant flow of analyte gases. One option is to employ a *mass flow controller* that provides a means of fixing and adjusting the flow of gas into the drift tube under electronic control. A potential disadvantage of mass flow controllers, aside from the cost, is that some VOCs can linger on the stainless steel surfaces in the interior of the flow controller, causing memory effects in time-resolved measurements. A way of avoiding this has been described by de Gouw *et al.* [22]. Their solution employs a gas line constructed of materials that are much less 'sticky' than stainless steel, namely Teflon or perfluoroalkoxy (PFA) polymer. The inlet to the PTR-MS is split off from this main inlet line and the flow rate is controlled through a combination of an upstream needle valve and a downstream pressure controller. Typical flow rates of analyte gas through the drift tube are in the region of 20–30 STP $cm^3\ min^{-1}$.

The second inlet option is to use a *critical orifice*, which is essentially a small hole drilled into a metal plate which allows air to enter the drift tube at a fixed rate. Providing the pressure on the downstream side of the orifice is more than a factor of two lower than on the upstream side, gas enters at the speed of sound and the flow rate cannot be increased any further. Such conditions are easily achieved in PTR-MS, where the gas usually enters at atmospheric pressure into a drift tube with a pressure near to 1 mbar. Clearly the diameter of the aperture in the critical orifice plate must be compatible with the pumping speed of the system in order to maintain the desired drift tube pressure, but an orifice of a few tens of microns is typical. A clear advantage of a critical orifice inlet is its simplicity but the flow rate through the orifice is sensitive to fluctuations in the temperature and pressure on the upstream side, so is not ideal in applications where such fluctuations might occur.

At the downstream end of the drift tube an aperture is required to allow a fraction of the ions to enter an ion transfer region leading to the mass spectrometer. Since the pressure inside the mass spectrometer must normally be many orders of magnitude lower than that of the drift tube, typically $\leq 10^{-5}$ mbar, the exit aperture of the drift tube is quite small. A typical exit aperture size will be 100–200 μm, but the precise size will be dictated by the flow rate of analyte gas into the drift tube as well as the pumping speed downstream of the drift tube, such that the mass spectrometer is maintained at a sufficiently low operating pressure.

Other drift tube designs are possible. For example, Figures 3.10 and 3.11 show one such arrangement originating from the laboratory of one of the authors. Here the drift tube is constructed from a single piece of static dissipative Teflon and the metal electrodes are located on the *outside* of the tube. This design is adapted from a drift tube developed for

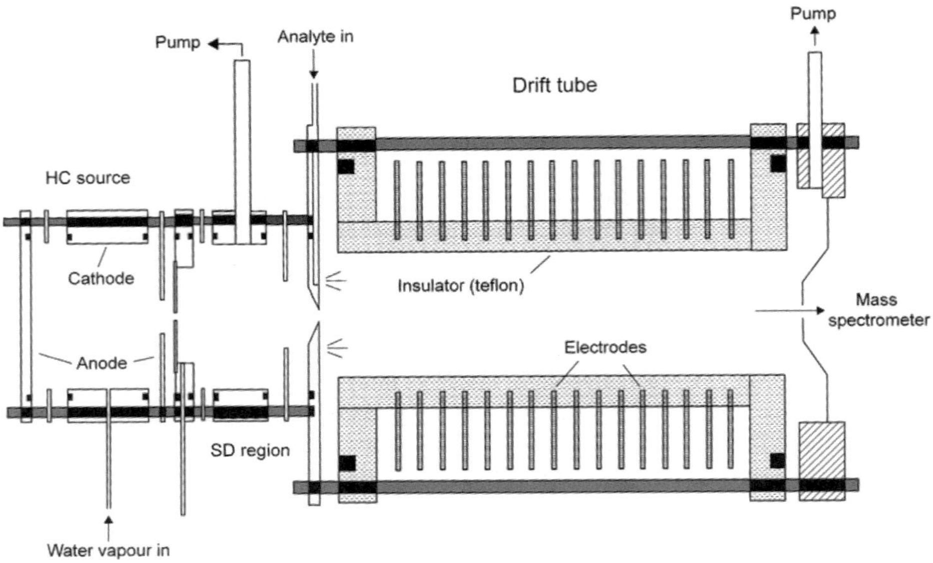

Figure 3.10 *Schematic cross section of an ion source and drift tube from the laboratory of one of the authors. Here the drift tube is constructed from a single block of Teflon and the metal electrodes are located in slots on the outside of the drift tube. A Venturi inlet was employed to introduce the analyte gas into the drift tube. HC refers to the hollow cathode discharge region and SD is the source drift region.*

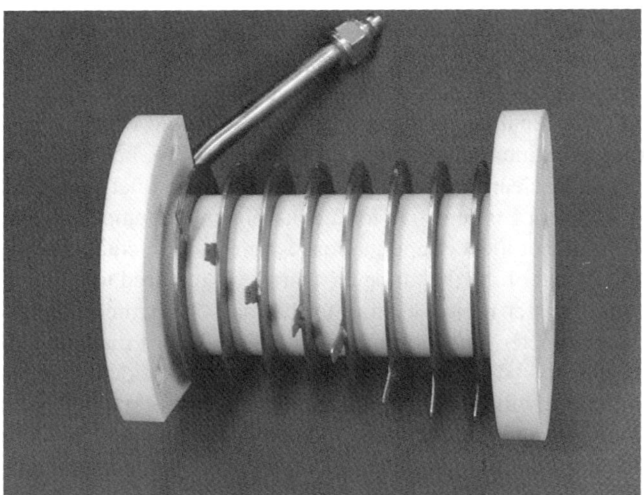

Figure 3.11 *Photograph of the drift tube shown schematically in Figure 3.10. Copyright © R. S. Blake, 2013.*

ion mobility spectrometry [23]. Although the electrodes no longer protrude into the drift tube, the electric field can penetrate inside the drift tube providing that the inner surface of the electrodes is not too far from the interior wall of the Teflon. In the case of Figure 3.10 this separation is 5 mm and at this effective thickness the Teflon is predicted to reduce the electric field inside the drift tube by roughly 25% when compared with the same electrode in the absence of the Teflon. An advantage of using a single piece of Teflon for the drift tube is that it avoids requiring multiple vacuum seals that are time consuming to assemble and may be prone to develop leaks if not seated correctly.

An alternative approach is to construct the drift tube out of resistive glass [24, 25]. Connection of a voltage across the two ends of this tube automatically provides a built-in continuous potential divider, and therefore a continuous electrical potential gradient along the tube. In addition to removing the need for multiple guard rings and insulating spacers this also ensures a highly homogeneous electric field along the tube because there are no localized field distortions provided by the presence of individual electrodes.

A typical drift tube in PTR-MS is ~10 cm in length, although both longer and shorter tubes have been reported. A long drift tube will provide a longer reaction time for a given value of the ratio *E*/*N* (see Equation 3.20), which will aid sensitivity but will also create a more bulky apparatus. The electric field is important because, as discussed elsewhere (Section 3.4.3), it delivers energy to ion–molecule collisions to prevent excessive clustering of neutral molecules onto ions. In particular we wish to avoid production of excessive quantities of $H_3O^+(H_2O)_n$ clusters that may show different chemistry with certain VOCs when compared with H_3O^+ (see Section 2.2.4.2). However, the energy delivered in these de-clustering collisions depends not only on the applied electric field, *E*, but also on the density of gas in the drift tube, i.e. the quantity *E*/*N*. This ratio is of critical importance in PTR-MS and is sometimes referred to by a specific name, the *reduced electric field*, *E*/*N*. The gas number density is defined as the number of gas particles per unit volume and this is readily calculated using the equation below, which is derived from the ideal gas equation:

$$N = \frac{N_A}{V_M}\frac{273.15}{T_d}\frac{P_d}{101.325} \tag{3.15}$$

N_A is Avogadro's number (6.022×10^{23} mol^{-1}) and V_M (22414 cm^3 mol^{-1}) is the molar volume of an ideal gas at 1 atmosphere of pressure (101.325 kPa) and at a temperature of 273.15 K. T_d is the operational temperature of the drift tube (in Kelvin), and P_d is the gas pressure in the drift tube (in kPa). Generally pressure gauges provide readings of pressure in units of millibar or Torr (equivalent to mmHg).[5] For conversion purposes 1 Pascal (Pa) is equal to 0.01 millibar or 0.0075 Torr. For example, using a PTR-MS instrument at an operating drift tube pressure of 2 mbar and a drift tube temperature of 60°C (or 333 K), the number density is $N = 4.35 \times 10^{16}$ cm^{-3}. The value for *E* is simple to calculate, being given by Equation 3.1. For a typical operating voltage of 500 V across a drift tube of length 10 cm, the electric field strength is 50 V cm^{-1}. Under these operating conditions, the reduced electric field is $E/N = 1.15 \times 10^{-15}$ V cm^2. Owing to the smallness of this number using conventional units, a different and more convenient unit is commonly adopted when

[5] The torr is a remnant from the days when pressure was routinely measured by a mercury manometer. Atmospheric pressure can support a 760 mm high column of mercury and 1 Torr was defined as 1/760 of atmospheric pressure. More usefully, 1 Torr = 1.333 mbar.

referring to *E/N* values. This unit is the townsend (Td) and is defined by the relation 1 Td $= 10^{-17}$ V cm^2. Thus in the above example *E/N* is 115 Td. When we refer to *E/N* values throughout the remainder of this book we will always use the Td unit.

As hinted at in the above calculation, the drift tube and associated gas inlet are almost invariably heated to try to prevent adsorption of VOCs on the vessel walls. Typical temperatures vary from 40°C to 100°C, with the upper temperature limited by the vacuum seals and other components used in the drift tube, which may be damaged and/or may degas excessively when raised to even higher temperatures. By avoiding these temperature-sensitive components, Mikoviny *et al.* have constructed a PTR-MS instrument where the ion source, drift tube and gas inlet are capable of operating continuously at temperatures as high as 250°C [25]. This rather extreme operating temperature was chosen to allow reliable measurements of semi-volatile trace gases.

Another advantage of operating at a high temperature is the improvement it offers to the *response time* of the instrument. The ability to be able to distinguish between distinct events in time using PTR-MS is limited partly by the time it takes for the analyte gas to pass through the inlet and drift tube and out into the vacuum system. In principle a 'packet' of gas entering the drift tube at one end can exit as a packet with the same spatial extent at the other end. However, in practice turbulent mixing as this gas packet collides with the existing gas as it enters the drift tube will spread this packet spatially. Furthermore, collisions with the walls and diffusion will both act in the same way. Nevertheless, the extent of this spatial delocalization is related to the time the gas spends in the drift tube and therefore it is useful to calculate this quantity. Suppose we have a drift tube operating at a pressure of 2 mbar and a temperature of 60°C which is receiving an air flow of 20 STP cm^3 min^{-1}, which is fairly typical for PTR-MS. Remember (see Section 3.3.1) that the STP flow rate is the volumetric flow rate that would occur if the system was operated under STP conditions (F_{STP}). However, since the drift tube is actually operated under very different conditions the actual volumetric flow rate (F_{drift}) can be obtained by applying the ideal gas equation and is given by

$$F_{\mathrm{drift}} = F_{\mathrm{STP}} \times \frac{P_{\mathrm{S}}}{P_{\mathrm{drift}}} \times \frac{T_{\mathrm{drift}}}{T_{\mathrm{S}}} \tag{3.16}$$

where P_{S} and T_{S} are the standard pressure and temperature, respectively. Application of this equation to the conditions listed above gives a flow rate of 12200 cm^3 min^{-1} or 200 cm^3 s^{-1}. For a drift tube of cylindrical cross section with a diameter of 4 cm and a length of 10 cm the internal volume is ~125 cm^3. Consequently, a slug of gas that is not subject to spatial delocalization should traverse the drift tube in ~0.6 s. Smaller drift tubes can be constructed to make the residence time smaller but one cannot push this quantity too much further.

We reiterate that the intrinsic time resolution of the PTR-MS instrument can be faster than the nominal gas residence time. However, other factors may work to lower the practical time resolution, including the adsorption and slow release of molecules on the surfaces of the gas inlet line and the walls of the drift tube (a so-called *memory effect*). Moreover, if the ion signal is weak then data may need to be accumulated for some time to achieve an acceptable signal-to-noise ratio, and therefore this may be the key factor dictating the ultimate time resolution of any measurement. With regard to memory effects, by operating the gas inlet and drift tube at a high temperature Mikoviny *et al.* were able to reduce the response times even for certain strongly adsorbing compounds, such as ammonia and dimethyl sulphoxide, to <0.5 s [25].

3.4.2 Ion Mobility and Transit Times

Positive ions entering the drift tube will be subject to an accelerating force supplied by the applied electric field. However, instead of a continuous increase in ion velocity as the ions move towards the electrode with the most negative potential, the ion swarm will quickly establish a constant drift velocity if the field applied is uniform. The constant drift velocity is a result of the significant pressure of gas in the drift tube, such that multiple ion–molecule collisions take place. The ions cannot continue to accelerate because an opposing force is generated through collisions with neutral gas molecules and eventually a steady-state drift velocity is reached.

The ion drift velocity, v_d, is linearly dependent on the electric field (E) at low electric fields[6] and can be expressed as

$$v_d = KE \tag{3.17}$$

where the coefficient K is known as the ion mobility.[7] This equation refers to the mean drift velocity of the ion swarm, not the velocities of individual ions. The ion mobility coefficient is a function of temperature and pressure, but it is also determined by the cross section for ion–neutral collisions, and therefore depends on the size (and structure) of the ion and the neutral gas molecules. It is therefore useful to introduce a pressure and temperature-independent quantity for a given ion/neutral combination known as the *reduced mobility*, K_0, where K_0 is related to K by the expression

$$K = \frac{760}{P}\frac{T}{273}K_0 \tag{3.18}$$

In Equation 3.18, the pressure is expressed in torr and the temperature is in kelvin. The important point about Equation 3.18 is that, once K_0 is known for a particular ion/neutral combination, the ion mobility can be calculated at any temperature and pressure. Equation 3.18 can be substituted into Equation 3.17 to yield

$$v_d = \frac{760}{P}\frac{T}{273}K_0E \tag{3.19}$$

---

[6] It is common practice in the PTR-MS literature to assume that K is independent of E. However, this is only true for low electric fields and more generally K should be expressed as

$$K(E) = K(0)[1 + \alpha_1(E/N)^2 + \alpha_2(E/N)^4 + \ldots.$$

It is important to recognize that the electric field conditions normally employed in PTR-MS are *not* low field conditions and thus the dependence of K on E cannot be ignored in an accurate quantitative treatment [24, 26]. However, this makes any calculations which require the use of K considerably more complex and in what follows we opt for simplicity and ignore any dependence of K on E. However, the wise reader should be wary of such calculations and not attach too much accuracy to the numbers extracted.

[7] The International Union of Pure and Applied Chemistry (IUPAC) recommend that the symbol μ be used to represent the mobility of a charged particle. However, this symbol clashes with that used to represent reduced mass (see Section 2.2.3.2) and the symbol K is more commonly used in the modern literature. We have therefore chosen to use K in this book.

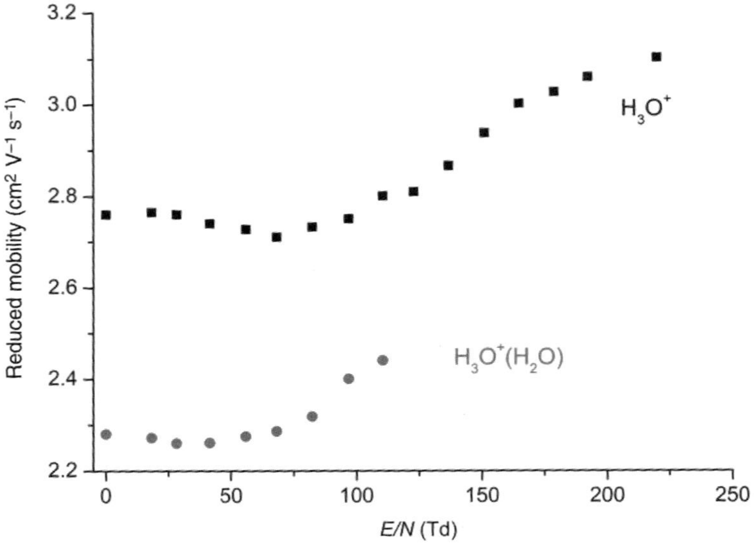

Figure 3.12 *Graph showing the reduced mobilities* (K_0) *of* H_3O^+ *and* $H_3O^+(H_2O)$ *in pure nitrogen as a function of* E/N. *Adapted with permission from Reference 30. Copyright 1976, American Institute of Physics.*

If we now define the parameter N_0 to be the gas number density under standard temperature and pressure (i.e. 273.16 K and 760 Torr = 1.01325 × 10^5 Pa), Equation 3.19 may be rewritten as follows:

$$v_d = K_0 N_0 \frac{E}{N} \tag{3.20}$$

Our expectation is that the drift velocity will decrease as the drift tube pressure increases. Equation 3.20 confirms this, although the effect of the gas is expressed here in terms of number density (N) rather than pressure.

If the velocity of the ion swarm along the drift tube is measured, which can be achieved by sending a short pulse of ions along the tube and measuring the time of arrival at the end, then the reduced mobility can be determined and thus K_0 can be calculated as a function of E/N. Extensive tabulations of reduced mobilities have been published [27–29]. As an example, Figure 3.12 shows reduced mobility data for both H_3O^+ and $H_3O^+(H_2O)$ in gaseous nitrogen obtained at a temperature of 298 K. $H_3O^+(H_2O)$ possesses a smaller reduced mobility than H_3O^+, which is expected given the larger size of the former ion. However, measurements have shown that the mean velocities of H_3O^+ and all $H_3O^+(H_2O)_n$ cluster ions in a PTR-MS drift tube are the same [31], which seems to contradict the comments above. The reason for this is that these clusters are not fixed cluster ions in air but can instead undergo multiple hydration and dehydration reactions as they make their way along the drift tube. The cluster ions therefore interchange as they progress along the drift tube and thus only an average mobility is observed. If it is necessary to be precise about the mobility of the cluster ions, then an effective mobility can be defined for a given E/N which will be a weighted average of the mobilities of the various cluster ions, the weighting coefficients being the relative abundances of each $H_3O^+(H_2O)_n$ cluster.

With the knowledge of its mobility, it is now possible to calculate the transit time for H_3O^+ through a drift tube. We will assume that the ion mobilities in air and nitrogen are the same, which should be a reasonably good approximation, and will carry out the calculation for an *E*/*N* of 120 Td. We will also assume that conditions have been chosen so that H_3O^+ is far more abundant than hydrated hydronium cluster ions. From Figure 3.12, the reduced mobility of H_3O^+ at this value of *E*/*N* is ~2.81 cm^2 V^{-1} s^{-1}. The time taken to travel a drift tube of length L is given by $t_d = L/v_d$ and thus, in combination with Equation 3.20, this gives

$$t_d = \frac{L}{K_0 N_0} \frac{N}{E} \tag{3.21}$$

Choosing a drift length of 10 cm and using the quantities $N_0 = 2.687 \times 10^{19}$ cm^{-3}, $K_0 = 2.81$ cm^2 V^{-1} s^{-1} and $E/N = 120$ Td (1.2×10^{-15} V cm^2) gives a drift time of $t_d = 110$ μs and a mean drift velocity of 906 m s^{-1}. This drift (reaction) time can be combined with ion count rate measurements and the reaction rate coefficient to calculate the concentration of a specific analyte molecule via Equation 1.10. Note that the drift time for ions is more than two orders of magnitude shorter than the typical time required for the neutral analyte gas to travel from one end of the drift tube to the other, that is, the ion drift time and the instrument response time are two very different quantities. It is also important to notice that, since t_d is inversely proportional to *E*/*N*, there is a loss in detection sensitivity as *E*/*N* is increased because the time available for proton transfer reactions is reduced.

Caution is warranted in employing calculated ion transit times. The ion transit time may not be the same as the reaction time, since analyte gas is usually added to the ion flow at a point some way along the drift tube. This can, of course, be allowed for by adjusting the value of L accordingly, but there are other complications to take into account, including the finite mixing time between ions and the analyte gas. This uncertainty in the effective length of the reaction zone can be circumvented by adding a compound at a known concentration and whose proton transfer reaction with H_3O^+ has a known rate coefficient. In effect, this well-defined reaction can then be used to empirically determine the reaction time, t_d. There is, however, another complication which arises from the different mobilities of H_3O^+ reagent ions versus those of the protonated recipient molecules, MH^+. Instead of these ions travelling along the drift tube at essentially the same velocity, as would be the case in SIFT-MS because their speeds are simply determined by the gas flow, large protonated organic cations will travel at a substantially slower velocity than H_3O^+ and this introduces an error into the calculation of concentrations using Equation 1.10. As discussed by Keck and co-workers, this error can be greater than 20% [32].

3.4.3 Ion–Molecule Collision Energies

In addition to a thermal translational energy, ions travelling along a drift tube are given an additional energy by the applied electric field. An expression for the total mean ion kinetic energy was derived by Wannier [33, 34] and takes the following form:

$$KE_{ion} = \frac{3}{2}k_B T + \frac{1}{2}m_{ion}v_d^2 + \frac{1}{2}m_b v_d^2 \tag{3.22}$$

The first term on the right-hand side of Equation 3.22 is simply the thermal contribution to the kinetic energy of the ion. The second term represents the kinetic energy of the ion as it is driven along the drift tube at velocity v_d. However, Wannier also suggested an additional contribution, the third and final term in Equation 3.22, which derives from collisions between the ions and buffer gas molecules (of mass m_b)[8] and gives rise to an additional, but randomly oriented, velocity contribution.

Although Equation 3.22 was derived as early as 1953, there was some doubt about its validity until the mid-1970s, when careful experiments were carried out on ion motion in various buffer gases, which showed the Wannier expression to be a good approximation to the real mean ion kinetic energy [35].

Equation 3.22 refers to the ion kinetic energy in the laboratory frame of reference. To understand the impact of the ion kinetic energy on an ion–molecule reaction outcome, it is the collision energy that is important, which means that the *relative* kinetic energy of the colliding species is required. Owing to the need to conserve linear momentum this is not the same as the kinetic energy given by Equation 3.22. Instead we need the kinetic energy relative to the centre-of-mass of the colliding system, KE_{CM}, because this is the energy that is available for a reaction process and it takes the form

$$KE_{CM} = \frac{1}{2}\left(\frac{m_n m_{ion}}{m_n + m_{ion}}\right)\left(v_{ion}^2 + v_n^2\right) \tag{3.23}$$

where

$$\frac{1}{2}m_{ion}v_{ion}^2 = KE_{ion} \tag{3.24}$$

and

$$\frac{1}{2}m_n v_n^2 = \frac{3}{2}k_B T \tag{3.25}$$

Notice that the molecule of mass m_n refers to the neutral molecule involved in the collision (which is an unreactive collision with a buffer gas molecule or a reactive collision with an analyte molecule).

Combining Equations 3.23, 3.24 and 3.25 gives

$$KE_{CM} = \left(\frac{m_n}{m_n + m_{ion}}\right)KE_{ion} + \left(\frac{m_{ion}}{m_n + m_{ion}}\right)\left(\frac{3}{2}k_B T\right) \tag{3.26}$$

Recognizing that

$$\left(\frac{m_{ion}}{m_n + m_{ion}}\right) = 1 - \left(\frac{m_n}{m_n + m_{ion}}\right) \tag{3.27}$$

leads to the final expression for KE_{CM}:

$$KE_{CM} = \frac{m_n}{m_{ion} + m_n}\left(KE_{ion} - \frac{3}{2}k_B T\right) + \frac{3}{2}k_B T \tag{3.28}$$

[8] In PTR-MS the buffer gas is normally air and so an average of the masses of N_2 and O_2, weighted by their relative abundances, will suffice to determine m_b, that is, a relative molecular mass of 28.8.

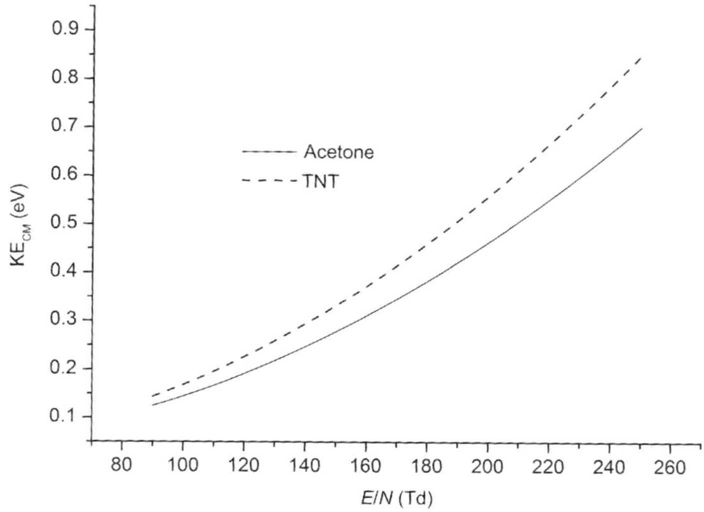

Figure 3.13 *Calculated centre-of-mass collision energies* (KE_{CM}) *for collisions of* H_3O^+ *with either acetone (solid line) or trinitrotoluene (TNT; dashed line) as a function of* E/N *for a drift tube operating at a temperature of 300 K with air as the buffer gas. A reduced mobility of 2.81* $cm^2\ s^{-1}\ V^{-1}$ *for* H_3O^+ *in air was used in the calculations. The nominal molecular masses of acetone and TNT are 58 Da and 227 Da, respectively.*

It is useful to illustrate the value of Equation 3.28 through a specific example. We build on the discussion in the preceding section, where the ion drift velocity for H_3O^+ was calculated to be 906 m s^{-1} at an *E/N* of 120 Td. Suppose the analyte molecule is acetone, which has a relative molecular mass of 58. To calculate the centre-of-mass collision energy between the H_3O^+ and acetone molecules we need to substitute the expression for KE_{ion}, given by Equation 3.22, into Equation 3.28. The relative molecular mass of the buffer gas will be assumed to be 28.8 (a weighted average of N_2 and O_2) and after performing the calculation we obtain $KE_{CM} = 3.09 \times 10^{-20}$ J = 0.19 eV at 300 K. This is almost an order of magnitude higher than the average thermal kinetic energy, $3k_BT/2$ at 300 K, so the electric field clearly provides substantial additional energy to the reactive collisions in PTR-MS. Notice that this is distinctly different from SIFT-MS, where no electric field is present and therefore ion–molecule collision energies are thermal.

The dependence of collision energies on *E/N* is illustrated further in Figure 3.13. This plot, obtained by application of Equation 3.28, shows the calculated KE_{CM} as a function of *E/N* for collisions between H_3O^+ and two different molecules, acetone and trinitrotoluene. As expected from Equation 3.28, the KE_{CM} is mass dependent, although the differences are modest at the lower end of the *E/N* scale.

3.4.4 Ion Cluster Distributions

In addition to providing the mechanism for the migration of reagent and product ions along the drift tube, the electric field also helps to prevent the formation of substantial quantities of cluster ions. A value of *E/N* should be chosen which adds sufficient energy to collisions between ions and neutral molecules to break up any ion clusters that might otherwise be formed while at the same time avoiding excessive fragmentation of unclustered analyte molecular ions. This is normally achieved by employing an *E*/*N* of between 100 and 140 Td.

To see the effect that the *E*/*N* has on cluster survival, it is useful to calculate the collision energy between a cluster ion and a buffer gas molecule. We will focus on the $H_3O^+(H_2O)$ cluster, since this is likely to be the most abundant of the hydrated hydronium cluster ions in any PTR-MS instrument. For reasons discussed in Section 3.4.2 the mean velocities of $H_3O^+(H_2O)$ and H_3O^+ in a drift tube will be the same despite their different reduced mobilities and therefore, since H_3O^+ will be far more abundant than $H_3O^+(H_2O)$ at an *E*/*N* of 120 Td (see below), we can use the drift velocity (906 m s^{-1}) of H_3O^+ calculated in Section 3.4.1 for illustration purposes. Employing Equation 3.28 and including the expression for KE_{ion} in Equation 3.22, the centre-of-mass collision energy reduces to

$$KE_{\text{CM(buffer-ion)}} = \frac{1}{2} m_{\text{b}} v_{\text{d}}^2 + \frac{3}{2} k_{\text{B}} T \tag{3.29}$$

when m_{b} is used in place of m_{n}. The mean collision energy is thus found to be 2.59×10^{-20} J or 0.16 eV. Experimental measurement has determined a dissociation energy for removing H_2O from $H_3O^+(H_2O)$ of 1.41 ± 0.06 eV [36], a value supported by independent high quality *ab initio* quantum chemical predictions [37, 38]. The average single collision will therefore not provide sufficient energy to fragment $H_3O^+(H_2O)$, but a multitude of such collisions could progressively deposit enough energy into the ion to bring about fragmentation. We can easily calculate the number of collisions between an H_3O^+ ion and the neutral buffer gas in a drift tube using the kinetic theory of gases. Since an estimate will suffice, the buffer gas will be assumed to be pure nitrogen at a pressure of 2 mbar and in a drift tube maintained at a temperature of 60°C, which corresponds to a number density for the nitrogen of $N = 4.4 \times 10^{16}$ molecules cm^{-3}. The collision frequency, z, for a single H_3O^+ ion is given by

$$z = \sigma \left(\frac{8 k_{\text{B}} T}{\pi \mu} \right)^{1/2} N \tag{3.30}$$

where σ is the collision cross section and μ is the reduced mass of the colliding system. The collision cross section between H_3O^+ and N_2 is in the region of 3×10^{-15} cm^2 [39] and the reduced mass is 1.9×10^{-26} kg to two significant figures. A temperature for this colliding system above that of the actual drift tube will be assumed, for reasons already seen above, and specifically we will use a value of 1200 K (see later for justification). Substituting the appropriate numerical values into Equation 3.30 and taking care to use consistent units gives a collision frequency of 2×10^7 collisions per second. Since the ions typically take on the order of 100 μs to pass through the drift tube, each H_3O^+ ion will therefore undergo in the region of 2000 collisions with the neutral gas molecules as it passes through the drift tube. Thus the requirement for multiple collisions for efficient energy transfer is met.

The calculations can be taken further by assuming an equilibrium distribution of cluster ions. Again we will concentrate on a single cluster ion and the equilibrium process of interest is

$$H_3O^+ + H_2O + M \rightleftharpoons H_3O^+(H_2O) + M \tag{3.31}$$

The procedure we follow here is similar to that described by Warneke and co-workers [31]. We specifically require the ratio $[H_3O^+(H_2O)]/[H_3O^+]$ that can be obtained from the equilibrium constant

$$K_{eq} = \frac{[H_3O^+(H_2O)]}{[H_3O^+][H_2O]} \tag{3.32}$$

if both $[H_2O]$ and K_{eq} can be found. The latter can be determined from a combination of Equations 2.9 and 2.12 where the enthalpy and entropy changes for Reaction 3.31 are taken from the work by Lau *et al.* [40] and have the values -132 kJ mol^{-1} and -102 J mol^{-1} K^{-1}, respectively. To calculate the Gibbs energy change for the reaction the temperature is required but this is not the same as the laboratory temperature of the drift tube, since the ions within the drift tube are given additional energy by the applied electric field (as mentioned above). An effective ion temperature, T_{eff}, can be defined by equating an effective ion thermal energy with the mean centre-of-mass collision energy between the ion and the buffer gas [35, 41–43], that is, by choosing

$$\frac{3}{2}k_B T_{eff} = KE_{CM(ion-buffer)} = \frac{3}{2}k_B T + \frac{1}{2}m_b v_d^2 \tag{3.33}$$

Assuming a temperature of 298 K for the drift tube, the effective ion temperature is calculated to be 1245 K and therefore $\Delta G^0_{1239} = -5.01$ kJ mol^{-1}. Consequently, $K_{eq} = 0.62$.

To complete the calculation of the concentration ratio $[H_3O^+(H_2O)]/[H_3O^+]$, the partial pressure of water vapour in the drift tube is required.[9] One source of water in the drift tube is the water deliberately added to the ion source. This is minimized in some instruments by adding a source drift region and pumping this to reduce the quantity of water vapour that can exit into the drift tube. A second source of water is the analyte gas. To keep the calculation simple, we will assume that the water in the drift tube originates from the analyte gas only, and this is assumed to enter with 100% relative humidity. The saturated vapour pressure of water at 25°C is 31.7 mbar and we can equate this to the partial pressure of water, p_{H_2O}, in the expression below:

$$\frac{[H_3O^+(H_2O)]}{[H_3O^+]} = \left(\frac{p_{H_2O}}{p^0}\right) \times K_{eq} \tag{3.34}$$

The saturated vapour pressure is used in the above expression rather than the actual pressure in the drift tube since the thermodynamic parameters are expressed relative to standard state conditions, in which the total pressure is 1 bar (p_O).

The ratios of $[H_3O^+(H_2O)]$ to $[H_3O^+]$ calculated using this procedure are shown in Table 3.1 for several different *E*/*N* values. These calculated values suggest that at 120 Td the proportion of $H_3O^+(H_2O)$ in the drift tube is relatively low. However, at smaller values of *E*/*N* the contribution of $H_3O^+(H_2O)$ grows substantially and below 100 Td hydrated hydronium becomes the dominant ion. Descending to still lower *E*/*N* values than those shown in Table 3.1 will see larger hydrated hydronium ions become the dominant species. As discussed in Chapter 2, proton transfer from $H_3O^+(H_2O)$ can be much slower than from

[9] To obtain a dimensionless equilibrium constant, the partial pressure required is a dimensionless partial pressure expressed relative to *standard pressure* (1 bar).

Table 3.1 *Calculated ratio of [$H_3O^+(H_2O)$] versus [H_3O^+] as a function of E/N*

E/N (Td)	$H_3O^+(H_2O)$:H_3O^+ abundance ratio
80	270
90	14
100	1.1
110	0.14
120	0.023

H_3O^+ for some compounds, or indeed in some cases does not occur. Furthermore, hydrated hydronium ions lead to other complications, such as ligand switching reactions (see Section 2.2.4.2) that can make mass spectral analysis more complex. It is for these reasons that the majority of PTR-MS experiments are carried out at an $E/N > 100$ Td. However, excessively high *E/N* values are also normally avoided since they can produce collision energies that can lead to extensive fragmentation of protonated organic ions. Furthermore, a higher *E/N* leads to a reduced reaction time, which in turn will decrease the sensitivity of the PTR-MS instrument. Consequently, the preferred reduced electric field is a compromise and most PTR-MS experiments operate at *E/N* values in the range 120–140 Td.

Predictions of hydrated hydronium cluster abundances of the type presented above are largely supported by experimental measurements of $H_3O^+(H_2O)_n$ relative abundances (see, e.g. Reference [44]). However, there are approximations involved in the calculations which mean that they should only be used as a guide to the relative quantities of hydrated hydronium ions. One such approximation is the assumption that the calculated ion temperature, T_{eff}, is a valid estimate of the temperature experienced by Reaction 3.31. T_{eff} is calculated as an effective *translational* temperature of the ions which does not directly account for the excitation of internal energy modes, that is, rotational and vibrational degrees of freedom, in either the ion or the neutral molecule. There is evidence of good equilibration taking place during collisions (see, e.g. Reference [45]), but the effective temperature used in the calculation of the equilibrium constant should still be regarded with some caution. Furthermore, the enthalpy and entropy changes employed in the equilibrium constant are only approximate and have also been assumed to be temperature independent, which in itself is only an approximation. There are yet more approximations involved and a more complete discussion has been provided in a review by de Gouw and Warneke [46].

3.5 Mass Spectrometry

The earliest PTR-MS studies used only quadrupole mass spectrometers (QMS) for ion mass analysis. However, more recently there has been increasing interest in exploiting some of the advantages of other types of mass analysers in PTR-MS instruments. Some of the more common and well-known types of mass analyser are listed in Table 3.2. With the exception of magnetic/electrostatic sector analysers, all have seen some application in PTR-MS and

Table 3.2 Commonly used types of mass analysers

Analyser	Main features
Quadrupole filter	Compact, relatively low cost, unit mass resolution, limited mass range
3D quadrupole ion trap	Extremely compact, fast scanning, unit mass resolution, option for MS^n experiments[a]
Magnetic and/or electrostatic sector	High resolution, no upper limit of mass
Time-of-flight	Good resolution, high throughput, no upper mass limit (in principle)

[a] MS^n experiments are described in Section 3.5.3.2.

it is likely in the coming years that there will be a considerable expansion in the use of alternatives to quadrupole analysers in PTR-MS.

In this section we will explain the operating principles of the three types of mass analyser used so far in PTR-MS studies, namely quadrupole, ion trap and TOF-MS. However, before doing so some basic definitions relevant to mass spectrometry are introduced.

3.5.1 Some Important Definitions

3.5.1.1 Ion Mass and Mass-to-Charge Ratio

The positions of peaks in a mass spectrum are determined by the ratio m/z, where m is the mass of a specific ion and z is the corresponding charge (as a multiple of the fundamental unit of electrical charge, e ($=1.60210 \times 10^{-19}$ C). This was discussed briefly in Section 1.5.2 but there is sometimes some confusion about units for mass in a mass spectrometry context and so in this section we deal with this in some detail.

It is common to see mass peaks identified by their m/z ratio, that is, a peak might be reported at m/z 69, which could alternatively be written as $m/z = 69$. Implicit in such reporting is that the absolute mass of the ion in the SI unit of kilograms is not used, since of course an ion mass expressed in kg is a very small quantity. Instead the mass in m/z is in unified atomic mass units, which is sometimes written as 'amu' but more correctly should be expressed as the unit 'u', where 1 u = one-twelfth the mass of one atom of ^{12}C and is equal to $1.6605402(10) \times 10^{-19}$ kg. The mass in some mass spectrometric data is quoted in amu or u, but actually it is common practice among many mass spectrometrists to employ the dalton (symbol Da) as the unit of mass, where 1 Da = 1 u.[10] With the exception of ^{12}C, the mass of any atom or molecule is not an exact integer number of daltons, but if the resolution of a mass analyser is relatively low then this distinction is unimportant and

[10] You may wonder why one bothers with the unit Da when it is the same as the unit u. This is a valid question, but they tend to be used in different contexts, with u being used to express more precise atomic and molecular masses whereas Da is more often used to express more approximate masses, for example, integer (nominal) values of u for particular molecules. To add to the not-inconsiderable confusion the 'Thomson', with the symbol Th and named after the famous physicist and discoverer of the electron, J. J. Thomson, has been suggested as the unit for m/z. However, this has not become accepted practice and is rarely if ever used in the PTR-MS literature.

it is common in that situation to approximate the mass of a given ion as an integer multiple of Da. This mass is often called the *nominal ion mass*.

In some types of mass spectrometry it is not unusual to encounter positively charged ions with $z > 1$, such as the doubly charged ions sometimes seen from electron impact ionization sources or the highly charged ions produced by the electrospray technique. However, in PTR-MS the charge on an ion is always $+\ 1e$, that is, $z = 1$. Consequently, it is acceptable to express a peak position in the mass spectrum in terms of its mass only. Both m/z and Da are commonly used in the PTR-MS literature and we use both in this book.

3.5.1.2 Mass Resolution

There are a number of parameters that specify the performance of a given type of mass analyser. One of the most important is the mass resolution. It is not always essential, but in some applications of PTR-MS it is desirable to be able to distinguish peaks arising from ions with similar masses, which means that the mass analyser must have sufficient resolution to achieve this aim. As if it were not enough to find some confusion with the units employed for ion mass, as seen in the previous section, there is also some ambiguity about the definition of resolution in mass spectrometry. In simple descriptive terms, the mass resolution provides a measure of the ability to distinguish the signals from ions with slightly different values of m/z. In other words, sharper peaks correspond to a higher mass resolution. To quantify this we can either use (i) the mass resolution, Δm, or (ii) the instrument resolving power, R. The resolving power can be expressed mathematically as

$$R = \frac{m}{\Delta m} \tag{3.35}$$

where m is the ion mass (corresponding to the peak maximum for an isolated peak). Δm is determined by the width of an isolated peak in the mass spectrum but there is no unique way of defining this and so it is necessary to introduce an appropriate definition. The usual way of defining Δm is as the full width at half maximum (FWHM) of a single peak, as shown in Figure 3.14(a). This definition of peak width is quite widely used in mass spectrometry, as well as being commonplace in other fields of science such as optical spectroscopy.

An alternative way of defining Δm is the '10% valley definition', which has been particularly prevalent in magnetic sector mass spectrometry. The basic idea is illustrated in Figure 3.14(b). This figure shows two peaks of the same heights that overlap in the centre such that the height of the valley is 10% of that of each peak maximum. When the relative valley height is this small or below, the two peaks are considered to be fully resolved. The quantity Δm in Equation 3.35 is defined as the separation of the two peak maxima when the valley height is exactly equal to 10% of the peak maximum. This definition of Δm is approximately the same as the width of a single, isolated, peak in the mass spectrum measured at a height of 5% relative to the peak maximum.

According to its definition in Equation 3.35, the mass resolving power is inversely proportional to the resolution, Δm. Thus, if $\Delta m = 0.01$ Da for a peak with a maximum located at 100 Da, then the resolving power would be 10^4. However, this definition is not universally used and in some definitions the resolving power and the resolution are essentially one and the same quantity. This is highly confusing and so in this book we will stick to the above definitions of resolving power as $m/\Delta m$ and resolution as Δm, where Δm will be considered to be the FWHM of an isolated peak in the mass spectrum.

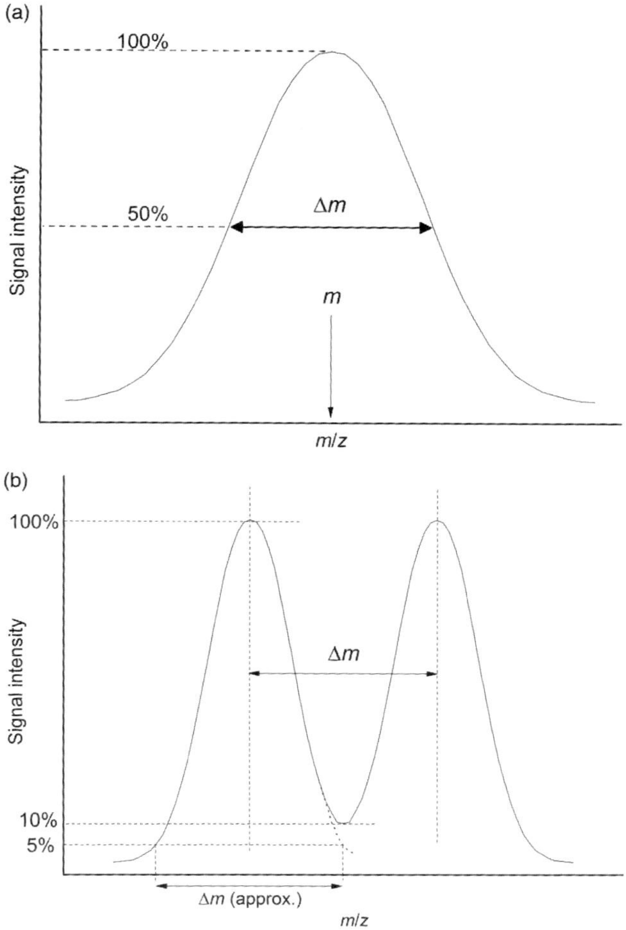

Figure 3.14 *Illustration of the two most common definitions of resolution in mass spectrometry: (a) full width at half maximum (FWHM) and (b) 10% valley definition. Note that the two partially overlapping peaks in (b) have been chosen to have the same heights but this definition would apply equally well to two peaks with different heights.*

3.5.1.3 Transmission and Dynamic Range

Other important quantities to consider when judging the suitability of a particular type of mass spectrometer are its *transmission* and its *dynamic range*. The term transmission reflects the fact that not all ions entering a mass spectrometer will make it all the way to the detector. Losses can be caused by a variety of factors, such as unstable ion trajectories, collisions of ions with the solid parts of grid electrodes, imperfect performance of ion lenses and the scattering of ions by residual background gas molecules. The transmission for some types of mass spectrometer can exceed 90% while in others it may be rather low (<10%). However, when transmission is referred to in mass spectrometry it is normally in the context of its dependence on m/z, rather than just the proportion of ions successfully transmitted all the way from one end of the mass spectrometer to the other. As well as the response of the ion optics, which may depend on m/z, we must also consider the response of the ion detector, which invariably depends on m/z and is therefore also included in the so-called transmission.

Clearly the transmission is an important quantity to take into account for any quantitative analysis. In particular it is important to know how the transmission varies with m/z for a given instrument. The transmission curve for a PTR-MS instrument can be determined in two ways. One option is to add selected compounds with different masses and known concentrations which are then detected following proton transfer. The relative ion signals from each compound can then be equated with the relative ion transmission values at the respective m/z values. However, this approach assumes no fragmentation and equal proton transfer rate coefficients. The latter issue can be avoided by separately calibrating the instrument for each of the compounds using one of the procedures discussed in Chapter 4.

There is an alternative procedure for determining the transmission of ion MH^+ relative to that of H_3O^+ that requires no knowledge of rate coefficients or any calibration. In this simple procedure a sufficient quantity of compound M is added to give a measurable decline in the H_3O^+ signal, for example, 10%. The corresponding gain in count rate is then measured for MH^+ and the ratio of the signal gained for MH^+ versus that lost for H_3O^+ gives the corresponding ion transmission ratio. Of course this can be repeated for multiple compounds to get the relative transmission ratios of the corresponding protonated ions, and from this one can construct a mass transmission curve for the instrument. In this kind of transmission measurement one chooses compounds that are known to undergo negligible fragmentation on protonation.

The dynamic range is the ratio of the largest to the smallest detectable ion signal. Thus a dynamic range of 10^5 implies that an ion signal 10^5 times smaller than the largest measurable ion signal can be discerned. More specifically it is common to refer to the *linear* dynamic range which is the range over which the ion signal is directly proportional to the analyte concentration. Clearly the dynamic range of the mass spectrometer is an important quantity in PTR-MS because the aim is often to determine the concentrations of compounds that may differ by several orders of magnitude. The dynamic range of a mass spectrometer is partly determined by the type of mass analyser and also by the detection method and associated electronics. If the noise level of the detector is high (see Section 3.6) then small ion signals will be swamped by this noise and a poor dynamic range will result.

Table 3.3 quantifies the basic performance of the three types of mass spectrometer that have seen significant use in PTR-MS: quadrupole mass filters, quadrupole ions traps, and

Table 3.3 *Illustrative performance specifications of different types of mass analysers*

Analyser type	Mass resolving power	Mass range	Linear dynamic range
Quadrupole filter	100–1000	Up to 4000[a]	10^6
3D quadrupole ion trap	10^3–10^4	Up to 4000	10^2–10^3
Time-of-flight MS	10^3–10^5	$>10^5$	10^3–10^5

[a] This is an indicative figure and it is possible to build specialized QMS systems capable of going to higher masses but the *quid pro quo* is very low mass resolution.

TOF-MSs. The operating principles behind each of these types of mass analyser will now be discussed in the next few sections.

3.5.2 Quadrupole Mass Spectrometry

The quadrupole mass filter was the type of mass analyser used in the earliest PTR-MS instruments and it is still the most popular choice at the time of writing. The popularity of the QMS stems from its relatively compact size and reasonable price. In this section the basic operating principles of QMS systems will be described along with a discussion of some of their performance characteristics.

3.5.2.1 Basic Principles of the Quadrupole Mass Spectrometer

The key structural component of a QMS is a set of four parallel rods, as shown in Figure 3.15. These rods, which act as electrodes, will ideally possess a hyperbolic cross section, although in practice they are normally of cylindrical cross section. To understand how mass selection is possible with this set of electrodes, we will work through a set of simplified scenarios that will explain aspects of the QMS operation. As will be seen, when these are combined we arrive at an explanation of how the QMS can act as a mass filter.

We start by considering a collection of applied electrical potentials (Φ) of equal magnitude applied to the four electrodes such that a quadrupolar electric field is generated in the xy plane. For a quadrupolar field the four rods will be at potentials $+\Phi/2$, $-\Phi/2$, $+\Phi/2$, and $-\Phi/2$, as shown in the lower image in Figure 3.15, that is, opposite rods have the same potential. To begin the discussion, the potentials will be assumed to be static, that is,

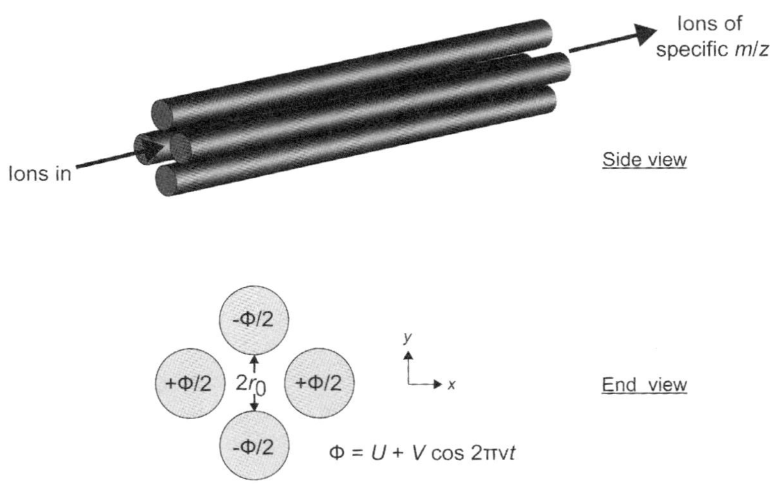

Figure 3.15 *Side and end views of rods employed in a quadrupole mass analyser. The rods are assumed to be of circular cross section rather than the idealized hyperbolic cross section, since the former are easier to fabricate. For cylindrical rods the best approximation to a hyperbolic field is obtained by positioning the rods, each of radius* r*, in a square array such that* $r = 1.148\ r_0$ *(see Reference 47, chapter VI by Austin, Holme and Leck).*

DC voltages will be applied to the rods. Now imagine a positively charged ion of mass m entering into the region between the electrodes from one end of the quadrupolar system. Furthermore, let us focus first on what happens in the xz plane. Providing the ion is not midway between the right hand and left hand rods, the repulsion from the nearest rod will drive the ion towards the opposite electrode. Naturally, the repulsion from the opposite rod will begin to grow as the ion crosses from one side to the other and eventually the direction of motion will reverse. This is a recipe for an oscillating trajectory in the x direction as the charged molecule moves along the z direction.

However, we now need to allow motion along the yz plane too. Since the potentials applied to the upper and lower rods are negative, these are attractive to a positive ion and any ion located nearer to one or other of these rods will be attracted to the nearest rod and may be lost by contact with this electrode. Consequently, the conclusion is that a static quadrupolar field can generate stable motion in one plane (xz in this case) but not in the other (yz).

Suppose, now that instead of a static quadrupolar field, an oscillating electric field component is also added operating at a radiofrequency (RF) such that

$$\Phi = U - V \cos 2\pi \nu t \tag{3.36}$$

In the above expression Φ is the net applied potential at time t, U is the static (DC) component, V is the maximum amplitude of the oscillating (AC) component, and ν is the oscillation frequency.

We have already seen that the static potentials applied to the upper and lower electrodes can lead to unstable (diverging) trajectories which, if not countered, will lead to the loss of positive ions. However, a stable trajectory is possible if an oscillating potential of sufficient magnitude and frequency is applied along the y direction, since the voltage applied to the upper and lower electrodes now alternates in sign. This will be most effective for light ions, since they are able to respond most rapidly to an oscillating electric field. Consequently, this part of the quadrupole system will transmit ions below a certain mass cut-off but heavier ions will not make it through to the end of the rods without colliding with a rod. Now consider motion along the x direction. We have already seen that the static field here can create a stable oscillating trajectory, but now this is disturbed if an oscillating component is added. Again it will be the lightest ions that are most affected by this oscillating potential, but if the amplitude and frequency of the applied field are suitably chosen the light ions can follow large amplitude trajectories that cause them to strike one of the rods. Thus motion along the x direction provides a means of removing light ions, that is, this part of the system acts as a barrier to low mass ions.

It can therefore be seen that the combination of static and oscillating components to the applied electrical potentials provides a mass filter that allows passage of ions with only a narrow range of masses. This set of transmitted masses can be changed by altering U, V and ν, and thus provides the basis for a mass spectrometer. To produce an actual mass spectrometer the quadrupole filter needs to be combined with a vacuum system, an ion source, and an ion detector at the downstream end of the mass filter.

The explanation above is very simplified. A more detailed account of QMS would consider the equations of ion motion and the permissible values of U, V and ν that transmit ions of a particular m/z to the detector. A full description of this topic is beyond the scope of this book and the interested reader is referred elsewhere for further details, such as the excellent book by Dawson [47]. Instead, in the next section a more practical description is provided which focuses on key performance characteristics of QMS systems.

3.5.2.2 *Practical Issues*

Solution of the equations of motion for ions moving in the *xy* plane make it possible to predict the trajectories for a given set of operating parameters, *U*, *V* and *ν*. For ions to be transmitted through the quadrupole rods and onwards to the detector, a stable trajectory is required such that any oscillatory motion is confined between the four rods. Furthermore, a means of selectively transmitting ions of a given *m/z* is required along with a means of scanning through the mass range such that a mass spectrum can be constructed.

In practice the RF frequency, *ν*, is normally maintained at a constant value, since variation of this quantity as a means of mass scanning requires more complex electronics than varying the two voltages, *U* and *V*. Figure 3.16 illustrates the effect of *U* and *V* on the ion trajectories, showing the regions for stable oscillatory motion (which is bounded by the solid lines for a given ion mass). Also shown superimposed on the diagram is a straight line corresponding to a fixed *U/V* ratio, which will be referred to as the *scan line*. If *U* and *V* are swept through values that follow the scan line, then the system will move in and out of stability regions for ions of different *m/z*. Consequently, in progressively following this line the quadrupole system acts a scanning mass filter. The scan line with the best resolution will cut through the stability regions close to their peaks. However, increasing the slope of the scan line will improve the mass resolution but if taken too far it will move beyond the stable trajectory regions for the ions, as will be evident from Figure 3.16. Although a linear scan line is preferred, a slightly non-linear scan line (not shown in Figure 3.16) can yield better results.

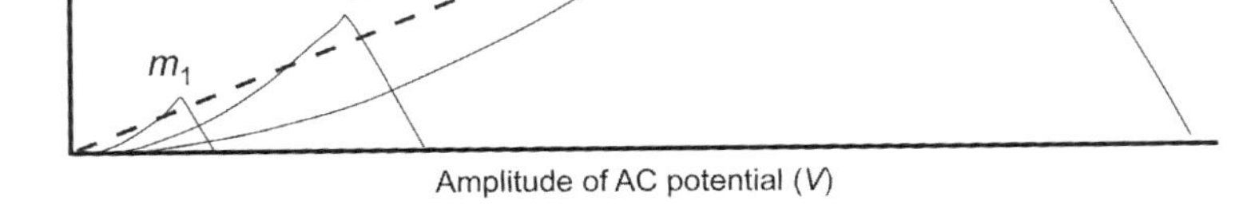

Figure 3.16 *Ion trajectory stability diagram for a linear quadrupole mass filter. The region inside the solid lines represents the possible values of U and V where stable trajectories are obtained for a given m/z. If the U/V ratio is varied such that it follows the scan line, shown by the dashed line in the figure, then at each time in the scanning period the filter will be optimized to transmit ions of a particular mass. This scanning process allows the quadrupole filter to act as a mass spectrometer by adding an ion detector at the downstream end of the filter.*

So far we have said little about motion along the z direction, other than the fact that there needs to be movement along z if the ions are to reach the detector at the far end of the rods. In principle the velocity along z is not all that critical, since the QMS operates on the basis of trajectory stability rather than requiring a low-divergence beam. However, in practice there are limitations. The best performance is achieved if the ions are focused to produce a beam of low divergence entering between the rods close to the principal axis of the quadrupole. To condition the ion beam before entry, an ion lens assembly is normally added before the quadrupole rods. The velocity given to the ion beam as it leaves the ion lens system should not be too fast in order to give the ions time to undergo a number of oscillations and therefore establish a stable (or unstable) trajectory inside the quadrupole: without this the resolution of the instrument will be diminished. The requirement for a number of complete oscillations in the ion trajectory in the xy sense is clearly linked to the frequency of the RF oscillations, since a low frequency will require a longer stay inside the quadrupole field (or a longer set of rods).

An example will serve to illustrate this point. Consider an ion of mass 50 Da accelerated to an ion energy of 5 eV directed along z. A simple calculation shows that this corresponds to a velocity v_z of $\sim$4400 m s^{-1}. For quadrupole rods of length 15 cm, which is fairly typical, the transit time through the rods will be $\sim$34 μs. A typical RF frequency applied in quadrupole mass spectrometry is 1 MHz, which corresponds to an oscillation period of 1 μs. Consequently, the 50 Da ions will have sufficient time to establish a stable or unstable oscillating trajectory. However, ions of mass 10 Da accelerated to 50 eV will pass through the rods in about 5 μs and so the establishment of a stable or unstable trajectory is more marginal. Note also that the transit time of ions through the quadrupole filter affects the mass scan rate, since a rapidly changing voltage occurring while ions of a given m/z are making their way through the system will change the trajectory conditions and thus worsen the mass resolution. Consequently, for optimum resolution the voltage scan speed should be reasonably slow.

Two key performance criteria for the quadrupole mass filter are the mass range and the mass resolution. Both can be calculated, assuming a perfect quadrupolar field. The maximum mass that can be transmitted is given by [47]

$$m_{\max} = \frac{7 \times 10^6 V}{\nu^2 r_0^2} \tag{3.37}$$

The quantity r_0 is half the distance between opposite electrodes, as shown in the lower part of Figure 3.15. Typical values of V are in the range 1–5 kV and r_0 can be as large as 10 mm. Choosing values of $V = 2$ kV, $r_0 = 3$ mm and $\nu = 1$ MHz, a value of $m_{\max} = 1560$ daltons is calculated.

The mass resolution is given by

$$\Delta m = \frac{4 \times 10^9 eV_z}{\nu^2 L^2} \tag{3.38}$$

where eV_z is the kinetic energy (in electron volts) of the incoming ion along the z direction and L is the length of the rods. The above relationship is independent of ion mass, and therefore in the ideal quadrupole filter the mass resolution is constant. For a quadrupole system of length 15 cm and an incoming ion energy of 5 eV, the mass resolution is calculated to be 0.89 Da. A higher RF frequency or longer rods can improve this resolution, but a higher frequency will lower the maximum mass cut-off. Furthermore, although longer rods may seem an appealing choice to improve the mass resolution, other factors may serve to degrade the actual performance which may actually worsen with longer rods, including inhomogeneous electric field effects and rod misalignments. Consequently, while an optimized quadrupole can achieve a resolution as low as 0.1 Da, most quadrupoles are operated as low-resolution mass filters with little better than a one Da resolution.

Perhaps the most favourable characteristic of QMSs is that they are relatively inexpensive, compact and robust, especially when compared with much higher resolution mass spectrometers such as magnetic sector devices. As a result, the QMS is an integrated part of many gas monitoring systems, being commonplace in gas chromatography/mass spectrometry and in residual gas analysers. It is therefore of no surprise that the first PTR-MS systems were also based on QMS. The principal disadvantages of QMS have been touched on above and include a relatively low mass resolution and a slow scan speed. As a mass filter, the fact that all ions except those of a specific m/z are rejected from the mass filter at any one instant in time makes the QMS unsuitable for investigating complex mass spectra in rapidly evolving situations.[11] Furthermore, the transmission efficiency through the filter depends on m/z because heavier ions spend more time in the fringing fields and may therefore be given unstable trajectories. A rise in transmission efficiency with m/z is seen for small masses ($m/z < 80$) but, as the mass increases further, the transmission starts to decline.

3.5.3 Quadrupole Ion Trap Mass Spectrometry

3.5.3.1 Basic Principles

The mass analyser described in the preceding section is sometimes referred to as a *linear* quadrupole, since the electrodes consist of parallel rods. Another type of mass spectrometer based on the use of a quadrupolar electric field is the three-dimensional (3D) quadrupole ion trap, which is sometimes also known as a Paul trap. This is a device that has been in existence for several decades, but it is only in the past twenty years or so that it has moved from a small number of research laboratories out into the commercial domain. Ion traps are now popular devices for a whole range of mass spectrometry applications, their popularity being enhanced by their very compact size and the ease with which they can be used in MS^n experiments (see Section 3.5.3.2). Several PTR-MS instruments have been constructed with a quadrupole ion trap and so a description of the basic operating principles

[11] When it is required to measure ions at several different but known masses using QMS it would clearly be highly inefficient to ramp through a full scan over m/z, since much of the time will be spent measuring signal at m/z values of no interest or even where there is no signal of any type. A smarter approach is to program the QMS so that it dwells for an appropriate length of time at specific m/z values. For example, if the aim in a PTR-MS experiment is to measure two VOCs using ion signals at two distinct m/z values (protonated ions), then one might set the data collection routine as follows: first, to be able to extract quantitative information on the VOCs, it is necessary to measure the count rate of H_3O^+. The H_3O^+ signal at m/z 19 is likely to be far too strong and will saturate (and shorten the lifetime of) the ion detector. Consequently, the signal at m/z 21, corresponding to $H_3{}^{18}O^+$, will be used instead and since the $^{16}O:^{18}O$ abundance ratio is known to be about 500:1, the total count rate for H_3O^+ can easily be deduced. The second stage is to switch the QMS to transmit the desired organic ions. If the signal at m/z 21 is much stronger than the signals from the protonated organic ions of interest, the dwell time for collecting data at m/z 21 could be set to be shorter than that for each organic ion. In this way, one can design a data acquisition scheme that is optimized for a particular application.

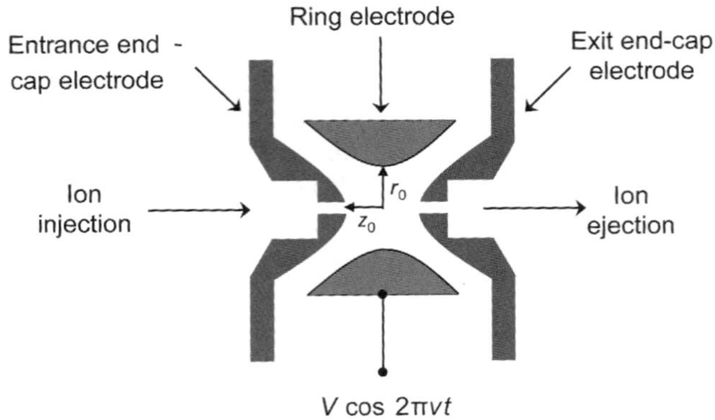

Figure 3.17 *Cross-sectional view of a quadrupole ion trap. An appropriate voltage is applied to the incoming end-cap electrode to allow ions to enter. The outgoing end-cap electrode is then switched to a potential to slow this batch of ions down so that they can be kept inside the trap. A radiofrequency voltage is applied to the ring electrode to trap and then select/eject ions of the desired m/z.*

is given here. For a more comprehensive account about ion trap mass spectrometers the reader is directed towards a book [48] and two review articles [49, 50].

The most commonly encountered type of quadrupole ion trap is the 3D ion trap which is composed of three distinct electrodes: a ring electrode and two end-cap electrodes. In contrast to the linear quadrupole encountered previously in this chapter, where ions pass through the filter (given the correct *m/z*), the effect of the ion trap is, as the name implies, to confine the ions. The arrangement of the electrodes is shown in cross-sectional view in Figure 3.17. The ideal end-caps will have a hyperbolic cross section while the ring electrode presents a donut-shaped surface (also possessing a hyperbolic profile) that effectively acts as a cylindrical wall for the ion trap. Holes in the centre of the two end-caps can be used to allow ions to enter and leave (see below). The idealized geometry (for the simplest mathematical analysis of ion motion) occurs when $r_0 = \sqrt{2}z_0$, where r_0 and z_0 are defined in Figure 3.17, although it is quite common to have stretched traps with a slightly larger z_0 to help reduce the effect of imperfect quadrupolar electric fields in the ion trap. In most quadrupole ion traps $r_0 \leq 1$ cm.

In the usual mode of operation an RF potential, with a typical frequency in the region of 1 MHz, is applied to the ring electrode while the end-cap electrodes are grounded. At first sight it seems strange to claim that this generates a quadrupolar electric field. However, analogy can be drawn with the more obvious quadrupolar field generated by the rod electrodes in the conventional mass filter, as shown in the lower image in Figure 3.15. The difference in the case of the ion trap is that a pair of opposing electrodes in the mass filter, which possess the same potential at any instant in time, have been merged into a single ring electrode in the ion trap. In the linear quadrupole mass filter an oscillating quadrupolar electric field, defined by the potentials $\Phi = U - V \cos 2\pi \nu t$, can produce stable and unstable ion trajectories in two dimensions, x and y. The stability is determined by a variety of factors including the DC and RF voltage amplitudes and the *m/z* of the ions. In the ion trap, stable ion motion requires stability in all three dimensions in order for the ions to remain confined. However, mathematically we can still employ a two-dimensional (2D) treatment since the system is cylindrically symmetrical and therefore the choice of the x and y directions is arbitrary. In other words, the axial coordinate z and the radical coordinate r $(= \sqrt{x^2 + y^2})$ is sufficient to define the oscillatory motion of the ions.

The stable motion of ions in an ion trap follows a figure-of-eight like trajectory. Stable oscillation is possible if the applied potentials rotate (oscillate) at the correct rate, thus periodically driving the ions out towards the electrodes and then switching polarity to provide a restoring force that pushes them back towards the centre. In practice these excursions should not be too large and so a background level of helium, at a pressure of typically 10^{-3} mbar, is added to the cell to quench any excess kinetic energy from the ions. This quenching 'herds' the ions into a compact region close to the centre of the trap, and improves the mass resolution.

There are two constituent (secular) frequencies to the oscillatory ion motion, the axial and radial frequencies. The amplitude of the axial motion can be raised by increasing the amplitude of the RF voltage, V, applied to the ring electrode. This affects ions with lower m/z values more than those with higher m/z, and provides the most effective means of operating the ion trap as a mass spectrometer. For a given V, ions below a certain m/z cut-off will undergo unstable excursions along z that lead to their eventual ejection through the apertures in the end-cap electrodes. Stable trajectories are defined by the criterion

$$\frac{zeV}{mr_0^2\pi^2\nu^2} \leq 0.908 \tag{3.39}$$

where e is the fundamental unit charge and all other quantities are as defined previously. Increasing the value of V leads to the ejection of ions with increasing m/z, thus providing a means for scanning mass. This means of operating the ion trap is the most popular and is known as the *mass-selective axial instability mode*. By coupling the ion trap with an ion detector beyond one of the end-caps, a mass spectrum can be constructed. A typical single mass scan via this means takes ~10–20 ms to complete. Improvements in the sensitivity and mass resolution can be achieved by applying a low amplitude supplementary oscillating voltage on the end caps at a lower frequency than the ring RF potential. This *axial modulation* serves to slowly 'stretch' the motion along the axial direction for ions near the low-mass threshold, thus reducing the impact of the space charge effect felt by the ions when confined near the centre of the cell.[12]

3.5.3.2 *Collision-Induced Dissociation*

In addition to its compact size, another beneficial feature of the quadrupole ion trap mass spectrometer is its convenience for tandem mass spectrometry (also known as MS/MS or MS^n). There are many instances where a single mass spectrum does not allow complete compound identification. Indeed this is a common problem in PTR-MS, as well as in numerous other mass spectrometry investigations. In particular, a peak at a given m/z might derive from one or more of several possible compounds. The inclusion of chromatographic

[12] When a substantial number of ions are gathered into a relatively small volume the electrostatic interactions between the ions can become comparable to the strength of interaction of each ion with applied external electric fields. The ion–ion interactions in such circumstances are referred to as a *space charge* effect.

separation prior to the ion trap can greatly assist compound identification but there are other means of compound discrimination.

Tandem mass spectrometry attempts to tackle this by first selecting ions of a given m/z, and then subjecting these ions to collision-induced dissociation (CID) to cause fragmentation. The identification of the fragment ions and, where helpful, the dependence of the fragmentation yields on the collision energy, can be used to aid in the identification of a specific compound. This tandem process can also be realized with conventional linear quadrupole filters, but two such devices are required and they must be separated by a collision cell. This is a bulky and expensive arrangement, whereas no additional equipment is required to do tandem mass spectrometry in a quadrupole ion trap.

To carry out MS/MS studies in an ion trap, the first requirement is to reject all ions from the trap except those with the m/z of interest. This can be achieved, for example, by applying a low-amplitude oscillating potential to the end-cap electrodes in order to resonantly excite the unwanted ions such that they are pushed out of the trap. Rather than applying a single frequency, a range of frequencies can be applied simultaneously to remove all of the unwanted ions. Various procedures have been described for doing this, the most well-known being the filtered noise field and the stored waveform inverse Fourier transform (SWIFT) methods [47]. The underlying principle at work here is that the applied voltage waveform contains multiple frequency components to eject all unwanted ions, but has a notch at the secular frequency of the ions of interest. This ejection process takes just a few milliseconds to complete, leaving the desired collection of ions behind in the trap. A second excitation process can then be used to excite the remaining ions by applying an oscillating potential to the end-cap electrodes that matches the axial secular frequency of the trapped ions. This resonant excitation process, sometimes referred to colloquially as 'tickling', increases the kinetic energies of the ions, yielding more energetic collisions with the helium atoms in the cell and leading, ultimately, to an increase in the internal energy within the ions. If this added energy is sufficient to cause fragmentation, then CID occurs. The degree of energy increase in the ions can be controlled by the amplitude and duration of the applied resonant excitation waveform (avoiding complete ejection of the ions). Following CID, another RF voltage ramp can be applied to the ring electrodes to record a mass spectrum.

The procedure above describes, somewhat simplistically, an MS^2 experiment. It is a straightforward matter to add further MS stages if desired. In principle the only limitation on the number of stages is brought about by the number of ions initially in the cell, which is usually limited to a maximum in the region of 10^5–10^6 ions in order to reduce coulombic repulsions (so-called *space charge* effects) between ions to an acceptable level. The ejection stages progressively reduce the number of ions in the trap and thus a large number of stages may ultimately leave too few ions in the trap for satisfactory detection.

3.5.3.3 Three-Dimensional Quadrupole Ion Traps in PTR-MS

The first description of an ion trap in a PTR-MS instrument was published in 2003. To date there are three distinct research groups that have constructed PTR ion trap mass spectrometers (PTR-IT-MS) [51–53]. Some of the driving forces for replacing a linear quadrupole mass filter with a 3D quadrupole trap have been described above, including the convenience for ion-selective CID and the small size of the trap. Another motivation is the potential for a much improved *duty cycle* achievable with an ion trap. Consider, for example, a trap that receives ions from a drift tube for a period of 1 s, after which it has the maximum number of ions stored within it commensurate with an acceptable space charge effect. All the ions stored in this period can then be ejected in the mass-selective axial instability mode to record a mass spectrum. Since this might take somewhere in the region of 20 ms, the only ions that are not accepted by the trap are those that arrive during this 20 ms data accumulation period. This means that the percentage of ions leaving the drift tube that can potentially be detected (the duty cycle) is

$$\frac{1}{1+0.02} \times 100 = 98\%$$

Contrast this with a linear quadrupole, which at any instant in time transmits ions with only a single *m/z* and rejects all others. The duty cycle achieved will depend on the chosen mass scan range and the scan speed but will at best be just a few percent. Consequently, the ion trap is capable of a much higher duty cycle than a linear quadrupole (assuming the latter is operated in the mass scanning mode). However, this need not necessarily translate into a higher detection sensitivity since other factors will also affect this, such as ion transmission efficiency from the drift tube to the quadrupole trap.

The transmission efficiency is perhaps the biggest single issue to tackle in the design of a PTR-MS instrument that employs an ion trap. However, external production of ions followed by injection into the ion trap was already well established before the first PTR-IT-MS instrument was reported, so adoption in PTR-MS requires similar principles. The approach used is to employ a multi-element ion lens system to collimate and focus the ions leaving the drift tube such that as many as possible pass through the small ($\sim$1 mm) aperture in the end-cap electrode. To prevent the trap overfilling, a specific fill time is set by gating the potential applied to one of the ion lens elements such that it switches from a transmissive to a blocking role at a predetermined time. Differential pumping is also required to ensure that the various parts of the instrument operate at acceptable pressures, that is, drift tube at $>$1 mbar, ion trap at $\sim 10^{-3}$ mbar and the ion detector at $\leq 10^{-5}$ mbar.

The use of an ion trap in PTR-MS gives rise to a problem that is not encountered when a linear quadrupole mass filter is used. Ordinarily, the vast majority of ions leaving the drift tube in PTR-MS are unreacted primary ions, most commonly H_3O^+. Since the number of H_3O^+ ions will typically exceed all other ions by several orders of magnitude, the upper limit of $\sim 10^5$ ions that can be stored in the trap poses a problem for the detection of the far less abundant protonated VOCs. Essentially, this is a problem resulting from the limited dynamic range of the quadrupole ion trap. A neat solution to this predicament has been described by Warneke and co-workers [52], who use a two-stage filling procedure that begins with a short fill of 50–100 ms in order to accumulate sufficient primary ions to gauge an accurate measure of the proportion of primary ions. After the RF voltage ramp has emptied the cell of primary ions, a second and much longer fill commences, up to 2 s, with the RF voltage kept above the threshold for trapping the primary ions. In this way a more statistically acceptable signal may be extracted for the secondary ions and then the $[RH^+]/[H_3O^+]$ ratio can be determined from the relative ion signals in the two distinct mass scans, with suitable weighting applied for the relative trap accumulation times.

The best limit of detection that has so far been achieved with PTR-IT-MS instruments based on a 3D ion trap is in the region of 50–100 pptv for specific VOCs such as toluene and benzene using 1 min of data accumulation [52]

3.5.3.4 The Linear Ion Trap in PTR-MS

An alternative to the 3D quadrupole ion trap (Paul trap) is the linear quadrupole ion trap. The linear ion trap is akin to a hybrid of the quadrupole mass filter and the 3D ion trap in that it consists of a four-rod assembly, like the quadrupole filter, but also it has entrance and end electrodes like the 3D ion trap. Confinement of ions along the axial direction is provided by DC potentials applied to the end electrodes. The quadrupole rods produce radial motion of the ions through application of an RF electric field, in a similar manner to that already described for the quadrupole mass filter. To record a mass spectrum axial ion ejection, initiated by RF excitation, can be used in a procedure similar to that used for the 3D ion trap.

Three-dimensional ion traps have problems with both poor ion collection efficiencies and space-charge effects, where the latter provides an upper limit on the number of ions that can be stored within the trap. The much larger spatial volume within a linear ion trap reduces the space-charge problem and therefore a larger quantity of ions can be stored in this trap when compared to the 3D quadrupole ion trap.

The incorporation of linear ion trap technology into PTR-MS was reported by Mielke and co-workers in 2008 [54]. The quantity of ions that could be stored within the ion trap was ~50 times larger than that in a 3D ion trap, which should confer a concomitant improvement in the limit of detection. Thus PTR-MS based on a linear ion trap as a mass analyser offers a detection sensitivity which has the potential to approach that achievable with systems based on quadrupole mass filters, although the best limit of detection so far achieved is near 100 pptv. Also, the linear ion trap preserves the easy use of MS/MS techniques exploited in 3D ion traps, which thus aids in compound identification.

3.5.4 Time-of-flight Mass Spectrometry

3.5.4.1 Basic Principles of TOF-MS

The simplest version of a TOF-MS is shown in Figure 3.18. It consists of a relatively short ion acceleration region and a much longer part known as the flight tube. Ions may be generated within the ion acceleration region (e.g. by electron impact or by photoionization) but in the case of PTR-MS the ions are generated externally by CI in a drift tube, which is coupled to the TOF-MS. These ions enter into the acceleration region along a direction perpendicular to the long axis of the flight tube, as shown in Figure 3.18, which will be referred to as the x direction: such an arrangement is known as orthogonal acceleration TOF-MS. The acceleration region is formed by two electrodes, known as the repeller and extractor electrodes. The extractor electrode is a high transmission grid to allow the majority of incident ions to pass through into the flight tube. The flight tube has no electrical field and therefore the velocity delivered to the ions in the acceleration region is maintained once they enter the flight tube.

Initially, the repeller and extractor electrodes are kept at the same electrical potential in order to prevent deflection of the incoming ions. To obtain a mass spectrum, this voltage

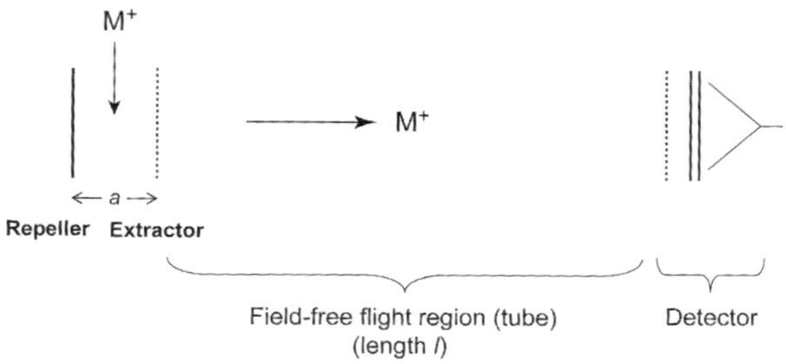

Figure 3.18 *Simplest possible arrangement for a time-of-flight mass spectrometer. The broken lines indicate fine mesh electrodes, which allow good ion transmission while also giving well defined electrical potentials.*

state is then altered rapidly such that the voltage on the repeller is higher than on the extractor. This potential gradient is used to send a pulse of ions into the flight tube and if a detector with sufficient sensitivity and sufficient speed of response is located near the end of the flight tube the arrival times of ions in the extracted packet can be determined. Heavier ions will take longer to reach the detector than lighter ions, and so measurement of the time-of-flight provides the basis for recording a mass spectrum. To be effective, collisions with background gases need to be avoided and so a good vacuum in the flight tube, and preferably $<10^{-5}$ mbar, is essential.[13]

Let us look more closely at the quantitative aspects of the simple time-of-flight analyser shown in Figure 3.18. The length of the acceleration region is denoted by a and the potential difference between the repeller and extractor plates is denoted by V. All positively charged ions of charge q (=ze) located close to the repeller surface will enter the flight tube with the same kinetic energy E given by

$$E = qV = \frac{1}{2}mv^2 = \frac{1}{2}m\left(\frac{l}{t}\right)^2 \tag{3.40}$$

where v is the ion velocity in the flight tube, m is the ion mass, l is the length of the flight tube, and t is the time taken to reach the detector. Assuming that $a \ll l$, the time taken for ions to travel to the detector after the voltage ramp is applied is given by

$$t = \sqrt{\frac{l^2}{2eV} \cdot \frac{m}{z}} \tag{3.41}$$

Consequently, measurement of the time-of-flight allows determination of m/z. The arrival times of the ions provide a spectrum in the time domain, which through application of Equation 3.41 is readily converted into the more usual mass spectrum.

[13] From the kinetic theory of gases we know that the mean free path of a gas particle is given by $k_BT/(2^{1/2}p\sigma)$, where k_B is the Boltzmann constant, T is the absolute temperature, p is the pressure (in Pa) and σ is the collision cross section (in m^2). A reasonable estimate of the collision cross section between a small molecular ion and N_2 or O_2 is 50 Å^2 = 5×10^{-19} m^2, and thus for a mass spectrometer operating at 300 K and a pressure of 10^{-5} mbar $\approx 10^{-3}$ Pa the mean free path is ~6 m. A typical length of the flight tube in a TOF-MS is 1 m and thus a vacuum of $\leq 10^{-5}$ mbar ensures that collisions between ions and any residual background gas are unlikely. Ideally one would want to avoid any realistic prospect of ion–molecule collisions, which means that the TOF-MS in the best systems will be at pressures of $<10^{-6}$ mbar.

It is not difficult to spot a flaw in this analysis. It has been assumed that all ions are accelerated from positions near to the surface of the repeller electrode, and are thus given the same energy. However, there is a potential gradient between the repeller and extractor electrodes, which means that ions starting their journey from near to the extractor electrode will experience a much lower voltage drop than those starting near the repeller, and therefore the former set of ions will receive a much lower energy than the latter as they journey into the drift region. The effect of this *spatial* distribution of ions is to reduce the mass resolution to such a degree that a simple mass analyser of the type shown in Figure 3.18 is of little practical value.

3.5.4.2 *Improving the Resolution: Spatial Focusing*

A major breakthrough in TOF technology, which led to the first practical TOF mass analysers, was achieved by Wiley and McLaren in 1955 [55]. Instead of a single acceleration region, a two-stage acceleration zone was employed, as shown in Figure 3.19. An extra electrode, known as the grid electrode and located between the repeller and extractor electrodes, is added. Ions produced or introduced into region 1 (repeller-grid) are initially accelerated through into region 2 (grid-extractor), where they receive a second tranche of acceleration that drives them onwards into the field-free flight tube. All ions passing through region 2 are given the same increase in energy because they experience the full voltage drop from grid to extractor but we know from our previous comments that ions leave region 1 with a spread of energies due to their different spatial positions along the x axis.

The contribution of Wiley and McLaren was to recognize that, by suitable choice of the lengths of regions 1 and 2, the length of the flight tube, and the voltages applied to the electrodes, ions of the same m/z starting from different points in region 1 could be made to arrive at the detector at virtually the same moment in time. The key here is to use region 2 as the main acceleration region. Region 1 therefore provides a relatively small part of the final ion energy and the spread of energies as a proportion of the total is therefore reduced. Those ions given the highest energy from region 1 also have the furthest to travel to the detector, since they start their journey closest to the repeller electrode, and by careful choice

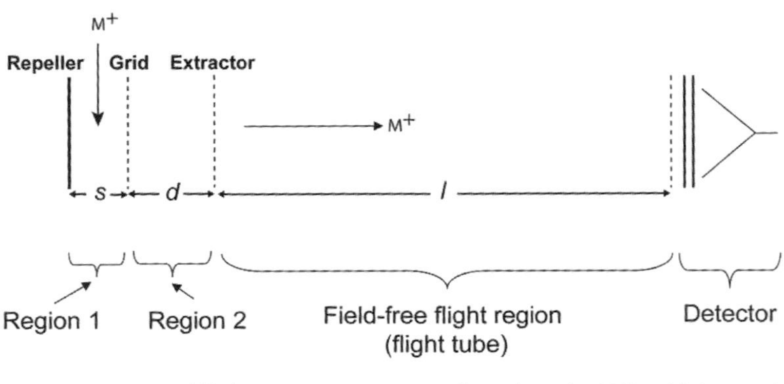

Figure 3.19 *Linear time-of-flight mass spectrometer based on the Wiley-McLaren design [55]. Reprinted with permission from [54]. Copyright 1955, American Institute of Physics.*

of operating conditions this group of ions can be made to arrive at the detector at nearly the same time as those that start their journey close to the grid electrode.

In Figure 3.19 the inter-electrode distances in regions 1 and 2 are denoted by s and d, but in terms of the mathematical treatment below, s is regarded as a variable that needs optimizing. The maximum final energy attained by the ions is given by

$$U_{\text{tot}} = q(V_s + V_d) \tag{3.42}$$

The overall flight time to the detector has contributions from the two acceleration regions and the flight tube, and can easily be shown to be

$$t_{\text{tot}} = \sqrt{\frac{m}{2U_{\text{tot}}}}\left(k^{1/2}s + \frac{2k^{1/2}d}{k^{1/2}+1} + l\right) \tag{3.43}$$

where k is the function

$$k = \frac{sE_s + 2dE_d}{sE_s} \tag{3.44}$$

and E_s and E_d are the electric fields in regions 1 and 2, respectively. If the condition $\partial t_{\text{tot}}/\partial s = 0$ is satisfied it implies that the time-of-flight of an ion is independent of its initial position s in region 1. Applying this condition to Equation 3.43 therefore allows the optimum position of the detector to be calculated such that singly charged ions of a given mass with different velocities due to their different spatial positions ‘focus’ at the detector. This leads to the following expression for the optimum detector position:

$$l_{\text{focus}} = sk^{3/2}\left(1 - \frac{2d}{(k + k^{1/2})s}\right) \tag{3.45}$$

In practice the detector position is fixed and spatial focusing is achieved instead by changing the ratio E_d/E_s. For a given choice of distances s and d, this means adjusting V_s and V_d to ensure that Equation 3.45 is satisfied. In reality one can avoid doing the calculation and take the practical approach instead, whereby voltages V_s and V_d are adjusted to give the best possible resolution, for example, by viewing the ion signal on an oscilloscope.

A Wiley-McLaren mass analyser of typical length of a few tens of centimetres can achieve a mass resolving power in the region of 200–300. The resolution can be improved by making the flight tube longer, but this will mean an added cost and a more bulky instrument. Furthermore, spatial focusing fails to correct another significant source of spectral broadening, so-called kinetic energy broadening. This is broadening caused by the variation in molecular speeds along axis x possessed by molecules in region 1. For a given starting point in region 1, if the ion is initially moving away from the detector along axis x then it must first slow down (due to the repulsive force offered by the higher potential of the repeller electrode) before turning around and moving in the reverse direction. This will result in a longer overall flight time than that of an ion of the same m/z starting from the same position but initially moving in the opposite direction. Any attempt to achieve a mass resolution much better than 0.1 Da must significantly reduce this kinetic energy contribution to the broadening.

3.5.4.3 *Reflectron TOF-MS*

The TOF analysers shown in Figures 3.18 and 3.19 are known as *linear* TOF analysers because the journey from acceleration region to the detector is a simple straight line. An

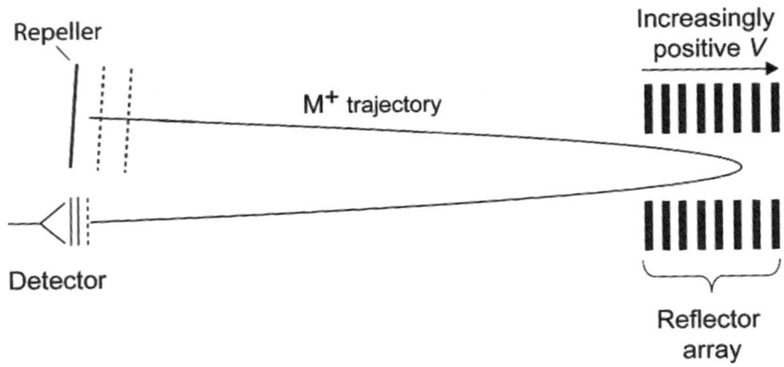

Figure 3.20 Schematic of a reflectron time-of-flight mass spectrometer.

alternative is the *reflectron* TOF-MS, in which the trajectories of all of the ions are deliberately reversed part way along their journey to the detector, as illustrated in Figure 3.20. A series of electrodes, known as the reflector array or simply as a reflectron, is employed towards the end of the flight tube to first slow, and then reverse the motion of the ions. This is achieved by having the front electrode at the same electrical potential as the flight tube and then subsequent electrodes at increasingly positive potentials. Ions moving fast along the x direction will penetrate further into the reflection zone than those of the same m/z but travelling slower, and thus the former will have longer overall flight paths to the detector. By appropriate choice of the reflector voltages, this additional path length for the faster ions can be set such that it compensates for their additional speeds and allows slow and fast ions of the same m/z to arrive at the detector at approximately the same time. This ingenious means of compensating for the variation of ion energies in region 1 due to thermal kinetic energy effects was first introduced by Mamyrin in 1973 and for an equivalent ion flight length it improves the mass resolution by about an order of magnitude when compared with the standard linear TOF-MS [56]. As a result, a resolving power well in excess of 1000 now becomes available in a reasonably compact configuration and most commercial TOF mass analysers now use a reflectron configuration.

3.5.4.4 Mass Calibration in TOF-MS

An important practical issue for TOF-MS concerns mass calibration. Equation 3.40 cannot be used to gain an accurate value for m/z for three reasons: (i) the ion travels at constant speed in the flight tube but not in the two-stage acceleration zone, nor in the reflecting array of a reflectron; (ii) the distance travelled by the ions will also vary for similar reasons, so l is not fixed for all ions; (iii) the starting time of the ion journey is difficult to measure precisely. For these and other reasons it is standard practice to calibrate the flight times for unknown ions using the measured flight times, t, of known ions. A calibration equation of the type shown below could then be used, which is a generalization of Equation 3.41:

$$t - t_0 = A\sqrt{\frac{m}{z}} \qquad (3.46)$$

Since Equation 3.46 has two unknowns, A and t_0, a minimum of two known peaks are required to determine their values.[14] The known ions could be, for example, H_3O^+ and $H_3O^+(H_2O)$, which would therefore not require the addition of any specific calibrant gases to the mixture. However, it is best to use calibrant ions that span the mass range of interest in order to achieve the best possible calibration, so the aforementioned pair of ions would be insufficient for most purposes since most VOCs of interest will have masses above 40 Da, and some may be much higher. One solution is to add one or more calibrant gases which span the appropriate range to calibrate the system and then return to the original measurements. The assumption would then be that, once a calibration has been performed, it will be reliable for subsequent measurements. This assumption may be sufficient for low resolution instruments or for a short period of time on a high resolution instrument, but it is worth noting that calibration factors can drift over time, most notably due to variations in the ambient temperature which can affect the electronics and can physically alter the length of the flight tube.

3.5.4.5 *Advantages and Limitations of TOF-MS*

The key strengths of TOF mass analysers can be summarized as follows:

- High ion transmission
- Complete mass spectrum for each ionization event
- Good mass resolution, particularly with a reflectron configuration
- Theoretically unlimited mass range

However, it is worth qualifying some of the above points. The second point in the above list seems highly desirable, since it reflects the fact that a TOF analyser acts not as a mass filter (as in a quadrupole mass analyser) but as a mass disperser (in a time domain sense). It is therefore tempting to claim that all available ions are detected regardless of mass. However, such a claim would neglect the fact that, for a continuously injected ion stream into region 1, only those ions that are between the repeller and grid electrodes at the instant the voltage ramp is applied will be injected into the flight tube. It is then necessary to wait until the heaviest ions in this packet of ions have reached the detector before the next packet is injected. If the transverse flow rate of ions through region 1 (see Figure 3.19) is fast compared to the orthogonal velocity of the heaviest ions injected into the flight tube, then most ions entering the mass spectrometer will have no opportunity to be injected into the analyser and will therefore go undetected. This corresponds to a less than ideal duty cycle and can be as low as 1–2%, thus limiting the detection sensitivity.[15] The duty cycle can be improved by either slowing down the incoming ion beam and/or by increasing the width of the acceleration flight tube regions (and by implication the detector), but even

---

[14] If more than two calibrant peaks are employed linear regression can be used to find the optimum values of A and t_0. For very high resolution TOF-MS data it has been suggested that a quadratic relationship be used for the calibration, for example, $Bt^2 + Ct + D = m/z$.

[15] The duty cycle is also mass dependent. The ion transfer optics should deliver the ions into the source region of the TOF-MS at a constant energy and therefore lighter ions will be moving faster than heavier ions. This motion is orthogonal to the motion given to the ion by the repeller pulse and will continue as the ions move along the flight tube. This will cause some ions at the edges of the injected ion packet to miss the detector, an effect which will be larger for lighter ions than for heavier ions. The duty cycle therefore increases with mass in a predictable way.

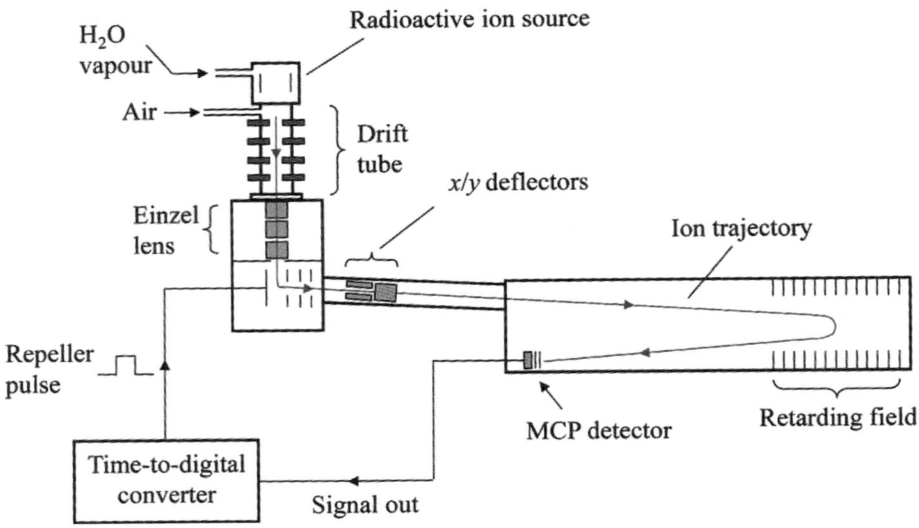

Figure 3.21 *Illustration of the instrument layout for the first published PTR-MS instrument containing a TOF-MS analyser [58]. Reproduced with permission from [57]. Copyright © 2004, American Chemical Society.*

then a duty cycle exceeding 10% is difficult to achieve. An alternative is to use a more complex ion multiplexing approach, in which many ion packets are injected into the flight tube in a pseudorandom sequence. This approach, known as Hadamard transform TOF mass spectrometry, can achieve a duty cycle as high as 100% but it is a technique that is still in its infancy [57].

Finally, we comment on the mass range. Although there is in principle no mass limit with a TOF analyser, in practice there are limitations. One such limitation arises from the ion detector which is usually an ion multiplier such as a microchannel plate (MCP) assembly (see Section 3.6.3). The detection efficiency of these detectors declines with mass.[16] Another mass limitation comes from the reflector array in reflectrons, which show a high mass cut-off effect because heavy ions can penetrate beyond the retarding electric field and will strike the walls of the instrument. Higher fields can of course be used to mitigate against this but with a corresponding detrimental impact on the mass resolution.

3.5.4.6 TOF-MS Analysers in PTR-MS

The first published report of a PTR-MS instrument based on TOF-MS was by Blake *et al.* in 2004 [58] and the experimental arrangement is shown in Figure 3.21. This instrument was constructed with the aim of monitoring complex trace VOC gas mixtures in the atmosphere, a task that is better suited to TOF-MS than QMS because the former collects data across

[16] The detection efficiency can be raised for high mass species by using electrodes to provide additional acceleration of ions prior to their hitting the detector, but without this extra energy the detection efficiency for ions with masses exceeding more than a few hundred Da is extremely low.

the entire mass range essentially instantaneously. The instrument coupled a radioactive ion source, which was described earlier in Section 3.3.3, with a commercial reflectron TOF-MS. This gave a limit of detection (see Section 4.7) at around the 10 ppbv level for 1 min of data accumulation. A similar instrument, this time with a hollow cathode ion source, was developed independently by Ennis *et al.* [59] and this delivered a somewhat better limit of detection near to 1 ppbv for a data acquisition time of 1 min. Although respectable, the sensitivities of these first generation instruments were well below those of existing quadrupole-based PTR-MS instruments. A major limitation here is the duty cycle of the TOF-MS, which was estimated to be as low as 1–2%. On the positive side, the mass resolving power was found to be better than 1000, although this is still not sufficient to resolve most compounds with the same nominal mass.

Improved sensitivities with TOF-MS have been reported by two groups. Tanimoto and co-workers constructed a PTR-MS instrument with a hollow cathode ion source and a linear TOF-MS and achieved detection sensitivities of <100 pptv for individual VOC components in a 1-min integration time [60]. This improvement in performance when compared with Ennis *et al.* is attributable to the higher operating pressure in the drift tube, which was in the region of 6 mbar versus 1 mbar in the case of Ennis *et al.* However, the most impressive achievement has been reported by Jordan *et al.*, who have developed a PTR-TOF-MS system with outstanding sensitivities and resolution [61]. The TOF-MS in this instrument, which is commercially sourced, was designed to achieve high duty cycles and this, along with careful collimation of the ions by an einzel lens into the region between the repeller and grid electrodes, has delivered a PTR-MS instrument with a sensitivity of around 20 pptv for data accumulation of roughly 1 min. Furthermore, the reflectron design means that the mass resolving power can be as high as 7000 while delivering high sensitivity, which now offers the realistic prospect of distinguishing between VOCs with the same nominal mass but different exact masses.[17] An example of this high-resolution capability is shown in Figure 3.22.

3.6 Ion Detectors

Suppose that a mass spectrometer in a PTR-MS instrument receives 10^6 ions per second at the detector, which is a fairly typical rate for H_3O^+ ions. Given that the charge on each ion is 1.602×10^{-19} C (to four significant figures), then the electrical current (which equals the charge flowing per unit time) incident on the detector is 1.6×10^{-13} A $= 160$ fA (f $=$ femto $= 10^{-15}$). In comparison the currents flowing for protonated organic ions present in an analyte gas are likely to be several orders of magnitude lower. Consequently, to be effective the ion detector must be capable of registering ions arriving at currents that are below the fA level.

There are two strategies that can be adopted to try and measure such low current levels: (1) use a very sensitive current measuring device (a picoammeter or, when suitably configured, an electrometer) or (2) devise some means to *amplify* the current, thus making it easier to

[17] This TOF-MS can operate in two modes, the so-called V and W modes. The V mode is equivalent to a standard single reflectron ion pathway whereas the W mode is essentially two consecutive reflectron pathways, such that the ion trajectory traces out a W-shape. The resolution in the W mode is even higher than the 7000 achieved in the V mode but with a substantial decrease in ion transmission, which lowers the instrument sensitivity.

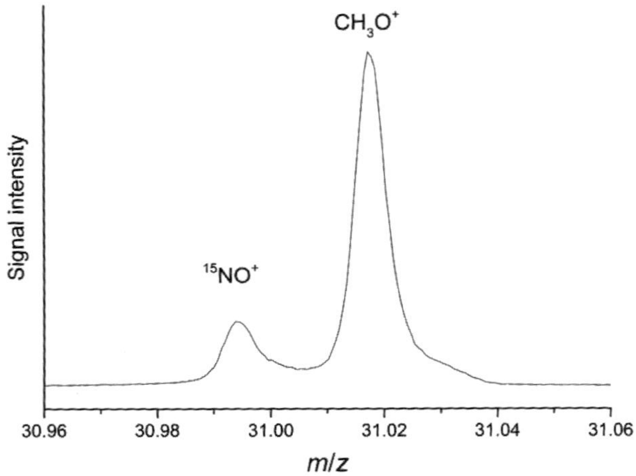

Figure 3.22 *High-resolution mass spectrum from a PTR-TOF-MS instrument showing two peaks at a nominal m/z of 31 arising from $^{15}NO^+$ and protonated formaldehyde (CH_3O^+). The exact masses of these two ions are 30.9950 and 31.0184 Da, respectively, and the high resolution of the instrument allows these two ions to be readily distinguished. This figure was kindly provided by Dr Philipp Sulzer of Ionicon Analytik GmbH and was recorded using an Ionicon PTR-TOF-8000 instrument.*

detect using more conventional devices. Procedure (1) is exemplified by the Faraday cup detector. This is a simple device which is little more than a metallic plate but which may be given the shape of a cup for more efficient collection of ions. However, the Faraday cup has a sensitivity that is limited by the current measuring device it is connected to and the very best picoammeters cannot be used for currents $<10^{-14}$ A. This means that, while a Faraday cup could be used to measure the flow of H_3O^+ ions, it lacks the sensitivity to be able to detect the much smaller quantities of protonated organic ions emanating from the drift tube in PTR-MS.

Detection strategy (2), namely current amplification, is therefore essential in PTR-MS. The underlying principle here is *electron multiplication*, which is used to convert an initially very small input current into an output current many orders of magnitude larger. The devices for doing this, so-called electron multipliers, come in several different guises, as explained in the sub-sections that follow. However, in each case an incoming ion strikes the surface of a suitably chosen material, which leads to the production of electrons known as secondary electrons. By configuring the device to allow multiple subsequent electron strikes, a single incoming ion can create an avalanche of electrons at the output of the detector.

3.6.1 Discrete Dynode Detector

One way of realizing an electron multiplier is through the discrete dynode design, which is illustrated in Figure 3.23. It consists of a series of electrodes, known as *dynodes*. A combination of a high voltage power supply and a potential divider delivers appropriate voltages to each individual dynode which facilitate the electron cascade (multiplication) process.

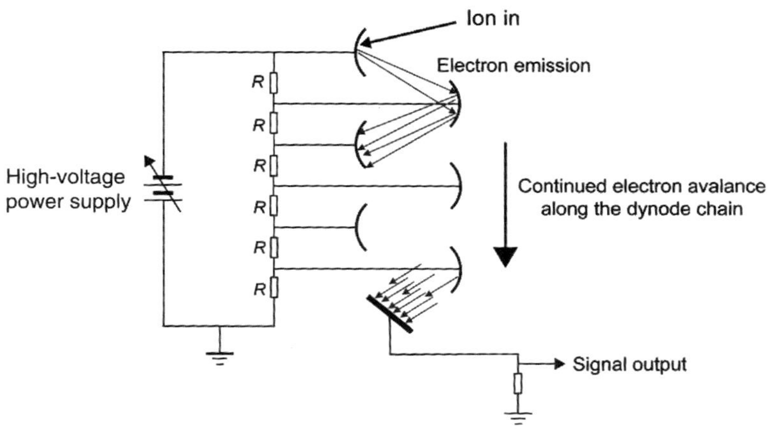

Figure 3.23 *Schematic layout of a discrete dynode detector. The DC power supply should be able to deliver a potential difference of 200–300 V between adjacent dynodes (using the potential divider circuit shown) and so for a 10-dynode detector ~2–3 kV needs to be delivered across the entire detector. Care must be taken to apply this voltage only when the detector is under high vacuum, otherwise damage due to electrical breakdown (arcing) can occur.*

For the detection of a positive ion the first dynode, sometimes called the conversion dynode, has the highest negative electrical potential in the dynode chain. Subsequent dynodes have potentials that are increasingly positive along the chain and it is common for the last dynode to have a potential near Earth (ground) potential. This series of progressively lower potentials at each dynode is most conveniently generated using a potential divider,[18] since this requires only a single DC power supply. Any secondary electrons produced at the first dynode will be accelerated towards the second dynode because the latter has a more positive potential, and therefore attracts electrons. In order to deliver amplification the number of electrons leaving each dynode must be greater than the incoming number. In other words, each dynode must be coated with a material that has a *secondary electron emission coefficient* >1. A material such as silica glass impregnated with lead oxide is typically used for this purpose. Modern discrete dynode detectors are generally stable in air, so frequent venting of the vacuum system or indeed storage on a shelf in a laboratory should not affect the performance of the detector.

The secondary electron emission coefficient is dependent not only on the material coated on the dynode surfaces, but also on the momentum of the incoming electrons. When the electrons are accelerated through a potential difference of ~200 V between adjacent

[18] The total resistance of a collection of n resistors connected in series is given by $R = R_1 + R_2 + R_3 + \ldots\ldots + R_n$. In a closed electrical circuit, the same current (I) must flow through all of these resistors and so from Ohm's law there is a voltage *drop* from the upstream end to the downstream end of each resistor given by IR_j such that the total voltage across the whole resistor chain is given by $V_{\text{applied}} = V_1 + V_2 + V_3 + \ldots\ldots V_n = I(R = R_1 + R_2 + R_3 + \ldots\ldots R_n)$. For example, if two resistors with equal resistances are placed in series and a voltage of 1000 V is applied across the two ends of this resistor pair then a connection to the wire between the two resistors will be at a potential of 500 V. An example of a potential divider circuit consisting of multiple resistors is shown in Figure 3.23.

dynodes each incoming electron might generate 2–3 secondary electrons (although even higher electron production levels are possible). Consequently, the amplification factor, which is known as the *gain*, would be between 2^N and 3^N in this case, where N is the number of dynodes. If the voltage is lowered, then the amplification decreases and thus the voltage applied across the dynode chain provides a means to control the gain of the ion detector. It is also worth noting that secondary electron emission at each dynode is a statistical process, and at the level of individual ions the output current from the detector for each incoming ion event follows a Poisson distribution (see Section 4.7). Consequently, when we refer to the gain of a detector it is a statistical average of these individual ion events.

It might seem reasonable to always operate at the highest possible gain, which can deliver amplification factors of 10^7 or even 10^8, but this is not necessarily so. Very large ion signals can lead to detector *saturation*, that is, the detector has reached its maximum output current and any further increase of the incoming ion current will not be registered. In addition, secondary electron production is a destructive process that leads to degradation of the dynode surface. Consequently, excessive electron production leads to premature ageing of the detector and necessitates more frequent replacement of the detectors, which are not cheap. If such excess currents are avoided by setting the gain at a compromise level then a discrete dynode electron multiplier can tolerate near-continuous operation for as long as 1–2 years before needing replacement.

Finally, it is important to recognize that the efficiency of detecting ions is dependent on their momentum as they strike the conversion dynode. Since most mass spectrometers accelerate all ions across a fixed voltage, the same kinetic energy is delivered to all ions regardless of mass. However, the momentum just prior to impact will be highest for the lightest ion and these are detected with higher efficiency than heavier ions. This decline in detection efficiency with mass occurs for all types of electron multipliers, including the channeltron and MCP detectors discussed in the next two sections. This discrimination against high-mass ions can be reduced if the conversion dynode is operated at a voltage such that ions of high mass are at high velocities.

3.6.2 Channel Electron Multiplier

Channel electron multipliers, sometimes known as *channeltrons*, replace the discrete dynode design with what amounts to an essentially continuous dynode chain. The concept is illustrated in Figure 3.24. The channeltron is essentially a tube, typically made from glass with a semiconducting interior surface. The resistance from one end to the other is high and if a voltage is applied across the two ends of the channeltron a continuous potential gradient runs along the length of the entire tube, that is, there is a 'built-in' potential divider. Ions striking the conversion dynode generate secondary electrons that enter the channeltron and strike the wall, generating further secondary electrons. The increasingly positive potential draws these new electrons deeper into the channeltron and new surface-strikes generate a cascade of electrons that progressively amplifies as the output region is approached. A large bunch of secondary electrons arrives at the electrode at the exit of the channeltron where the resulting current is processed in much the same way as for a discrete dynode detector (see later for details).

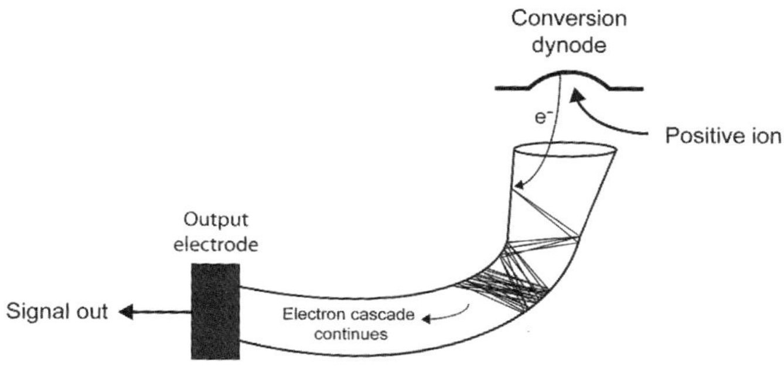

Figure 3.24 *Schematic diagram of a channel electron multiplier (channeltron).*

A channeltron with a straight channel has a maximum gain of around 10^4 to 10^5. This is caused by positive ion production at various points along the tube, for example, by desorption of adsorbed gases which are then ionized by the secondary electrons. The positive ions travel in an opposite direction to the electrons, heading back towards the conversion dynode and striking the walls, thus producing more secondary electrons which then undergo amplification in the usual manner. This additional and unwanted source of electrons leads to a build-up of space charge near the output end of the channeltron which limits the output current. By making the channeltron curved, as illustrated in Figure 3.24, the ions strike the wall before they can build up sufficient kinetic energy to produce significant quantities of secondary electrons. A curved channeltron can deliver signal gains in excess of 10^7.

Discrete dynode and channeltron detectors have many similar characteristics, which is not surprising given the similarity of their operating mechanisms. Discrete dynode detectors generally have longer lifetimes and higher detection sensitivities than channeltrons because they possess a larger active surface area. The discrete dynode detector is also more physically robust than a channeltron but the latter can be obtained in a more compact form and is usually cheaper than a discrete dynode detector.

3.6.3 Microchannel Plate Detector

An MCP is a resistive (lead oxide) glass plate containing numerous narrow channels arranged in a honeycomb structure, each channel acting as a miniature electron multiplier. It differs from most discrete dynode and channeltron detectors in that it has a larger surface area for incoming ions, thus allowing efficient ion collection from a spatially broad source, while still delivering a high gain. Furthermore, because the transit distance through the MCP is short (0.5–2.0 mm) when compared to a channeltron or a discrete dynode detector, the MCP is capable of high time resolution (<1 ns), which makes it particularly suitable for use in TOF-MS.

Most MCP detectors employ two plates in series to achieve a high gain. A typical configuration is shown in Figure 3.25. Each individual channel has an entrance diameter in the region of 10 μm and the channel walls are oriented at an angle to the surface of

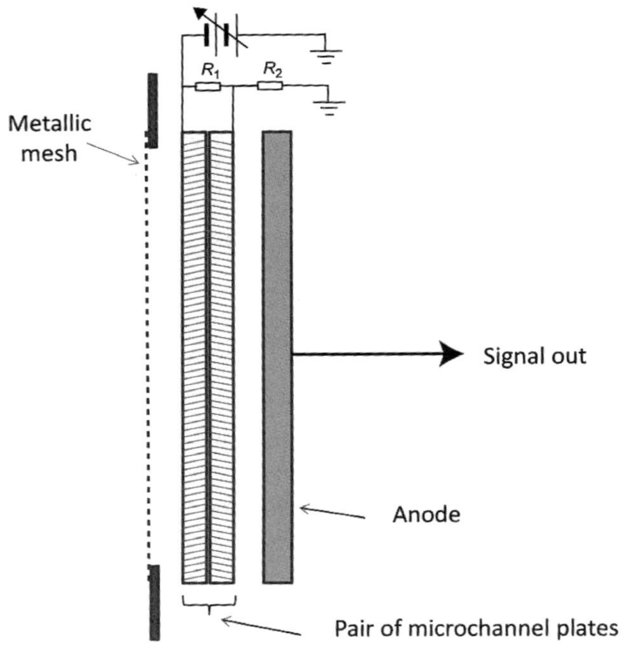

Figure 3.25 *Schematic diagram of a microchannel plate (MCP) detector. This particular detector shows two MCPs in series, mounted with a Chevron configuration to minimize ion feedback (see text). The front of the first MCP is at a high negative voltage, typically −2 kV with respect to Earth (ground) potential.* $R_1 \gg R_2$ *so that most of the voltage drop occurs across the two plates, with a roughly 1 kV voltage drop across each plate. This voltage drop continuously accelerates the cascade of secondary electrons within each microchannel to produce high amplification. The electron current leaving the back of the second MCP is collected by the anode and processed in the same way as, for example, that from the discrete dynode detector. The metallic mesh in front of the detector is at the electrical potential of the flight tube (usually Earth potential) and is used to shield the ions in the flight tube from the high voltage on the front MCP.*

the plate so that incoming charged particles are guaranteed to strike the wall. For positive ion detection the ions enter the MCP detector through a conducting mesh, often at Earth potential, and are then accelerated through a voltage of ~2 kV, whereupon they strike the channel walls and can generate one or more secondary electrons. This starts a cascade process, as this first batch of electrons is accelerated towards the back of the second plate by the increasingly positive electrical potential in moving from the left to right in Figure 3.25. Just as with the channeltron, ion feedback is a potential problem and this is minimized in two-plate MCP detectors by creating a V-like orientation of the channels in the two plates. A dual-MCP in such an arrangement is said to be *chevron* mounted and is easily capable of an amplification of 10^8 from a single incoming ion event. The gain can be controlled by adjusting the voltage across the MCPs, which in Figure 3.25 means controlling the voltage applied to the front of the first MCP.

Care must be taken in the handling of MCP detectors, since the plates are relatively fragile and can crack or can even shatter if subject to a significant mechanical shock. Voltages significantly in excess of ~1.1 kV across each plate can lead to electrical breakdown and consequent irreparable damage to the MCP, so excessively high voltages must be avoided. The MCP surfaces are degraded by contact with water vapour and, while they can tolerate short periods of up to a few hours exposed to air, long term storage must be under vacuum or in an exceptionally dry environment.

3.7 Analogue versus Digital Signal Processing

Ultimately, the signal output from the ion detector is always converted into a digital signal for computer presentation, analysis and storage. Before we discuss the ways in which this can be achieved, we first need to be clear about the difference between an analogue signal and a digital signal. When the ion current emanating from the detector is relatively large it is intrinsically an *analogue* signal, since it can have an essentially continuous range of values. On the other hand, digital signals have only a finite set of values, that is, they are discrete rather than continuous. Digital signals allow information to be stored in binary form in a computer, where each number in the binary sequence is stored as a *bit*. For example, if we have an 8-bit signal then the maximum value this signal can have in binary form is 11111111, which corresponds to 255 in decimal. If we therefore measure a signal and somehow convert it into digital form it can have 256 possible values if stored as an 8-bit integer. When restricted to 8-bit numbers a value of 254 cannot be distinguished from 254.4, so a possible consequence of digitization is a loss of information (and in particular a restriction on the dynamic range of the instrument). The information content can be increased by allocating more bits to each number, but this comes at the expense of making the electronics more complex and therefore increases the cost.

There are two ways in which a digital signal can be generated from an ion detector. The first is by analogue processing, in which the output signal (current) is converted to a voltage, amplified if necessary, and is then transformed into a digital quantity by an electronic device known as an analogue-to-digital converter (ADC), as illustrated in Figure 3.26. This is referred to as analogue signal processing because we are essentially measuring the ion current and the A to D converter is simply there to allow the signal to be stored in a digital format within a computer.

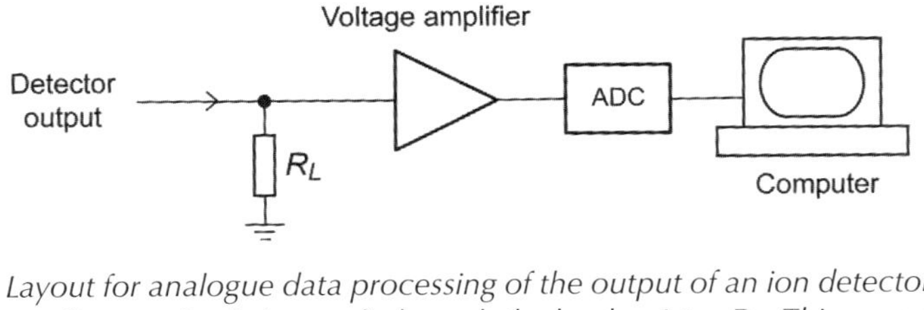

Figure 3.26 *Layout for analogue data processing of the output of an ion detector. The current from the detector flows to Earth (ground) through the load resistor, R_L. This generates a voltage at the upstream end of R_L which can be amplified by a voltage amplifier and then converted into a digital integer before arriving at a computer.*

Analogue signal processing can be used effectively when signal levels are relatively high. However, if ions arrive only sporadically at the detector, as would be the case when trying to detect ions from a compound present at very low concentrations, the analogue detection mode is unsuitable. This is because the noise from the detector, which is unavoidable in all detectors, is continuously being accumulated and will overwhelm genuine signals when present at very low levels. This is the scenario most often encountered in PTR-MS, so for PTR-MS applications a purely digital signal processing scheme is invariably employed.

Digital signal processing employs ion (pulse) counting. Essentially, each ion incident on the detector can lead to a current pulse at the detector. With sufficient amplification it is therefore possible to register the arrival of a single ion at the detector. However, there is a complication: all detectors receive background, or dark, signals that would be observed even if the drift tube of the PTR-MS instrument is completely blocked so that no ions can enter the mass spectrometer. Sporadic signals from cosmic rays passing through the instrument is one source of dark counts, as is thermionic emission of electrons from the detector surfaces. These dark counts amount to detector noise and it is necessary to discriminate between a genuine signal pulse and dark noise pulses, otherwise the genuine signal may be overwhelmed by the dark noise. Fortunately, the electrons generated by dark noise tend not to undergo the full amplification available within the detector because they are generated at an arbitrary position within the dynode chain. Consequently, the genuine incoming ions almost always generate a larger signal pulse at the detector output than typical dark noise pulses. This difference can be put to use by operating the detector at a relatively high gain and the subsequent electronics are then configured to count only those pulses arising from genuine ions, as illustrated in Figure 3.27.

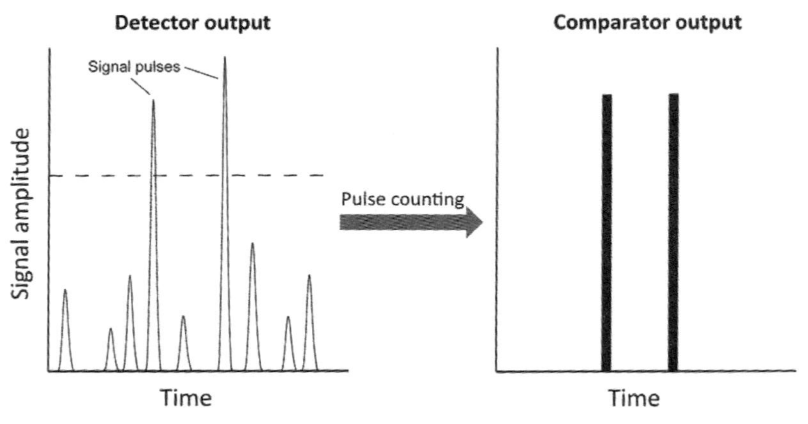

Figure 3.27 *Illustration of the function of discriminator electronics for pulse counting (see also Figure 3.28). The output of the detector delivers signal pulses which exceed the amplitude of most noise pulses. The signal pulses can then be selectively counted by use of a comparator unit located within the discriminator. A preset voltage level is chosen by the user such that most signal pulses exceed this voltage whereas most noise pulses do not. The comparator then compares the detector pulse with the preset discriminator voltage and only those pulses that exceed this threshold are passed on to a pulse counting unit.*

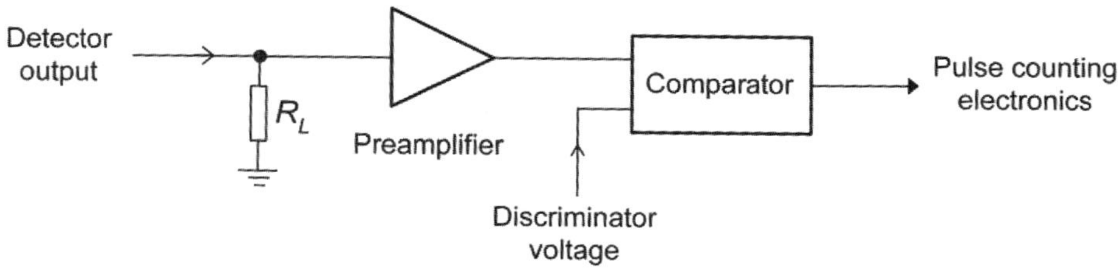

Figure 3.28 A typical arrangement for ion pulse counting. The signal from the detector generates a voltage across a load resistor, R_L, as it flows to Earth (ground). The voltage can be amplified (optional) by a preamplifier and is then fed into a comparator, which outputs a voltage pulse for subsequent counting providing the input signal exceeds a user-defined threshold set by a discriminator unit. In this way, the user can set the discriminator such that dark counts generate a voltage that lies below the threshold while genuine ion signal pulses usually lie above the threshold, and so only the latter are counted. A fast current pulse is desirable so a small resistance (50 Ω is typical) for the load resistor ensures that the current is discharged to Earth (ground) relatively rapidly.

The practical implementation of the pulse counting mode uses the arrangement shown in Figure 3.28. The current from the detector flows to Earth (ground) through the load resistor, R_L, generating a voltage pulse across the resistor. A low load resistance is necessary to allow rapid current transit, since the current flow is impeded by a higher resistance.[19] The standard choice for fast pulse counting is a 50 Ω load resistor connected to the detector by a coaxial cable with a 50 Ω impedance.[20] The voltage pulse can be fed directly into a comparator, or else it can first be amplified and then fed into the comparator. The purpose of the comparator is to allow the signal voltage to be compared with another voltage preset by the user, which is provided by a *discriminator*. If the discriminator threshold is exceeded, a voltage pulse is output from the comparator, whereas all voltage pulses lying below the discriminator threshold yield no output pulse from the comparator. The circuitry inside the comparator is normally designed to deliver an output voltage pulse in the form of one of the standard logic pulses used in digital signal processing, which come with designations such as NIM, TTL and ECL. The important point is that the logic pulse will be in a format recognized by subsequent pulse counting electronics. The pulse counter divides the data accumulation process into a series of short time bins (typically a few ns or less) and will either register a single count or no count in each time bin. Accumulation of data over a significantly long period of time can be expressed as a count rate, for example, counts per second.

Very high count rates cause a potential problem for pulse counting. If the count rate becomes too high then it is possible that two ions may arrive at the detector at essentially the same time, that is, within the time period of a single time bin. Since only one ion count

[19] The speed at which the current pulse is discharged to Earth (ground) is proportional to the *time constant*, *RC*, where *R* is the resistance and *C* is the capacitance. Although we have no capacitor deliberately placed in the circuit, there is always some capacitance arising in the detector circuitry. Since the circuit resistance can easily be controlled, low *R* is chosen to generate an acceptably small time constant.

[20] This is an example of *impedance matching*. If the electrical current pulse encounters discontinuities in impedance then part of the signal is reflected back and forth along the detector-amplifier part of the circuit, leading to signal 'ringing'.

can be registered within a single time bin, the potential exists for underestimating the ion count rate. This is one of the factors limiting the dynamic range of a TOF-MS instrument. Mathematical corrections can be applied approximately to allow for these *dead time* losses but an alternative option is to switch to the analogue mode of data collection if the count rate is too high for satisfactory pulse counting.

References

1. Chambers, A., Fitch, R. F., Halliday, B. S. (1998) *Basic Vacuum Technology*, IoP Publishing, Bristol.
2. O'Hanlon, J. F. (2003) *A User's Guide to Vacuum Technology*, John Wiley & Sons, Inc.
3. Yoshimura, N. (2007) *Vacuum Technology: Practice for Scientific Instruments*, Springer-Verlag, Berlin.
4. Dahl, P. (1973) *Introduction to Electron and Ion Optics*, Academic Press, New York.
5. Szilagyi, M. (1988) *Electron and Ion Optics*, Plenum Press, New York.
6. Liebl, H. (2008) *Applied Charged Particle Optics*, Springer, Berlin.
7. Marcus, R. K. (Ed.) (1993) *Glow Discharge Spectroscopies*, Plenum Publishing, New York.
8. Pillow, M. E. (1981) A critical review of spectral and related physical properties of the hollow cathode discharge. *Spectrochim. Acta* **36B**, 821.
9. Mann, M. M., Hustrulid, A., Tate, J. T. (1940) The ionization and dissociation of water vapour and ammonia by electron impact. *Phys. Rev.* **58**, 340.
10. Ikezoe, Y., Matsuoka, S., Viggiano, A. (1987) *Gas Phase Ion-Molecule Reaction Rate Constants through 1986*, Maruzen Company Ltd, Tokyo.
11. Hansel, A., Jordan, A., Holzinger, R. *et al.* (1995) Proton transfer reaction mass spectrometry: on-line trace gas analysis at the ppb level. *Int. J. Mass Spectrom. Ion Processes* **149–150**, 609.
12. Smith, D., Adams, N. G., Miller, T. M. (1978) A laboratory study of the reactions of N^+, $N_2{}^+$, $N_3{}^+$, $N_4{}^+$, O^+, $O_2{}^+$, and NO^+ ions with several molecules at 300 K. *J. Chem. Phys.* **69**, 308.
13. Inomata, S., Tanimoto, H., Aoki, N. *et al.* (2006) A novel discharge source of hydronium ions for proton transfer reaction ionization: design, characterization, and performance. *Rapid Commun. Mass Spectrom.* **20**, 1025.
14. Hanson, D. R., Koppes, M., Stoffers, A. *et al.* (2009) Proton transfer mass spectrometry at 11 hPa with a circular glow discharge: sensitivities and applications. *Int. J. Mass Spectrom.* **282**, 28.
15. Hanson, D. R., Greenberg, J., Henry, B. E., Kosciuch, E. (2003) Proton transfer reaction mass spectrometry at high drift tube pressure. *Int. J. Mass Spectrom.* **223**, 507.
16. Blake, R. S., Wyche, K. P., Ellis, A. M., Monks, P. S. (2006) Chemical ionization reaction time-of-flight mass spectrometry: multi-reagent analysis for determination of trace gas composition. *Int. J. Mass Spectrom.* **254**, 85.
17. Wyche, K. P., Blake, R. S., Willis, K. A. *et al.* (2005) Differentiation of isobaric compounds using chemical ionization reaction mass spectrometry. *Rapid Commun. Mass Spectrom.* **19**, 3356.

18. Norman, M., Hansel, A., Wisthaler, A. (2007) $O_2{}^+$ as reagent ion in the PTR-MS instrument: detection of gas-phase ammonia. *Int. J. Mass Spectrom.* **265**, 382.
19. Jordan, A., Haidacher, S., Hanel, G. *et al.* (2009) An online ultra-high sensitivity proton-transfer-reaction mass-spectrometer combined with switchable reagent ion capability (PTR + SRI – MS). *Int. J. Mass Spectrom.* **286**, 32.
20. Knighton, W. B., Fortner, E. C., Herndon, S. C. *et al.* (2009) Adaptation of a proton transfer reaction mass spectrometer instrument to employ NO^+ as reagent ion for the detection of 1,3-butadiene in the ambient atmosphere. *Rapid Commun. Mass Spectrom.* **23**, 3301.
21. Sulzer, P., Edtbauer, A., Hartungen, E. *et al.* (2012) From conventional proton-transfer-reaction mass spectrometry (PTR-MS) to universal trace gas analysis. *Int. J. Mass Spectrom.* **321–322**, 66.
22. de Gouw, J. A., Goldan, P. D., Warneke, C. *et al.* (2003) Validation of proton transfer reaction-mass spectrometry (PTR-MS) measurements of gas-phase organic compounds in the atmosphere during the New England Air Quality Study (NEAQS) in 2002. *J. Geophys. Res.* **108**(D21), 4682.
23. Soppart, O., Baumbach, J. (2000) Comparison of electric fields within drift tubes for ion mobility spectrometry. *Meas. Sci. Technol.* **11**, 1473.
24. Thornberry, T., Murphy, D. M., Thomson, D. D. *et al.* (2009) Measurement of aerosol organic compounds using a novel collection/thermal-desorption PTR-ITMS instrument. *Aerosol Sci. Technol.* **43**, 486.
25. Mikoviny, T., Kaser, L., Wisthaler, A. (2010) Development and characterization of a high-temperature proton-transfer-reaction mass spectrometer (HT-PTR-MS). *Atmos. Meas. Tech.* **3**, 537.
26. Revercomb, H. E., Mason, E. A. (1975) Theory of plasma chromatography/gaseous electrophoresis – a review. *Anal. Chem.* **47**, 970.
27. Ellis, H. W., Pai, R. Y., McDaniel, E. W. *et al.* (1976) Transport properties of gaseous ions over a wide energy range. *At. Data Nucl. Data Tables.* **17**, 177.
28. Ellis, H. W., McDaniel, E. W., Albritton, D. L. *et al.* (1978) Transport properties of gaseous ions over a wide energy range. Part II. *At. Data Nucl. Data Tables.* **22**, 179.
29. Ellis, H. W., Thackston, M. G., McDaniel, E. W., Mason, E. A. (1984) Transport properties of gaseous ions over a wide energy range. Part III. *At. Data Nucl. Data Tables.* **31**, 113.
30. Dotan, I., Albritton, D. L., Lindinger, W., Pahl, M. (1976) Mobilities of $CO_2{}^+$, N_2H^+, H_3O^+, $H_3O^+ \cdot H_2O$, and $H_3O^+ \cdot (H_2O)_2$ ions in N_2. *J. Chem. Phys.* **65**, 5028.
31. Warneke, C., van der Veen, C., Luxembourg, S. *et al.* (2001) Measurements of benzene and toluene in ambient air using proton-transfer-reaction mass spectrometry: calibration, humidity dependence, and field intercomparison. *Int. J. Mass Spectrom.* **207**, 167.
32. Keck, L., Oeh, U., Hoeschen, C. (2007) Corrected equation for the concentrations in the drift tube of a proton transfer reaction-mass spectrometer (PTR-MS). *Int. J. Mass Spectrom.* **264**, 92.
33. Wannier, G. H. (1951) On the motion of gaseous ions in a strong electric field. I. *Phys. Rev.* **83**, 281.
34. Wannier, G. H. (1953) Motion of gaseous ions in strong electric fields. *Bell Syst. Tech. J.* **32**, 170.

35. McFarland, M., Albritton, D. L., Fehsenfeld, F. C. *et al.* (1973) Flow-drift technique for ion mobility and ion-molecule reaction rate constant measurements. II. Positive ion reactions of N^+, O^+, and $H_2{}^+$ with O_2 and O^+ with N_2 from thermal to 2 eV. *J. Chem. Phys.* **59**, 6620.
36. Dalleska, N. F., Honma, K., Armentrout, P. B. (1993) Stepwise solvation enthalpies of protonated water clusters: collision-induced dissociation as an alternative to equilibrium studies. *J. Am. Chem. Soc.* **115**, 12125.
37. Xie, Y., Remington, R. B., Schaefer III, H. F. (1994) The protonated water dimer: extensive theoretical studies of $H_5O_2{}^+$. *J. Chem. Phys.* **101**, 4878.
38. Huang, X., Braams, B. J., Bowman, J. M. (2005) *Ab initio* potential energy and dipole moment surfaces for $H_5O_2{}^+$. *J. Chem. Phys.* **122**, 044308.
39. Bekstein, A., Benhenni, M., Yousfi, M. (2011) Collisions cross sections and swarm coefficients of water vapour ion clusters $(H_2O)_nH^+$ with $n = 1, 2$ and 3 in N_2, O_2 and air. *Eur. Phys. J. D* **61**, 153.
40. Lau, Y. K., Ikuta, S., Kebarle, P. (1982) Thermodynamics and kinetics of the gas-phase reactions $H_3O^+(H_2O)_{n-1}$ + water = $H_3O^+(H_2O)_n$. *J. Am. Chem. Soc.* **104**, 1462.
41. Albritton, D. L., Dotan, I., Lindinger, W. *et al.* (1977) Effects of ion speed distributions in flow-drift tube studies of ion–neutral reactions. *J. Chem. Phys.* **66**, 410.
42. Viehland, L. A., Mason, E. A. (1977) Statistical–mechanical theory of gaseous ion–molecule reactions in an electrostatic field. *J. Chem. Phys.* **66**, 422.
43. Fairly, D. A., Milligan, D. B., Freeman, C. G. *et al.* (1999) Competitive association and charge transfer in the reactions of NO^+ with some ketones: a selected ion flow drift tube study. *Int. J. Mass Spectrom.* **193**, 35.
44. de Gouw, J. A., Warneke, C., Karl, T. *et al.* (2003) Sensitivity and specificity of atmospheric trace gas detection by proton-transfer-reaction mass spectrometry. *Int. J. Mass Spectrom.* **223–224**, 365.
45. Adams, N. G., Smith, D., Ferguson, E. E. (1985) Comparative effects of temperature and kinetic energy change on the reaction of $O_2{}^+$ with CH_4 and CD_4. *Int. J. Mass Spectrom. Ion Processes.* **67**, 67.
46. de Gouw, J., Warneke, C. (2007) Measurements of volatile organic compounds in the earth's atmosphere using proton-transfer-reaction mass spectrometry. *Mass Spectrom. Rev.* **26**, 223.
47. Dawson P. H. (Ed.) (1997) *Quadrupole Mass Spectrometry and its Applications*, American Institute of Physics.
48. March, R. E., Todd, J. F. (2005) *Quadrupole Ion Trap Mass Spectrometry*, 2nd edn, John Wiley & Sons, Inc.
49. March, R. E. (1997) An introduction to quadrupole ion trap mass spectrometry. *J. Mass Spectrom.* **32**, 351.
50. March, R. E. (1998) Quadrupole ion trap mass spectrometry: theory, simulation, recent developments and applications. *Rapid Commun. Mass Spectrom.* **12**, 1543.
51. Prazeller, P., Palmer, P. T., Boscaini, E. *et al.* (2003) Proton transfer reaction ion trap mass spectrometer. *Rapid Commun. Mass Spectrom.* **17**, 1593.
52. Warneke, C., de Gouw, J. A., Lovejoy, E. R. *et al.* (2005) Development of proton-transfer ion trap-mass spectrometry: on-line detection and identification of volatile organic compounds in air. *J. Am. Soc. Mass Spectrom.* **16**, 1316.

53. Steeghs, M. M. L., Sikkens, C., Crespo, E. *et al.* (2007) Development of a proton-transfer reaction ion trap mass spectrometer: online detection and analysis of volatile organic compounds. *Int. J. Mass Spectrom.* **262**, 16.
54. Mielke, L. H., Erickson, D. E., McLuckey, S. A. *et al.* (2008) Development of a proton-transfer reaction-linear ion trap mass spectrometer for quantitative determination of volatile organic compounds. *Anal. Chem.* **80**, 8171.
55. Wiley, W. C., McLaren, I. H. (1955) Time-of-flight mass spectrometer with improved resolution. *Rev. Sci. Inst.* **26**, 1150.
56. Mamyrin, B. A., Karataev, V. I., Shmikk, D. V., Zagulin, V. A. (1973) The mass-reflectron, a new nonmagnetic time-of-flight mass spectrometer with high resolution. *Sov. Phys. JETP* **37**, 45.
57. Zare, R. N., Fernandez, F. M., Kimmel, J. R. (2003) Hadamard transform time-of-flight mass spectrometry: more signal, more of the time. *Angew. Chemie Intl. Ed.* **42**, 30.
58. Blake, R. S., Whyte, C., Hughes, C. O. *et al.* (2004) Demonstration of proton-transfer reaction time-of-flight mass spectrometry for real-time analysis of trace volatile organic compounds. *Anal. Chem.* **76**, 3841.
59. Ennis, C. J., Reynolds, J. C., Keely, B. J., Carpenter, L. J. (2005) A hollow cathode proton transfer reaction time of flight mass spectrometer. *Int. J. Mass Spectrom.* **247**, 72.
60. Tanimoto, H., Aoki, N., Inomata, S. *et al.* (2007) Development of a PTR-TOFMS instrument for real-time measurements of volatile organic compounds in air. *Int. J. Mass Spectrom.* **263**, 1.
61. Jordan, A., Haidacher, S., Hanel, G. *et al.* (2009) A high resolution and high sensitivity proton-transfer-reaction time-of-flight mass spectrometer (PTR-TOF-MS). *Int. J. Mass Spectrom.* **286**, 122.

4

Quantitative Analysis

4.1 Introduction

An aim of many applications of PTR-MS is to determine the absolute concentration of one or more trace volatile organic compounds (VOCs) in air. PTR-MS is, in principle, capable of determining absolute compound concentrations without recourse to instrument calibration. However, as detailed below, there are various sources of uncertainty which can make such non-calibrated determinations relatively unreliable. Consequently, for many applications some form of instrument calibration is at least desirable, and perhaps essential.

This chapter deals with issues encountered when using PTR-MS as a *quantitative* technique. It starts by showing how the concentration of a gas constituent can be calculated from a PTR-MS measurement without calibration, and then moves on to consider why calibration can be important. The most commonly used methods for trace gas calibration are then described. The chapter closes with a discussion of the accuracy, precision and limit of detection for PTR-MS measurements.

4.2 Extracting the Concentration of a Trace Gas from PTR-MS

The simplicity of PTR-MS is embodied in Equation 1.10, which we repeat below:

$$\frac{[\mathrm{MH^+}]}{[\mathrm{H_3O^+}]} = \frac{i(\mathrm{MH^+})}{i(\mathrm{H_3O^+})} = k[\mathrm{M}]t \tag{4.1}$$

According to this equation, it is possible to determine the absolute concentration of compound M if the ratio $i(\mathrm{MH^+})/i(\mathrm{H_3O^+})$ is measured and the reaction time t and the rate coefficient k are known. In Section 4.4, we will consider some of the limitations arising from the application of Equation 4.1. However, in this short section our aim is to demonstrate how the concentration of compound M can be calculated from the acquired PTR-MS measurements.

Proton Transfer Reaction Mass Spectrometry: Principles and Applications, First Edition. Andrew M. Ellis and Christopher A. Mayhew.

When t is expressed in seconds (s) and rate coefficients for gaseous reactions are in cm^3 s^{-1} then [M] would be obtained as a number density from a calculation using Equation 4.1, that is, molecules cm^{-3} inside the drift tube. However, the user invariably wishes to know the concentration of the compound *outside* of the drift tube in the original gas. Instead of using the concentration (moles/volume) or number density (molecules/volume), we have seen in Chapter 1 that it is convenient to employ the volume mixing ratio (VMR), expressed in parts per million by volume (ppmv), parts per billion by volume (ppbv) or parts per trillion by volume (pptv). One advantage of using VMR is that it is the same inside the drift tube as it is outside in the analyte. Furthermore, unlike number density, the VMR is temperature independent. VMRs will be encountered many times in the remainder of this book, although we may sometimes refer to them as concentrations. It will be evident to the reader that concentration and VMR are not the same quantities, but one is proportional to the other and converting between the two is straightforward. We therefore consider references to concentrations in units of ppbv or pptv as a tolerable misdemeanour.

To extract a VMR from Equation 4.1, it is necessary to include the total number density, $N_{\underline{d}}$, of the gas in the drift tube, which can be determined from the pressure and the temperature. For example, if the drift tube pressure is 2 mbar and the temperature is 40°C, then application of the ideal gas equation gives $N_d = pN_A/RT = 4.69 \times 10^{22}\ m^{-3} = 4.69 \times 10^{16}\ cm^{-3}$ to three significant figures. The VMR for substance M is obtained from the following adaptation of Equation 4.1:

$$\mathrm{VMR_M} = \frac{i(\mathrm{MH^+})}{i(\mathrm{H_3O^+})} \frac{1}{kt} \frac{10^9}{N_d} \tag{4.2}$$

Equation 4.2 gives the VMR for compound M in ppbv if N_d is given in cm^{-3}. As an illustration, suppose we measure $i(MH^+)/i(H_3O^+) = 10^{-4}$. If the reaction time is 100 μs and the proton transfer rate coefficient is $2 \times 10^{-9}\ cm^3\ s^{-1}$, both of which are fairly typical values, then we calculate a mixing ratio of 10.7 ppbv for compound M. It should be noted that the above calculation assumes that there is a negligible contribution to the drift tube pressure from ingress of gas from the ion source. If this is not the case then this dilution effect must be allowed for in the calculation.

In applying Equation 4.2, it is important to recognize that direct measurement of the total count rate of H_3O^+ is not possible. Depending on the instrument used, count rates for H_3O^+ can be as high as 10^7 counts per second and at this high rate detector saturation occurs. Furthermore, high count rates can lead to premature ageing of an ion detector, such as a channeltron electron multiplier or MCP (see Section 3.6 for details about these types of detector). For this reason, it is common to exploit the naturally occurring isotope of oxygen, ^{18}O, and measure the count rate of $H_3{}^{18}O^+$ at m/z 21. Since the natural abundance of ^{18}O relative to ^{16}O is, to a good approximation, 1:500, the ion count rate at m/z 21 is multiplied by 500 to obtain the actual count rate for H_3O^+ in Equation 4.1. Stopping the H_3O^+ ions at m/z 19 from reaching the detector is straightforward with a quadrupole mass spectrometer but a time-of-flight mass spectrometer is not a mass filter and is normally configured to detect all ions. A solution to this is to employ a *mass gate*, which provides a voltage pulse applied to the instrument at the pre-determined ion flight time to deflect the m/z 19 ions away from the detector.

4.3 Normalized Counts per Second

It is common for ion count rates to be quoted as values relative to the H_3O^+ count rate rather than as absolute count rates. This is because, even within a single instrument, the H_3O^+ count rate can change over the course of a day and may differ significantly from one day to the next. However, the relative response for different compounds should be unchanged by such fluctuations in the absolute signal levels generated by the instrument. Since most PTR-MS instruments generate absolute H_3O^+ counts near to $10^6\ s^{-1}$, to within an order of magnitude, it has become common practice to quote signal levels in *normalized counts per second*, or ncps, where

$$\text{ncps} = 10^6 \times \frac{i(\text{MH}^+)}{i(\text{H}_3\text{O}^+)} \tag{4.3}$$

In other words, the ion count rate for MH^+ is quoted relative to an assumed H_3O^+ signal of one million counts per second.

The use of normalized count rates is not universally employed and of course it does not allow direct comparisons of absolute detection sensitivities of different instruments. Instruments with dramatically different H_3O^+ count rates will have very different absolute detection sensitivities (see next section) and yet may have similar normalized detection sensitivities.

4.4 Why Calibrate?

In the practical exploitation of Equation 4.1 for calculating trace gas concentrations, there are several potential uncertainties, some of which have already been mentioned in the earlier chapters. For example, an accurate determination of $[MH^+]/[H_3O^+]$ would require knowledge of the relative *extraction* efficiencies of MH^+ and H_3O^+ ions into the mass spectrometer, as well as their relative *transmission* efficiencies through the mass spectrometer. Furthermore, the *detection* efficiencies of ion detectors are rarely flat as a function of m/z. Added to this is the possibility that some protonated parent molecules will fragment either because they are unstable or because they fragment under the collisional regime brought about by the specific E/N in the drift tube. This is more of a problem for some chemical compounds than others, and notable examples are alcohols and some aldehydes, where elimination of H_2O on receipt of a proton is a significant, and sometimes dominant, product channel (see Section 2.2.4.1). Whenever fragmentation occurs, uncertainty in the determination of concentration will follow, since the assumption behind Equation 4.1 is that protonation of molecule M leads only to formation of MH^+. In reality, where fragment ions are observed, the numerator in Equation 4.1 should become the sum of all product ions derived from the original neutral molecule, M (protonated parent and fragment ions). However, this is impossible to do reliably when analyzing a complex chemical environment owing to problems in knowing what specific ions contribute to signal at a specific m/z. These overlapping contributions can come from product ions of an unrelated VOC (which can be the protonated parent or fragments), isotopologues containing, for example, ^{13}C, or product ions resulting from reactions with impurity reagent ions such as O_2^+ and NO^+. Consequently, the determination of concentration from the ion count rate ratio $i(MH^+)/i(H_3O^+)$ or a modified version of this which allows for fragmentation is a source of uncertainty.

Unfortunately, there are other uncertainties that will further limit the accuracy of any quantitative analysis. A difference in ion mobilities for H_3O^+ versus MH^+ is one such factor, as discussed already in Section 3.4.1. Radial diffusion is another issue, which will tend to reduce the proportion of ions passing through the aperture at the end of the drift tube. The radial diffusion will vary from one ion to another because of differences in diffusion coefficient, but the effect of radial diffusion is expected to be fairly small. Principal among the remaining uncertainties is the likely error margin for the rate coefficient, k. Quoted error margins for proton transfer rate coefficients are rarely better than $\pm 25\%$, and sometimes as large as $\pm 50\%$. Alternatively, one can use theoretical estimates of k, as described in Chapter 2, but here too there are limitations. For example, to calculate proton transfer rate coefficients one needs to know the permanent electric dipole moment (if non-zero) and the polarizability of the neutral molecule, and while experimental values may be known for small molecules they may not be available for many larger molecules. Even if experimental values are known, there will be some margin of error associated with them, and if they are not known then an estimate will be required, such as recourse to *ab initio* quantum chemical calculations. Added to this uncertainty about the dipole moment and polarizability is the fact that the theoretical models developed to calculate k involve approximations, such as the assumption of a point charge for the proton source. Consequently, the error margins for calculated proton transfer rate coefficients may not be significantly better than available from experiment and could be worse. Furthermore, rate coefficients are temperature dependent when a polar neutral molecule is involved in the ion–molecule reaction (see Section 2.2.3.2) and the effective temperature occurring inside the PTR-MS drift tube for proton transfer reactions is also an uncertain quantity, as addressed in Section 3.4.4. Consequently, whether the rate coefficient is taken from experiment or from theory, the ill-defined reaction temperature adds further uncertainty to the margin of error.

Given the many potential sources of error intrinsic to application of Equation 4.1, and there are others we could have added but have not mentioned, the bottom line is that, if one is aiming for accurate concentrations, calibration of the instrument against appropriate gas standards is essential. An illustration of compound calibration in PTR-MS is shown in Figure 4.1, which shows a calibration plot for methanol obtained by one of the authors. Known concentrations of gaseous methanol have been added to the inlet of a PTR-MS instrument using one of the techniques described in section 4.5.2 (in this case a permeation

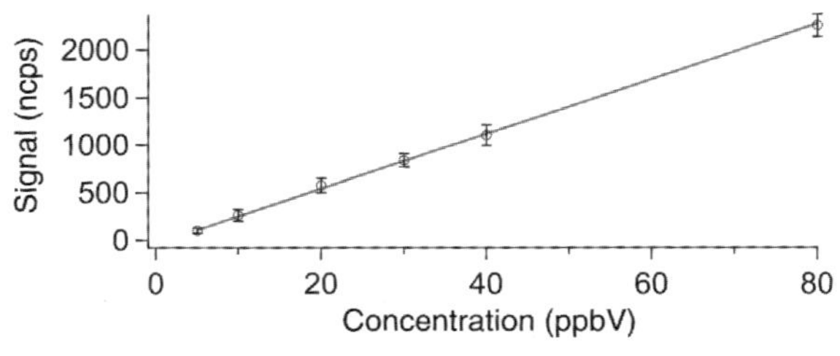

Figure 4.1 *Calibration plot for methanol derived from PTR-MS measurements of the protonated molecule. The calibration was performed using a permeation tube system. Several repeat measurements were made at each calibrant concentration and the error bars shown represent the standard deviation of each set of measurements. Notice that the signal of protonated methanol is expressed in normalized counts per second (see Section 4.3).*

tube). The ion signal for protonated methanol has been normalized to an H_3O^+ count rate of 10^6 s^{-1} and so is plotted on the vertical scale in units of ncps. The horizontal scale plots the concentration in ppbv. The instrument response shows good linearity and the calibration factor can be extracted from the slope of this plot (28.3 ncps/ppbv). This slope represents the *sensitivity* of the instrument towards methanol and because it is expressed in terms of ncps rather than an absolute count rate it is sometimes called the *normalized detection sensitivity*. An obvious advantage of expressing the ion signal in terms of ncps, instead of absolute ion count rates, is that the calibration factor should remain valid even if the absolute H_3O^+ count rate changes. However, the normalized detection sensitivity can change if the mass-dependent transmission characteristics of the instrument change, for example, through inadvertent contamination of ion optics or indeed through intentional tweaking of the lens voltages to improve the overall signal levels. The effective transmission of the mass spectrometer can also change as the ion detector ages. Substantial differences can be observed between normalized detection sensitivities from one type of PTR-MS instrument to another. For example, a PTR-MS instrument with a quadrupole mass spectrometer may show very different transmission behaviour as a function of m/z compared with a time-of-flight mass spectrometer and so there is no reason to expect their normalized detection sensitivities to be identical.

The linear response of PTR-MS can extend over a much wider range of mixing ratios than shown in Figure 4.1, ranging from typically a few tens of ppmv at the high end through to the pptv range at the low end. The limitation at high mixing ratios is that large quantities of VOCs will begin to substantially deplete the quantity of H_3O^+ in the drift tube. It was an assumption in developing the basic kinetic model for PTR-MS, as described in Section 1.4.3, that the concentration of H_3O^+ is fixed. If this condition is not met then a deviation from linearity will occur.

At the low mixing ratio end of the scale problems arise not from a deviation from linearity, but rather because of an offset introduced by interferences from other compounds at a given m/z. Indeed, this is a problem which is not restricted to compounds at very low mixing ratios, since it is possible that the signal at a given m/z, which is ascribed to the protonation of a specific molecule, may contain contributions from another compound present in the analyte, whether known or unknown. However, even if such direct interferences from the analyte do not occur, background signals can emanate from within the instrument. This background can derive from residual quantities of previously studied compounds trapped within the instrument, as well as VOC emissions from the constituent components (O-rings, insulators, etc.). Although usually small, amounting to no more than tens of pptv in a well-designed and maintained instrument, these interfering signals are difficult to eliminate completely. To deal with their effect on quantitative analysis, it is good practice to record a background spectrum in the absence of the analyte VOCs and then subtract this background mass spectrum from each analyte spectrum to deliver a mass spectrum that is representative of the analyte only. To record a background spectrum, pure nitrogen or a purified air mixture known as zero air can be used in place of the analyte gas. However, this is likely to change the water content of the gas entering the drift tube and therefore it is essential to ensure that the analyte and background spectra are expressed in ncps units so that a meaningful subtraction can be performed.

A better way of obtaining a background mass spectrum is to employ a catalytic converter, which can scrub the analyte gas clean of VOCs while leaving the water level unchanged [1]. This can be fitted to the inlet system as a bypass unit. Since some background signals depend on the humidity present in the analyte stream, and specifically the presence or absence of $H_3O^+(H_2O)_n$ clusters, the ability to leave the humidity level intact is a clear advantage for accurate background subtraction.

4.5 Calibration Techniques

To calibrate the performance of a PTR-MS instrument for a given analyte, this analyte needs to be delivered to the instrument at a known concentration with no other VOCs being introduced that could interfere with the ion signals that are associated with that analyte. We shall refer to the analyte as the *calibrant* when it is in the calibration gas. One of the challenges in calibrating a PTR-MS instrument is to deliver the calibrant at one or more known concentrations which are typical of those encountered in actual measurements. Since PTR-MS is routinely employed to detect gases at ppbv or pptv concentration levels, a calibration will only be reliable if it is able to deliver the calibrant at these sorts of levels. This is a challenging requirement and necessitates the use of specialist equipment and/or the purchase of carefully prepared gas standards. Essentially, there are two ways of generating gas mixtures with known concentrations of specific components: *static* or *dynamic* methods.

4.5.1 Static Gas Calibration

In static methods, a gas mixture is prepared prior to its use in calibration. If the aim is to make a mixture with one or more VOCs, perhaps diluted in nitrogen, the mixture can be made by introducing a known quantity of the desired VOC(s) into an appropriate container. The delivered partial pressure of the VOC must be below the vapour pressure of the corresponding liquid, otherwise it will begin to condense. The amount of VOC added can be determined from an accurate measure of the resulting partial pressure. To prepare such mixtures and to establish a reliable concentration requires considerable effort [2]. Such an approach is rarely cost-effective for individual researchers and research teams and therefore it is normally specialist gas suppliers who provide such mixtures. These are typically provided in pressurized cylinders and good suppliers will list the mixture composition along with some indication of the margin of error.

Although the use of a static mixture is relatively simple for calibration purposes, it has a number of weaknesses. First, certain compounds may have a tendency to stick to the walls of the container, or may even undergo gradual chemical decomposition, hence reducing their concentrations by an unknown amount when delivered into the PTR-MS instrument. Wall adsorption is more of a problem for polar compounds and/or low vapour pressure compounds than for high vapour pressure hydrocarbons. This 'stickiness' of certain compounds can be mitigated to some extent by ensuring that the inner walls of the cylinder are constructed from or coated with a material which shows less tendency for compound adsorption. One such material is Silcosteel®, in which a passivation layer is added to stainless steel in order to reduce VOC adsorption. Another option is a Teflon® coating, which also shows reduced adsorption tendencies compared with bare stainless steel.

There are other problems associated with static calibration standards. One problem is the use of a single calibration point to determine the response of an instrument. Quite often one would like to establish that the instrument response is truly *linear* with respect to concentration over a wide range, but a fixed concentration in the gas standard will not allow this. Of course the gas standard mixture could be subjected to a dilution process, whereby an inert balance gas is added to lower the compound concentrations, but in doing this there is added complexity along with the introduction of new uncertainties. Another issue to consider is cost. Carefully prepared gas standards are quite expensive to buy. If the interest is in one or two specific compounds at well-defined concentrations, then the purchase of a gas standard cylinder which has been commercially sourced may be cost-effective. However, if interest is in a wide and changeable set of compounds then *dynamic* calibration techniques may be a better choice.

4.5.2 Dynamic Methods

In dynamic gas calibration, a mixture of the desired concentration is generated *in situ* by the continuous flow of one or more calibrant gases into a stream of an unreactive diluent gas. The diluent is typically ultrapure nitrogen or so-called zero air (a mixture of oxygen and nitrogen in their usual atmospheric ratio but which has been purified such that the total hydrocarbon content is typically <1 ppmv and the moisture level is usually <5 ppmv).

To generate the very low concentration levels typically required in trace gas analysis, which may be ppbv or pptv, it is not practical to do this by simply using dynamic blending of two gas streams using two independent flow controllers. As an example, consider the possibility of making a 1 ppmv mixture, which is a relatively high concentration in trace gas terms. We could do this by blending, say, a 1000 L min^{-1} flow of diluent gas with 0.001 L min^{-1} flow of calibrant gas. However, this would generate a large margin of error in the composition of the resulting mixture because of the huge disparity in flow rates.

Consequently, some means of continuously delivering very small but reliable quantities of calibrant gas with minimal fluctuation is required. The most commonly used method employs *permeation tubes* [2,3]. These contain the calibrant, often in liquid form, and allow a controlled leak into the external environment via diffusion through permeable tube walls. A typical permeation tube is a few centimetres in length and <1 cm in diameter and its walls are constructed from a permeable material such as Teflon, although other polymeric materials are sometimes employed. The tube is filled with a quantity of the calibrant and completely sealed. The emission rate of a compound through the walls of a permeation tube depends on a number of factors including the wall material, the wall thickness, the surface area, the temperature and the calibrant material itself. This loss of material is determined gravimetrically, by periodic weighing, to determine the rate of mass loss of the contents of the tube. This measurement provides the emission rate and, once this quantity is known, the permeation tube can be employed in combination with a diluent gas at a known flow rate to generate a mixture with a known concentration of the calibrant, as illustrated in Figure 4.2.

It is possible to make permeation tubes from scratch, or they can be bought empty from commercial suppliers and can then be filled by the user. However, this requires the user to be equipped with the appropriate gravimetric equipment. Furthermore, gravimetric measurements must be made over relatively long periods of time (several weeks might be typical) to register a sufficient mass loss, so that the permeation rate can be determined with adequate accuracy. For many users, it is far more convenient to purchase tubes already filled

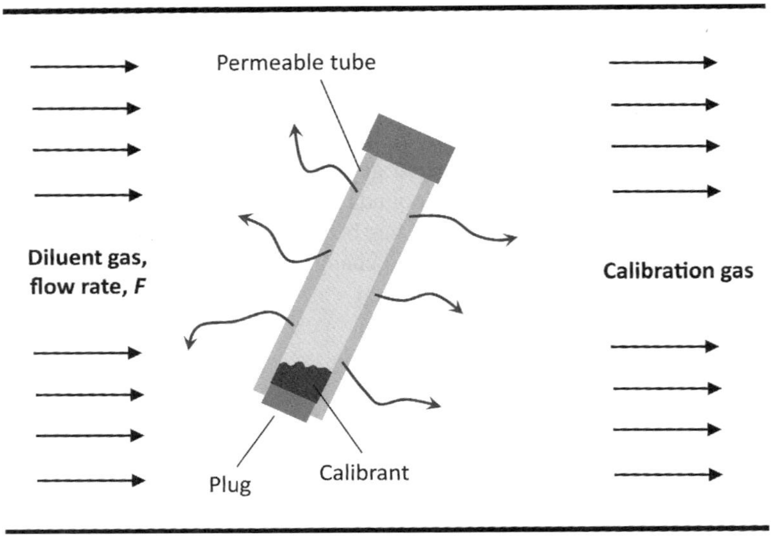

Figure 4.2 *Illustration of the action of a permeation tube. To ensure constant emission of the calibrant the permeation tube must be located inside an oven capable of delivering a stable temperature.*

with the compound of interest and which have emission rates specified by the manufacturer. Permeation tubes sourced commercially are usually accompanied by a certificate stating the permeation rate, for example, in ng min^{-1}. Furthermore, this emission rate is usually traceable to specific standards, such as those provided by the National Institute for Standards and Technology in the United States. Consequently, non-experts in gas calibration are likely to have more confidence in their calibration procedure when using commercially supplied permeation tubes than home-made ones. It should be noted that the quoted emission rate is only valid at a specific temperature, which is recorded on the certificate. Thus, in order to deliver the calibrant at the supplier-specified rate, the permeation tube must be operated at this particular temperature and the temperature should be stable to at least $\pm 0.1^{\circ}$C. Typical emission rate accuracies are $\pm 2\%$ or better. In combination with error margins associated with flow controllers, which deliver the diluent gas, the overall accuracy of a calibration using a permeation tube is typically $\pm 5\%$.

Once the emission rate is known, the concentration of the calibrant gas in ppmv can be calculated from the expression

$$C = \frac{V_{\mathrm{m}} E}{MF} \tag{4.4}$$

where $E =$ emission rate (ng min^{-1}), F is the diluent gas flow rate (mL min^{-1}), M is the molecular mass (g mol^{-1}) of the calibrant and V_{m} is the molar volume (in dm^3 mol^{-1}) of the diluent gas. Since flow rates are normally quoted for standard conditions of 0°C and 1 atmosphere of pressure, the molar volume used is for an ideal gas under these conditions, 22.414 dm^3.

Permeation tube methodology confers several important advantages when compared with the use of static gas calibration mixtures. First, because the mixture is generated dynamically, there is no long-term contact of the gas constituents with a metal wall, as there is in a gas cylinder standard, and this helps to minimize surface adsorption and/or surface-catalysed compound decomposition. Second, the range of usable compounds is very large and could include gases, liquids and solids under ambient conditions. Third, as already mentioned, the release of small quantities of calibrant makes it possible to achieve very low concentrations in the calibration mixture. Furthermore, simply by changing the flow rate of the diluent, the concentration can be changed in a controllable manner, thus allowing the instrument response to be assessed over a range of calibrant concentrations, for example, for checking for a satisfactory linear response. With a single dilution stage, the concentration levels can be set to as low as a few ppbv. To enter into the pptv range a secondary dilution stage is required. With multiple ovens, mixtures containing defined quantities of multiple calibrants can be generated.

A significant disadvantage of dynamic calibration using permeation tubes is the initial investment required in suitable equipment. Commercially supplied permeation tubes are not cheap. However, beyond that, an experimental arrangement is required which must include a temperature-stabilized oven to maintain an accurate and precise temperature for the permeation tube. Furthermore a gas manifold, with one or more flow controllers, is essential. Users themselves can buy and put these elements together, but commercial suppliers of complete calibration units are also available. Some of the commercial units can be fitted with a humidifier to generate a calibrant mixture at known humidity as well as a known calibrant concentration, which is excellent if required but adds to the cost. A relatively long start-up time is another disadvantage of calibration using permeation tubes. It is essential that the permeation tube has reached equilibrium at the chosen temperature in order for the quoted emission rate to be valid. Unfortunately, this can take several hours to achieve when a new permeation tube is inserted into the calibration system. It should also be noted that permeation tubes have a finite lifetime, ranging from a few months to several years. This is determined partly by the quantity of material originally placed within the tube and the permeation rate through the walls: eventually the tube will become empty. To prolong the fill lifetime, permeation tubes are normally stored within a refrigerator when not in use since this lowers the permeation rate. Storage at low temperature can also prolong the lifetime of certain sensitive compounds which may undergo some chemical decomposition if left at or above room temperature for extended periods.

Finally, we note that *diffusion tubes* can be used as an alternative to permeation tubes. These involve the use of a liquid reservoir which is connected to the external environment by a capillary. Vapour will diffuse along this tube at a fixed rate at a fixed temperature. Consequently, a diffusion tube can then be used in a similar manner to a permeation tube. However, diffusion tubes require more operator skill and are suitable for fewer compounds than permeation tubes. Permeation tubes are therefore used far more widely than diffusion tubes in dynamic calibration techniques.

4.5.3 Alternative Dynamic Calibration Procedures

It is not the purpose of this chapter to provide an exhaustive account of gas calibration techniques. However, we note that other techniques have been developed for calibrating

PTR-MS systems, some of which are variants of the techniques already described above. These alternative methods aim for portability, a reasonable cost and applicability to many volatile compounds.

One such approach is the so-called mobile organic carbon calibration system (MOCCS) [4]. Instead of purchasing certified permeation tubes, the tubes are filled by the user with the desired compound. A lengthy gravimetric calibration of the emission rates of these tubes is avoided by calibrating them within the MOCCS system via a total organic carbon measurement, in which CO_2 production is determined by infrared spectroscopic analysis following catalytic oxidation of the calibrant emitted by the permeation tube. Since the carbon comes from the organic content emitted by the permeation tube, it is then possible to calculate the emission rate of the chosen compound. In the long term, cost savings from this technique will come from avoiding purchasing fully certified permeation tubes.

An alternative to working with permeation tubes is dynamic solution injection (DSI) [5]. As the name implies, this involves the injection of the calibrant organic compounds in liquid form into the gas analysis system. The calibrant is dissolved in a suitable solvent, a good choice being cyclohexane since it has a lower proton affinity than water, and therefore will not interfere with PTR-MS analysis. A solution of known concentration is prepared and then a precise mechanical injection system periodically adds small aliquots of the chosen liquid into a mixing chamber, where it is combined with an excess of nitrogen gas. Given that only small amounts of liquid are injected into this flow, full volatilization occurs without the need for any heating. This generates a calibrant mixture of known concentration, providing the initial solution has a known calibrant concentration and also the injection volumes and injection repetition rates are known. Calibrant concentrations as low as a few ppbv have been generated by this procedure but this technique is relatively new and requires further investigation to establish its limitations.

4.6 Effect of Humidity

The determination of the concentrations of compounds by PTR-MS can be affected by humidity. The humidity effect derives from the increase in the quantity of hydrated hydronium clusters, $H_3O^+(H_2O)_n$, within the drift tube as the humidity of the analyte gas increases. Some compounds undergo slower proton transfer from hydrated hydronium clusters when compared with H_3O^+, while some will not react at all (see Section 2.2.4.2). Raising the *E*/*N* of the drift tube can reduce the proportion of hydrated hydronium clusters but there are limits as to how far this can be pushed before other, unwanted, fragmentation effects occur. Raising the *E*/*N* excessively is also undesirable from the point of view of detection sensitivity, since it reduces the reaction time for the analyte gas in the drift tube.

Sample humidity will affect those compounds that do not react with $H_3O^+(H_2O)$. An illustration of this has been provided by Warneke *et al.*, who used benzene and toluene as the target compounds [6]. Both benzene and toluene will accept a proton from H_3O^+ but proton transfer from $H_3O^+(H_2O)$ is thermodynamically forbidden. The sensitivity towards these compounds declined at a fixed *E*/*N* (106 Td in this specific case) as the relative humidity was increased from 20% to 100%. This occurs because more of the H_3O^+ is tied up in the form of unreactive $H_3O^+(H_2O)$ clusters as the humidity is increased. Deduction of the $H_3O^+/H_3O^+(H_2O)_n$ abundance ratio, which can be used to assess the impact of humidity on the detection sensitivity, is complicated because some PTR-MS instruments, such as the one employed by Warneke *et al.*, include a collision-induced dissociation (CID) region at the downstream end of the drift tube. The purpose of this CID region is to raise the *E/N* over a short distance and thus help to simplify the mass spectrum by removing cluster ions. Not surprisingly, Warneke and co-workers showed that an elevated *E/N* in the CID region relative to the rest of the drift tube leads to an overestimate of the amount of H_3O^+, since it includes contributions from fragmented $H_3O^+(H_2O)_n$ ions, which if not taken into account will result in an underestimate of the concentration of the VOC(s) under investigation.

Variation in sample humidity is also a problem for those compounds whose proton affinities are only marginally above that of water. It is tempting to think of proton transfer as a one-way reaction, but in reality Reaction 1.5 is reversible and thus deprotonation of a VOC is also possible by reaction with water. For most VOCs, the high exothermicity of the proton transfer Reaction 1.5 means that the reverse reaction ($MH^+ + H_2O \rightarrow H_3O^+ + M$) is extremely slow, and thus equilibrium overwhelmingly favours the protonated VOC. However, there are a few examples where this is not the case. One of the best known is formaldehyde, whose proton affinity is only 22 kJ mol^{-1} above that of H_2O. The proton transfer reaction between H_3O^+ and formaldehyde is fast, with a rate coefficient of 3×10^{-9} cm^3 s^{-1} at 300 K [7]. At the same temperature, the rate coefficient for the reverse reaction is some two orders of magnitude slower, so the reversibility of the reaction might be thought to be unimportant. However, the rate coefficient for the reverse reaction is strongly temperature dependent and increases as the temperature is raised. Since the effective ion temperature is elevated in PTR-MS by the applied electric field in the drift tube (see Sections 3.4.3 and 3.4.4), the rate coefficient for the reverse reaction will increase with increasing *E/N* [8]. Furthermore, increased humidity will also enhance the rate of the reverse reaction since it increases the concentration of water vapour. Consequently, it is vital to consider the effect of *E/N* and humidity when attempting a determination of the concentration of formaldehyde via PTR-MS [9].

The effect of humidity on the determination of concentrations is compound specific. To incorporate the humidity factor into the calibration process de Gouw and co-workers [10] have suggested that the ratio

$$\frac{i(\mathrm{MH^+})}{i(\mathrm{H_3O^+}) + X_\mathrm{M} \times i(\mathrm{H_3O^+(H_2O)})}$$

be used in place of $i(MH^+)/i(H_3O^+)$ in Equations 4.1 and 4.2. The quantity X_M is a multiplier whose value can be determined from calibration measurements at low and high humidities for a given compound for a fixed temperature and *E/N*. The rationale here is that the dependence on humidity is roughly linear and the dominant contributor to the humidity effect is the $H_3O^+(H_2O)$ ion, rather than larger hydrated hydronium clusters. In effect, by determining the value of X_M for a given compound, the exploitation of the above ratio largely removes any humidity contribution. The value of X_M is determined by both the efficiency of proton transfer from $H_3O^+(H_2O)$ to the analyte molecule M and the transmission of $H_3O^+(H_2O)$ through the mass spectrometer relative to that of H_3O^+. In a quadrupole mass spectrometer, the transmission of $H_3O^+(H_2O)$ is typically a factor of two higher than that of H_3O^+, and so if M is equally as reactive with $H_3O^+(H_2O)$ as it is with H_3O^+, then we would expect $X_M \approx 0.5$. However, instruments with different types of quadrupole mass filters, or some other type of mass spectrometer such as a time-of-flight system, may have different values of X_M because their transmission as a function of *m/z* can differ, and therefore this factor X_M needs to be determined for individual instruments.

In terms of the reactivity contribution, significant differences can occur from one type of analyte molecule to another. For example, owing to its high proton affinity (812 kJ mol^{-1}) acetone reacts via a proton transfer process as readily with $H_3O^+(H_2O)$ as it does with H_3O^+ and therefore on a quadrupole-based PTR-MS system a value of $X_M \approx 0.5$ will be appropriate. However, X_M has a much lower value for toluene and benzene, since both possess a higher proton affinity than H_2O but a lower proton affinity than $(H_2O)_2$ (see Section 2.2.4.2 for a more detailed discussion of the reactivity of toluene with $H_3O^+(H_2O)$). Note that the value of X_M may also depend on the value of *E/N* and the temperature of the drift tube. Toluene is a good example of this since the rate of its endothermic reaction with $H_3O^+(H_2O)$ will increase as the ion–molecule collision energy increases. An illustration of the effect of humidity is shown in Figure 4.3.

4.7 Accuracy, Precision and Limit of Detection

In terms of characterizing, the ultimate performance of PTR-MS as a measurement technique, it is useful to quantify the *accuracy* and *precision* of any quantitative determination. The reader is reminded of the definitions of these two terms: accuracy reveals how close a series of measurements are to the true value of the desired quantity, while precision is a measure of how reproducible each consecutive measurement is. Thus if multiple measurements of a trace gas concentration are made under identical conditions, the best estimate of the concentration will be the *mean* of these values. However, there is no guarantee that the mean value will be close to the true value, since *systematic* errors may be incorporated in this determination. As an example, if the rate coefficient used in the application of Equation 4.1 differs from the true value by a factor of two, then this relatively large error will be incorporated into the determination of the gas concentration. All of the potential sources of error for concentration determinations discussed in Section 4.4 are sources of systematic error.

The reproducibility of measurements, as reflected in the precision, is the product of random variations. In PTR-MS, the primary source of random variations is instrumental noise arising from fluctuations in the ion count rate. Imagine a situation where the conditions within the instrument, such as gas pressure and electrode voltages, are perfectly steady, but the ion count rate at the detector is relatively low. One can imagine that there is an intrinsic production rate of protonated organic ions, MH^+, which could be measured reproducibly over a long period of time. However, measurements made over a short period of time will show fluctuations. Thus one measurement might detect, say, 10 ions in a fixed acquisition period while the next might measure 8 ions. This kind of statistical fluctuation is well understood and is described by Poisson statistics.[1] A special case of Poisson statistics is

[1] Poisson statistics applies when an event occurs at a known average rate but each individual occurrence is independent of the time elapsed between any preceding event(s). Radioactive decay is another example where Poisson statistics applies.

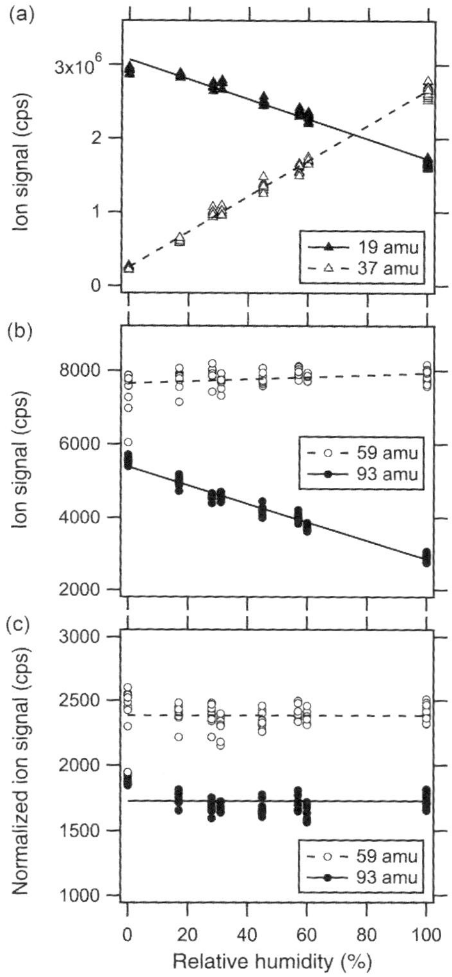

Figure 4.3 *An illustration of the impact of humidity on the detection of acetone (protonated signal at m/z 59) and toluene (protonated signal at m/z 93). Panel (a) shows an approximately linear decrease in [H_3O^+] with humidity and a corresponding linear increase in [$H_3O^+(H_2O)$]. Panel (b) shows the corresponding effects on the protonated acetone and toluene signals and panel (c) shows the response when the humidity effect was included via the factor X_M. A value of $X_M \approx 0.5$, which reflects the relative transmission of $H_3O^+(H_2O)$ to H_3O^+, was appropriate for acetone since it reacts with both H_3O^+ and $H_3O^+(H_2O)$ at roughly the same rate. On the other hand, toluene is much less reactive with $H_3O^+(H_2O)$ than with H_3O^+ and taken together with the transmission factor an $X_M \approx 0.1$ was found for this molecule. The original figure was kindly supplied by J. A. de Gouw. Reproduced with permission from Reference 11. Reproduced with permission from [11]. Copyright © 2006 John Wiley & Sons, Ltd.*

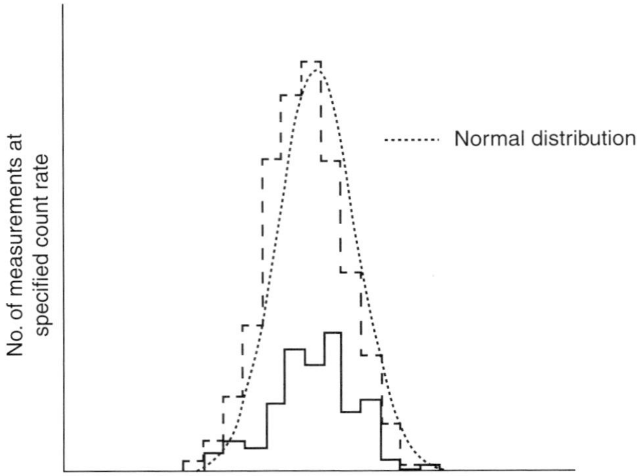

Figure 4.4 *An illustration of the statistical (Poissonian) nature of ion counting for moderate count rates. The data shown by the dashed line correspond to a higher overall signal level than for the solid line. The former is a reasonable match to a Gaussian (normal) distribution whereas the latter is a relatively poor approximation.*

where the count rate is sufficiently large such that a *Gaussian distribution*, also known as a *normal distribution*, applies, as illustrated in Figure 4.4.

For a Poisson distribution, the degree of random scatter in the measured count rate is reflected in the *standard deviation*, σ, which is given by

$$\sigma = \frac{\sqrt{N}}{t} = \sqrt{\frac{\text{cps}}{t}} \tag{4.5}$$

where $N =$ total number of counts, t is the measurement time and 'cps' is the count rate per second. The standard deviation is an indication of the precision of a set of measurements, and in the usual language of statistical distributions we can specify *confidence limits*, which express the probability that, in the hypothetical case where there is no systematic error, the mean value lies within a certain range of the true value. A detailed theoretical analysis shows that $\pm\sigma$ represents a 68% confidence limit and $\pm 2\sigma$ corresponds to a 95% confidence limit [12].

As an illustration, suppose that a PTR-MS measurement is carried out continuously over 10 s and yields a cps of 50. From Equation 4.5, we calculate $\sigma = 2.2$. This reveals that, given no other changes to the experiment, there is a 95% probability that in another 10 s measurement the signal would lie within the range 45.6–54.4 cps, providing Poissonian noise dominates. Thus a measurement which gave a cps of 40 cps would lie well outside of this range. In such circumstances, it would be valid to claim that this is the consequence of a genuine decline in the concentration of the trace gas, rather than a result of insufficient precision. A simple statistical analysis of this type can therefore define the effective time resolution of a continuous sequence of measurements. Clearly if the count rate is high then the effective time resolution improves accordingly (although in PTR-MS we would still ultimately be limited by the intrinsic response time of the drift tube, which was discussed at the end of Section 3.4.1).

For a given set of data acquisition conditions, each compound in a gas mixture will have a minimum possible detectable concentration, also known as the limit of detection (LOD). This is normally defined as the case where the signal amplitude in the appropriate mass channel exceeds the background noise level by a factor of three, that is, the LOD corresponds to a signal/noise ratio of 3.[2] Compounds with similar masses and similar proton transfer kinetics should possess similar LODs. However, this does not always transpire in practice. In particular, once very low concentrations are approached, such as a VMR at a low pptv level, then background signals due to chemical impurities present in the instrument can be found in almost every mass channel, as already discussed in Section 4.4. As a result, the fluctuating analyte signal needs to be distinguished from the fluctuating background signal.

The LOD for compound M can be expressed mathematically as

$$\mathrm{LOD_M} = \frac{3 \times \mathrm{VMR_M}}{(S/N)} \tag{4.6}$$

where S is the signal level under background conditions (no added analyte) and N is the corresponding noise level, which can be equated with the standard deviation ($\sigma_{\text{background}}$) in the background signal level from a series of measurements. An alternative way of expressing Equation 4.6 is as

$$\mathrm{LOD_M} = \frac{3\sigma_{\text{Background}}}{\varepsilon_\mathrm{M}} \tag{4.7}$$

where ε_M is the instrument sensitivity for compound M in, for example, counts per second per ppbv, as derived from a calibration plot of the type shown in Figure 4.1: this would yield the LOD in ppbv. The quantity $\sigma_{\text{background}}$ is the error margin associated with the y-axis intercept in a calibration plot of the type shown in Figure 4.1. This error margin, the standard deviation in the y-axis intercept, can be obtained from a linear regression analysis.

For reasonable data acquisition times of a few seconds, the best attainable LODs via PTR-MS are on the order of a few tens of pptv, with significant variations from compound to compound [10].

4.8 Validation of PTR-MS

As a means of monitoring gas compositions, some independent verification of the performance of PTR-MS would help to build confidence in the technique. Such verification studies have been carried out and involve comparisons with other gas analysis techniques, particularly gas chromatography (GC). It is important to recognize at the outset that PTR-MS and GC techniques will not always agree, since PTR-MS using a mass spectrometer may not be able to separate compounds with the same nominal mass, whereas GC–MS almost always can because of the additional discrimination provided by compound separation on the chromatographic column. An additional complication is that fragment ions from

[2] A signal/noise ratio of two is sometimes chosen in the definition of LOD, but we prefer the more conservative value of three.

heavier compounds may contribute to the protonated parent signal of a lighter compound in PTR-MS. Nevertheless, for many important trace constituents of air the news is good.

The most comprehensive validation studies have been reported by the NOAA group in Boulder, Colorado, who have exhaustively compared GC and PTR-MS measurements for a large number of compounds [10, 13]. Reinforcing the point made in the previous paragraph, the performance is compound specific but for many important molecules, such as methanol, acetone, isoprene and benzene, the PTR-MS and GC–MS data agreed within the measurement uncertainties. The reader is encouraged to consult the original papers if they wish to check on the response for other compounds, where interferences may become more significant. Clearly, there is no set accuracy that can be assigned to PTR-MS measurements given the compound-specific nature, but as a general statement, when severe interferences are avoided, the accuracy of measurements is likely to be in the range of ±20% or better when calibration is employed. Even with compounds which show little or no fragmentation in PTR-MS, such as benzene, there are limitations in improving the precision, particularly when working at very low concentrations, where background signal counts are highly significant. For compounds present at the low ppbv level, the best achievable precision will typically be in the range of 10–15%, although this can be improved to some extent by increasing the data acquisition period in the appropriate mass channel.

References

1. de Gouw, J. A., Warneke, C., Karl, T. *et al.* (2003) Sensitivity and specificity of atmospheric trace gas detection by proton-transfer-reaction mass spectrometry. *Int. J. Mass Spectrom.* **223–224**, 365.
2. Barratt, R. S. (1981) The preparation of standard gas mixtures. A review. *The Analyst* **106**, 817.
3. Scaringelli, F. P., O'Keeffe, A. E., Rosenberg, E., Bell, J. P. (1970) Preparation of known concentrations of gases and vapors with permeation devices calibrated gravimetrically. *Anal. Chem.* **42**, 871.
4. Veres, P., Gilman, J. N., Roberts, J. M. *et al.* (2010) Development and validation of a portable gas phase standard generation and calibration system for volatile organic compounds. *Atmos. Meas. Tech.* **3**, 683.
5. Jardine, K. J., Henderson, W. M., Huxman, T. E., Abrell, L. (2010) Dynamic solution injection: a new method for preparing pptv–ppbv standard atmospheres of volatile organic compounds. *Atmos. Meas. Tech.* **3**, 1569.
6. Warneke, C., van der Veen, C., Luxembourg, S. *et al.* (2001) Measurements of benzene and toluene in ambient air using proton-transfer-reaction mass spectrometry: calibration, humidity dependence, and field intercomparison. *Int. J. Mass Spectrom.* **207**, 167.
7. Bohme, D. K., Mackay, G. I., Tanner, S. D. (1979) An experimental study of the gas-phase kinetics of reactions with hydrated hydronium(1+) ions ($n = 1$–3) at 298 K. *J. Am. Chem. Soc.* **101**, 3724.
8. Hansel, A., Singer, W., Wisthaler, A. *et al.* (1997) Energy dependencies of the proton transfer reaction $H_3O^+ + CH_2O \rightleftharpoons CH_2OH^+ + H_2O$. *Int. J. Mass Spectrom. Ion Processes* **167–168**, 697.

9. Karl, T., Jobson, T., Kuster, W. C. *et al.* (2003) Use of proton-transfer-reaction mass spectrometry to characterize volatile organic compound sources at the La Porte super site during the Texas Air Quality Study 2000. *J. Geophys. Res.* **108**, D4508.
10. de Gouw, J. A., Goldan, P. D., Warneke, C. *et al.* (2003) Validation of proton transfer reaction-mass spectrometry (PTR-MS) measurements of gas-phase organic compounds in the atmosphere during the New England Air Quality Study (NEAQS) in 2002. *J. Geophys. Res.* **108**, 4682.
11. de Gouw, J. A., Warneke, C. (2007) Measurements of volatile organic compounds in the earth's atmosphere using proton-transfer-reaction mass spectrometry. *Mass Spectrom. Rev.* **26**, 223.
12. Mandel, J. (1985) *The Statistical Analysis of Experimental Data*, Dover Publications.
13. Warneke, C., de Gouw, J. A., Kuster, W. C. *et al.* (2003) Validation of atmospheric VOC measurements by proton-transfer-reaction mass spectrometry using a gas-chromatographic preseparation method. *Environ. Sci. Technol.* **37**, 2494.

Section 2
Applications

5

PTR-MS in the Environmental Sciences

5.1 Background

A wide variety of VOCs are emitted into the atmosphere by natural and anthropogenic processes and these can influence the chemistry of the atmosphere. For example, tropospheric VOCs play a role in the photochemical formation of ozone and their oxidation results in low vapour pressure compounds that can partition onto particles to form secondary organic aerosols.[1] Globally, biogenic emissions are far more dominant than anthropogenic emissions and the major sources of these VOCs are the world's forests and oceans. It is only in densely populated urban areas that anthropogenic VOC emissions can dominate over those of biogenic origin. Some VOCs (e.g. methanol) have both anthropogenic and natural sources and for these compounds it may be difficult to determine their origin.

Given their importance in the atmosphere, accurate atmospheric measurements of VOCs, in terms of their identification, concentrations and evolution, covering a wide range of environments and altitudes, are important for a quantitative understanding of atmospheric chemical processing.

Owing predominantly to its fast response, high sensitivity (sub ppbv in seconds), the lack of reaction of H_3O^+ with the major constituents of air, and the large range of VOCs that it can detect, it is not surprising that soon after its development PTR-MS was applied to the detection of VOCs in the Earth's atmosphere. Substantial reviews of atmospheric and environmental applications of PTR-MS and details on its capabilities as an atmospheric sensor have been provided by Hewitt *et al.* [1], de Gouw and Warneke [2] and Blake *et al.* [3], which cover atmospheric research with PTR-MS from the first studies in the mid-1990s

[1] Inorganic and organic aerosols that are directly emitted into the atmosphere are known as *primary* aerosols, for example, aerosols from ocean spray, volcanic eruptions or combustion. Those aerosols that are formed in the atmosphere through a sequence of chemical reactions are referred to as *secondary* aerosols, and they account for a major component of tropospheric particulates.

Proton Transfer Reaction Mass Spectrometry: Principles and Applications, First Edition. Andrew M. Ellis and Christopher A. Mayhew.

up to approximately the end of 2008. Drawing extensively from publications, the aim of this chapter is to provide a comprehensive review of PTR-MS research in the environmental sciences which will cover studies from the early 1990s up to the present day.

PTR-MS is now a well-established analytical technique in the environmental sciences, having a major impact on the detection and understanding of, for example, pollution outflow, biomass burning and sources of anthropogenic and biogenic VOCs in the atmosphere. However, it should be appreciated that when it was first introduced the PTR-MS findings were revolutionary, particularly since the fast time response enabled rapid changes in VOC emissions to be monitored. GC–MS techniques, which had been previously used to detect VOCs in the environment, are simply too slow to capture transient events on the minute-by-minute or faster timescale.

PTR-MS environmental research started with ground-based measurements, the details of which were first published in 1997. These first studies investigated the diurnal variations of various VOCs in ambient air on the western outskirts of the city of Innsbruck in Austria [4, 5]. These early atmospheric demonstrations, showing the applicability of PTR-MS to monitor temporal variations of VOC concentrations, were quickly followed up by pioneering airborne studies, starting in 1998 with measurements of primary organics and their oxidation products over the tropical forests of Surinam [6–8]. Since those groundbreaking studies there have been a number of other airborne measurements, including the monitoring of aromatic and oxygenated VOCs over the Deepwater Horizon oil spill in the Gulf of Mexico, which occurred from April to August 2010 [9]. In that study, a narrow plume of hydrocarbons was observed downwind of the spill, which was attributed to the evaporation of oil from the sea's surface. Mixing ratios of up to 10 ppbv of volatile compounds, which included C_9 aromatics, were determined.

As well as providing an excellent field instrument, PTR-MS can also be coupled with simulation chambers to explore the underlying atmospheric chemistry. One illustrative study was reported by Wisthaler *et al.*, who used PTR-MS to measure acetone and other gas-phase products from the oxidation reaction of OH with α- and β-pinene in air and in the presence of NO_x [10]. Another example which is related to atmospheric chemistry, is the work by Northway *et al.*, who found unexpectedly high levels of acetaldehyde during aircraft measurements which increased with the ambient ozone level [11]. This was attributed to reactive organic compounds present in the PTR-MS sampling inlet. From this the researchers suggested that any field measurements of acetaldehyde should be checked for possible ozone-induced artefacts in both clean and polluted environments. Whilst not strictly using PTR-MS to investigate atmospheric processes, their study highlights the importance of understanding the chemistry that may occur in the PTR-MS instrument if atmospheric data are to be correctly interpreted.

As mentioned earlier, an important atmospheric quantity to determine is the concentration of a VOC of interest. However, unless special techniques are utilized PTR-MS does not provide a unique indicator for the identity of trace gases in an air sample, because it can only provide a nominal m/z value, that is, isomeric (and isobaric if high resolution mass spectrometers are not used) compounds cannot usually be distinguished.[2] Furthermore,

[2] Of course isomers have identical masses and therefore cannot be distinguished using the m/z of the protonated species. It may be possible to differentiate isomers in PTR-MS if they have different fragmentation patterns for a given E/N or over a range of E/N. Alternatively, a different reagent ion other than H_3O^+ could be used, for example, NO^+ or $O_2{}^+$, which may produce different product ions upon reaction with different isomers.

product ions resulting from fragmentation, clustering, and secondary ion–molecule reactions can inhibit accurate assignments and thereby make it challenging to determine the concentrations of some VOCs in the diverse chemical environment of the atmosphere.

The problems mentioned above limit the analytical use of PTR-MS. However, these are more than compensated by the many good qualities of the technique, such as its high reliability (allowing measurements to be taken over extended periods of time), high sensitivity, the large range of detectable VOCs, its compactness and robustness, and its rapid response time, all of which have made PTR-MS ideal for deployment on a range of atmospheric measurement platforms, including ground-based and field site measurement stations [12–14], vehicles [15], ships [16] and, as already mentioned, aircraft [9, 17–20]. Nevertheless, to acquire the actual atmospheric composition of trace compounds, to have an improved knowledge of their effects on the atmospheric chemistry and to provide a greater understanding about the sources and sinks of atmospheric VOCs, it is important that VOCs are identified with a high level of confidence and that their concentrations as determined by the PTR-MS technique are reliable.

Towards this goal, there have been extensive studies that have compared PTR-MS measurements of atmospheric VOCs with those obtained by other atmospheric analytical techniques, such as GC–MS [17], GC-FID [21–26], atmospheric pressure chemical ionization mass spectrometry (AP-CIMS) [27], differential optical absorption spectroscopy (DOAS) [24, 28] and Fourier transform infrared (FTIR) spectroscopy [17, 29], in addition to offline sampling methods coupled to GC analysis [30–34]. These studies have shown that PTR-MS is capable of accurately measuring concentrations of VOCs providing that there is no contribution to the *m/z* of interest in a mass spectrum by interfering species. If other compounds are present in the atmospheric sample which can lead to ions (protonated parent compounds, cluster ions or fragment ion species) at the nominal *m/z* of the protonated VOC of interest, then the lack of specificity associated with PTR-MS requires that the actual identity of the compound still needs to be confirmed by other analytical techniques, such as GC–MS.

There have been numerous PTR-MS studies of the atmosphere. A convenient way to provide a comprehensive summary of these activities is to adapt and extend tables given in the review articles by de Gouw and Warneke [2] and Blake *et al.* [3], which provide a general but incomplete list of the compounds identified in ambient air. This combined table is provided in Table 5.1.

The compounds listed in the table highlight the wide range of VOCs that can be detected with PTR-MS, including nitrogen-containing and oxygenated VOCs when using H_3O^+ as the probe ion. As discussed above, it should be noted that unequivocal identification of the compounds given in Table 5.1 using PTR-MS alone is not possible, but a number of studies have used other analytical methods, such as combined GC-PTR-MS, to help with the assignment [35, 36]. Of the compounds listed in Table 5.1, nitrogen-containing compounds play a significant role in a range of oxidation processes in the atmosphere [37]. Oxygenated VOCs, such as alcohols, aldehydes, ketones and carboxylic acids, are also ubiquitous in the atmosphere [38], with many resulting from the photooxidation of VOCs. These compounds contribute to the formation of secondary organic aerosols and a number of research groups have demonstrated the use of PTR-MS for the detection of their precursors [28, 39–45].

While Table 5.1 provides an illustration of the wide range of atmospheric VOCs that can be detected, a glance at the species listed shows that several important atmospheric compounds cannot be detected using H_3O^+ as the reagent ion. These include methane, the

Table 5.1 *Compounds detected in the atmosphere by PTR-MS using H_3O^+ as the reagent ion and assigned by the m/z used for identification (not necessarily the protonated parent value)*

m/z[a]	Proposed compound (molecular formula)	References
18	Ammonia (NH_3)	[17, 26, 29]
28	Hydrogen cyanide (HCN)	[17, 29, 49, 118, 137]
31	Formaldehyde (CH_2O)	[17, 20, 26, 28, 29, 48, 67, 118, 136, 137]
33	Methanol (CH_4O)	[6, 7, 12–17, 19, 23, 26, 29, 32, 33, 41, 42, 44, 62, 64, 66, 68, 76, 79, 82, 84–86, 89, 90, 94, 99, 100, 104–106, 108, 109, 111, 115, 116, 118, 120, 123, 125, 126, 128–130, 136, 137, 156, 162–164, 176]
42	Acetonitrile (C_2H_3N)	[6, 7, 12, 16, 17, 19, 27, 29, 32, 33, 44, 61, 62, 64–66, 94, 111, 115, 116, 118, 126, 129, 137, 156, 161–166]
43	Propene (C_3H_6)	[22, 26, 29, 118]
	Fragment from 2-propanol (C_3H_8O)	[6]
45	Acetaldehyde (C_2H_4O)	[6, 12, 14–17, 22, 23, 26, 29, 33, 42, 45, 66, 67, 70, 76, 77, 79, 84–86, 89, 90, 94, 99, 100, 104, 105, 109, 111, 115, 118, 120, 123, 125, 126, 128, 129, 136, 156, 163, 164, 176]
47	Formic acid (CH_2O_2)	[7, 17, 26, 29, 79, 128]
	Ethanol (C_2H_6O)	[14, 26, 86]
54	Acrylonitrile (C_3H_3N)	[17, 29, 118]
56	Propanenitrile (C_3H_5N)	[17, 29]
57	Acrolein (C_3H_4O)	[17]
	Hexenal fragments	[104, 105]
	Butanol (C_4H_9OH)	[105, 118]
	Butenes (C_4H_8)	[15, 119]
	Methyl *tert*-butyl ether ($C_5H_{12}O$)	[15, 119]
59	Acetone (C_3H_6O)	[6, 7, 12, 14–17, 19, 22, 23, 26, 27, 29, 32, 33, 42–44, 61, 62, 65–67, 76, 77, 79, 84–86, 89, 90, 94, 99, 100, 104–106, 110, 111, 115, 116, 120, 123, 125, 126, 128, 129, 137, 156, 162–164, 195]
61	Acetic acid ($C_2H_4O_2$)	[7,14–17,26,29,33,76,77,79,94,118,128,136,156]
	Glycolaldehyde ($C_2H_4O_2$)	[17, 29]
	Methyl formate ($C_2H_4O_2$)	[26]
63	Dimethyl sulfide (C_2H_6S)	[2, 6, 12, 17, 31, 33, 65, 115, 116, 164]
68	Pyrrole (C_4H_5N)	[17]
69	Isoprene (C_5H_8), furan (C_4H_4O)	[6–8, 12, 14, 16, 17, 21–23, 26, 29, 31, 61, 64, 68, 79, 82, 92–94, 99, 100, 103, 104, 111, 115, 118, 123, 125, 126, 129, 130, 136, 156, 162–164, 176]
	1-penten-3-ol ($C_5H_{10}O$)	[105]
	2-methyl-3-butene-2-ol ($C_5H_{10}O$)	[56, 73, 98]
71	Methyl vinyl ketone, Methacrolein (C_4H_6O)	[2, 6–8, 12, 14, 16, 17, 23, 26, 33, 42, 61, 64, 76, 79, 82, 85, 93, 94, 100, 103, 111, 115, 118, 156, 162]
	Pentanol ($C_5H_{12}O$)	[105]
73	Methyl ethyl ketone (C_4H_8O)	[6, 7, 12, 16, 17, 26, 33, 67, 89, 105, 111, 128, 129, 136, 156]
	Methyl propanaol ($C_4H_{10}O$)	[17]

Table 5.1 *(Continued)*

m/z^a	Proposed compound (molecular formula)	References
75	Hydroxyacetone ($C_3H_6O_2$)	[7, 17, 29, 63]
	Methyl acetate ($C_3H_6O_2$)	[17, 105]
77	Peroxyacetyl nitrate (PAN) ($C_2H_3NO_5$). (A secondary ion–molecule reaction of protonated PAN with water leads to the product ion $CH_3C(O)OOH_2^+$)	[12, 16, 44, 50, 156]
79	Benzene (C_6H_6)	[12, 15–17, 19, 21–24, 26, 29, 30, 33, 45, 61, 62, 64, 111, 115, 117, 118, 120, 123, 125, 136, 156, 164]
	Ethylbenzene ($C_6H_5C_2H_5$)	[17]
81	Monoterpenes ($C_{10}H_{16}$) detected via fragment ($C_6H_9^+$), hexenal ($C_6H_{10}O$) detected via fragment ion or fragment ion from other C_6-compounds	[2, 12, 14, 16, 23, 25, 33, 34, 42, 75, 77, 80, 86, 90, 91, 102, 104, 156, 250]
83	Fragment of $C_{6/8}$-compounds: hexenol, hexanal, ($C_6H_{12}O$); hexenyl acetate ($C_8H_{14}O_2$) Methylfuran (C_5H_6O), isoprene hydroxyl carbonyls	[6, 7, 12, 14, 17, 64, 78, 94, 100, 104, 105, 118, 250]
85	Pentanal ($C_5H_{10}O$)	[14]
	Ethyl vinyl ketone (C_5H_8O)	[64, 91]
	Pentenone ($C_5H_{10}O$)	[104]
87	2-methyl-3-buten-2-ol ($C_5H_{10}O$), C_5 carbonyls, methacrylic acid ($C_4H_6O_2$)	[7, 12, 14, 17, 26, 42, 56, 73, 76, 84, 104]
89	Ethyl acetate ($C_4H_8O_2$)	[15]
91	Peroxypropionyl nitrate (PPN) ($C_3H_5NO_5$) (fragment ion $CH_3CH_2C(O)OOHH^+$)	[50]
93	Toluene (C_7H_8)	[12, 15–17, 19, 21–26, 29, 30, 33, 45, 62, 64, 100, 117, 118, 120, 123, 125, 126, 128–130, 136, 156, 164]
95	2-vinylfuran	[7]
	Phenol (C_6H_6O)	[7, 17, 24, 29, 68]
97	Fufural ($C_5H_4O_2$)	[14, 17]
99	Hexanal ($C_6H_{12}O$)	[7, 12, 14, 64, 80, 90, 102, 104]
	3-methyl-3-peten-2-one ($C_6H_{10}O$)	[105]
101	Isoprene hydroperoxides ($C_5H_{10}O_3$)	[6–8, 63]
	cis-3-hexenol/hexenol ($C_6H_{12}O$)	[12, 104]
	Hexanal ($C_6H_{12}O$)	[12, 90, 104]

(continued)

Table 5.1 (Continued)

m/z[a]	Proposed compound (molecular formula)	References
103	Peroxymethacrylic nitric anhydride (MPAN, $C_4H_5NO_5$)	[50]
105	Styrene (C_8H_8)	[16, 22, 24, 45, 118, 129]
	Peroxy isobutyryl nitrate (piBN, $C_4H_7NO_5$)	[16]
107	C_8 aromatics (C_8H_{10})	[12, 15–17, 21, 22, 24, 26, 29, 33, 45, 117, 118, 120, 126, 128, 130, 136, 156]
109	Cresols (C_7H_8O)	[24]
115	Heptanal ($C_7H_{14}O$)	[26, 64]
121	C_9 aromatics (C_9H_{12})	[12, 15–17, 21, 22, 24, 26, 33, 45, 117, 120, 128, 136, 156]
129	Octanal ($C_8H_{16}O$)	[64]
	Naphthalene ($C_{10}H_8$)	[118, 136]
135	C_{10} aromatics ($C_{10}H_{14}$)	[45, 96, 120,]
137	Monoterpenes ($C_{10}H_{16}$)	[2,12,14,16,23,25,26,33,34,42,70,75,77,79,82, 85,86,90,91,94,98–100,104,115,123,156,176]
139	Nopinone ($C_9H_{14}O$)	[42]
143	Nonanal ($C_9H_{18}O$)	[26, 64]
	Hexenyl acetate ($C_8H_{14}O_2$)	[104]
149	Methyl chavicol ($C_{10}H_{12}O$)	[42, 74, 77, 94, 100, 101]
151	Pinonaldehyde ($C_{10}H_{16}O_2$)	[42]
153	Terpenoid ($C_{10}H_{16}O$), e.g. camphor	[14, 176]
163	C_{12} aromatics ($C_{12}H_{18}$)	[42]
205	Sesquiterpenes ($C_{15}H_{24}$)	[14, 77, 94, 100, 176]

Source: Adapted from References 2 and 3.
[a]On transfer of a proton from H_3O^+ to a VOC the dominant peak in the mass spectrum often arises from the protonated parent molecule. However, fragmentation may be significant for some compounds upon protonation, for example, monoterpenes, and in such cases a fragment ion is used as a marker for the particular compound(s). Where this occurs the fragment ion m/z has been provided in the table.

alkanes C_nH_{2n+2} for $n \leq 7$, and the unsaturated hydrocarbons ethene and acetylene. This lack of reactivity with H_3O^+ is simply because the proton affinities of these compounds are smaller than that of water, as was discussed in Section 2.2.4.1. To observe compounds which are unreactive with H_3O^+ using a PTR-MS instrument, a different gas/vapour must be introduced into the hollow cathode ion source (see Section 2.3).

Problems of detection also arise for compounds whose proton affinities are close to that of water (PA(H_2O) = 691 kJ mol^{-1}). This occurs for several compounds that may be significant in an atmospheric environment, including hydrogen sulfide (PA(H_2S) = 705 kJ mol^{-1}), hydrogen cyanide (PA(HCN) = 713 kJ mol^{-1}) and formaldehyde (PA(HCHO) = 712 kJ mol^{-1}) (the PAs for the above molecules have been taken from reference [46]). (Generally, inorganic molecules have proton affinities below that of water but H_2S and HCN are clear exceptions to this rule.) The closeness of the proton affinities of H_2S, HCN and HCHO to that of water and the considerably greater concentration of water than those of trace compounds in the drift tube reactor of a PTR-MS lead to efficient back reactions, that is, the reverse of Reaction 1.5. The reverse reaction complicates the analysis, and in particular the determination of the concentration of the compound of interest, because the signal intensities of the protonated parent molecule will be dependent on the temperature and humidity in the drift tube reactor (see also Section 4.2 for a discussion of the effect of humidity). For example, a series of laboratory studies of biomass-burning emissions found that PTR-MS measurements of HCHO and HCN led to concentrations which were factors of 3 and 12 lower, respectively, than those obtained using a FTIR spectrometer as a result of the back reaction channel [29].

In a later study investigating biomass emissions from controlled laboratory fires it was again concluded that PTR-MS measurements underestimated the concentration of HCN, and in this study by roughly a factor of 5 in comparison to values determined by FTIR spectroscopy [17]. From the collected field data, it was found that the HCHO ratios between PTR-MS and FTIR measurements varied significantly with ambient atmospheric humidity, ranging from ~0.2 at the lowest humidity to 0.05 at the highest humidity. It was concluded that the addition of water from the hollow cathode discharge ion source into the drift tube needs to be taken into account in addition to the humidity of the inlet air if HCN and HCHO concentrations are to be accurately determined.

An estimate of the drift tube humidity can be derived from the signal strength of the protonated water dimer, $H_3O^+(H_2O)$, at *m/z* 37 [47]. A non-linear but reproducible calibration factor between the signal intensity at *m/z* 37 and a deliberately added reference water vapour concentration provided a measure of water vapour concentration in the drift tube. However, considerable care must be taken when using the signal intensity at *m/z* 37 for this purpose, and particularly at low *E/N*, because the signal level may be very large and the user must be aware of possible detector saturation. For high humidity conditions and/or low *E/N* values it is safer to use the signal intensity of the ^{18}O isotopologue of the protonated water dimer at *m/z* 39 to determine humidity.

For an accurate measurement of HCHO concentration, Inomata *et al.* derived an instrumental calibration factor for their PTR-MS instrument as a function of drift tube humidity using a dynamic dilution method, that is, they determined the decrease in the sensitivity for the detection of ions at *m/z* 31 with increasing humidity [28]. This calibration factor was then deployed in a field campaign. Once corrections had been made for interferences from other compounds to the protonated ion signal at *m/z* 31, the measured HCHO concentrations obtained with PTR-MS were found to be in good agreement with those determined by multi-axis differential optical absorption spectroscopy (MAX-DOAS). In another reported study, Steinbacher *et al.* compared measurements of HCHO using a Hantzsch monitor (a fluorescence-based technique) with those from PTR-MS [48]. As expected, the calculated mixing ratio for HCHO was found to be lower from the PTR-MS measurement compared to the value obtained from the fluorescence data. However, it was found that the PTR-MS response was lower when using dry air (21%) than when using the more humid laboratory air (38%). Based on the discussion above, the apparent higher PTR-MS sensitivity with the more humid air is opposite to what should have been expected. The researchers attributed this discrepancy to a volatile interferent in the laboratory air contributing to the signal intensity at *m/z* 31, possibly a fragment from protonated methyl hydoperoxide (CH_3OOH). This again emphasizes the point we have already raised several times, namely that interferents which contribute to a given *m*/*z* of interest are always a concern that needs to be considered when relying only on PTR-MS for any analytical measurement.

Another example where interferents are significant is the case of HCN, whose protonated signal is at *m*/*z* 28. O_2^+ is always present to some extent in the drift tube when air is used as the inlet gas, as discussed in Section 3.3.2, and this ion can transfer its charge in collisions with ethene leading to an ion with the same nominal *m*/*z* as protonated HCN. If this contribution is not taken into account the HCN measurements will be compromised. Knighton *et al.* have estimated that if ethene is present it can contribute 0.1 ppbv HCN equivalent per ppbv of ethene when the O_2^+ intensity is at ~2% of the H_3O^+ level [49]. Knighton *et al.* also calibrated their PTR-MS instrument at various HCN mixing ratios, temperatures and relative humidities, and developed a model to accurately account for the temperature and humidity dependences of the PTR-MS sensitivities for the detection of HCN.

It is not only reverse reactions of protonated species with water that can complicate the mass spectral analysis of PTR-MS data. It is possible that, once formed, a protonated parent molecule could go on to react with water in a bimolecular reaction, resulting in a secondary product ion. An example of this is protonated peroxyacetyl nitrate, which is commonly referred to as PAN ($CH_3C(O)OONO_2$), and is a pollutant present in photochemical smog. Using a selected ion flow drift tube apparatus (see Section 1.4.1) Hansel and Wisthaler found that protonated PAN bimolecularly reacts with water to form a product ion at *m*/*z* 77 ($CH_3C(O)OOH_2^+$) together with neutral HNO_3 [50]:

$$CH_3C(O)OONO_2H^+ + H_2O \rightarrow CH_3C(O)OOH_2^+ + HNO_3 \quad (5.1)$$

It was estimated that if more than 0.2% water vapour is present in the drift tube then more than 95% of the protonated PAN reacts with water to form $CH_3C(O)OOH_2^+$. This study serves to illustrate the need for fundamental investigations of ion–molecule reactions, which may involve other techniques such as SIFT-MS or SIFDT-MS, in order to aid in the interpretation and use of PTR-MS data.

5.2 Use of Reagent Ions Other Than H_3O^+

Some of the problems, and hence limitations, associated with PTR-MS for atmospheric measurements may be overcome by operating the PTR-MS with a reagent ion other than H_3O^+. With a different reagent ion different types of ionization processes, such as charge transfer or hydride abstraction, could become dominant. Such reactions and their energetics were discussed in Chapter 2. Often it is observed that if a bimolecular ion–molecule reaction is exothermic it occurs at a collision-limited rate, but this need not always be the case. This is particularly true for charge transfer reactions. Fortunately, there are many compilations of experimentally determined reaction rate coefficients which may be referred to in order to determine the efficiency (and type) of many ion–molecule reactions [51, 52]. However, it is important to appreciate that the majority of the data provided in the compilations have been determined under thermal conditions (usually 300 K) using instruments such as the Flowing Afterglow (Section 1.3.3) or Selected Ion Flow Tube (Section 1.3.4).

An important class of ion–molecule processes in a PTR-MS context are charge transfer reactions, particularly when the recombination energy (RE)[3] of a reagent ion (X^+) is less than the ionization energy (IE) of H_2O (12.62 eV). If RE(X^+) > IE(H_2O) rapid switching from H_3O^+ to X^+ will not be possible. This is because charge transfer to any residual H_2O will occur leading to H_2O^+, which itself will rapidly react with H_2O, resulting in H_3O^+ as the dominant reagent ion. In other words any attempt to generate X^+ will simply lead to H_3O^+ if there is significant residual water.

Of the reagent ions that could potentially be used with PTR-MS for analytical purposes, O_2^+ and NO^+ have been particularly useful because (i) they are easily produced in high abundance without the need for expensive gases, (ii) the enthalpy change for reaction with many organic molecules is such that the process can still often be regarded as soft and (iii) their recombination energies are less than the ionization energy of water. O_2^+ has a recombination energy of only 12.07 eV, which is sufficiently high to permit charge transfer reactions with many light hydrocarbons and other VOCs, but not so high as to result in major fragmentation (dissociative charge transfer). For example, O_2^+ reacts efficiently with ethane by non-dissociative charge transfer:

$$O_2^+ + C_2H_6 \rightarrow C_2H_6^+ + O_2\ (\Delta H = -53\ \text{kJ mol}^{-1}) \tag{5.2}$$

The exothermicity given for this reaction has been determined from the difference in the ionization energies of O_2 (12.07 eV) and C_2H_6 (11.53 eV). This enthalpy change ignores the kinetic energy gained by the O_2^+ in the drift tube but at normal *E/N* values this will result in insufficient additional energy to cause any significant fragmentation of $C_2H_6^+$.

As already discussed in Section 2.3 of Chapter 2, Norman *et al.* have demonstrated the use of O_2^+ for the detection of NH_3 [53]. It is difficult to quantify atmospheric ammonia when using H_3O^+ as the reagent ion because significant quantities of NH_4^+ are believed to be formed by reaction with N_2 diffusing into the ion source. However, use of O_2^+ as the reagent ions means that non-dissociative charge transfer can be employed leading to an NH_3^+ ion, which will appear in a mass spectrum at *m/z* 17 and is therefore free from any interference from NH_4^+.

NO^+ can react with VOCs in a bimolecular manner via various routes, including charge transfer (providing molecule M has an ionization energy less than the recombination energy of NO^+ (9.26 eV)), or hydride abstraction resulting in a $[M\text{-}H]^+$ product ion, and hydroxide ion transfer (resulting in $[M\text{-}OH]^+$ product ions). It is also possible for NO^+ to react with a compound through a three-body association process resulting in an adduct ion, $M.NO^+$. However, if these adduct ions are weakly bound they will undergo fragmentation in the drift tube and will not be observed unless the *E/N* is reduced to a sufficiently low value, usually below 100 Td. Wyche *et al.* were the first to demonstrate the potential of NO^+ in a PTR-MS instrument by using this reagent ion to distinguish between isobaric and isomeric compounds [54]. Aldehydes and ketones, including the atmospherically important compounds methacrolein and methyl vinyl ketone, were chosen for this illustration. Methacrolein ($CH_3C{=}CH_2CHO$) and methyl vinyl ketone ($CH_2{=}CHCOCH_3$) are isomers (molecular formula C_4H_6O) and so they cannot be distinguished on the basis of the mass

[3] The *recombination energy* is the energy released when an electron 'recombines' with an ion to form a neutral atom or molecule. It is numerically equal to the ionization energy of the corresponding neutral species that is formed.

of the protonated parent molecule. However, it was found that whereas methyl vinyl ketone showed little reaction with NO^+, the reaction with methacrolein resulted in a number of product ions at *m/z* 69, 41 and 39. The ion at *m/z* 69 was assigned to a product resulting from hydride abstraction. The other two product ions were considered to result from dissociative charge transfer, with *m/z* 41 resulting from CHO or CH_3CH_2 elimination and *m/z* 39 from loss of CH_3O. Given that the ionization energy of methacrolein is 9.9 eV, charge transfer to NO^+ is endothermic. Consequently, for the dissociative charge transfer channels to be possible a substantial fraction of the NO^+ reagent ions used in the study by Wyche *et al.* must have been in an excited state. Independent of this potential complication, importantly this study showed the potential for discriminating isomeric compounds using a PTR-MS with reagent ions other than H_3O^+.

Another example of the potential use of NO^+ is in the detection of 1,3-butadiene (C_4H_6). Although H_3O^+ will react efficiently with 1,3-butadiene (PA = 783 kJ mol^{-1}), the protonated parent species produced has the same nominal *m/z* of 55 as the protonated water trimer, $H_3O^+(H_2O)_2$, which is always present in the drift tube of a PTR-MS instrument (the actual amount depends on the humidity of the air in the drift tube and the *E/N* value being used). To overcome this complication, Knighton *et al.* used NO^+ to detect 1,3-butadiene in the atmosphere via non-dissociative charge transfer to produce $C_4H_6{}^+$ at *m/z* 54 as the only product ion [55]. Detection sensitivities as low as 45 pptv were demonstrated within just a second of signal accumulation. Given that the typical ambient concentration of 1,3-butadiene is expected to be of the order of 10–100 pptv, the sensitivity demonstrated by Knighton *et al.* is sufficient to allow 1,3-butadiene to be detected rapidly in the atmosphere using NO^+ as the reagent ion.

More recently NO^+ has been used to show how low concentrations of isoprene (C_5H_8) can be determined without any interference from 2-methyl-3-buten-2-ol (MBO) [56]. MBO (CH_2=CHC(CH_3)$_2$OH) has been found to be the dominant VOC emitted from ponderosa pine and reaction of MBO with H_3O^+ leads to two product ions. One of these product ions is observed at *m/z* 87, which is the protonated parent molecule. The second and dominant product ion is detected at *m/z* 69, which results from the loss of H_2O from the protonated parent molecule, that is, the product ion is $C_5H_9{}^+$. Unfortunately, this is exactly the same molecular formula as protonated isoprene. When isoprene is the dominant VOC, contributions to *m/z* 69 from MBO can generally be ignored. For coniferous ecosystems this is not the case, and therefore H_3O^+ is not an ideal reagent ion to be used for determining isoprene mixing ratios. However, by using a PTR-MS instrument under standard operating conditions (*E/N* = 115 Td) it has been found that NO^+ reacts with isoprene (IE = 8.86 eV) mainly by non-dissociative charge transfer leading to the product ion $C_5H_8{}^+$ at *m/z* 68. On the other hand MBO undergoes a hydroxide ion transfer reaction with NO^+ leading to the production of $C_5H_9{}^+$ as the dominant product ion and HNO_2 as the neutral species. This selective ionization allowed the researchers to determine the low daytime concentrations of isoprene (200–250 pptv) without any interference from MBO (whose daytime concentration was found to be ~1.5 ppbv) in a Colorado ponderosa pine forest.

A significant advantage to note is that $O_2{}^+$ and NO^+ do not react with water and it is therefore possible to switch rapidly to either one of these reagent ions after operating the PTR-MS instrument in H_3O^+ reagent ion mode. This ability to switch fairly rapidly (within tens of seconds) between reagent ions offers the potential to greatly enhance the analytical capabilities of PTR-MS. However, rapid switching is not possible for reagent ions which can react with H_2O. For example, the recombination energy of Kr^+ is ~1.4 eV higher than the ionization energy of water. Fast switching from H_3O^+ to Kr^+ as reagent ions from a hollow cathode discharge ion source is therefore not possible because Kr^+ will always transfer its charge to traces of H_2O when the source gas is switched. The resulting H_2O^+ will then react with H_2O to form H_3O^+. Furthermore, Kr^+ will react efficiently with O_2 in the PTR instrument, leading to a large $O_2{}^+$ signal, unless conditions are changed within the drift tube. By operating the PTR-MS for many tens of minutes without any water being introduced into the ion source and by using inlet air that was heavily diluted with helium, Sulzer *et al.* showed that a large Kr^+ ion concentration could be generated in the drift tube, resulting in a signal intensity of $\sim 10^7$ counts per second of Kr^+ at the ion detector [57]. Other ions generated in the ion source or drift tube had substantially lower ion intensities, the most abundant being H_3O^+ at 3–8% (depending on the humidity of the air), KrH^+ at ~4% and He^+ at about 2%. The important point is that the higher recombination energy of Kr^+ compared to $O_2{}^+$ allows it to react via a charge transfer process with many of the important (inorganic) atmospheric species, such as CO, CO_2, CH_4, NO_x and SO_2, which have previously been unavailable for detection using a PTR-MS instrument.

It can of course be argued that when using reagent ions that are not acting as proton donating species they should not be included under the auspices of PTR-MS, and in fact some researchers prefer to use terms such as 'Soft Chemical Ionization Mass Spectrometry (SCIMS)' or 'Chemical Ionization Reaction Mass Spectrometry (CIR-MS)' [54]. However, for convenience we shall not adopt that terminology here and will keep referring to the technology as PTR-MS.

5.3 Biogenic VOCs

5.3.1 General Details

Biogenic VOCs are a result of biological processes in marine and terrestrial environments. They play a significant role in the chemistry of the biosphere: for example, through the production of tropospheric ozone and their contribution to the formation of aerosols, particulate matter and the global carbon cycle. After methane the most common biogenic VOCs released by terrestrial vegetation into the atmosphere are isoprene (C_5H_8) and the monoterpenes ($C_{10}H_{16}$), but other volatile compounds such as sesquiterpenes, alcohols, aldehydes, esters and acids are also involved. Isoprene is a particularly important compound to detect, mainly because it is highly reactive with various other compounds in the atmosphere.

Studies of biogenic VOCs and how they evolve in time provide crucial information about various processes in plants, including their response to external stimuli (e.g. stress or wounding), and hence potential information on biological pathways can be extracted. The investigation of biogenic VOC emissions is arguably the most dominant and most successful environmental application of PTR-MS to date. PTR-MS studies have provided unique data which have considerably enhanced our understanding of biogenic VOC formation, emission and evolution. In what follows, we will focus our attention on the results of the various studies rather than providing details on the experimental arrangements and procedures. For this information the reader is referred to the original papers.

The first published, and essentially pilot, studies were by Lindinger and colleagues, who reported measurements of oxidized VOCs being emitted from decaying plant material and the diurnal variation of the concentration of various VOCs in the atmosphere at the western outskirts of Innsbruck, Austria, measured during a 6 day period in the month of

August [4]. The next on-line atmospheric PTR-MS measurements were of isoprene and methyl vinyl ketone/methacrolein, made during a flight above the Amazon region in March 1998 [4, 58, 59]. During the same period, laboratory measurements revealed the role of temperature and wetness on VOC emissions from decaying dead vegetation [60], which hitherto had been unrecognized as a source of partially oxidized VOCs and is therefore of direct relevance to atmospheric chemistry.

Since those preliminary and pioneering investigations, PTR-MS has been successfully and comprehensively used to study biogenic emissions from various sources. These include various forests (e.g. tropical rainforests, coniferous, pine and boreal), woodland and grassland savannahs, plantations and oceans. Some studies have taken advantage of airborne measurements to include investigations from more than one type of biogenic source. An example of this is the airborne measurements of VOCs taken over West Africa by Murphy *et al.*, which encompassed urban areas (Lagos, Cotonou and Niamey), an ocean, a mosaic forest (land containing a mixture of forest, savannah and grassland) and the Sahel desert [61]. In that study, PTR-MS measurements of ion intensities at m/z values of 42, 59, 69, 71 and 79 were used to calculate mixing ratios of acetonitrile, acetone, isoprene, the sum of methyl vinyl ketone and methacrolein, and benzene, respectively. Consistent with the rapid atmospheric oxidation of biogenic VOCs, strong spatial associations between vegetation and isoprene and its oxidation products (methyl vinyl ketone and methacrolein) were found. Evidence for enhanced levels of acetonitrile, benzene and acetone in air masses that had been affected by biomass burning was also provided. In addition to this, elevated levels of benzene and acetone in air masses influenced by anthropogenic VOCs coming from cities were also observed. The study also showed that a vertical redistribution of VOCs emitted from the surface occurred as a result of deep convection and mixing associated with fair-weather cumulus clouds. Furthermore, the data from over the ocean implied that acetone is depleted in the marine boundary layer, whereas acetonitrile is not.

A series of interesting ground-based PTR-MS measurements of reactive organic compounds, particularly methanol and acetone, were undertaken during July–August 2001 by Salisbury *et al.* as part of the Mediterranean INtensive Oxidant Study (MINOS) campaign in Crete [62]. This study is discussed in more detail in section 5.5. We just comment here than an important conclusion from that work is that biogenic sources of acetone are more significant than anthropogenic sources.

The identification of a VOC to a specific biogenic source is sometimes difficult if this VOC also comes from anthropogenic sources. However, from roadside measurements taken in Innsbruck, Austria, Holzinger *et al.* were able to separate biogenic and anthropogenic (traffic) sources of methanol released into the troposphere by using benzene as a marker for traffic emissions, and from this it proved possible to determine seasonal variations in methanol concentrations [41]. This study also provided evidence which suggested that methanol is removed from the atmosphere at night via deposition onto dew-wetted surfaces.

5.3.2 Forest Emissions

5.3.2.1 Tropical Rainforests

An investigation of VOCs over the tropical rainforest of Surinam by Crutzen and colleagues was the first time a PTR-MS instrument had been employed in flight [6–8]. This work was carried out as part of the Large-Scale Biosphere–Atmosphere experiment in

Amazonia (LBA)–Cooperative LBA Airborne Regional Experiment (CLAIRE) campaign during the period 12–19 March 1998, with measurements taken between the heights of 60 and 12 500 m. The data obtained showed large emissions of isoprene, which was measured together with its oxidation (degradation) products, methylvinyl ketone and methacrolein, by PTR-MS. Average boundary layer mixing ratios were reported with some illustrative values being as follows [6]: methanol (1.1 ppbv; detected at *m/z* 33), acetonitrile (190 pptv; *m/z* 42), acetaldehyde (1.7 ppbv; *m/z* 45) and isoprene (1.7 ppbv; *m/z* 69). For some VOCs, for example, formic and acetic acids, mixing ratios could not be determined owing to sampling difficulties, attributed to line losses and memory effects [7]. For *m/z* 87 (C_5 carbonyls and methacrylic acid) the mixing ratios were found to vary between 500 pptv at the upper end and below the detection limit (which occurred at high altitudes) at the low end. Reference [7] also reported GC-FID analysis of air samples collected in canisters during flight and provided more details on isoprene and its oxidation products and isoprene-related peroxides. These early on-line airborne studies provided information about the dependence of volume mixing ratios on height and the time of day, the results of which were found to be consistent with the emissions estimated from model calculations [8]. Interestingly, an ion signal attributed to dimethyl sulfide was also observed. Given that this compound is associated with marine environments, Crutzen *et al.* postulated that this compound is transported deep into the tropical forest from the coast and that it is not rapidly lost through reactions with OH owing to the strong competition from OH reactions with hydrocarbons [6].

Data from long-term PTR-MS field measurements have been, coupled to the procedure known as the variability-lifetime method to assess the dominant cause of variability in atmospheric VOC concentrations. These have been employed to understand and identify VOC sources and VOC photochemical processing, that is, source-receptor relationships of VOCs detected in ambient air associated with tropical rain forest sites [63]. Notably, Williams *et al.* used the aircraft-measurement data taken over the Surinam rainforest in the "mixed-layer" region (0-1 km) to provide an estimate of the average OH concentration in air massed along the aircraft back trajectory and found the value to be in the range of 2.0×10^5 molecules cm^{-3}. Based on other measurements, this is approximately 5 times less than expected. A possible reason given for this is the presence of photochemical sources of acetone and methyl vinyl ketone along the back trajectory of the air encountered. (Other studies using variability-lifetime methods included VOC measurements at a site surrounded by forests in Austria allowing an average OH number density of $(1.5 \pm 0.4) \times 10^5$ molecules cm^{-3} to be determined [64], and from the remote marine boundary layer of an ocean [65] (see section 5.3.5).)

Eerdekens *et al.* reported airborne and ground-based measurements of biogenic VOCs taken during the long dry season covering a large area of the northern Amazonian rainforest [66], taken as part of the October 2005 Guyana's Atmosphere–Biosphere exchange and Radicals Intensive Experiment with the Lear jet (GABRIEL) campaign. PTR-MS was used to measure the vertical (from 35 m to 10 km) and diurnal (from 9 a.m. to 4 p.m.) profiles of isoprene, isoprene's oxidation products (methacrolein and methyl vinyl ketone), methanol and acetone, which were then used by the researchers to empirically estimate emission fluxes from the forest canopy on a regional scale. The empirically derived isoprene fluxes were found to be significantly smaller (by approximately a factor of two) than those predicted from the biogenic Model of Emissions of Gases and Aerosols from Nature in the single-column chemistry and climate model.

5.3.2.2 Coniferous Forests

VOCs being emitted from a spruce forest in Bavaria have been investigated by Müller *et al.* [67]. By using a 30 m research tower, the diurnal mixing ratios of aldehydes and ketones were determined at two heights. One was in the treetops at a height of 12 m and the other was taken above the canopy at a height of 24 m. The volatiles that were found to have the highest concentrations were acetone (up to 9.1 ppbv), formaldehyde (up to 6.5 ppbv), acetaldehyde (up to 5.5 ppbv) and methyl ethyl ketone (up to 1.8 ppbv). Pinonaldehyde, which results from the oxidation of α-pinene, was found to reach its highest concentration (0.15 ppbv) in the early morning, whereas for most other compounds their maximum abundance was found to occur near to midday, as determined in 2,4-dinitrophenylhydrazine (DNPH)[4] measurements. PTR-MS measurements verified this, with the exception of the daytime measurements of acetaldehyde and acetone, for which significant and unexplained differences were observed between the PTR-MS and DNPH results, that is, little variation in the PTR-MS derived mixing ratios was found throughout 24 h of measurements. For the measurements at night the mixing ratios obtained for acetaldehyde were almost identical, whereas for acetone lower mixing ratios were obtained from the DNPH method. Formaldehyde did not show any distinct diurnal variations in their studies. Often it was found that the concentrations of VOCs determined above the canopy were higher than found in the treetops, which was attributed to secondary formation routes to these compounds in the atmosphere above the forest.

Research groups at the University of Helsinki and the Finnish Meteorological Institute have studied atmospheric concentrations of VOCs emitted by a boreal forest in Finland [68–70]. This work showed that it was possible to place biogenic VOCs into one of three classifications depending on concentrations and any associated time dependence: the first class contained VOCs that have a high diurnal variation, with maximum mixing ratio values being observed in the afternoon (e.g. methanol, acetone, methyl vinyl ketone and hexanal); the second class included compounds such as monoterpenes and phenol, which were found to have their highest concentrations at night when the mixing of the atmospheric surface layer was weak; the third class of compounds incorporated VOCs that have rather long atmospheric lifetimes, such as benzene, and for which no obvious diurnal cycle and no dependence on local meteorological parameters were observed [68]. Further PTR-MS studies associated with the Finnish boreal forest were undertaken as part of the Hyytiälä United Measurements of Photochemistry and Particles in Air – Comprehensive Organic Precursor Emission and Concentration study (HUMPPA–COPEC) campaign during the summer of 2010. The aim of this campaign was to characterize the atmospheric chemistry associated with a Boreal forest site in summer, with the objective to measure VOCs from both within and at various heights above the canopy so that spatial and temporal distributions could be determined. A paper by Williams *et al.* provides a detailed overview of the campaign, including an examination of the meteorological and chemical influences observed [71]. In addition to PTR-MS measurements of VOCs, the researchers used a modified (essentially a more fully automated) version of the comparative reactivity method (see Section 5.6.2)

[4] DNPH derivatization is a method used to measure aldehydes and ketones. Nucleophilic addition is used to convert the carbonyl compounds to stable hydrazine derivatives prior to analysis. Either GC methods or HPLC/UV analysis can be used to detect the derivatives.

to determine a total OH reactivity (i.e. the quantity of total atmospheric constituents that can react with OH) of ~12 s^{-1} both above and within the canopy [72]. This is higher than values of total OH reactivity based on calculations using individual VOC measurements. Consequently, there is an unexplained and significant "missing" sink of OH.

PTR-MS measurements have been undertaken at a ponderosa pine forest in central California by various groups [34, 42, 73–75]. One of these studies was the first to identify sesquiterpenes ($C_{15}H_{24}$) and methyl chavicol (estragole) ($C_{10}H_{12}O$) in the atmosphere by PTR-MS [42]. Exploration of the vertical gradients of VOC mixing ratios revealed that large quantities of previously unobserved oxidation products of short-lived biogenic precursors were present in the atmosphere. The implication of this is that global terpene emission and secondary organic aerosol formation may be larger than had been assumed up to that point in time, because the very reactive precursors would be severely underestimated if the observed oxidation products are not taken into account. Furthermore these precursors, once oxidized by ozone, should produce secondary organic aerosols and hydroxyl radicals. Another study measured the total monoterpenes (*m/z* 81 and 137), total sesquiterpenes (*m/z* 205) and methyl chavicol (*m/z* 149) from enclosed branches using zero air flowing through a chamber system [73]. This study included not only ponderosa pine, but also shrub species (manzanita) and whitethorn (*Ceanothus*). In addition to PTR-MS, measurements were undertaken with GC-FID and SPME/GC–MS. The researchers demonstrated that sesquiterpenes and methyl chavicol were major contributors to the overall organic flux observed at their site of investigation. In another study the mixing ratio and biogenic emission rate for methyl chavicol were determined. The daytime basal emission rate was found to be in the range of 0.72–10.2 μg C g^{-1} h^{-1} and is estimated to account for 15–36% of the total gas-phase terpene (sum of total monoterpenes, sesquiterpenes and oxygenated terpenes) mass. (The actual rate was found to be dependent on the age of the pine needle and on the time of year.)

Holzinger *et al.* recorded the seasonal variability of monoterpene emission factors for a ponderosa pine plantation in California over an 11-month period (June 2003–April 2004) [75]. (As an aside, although such long-term studies highlight the reliability of PTR-MS, it is important that the PTR-MS apparatus is regularly calibrated to ensure that there has been no change in its sensitivity, or if there is that this is taken into account when analyzing data taken over a long time period.) From these measurements, Holzinger *et al.* suggested that the total annual monoterpene emission had been underestimated by ~50% when using a model optimized to reproduce monoterpene emissions in summer.

Karl *et al.* used PTR-MS to investigate the bidirectional exchange of oxygenated VOCs between a loblolly pine (*Pinus taeda*) plantation with a sweetgum (*Liquidambar styraciflua*) understory[5] and the atmosphere [76]. The emission and deposition patterns for a number of VOCs, including methanol, acetaldehyde, acetone, C_3/C_4 carbonyls (such as methyl ethyl ketone and methylglyoxal, as determined by ions at *m/z* 73), methyl vinyl ketone/methacrolein and acetic, acid were determined. During the day it was found that the canopy scale measurements showed significant emission of methanol and acetone, while in comparison acetaldehyde, methyl vinyl ketone/methacrolein, and acetic acid were mainly deposited. At night all of the oxygenated compounds exhibited strong losses. The exchange

[5] An understory refers to those plants that are growing beneath the canopy.

of acetaldehyde between the trees and the atmosphere was found to be complicated. From laboratory measurements it was observed that, depending on ambient conditions, acetaldehyde can either show a net emission from the trees or be taken up by the trees, and that mature pine needles emitted more acetaldehyde than younger ones.

Measurements of biogenic VOCs from a ponderosa pine ecosystem were also made by Kim *et al.*, who deployed two PTR-MS systems to measure VOCs near Woodland Park, Colorado, United States [77]. Biogenic VOC emissions and their ambient distributions were simultaneously measured with the two instruments. GC–MS was used to aid in the VOC assignments. 2-methyl-3-butene-2-ol (50%) and several monoterpenes (34%) were found to be the dominant biogenic VOCs emissions as determined from *m/z* transmission-corrected data averaged over 3 days of measurements. No isoprene associated with the forest was identified because isoprene is known to be a minor VOC emitted from such ecosystems and what was observed was attributed to its transport from other ecosystems (e.g. aspen groves and oak shrubland). This lack of isoprene agrees with an observation by Karl *et al.* from the same site, where average daytime concentrations of 2-methyl-3-butene-2-ol and isoprene were determined to be 0.2–1.4 ppbv and 90–250 pptv, respectively [56]. This study avoided using H_3O^+ as a reagent ion because this predominantly reacts dissociatively with 2-methyl-3-butene-2-ol, resulting in an ion at *m/z* 69 (corresponding to the loss of H_2O from the protonated parent molecule), and hence it is difficult to separate this from any *m/z* 69 associated with protonated isoprene. Instead NO^+ was used as the reagent ion since this reacts with isoprene via non-dissociative charge transfer leading to an ion at *m/z* 68 ($C_5H_8{}^+$), whereas 2-methyl-3-butene-2-ol undergoes hydroxide ion transfer resulting in *m/z* 69 ($C_5H_9{}^+$) as the dominant product ion (see Section 5.2).

5.3.2.3 *Deciduous Forests*

Ammann *et al.* demonstrated the ability of PTR-MS for measuring continuous vertical biogenic VOC concentration profiles throughout a mixed deciduous forest canopy [23]. As expected, considerable diurnal variations in VOC concentrations were found, with a distinct concentration maximum being observed during the daytime in the so-called crown region (14–28 m above ground) of the forest canopy. In this region of the canopy, the main emission results from the interaction of sunlight on the leaves. Ammann *et al.* showed that their data sets provided useful information on the interactions between emission, photochemical transformation and transport processes within and above the forest canopy.

In parallel with laboratory studies investigating VOC emission from leaves during leaf drying, senescence (biological aging) and following freeze–thaw damage, Fall and colleagues undertook on-line PTR-MS measurements of ambient air at the Sonnblick Observatory in the Austrian Alps following a hard freeze [35,78]. C_5 and C_6 VOCs were observed in the field study which, by comparison with the laboratory study, were attributed to emissions resulting from frost-damaged local vegetation. The concentrations of the C_5 and C_6 VOCs (including 1-penten-3-ol, methylbutanals, hexenals, hexanal, pentenol and pentenone) were found to range from 300 pptv to 6 ppbv. It was concluded that the damage to the leaves of the trees resulting from freezing will release VOCs that should contribute to regional tropospheric photochemistry in the autumn. Further details dealing with VOC emissions from damaged plants are provided later in Section 5.7.7.

5.3.2.4 Eddy Covariance Measuring Methodologies

We have already mentioned that a major reason why PTR-MS is valuable for studying biogenic emissions is because of its fast time response, making it suitable, for example, to measure biogenic VOC fluxes using techniques such as eddy covariance,[6] relaxed eddy accumulation, and disjunct eddy covariance as well as more conventional enclosure methods [34, 79–86]. Eddy covariance provides a direct approach for measuring the exchange of biogenic VOCs between the terrestrial ecosystems and the atmosphere and for measuring vertical fluxes in the turbulently mixed planetary boundary layer.

An experimental validation of eddy covariance techniques was provided by Ammann *et al.* [47]. They compared the water vapour concentrations and eddy covariance fluxes using PTR-MS, determined by monitoring $H_3O^+(H_2O)$, to a common reference system comprising an infra-red gas analyzer (IRGA). The water vapour flux obtained from PTR-MS measurements was found to be in good agreement with the value obtained using IRGA providing the first credible evidence that PTR-MS can be used for eddy flux measurements of VOCs.

Table 5.2 provides a summary of the environments in which eddy covariance measurements of VOC fluxes using PTR-MS have been made. For convenience and for completeness other methodologies for determining biogenic flux using PTR-MS and ecosystem types other than forests, such as alfalfa fields [81], are included in the table.

A relaxed-eddy-accumulation set-up was applied to determine isoprenoid fluxes above a Norway spruce forest in July 2001/2002 [85]. Disjunct eddy covariance measurements of monoterpene fluxes from a Norway spruce forest consisting of *Picea abies* L. [Karst.] trees using PTR-MS have been described by Grabmer *et al.* [87], whose results compare well with the data obtained using relaxed eddy accumulation in a follow-up study which investigated VOC emissions from Norway spruce (*Picea abies* L. [Karst]) twigs in the field using a more conventional dynamic enclosure method [85]. Emissions of isoprenoids (i.e. isoprene and monoterpenes) and oxygenated VOCs (i.e. acetaldehyde, acetone, methanol and ethanol) were investigated in this study. The data revealed that while compounds such as isoprene were well represented by temperature and solar radiation emission models (e.g. see Reference 88), the oxygenated VOC emissions were not well described by such models.

Lee *et al.* used PTR-MS and eddy covariance methods to determine mixing ratios and fluxes of total monoterpenes above a ponderosa pine plantation [34]. The results from these studies were compared to those obtained using GC-FID and a relaxed eddy accumulation system. It was found that the daytime measurements resulted in values that agreed well with each other. However, at night PTR-MS detected additional reactive terpenes which

[6] Eddy covariance is an established statistical method used in micrometeorology to determine the movement/exchange of air as a result of turbulence (eddies) within atmospheric surface boundary layers from various ecosystems (including forests, crops, grasslands, oceans, lakes, urban areas etc.). It provides measurements of gas exchange rates (emissions and fluxes) of trace gases. Eddy covariance techniques use details on air speed, gas concentration, pressure, temperature, and humidity. Vertical fluxes of biogenic VOCs are determined as a covariance of the vertical wind velocity component and the biogenic VOC concentrations, both of which require rapid measurements. Thus the real-time capabilities and fast instrumental response time makes PTR-MS an ideal instrument for estimating concentrations to be used in eddy covariance techniques. More details on eddy covariance measurements may be found in "Eddy Covariance: A Practical Guide to Measurement and Data Analysis", Springer Atmospheric Sciences (2012), Eds. M. Aubinet, T. Vesala, and D. Papale.

Table 5.2 *Summary of measurement methodologies for determining biogenic VOC fluxes using PTR-MS*

Ecosystem type	Method[a]	References
Mixed deciduous forest	EC	[82]
Mixed temperate forest	DEC	[98]
Tropical rain forest	EC, DEC, vDEC, CBL, MLG, MLV	[66, 79, 92, 94]
Ponderosa pine	EC	[34, 42, 75]
Scots pine	DEC, EM	[90, 91]
Subalpine forest	vDEC	[84]
Norway spruce	REA, EM, DEC	[85–87]
Beech forest	REA	[89]
Mediterranean macchia	DEC	[12]
Pine plantation	EC, DEC, REA	[76]
Oil palm plantation	EC, vDEC	[100, 101]
Mountain grassland	EC, DEC	[13, 14, 104]
Agricultural grassland	EC	[105, 108, 109]
Alfalfa	DEC	[81, 102]
Hay field (harvesting)	EC	[80, 247]
Agricultural soil	REA	[106]

Source: Adapted from Reference 3.
[a] EC = Eddy covariance; DEC = disjunct eddy covariance; vDEC = virtual disjunct eddy covariance; REA = relaxed eddy accumulation; EM = enclosure method; CBL = convective boundary layer budgeting; MLG = mixed layer gradient; MLV = mixed layer variance.

were not resolved by GC-FID. This resulted in PTR-MS providing mixing ratios for these terpenes which were ~30% higher than those determined from GC-FID.

Schade *et al.* used PTR-MS to measure in-canopy mixing ratio gradients and above-canopy fluxes of several VOCs (methanol, acetone, monoterpenes) in a European beech (*Fagus sylvatica*) forest in Denmark [89]. A relaxed eddy accumulation method was used for the determination of the above-canopy fluxes. The monoterpene emissions and in-canopy mixing ratios showed a diurnal cycle which was found to be consistent with light-dependent emissions, in contrast to temperature-only-driven emissions that had been observed for most conifer species. The measurements also showed that deposition of methanol occurred mostly during the day and that acetone was only emitted once ambient temperatures went above 20°C.

Spirig *et al.* have reported PTR-MS measurements taken in a mixed deciduous forest in north-west Germany coupled to eddy covariance methodology for the determination of biogenic VOC fluxes [82]. Fluxes for isoprene and the monoterpenes were shown to be consistent with light and temperature dependent emissions of these compounds. Average daytime emissions of methanol, isoprene and monoterpenes were determined to be 0.087 μg m^{-2} s^{-1}, 0.94 μg m^{-2} s^{-1} and 0.3 μg m^{-2} s^{-1}, respectively.

Rinne *et al.* used PTR-MS with a disjunct eddy covariance method to measure the fluxes of methanol, acetaldehyde and monoterpenes above a Scots pine (*Pinus sylvestris*) forest in Finland [90]. Chemical modeling showed that there was little loss owing to reaction and that the VOC fluxes followed the traditional exponential temperature-dependent emission algorithm. On-line field measurements of daily monoterpene emission patterns from the same forest using PTR-MS coupled to an *in situ* sampling chamber in the canopy have been reported by Ruuskanen [91]. That study showed that canopy-scale emissions of monoterpenes and oxygenated VOCs were of the same magnitude.

Karl *et al.* tested a virtual disjunct eddy covariance device with field measurements of biogenic VOC fluxes at a subalpine forest site in the Rocky Mountains of the United States [84], where a short grazing season is only possible owing to the altitude. Significant fluxes of MBO, methanol, acetone and acetaldehyde were observed.

Karl and Guenther reported fast time response measurements of biogenic VOCs above a deciduous temperate forest, a tropical rain forest and an evergreen needle forest in order to assess atmospheric variability, lifetimes and fluxes of biogenic VOCs [83]. This work demonstrated the use of a variance method as an alternative to eddy covariance and disjunct eddy covariance methods for estimating surface layer VOC fluxes. The key conclusion was that the fast monitoring capabilities of PTR-MS could be taken advantage of for variability measurements of biogenic VOCs so that the magnitude and homogeneity of surface fluxes can be assessed and the atmospheric lifetimes of VOCs can be characterized in remote places.

Karl *et al.* have undertaken airborne and ground-based mixing ratio and flux measurements using eddy covariance and were the first to use mixed layer gradient and mixed layer variance methods to determine the influence of isoprene and monoterpene emissions on the atmospheric chemistry occurring in the Amazon basin [92]. This airborne study covered several different environments including (1) mixed forest/plantation, (2) primary tropical forest, (3) soybean plantation, (4) water and (5) urban areas (the city of Manaus). The results obtained were compared to those from the biogenic Model of Emissions of Gases and Aerosols from Nature (MEGAN). This work suggested that isoprene and the monoterpenes account for $\sim$75% of the total OH reactivity, and therefore that these are important VOCs that need to be taken into account for modelling atmospheric chemistry in the Amazon. As found in a later study [61], the presence of fair-weather clouds (cumulus humilis) was also shown to play a significant role in the vertical distribution of VOCs through the planetary boundary layer, the cloud layer and the free troposphere. Karl *et al.* also used their PTR-MS field measurements of isoprene, the sum of methyl vinyl ketone and methyl ethyl ketone and hydroxyacetone taken during the Amazonian Aerosol Characterization Experiment in 2008 to show that the production of certain oxygenated VOCs (e.g. hydroxyacetone), which result from the photooxidation of isoprene in the lower atmosphere, was significantly underestimated by standard chemistry schemes available at that time [93].

Using a high-sensitivity PTR-MS instrument Langford *et al.* measured fluxes and concentrations of VOCs emitted by a south-east Asian tropical rainforest in Malaysia using data collected over 48 days throughout the period April–July 2008 as part of the Oxidant and Particle Photochemical Processes (OP3) field study of rainforest atmospheric chemistry [94]. The above-canopy fluxes of isoprene, monoterpenes and oxygenated VOCs (including methanol, acetone and acetaldehyde) were determined by virtual disjunct eddy covariance. As the researchers expected, isoprene was found to be the dominant non-methane hydrocarbon (NMHC) emitted, accounting for $\sim$80% of the measured emission of reactive carbon fluxes. However, the emission rates determined for isoprene and the monoterpenes were lower by factors of $\sim$4 and 1.8, respectively, than the default values used in the MEGAN model of biogenic VOC emissions. The researchers concluded that this highlights the need for direct canopy-flux measurements of VOCs from the world's tropical forests. VOC flux measurements from this rainforest using PTR-MS were further reported in studies by Pugh *et al.* [95,96]. The first paper provides a box modelling study of the chemistry occurring in the tropical rainforest planetary boundary layer [95]. The second paper provides high temporal resolution PTR-MS isoprene concentration measurements together with a box model of atmospheric chemistry to estimate [OH]. Another study associated with this region was by Robinson *et al.* [97], who used PTR-MS to measure first generation isoprene oxidation products, namely methyl vinyl ketone and methacrolein.

Continuous measurements of isoprene and monoterpene fluxes were undertaken above a mixed forest site in eastern Belgium over a period of several months using a disjunct eddy covariance technique together with PTR-MS [98]. CO_2 fluxes were measured simultaneously in order to deduce the gross primary production of the VOCs. This study confirmed that temperature and light intensity are the most important factors driving the production of biogenic VOCs, with temperature being found to be the more significant of the two.

5.3.2.5 Forest VOCs and m/z Assignments

Finally, we conclude our discussion of VOC emissions from forests by returning to a key issue mentioned earlier in this chapter, namely the difficulty of assigning a given nominal *m/z* in a PTR-MS mass spectrum to a specific VOC. In the chemically complex atmosphere of a forest, this is almost impossible without detailed validation. To date this is still lacking, and major investigations are required to provide the details, algorithms and operational procedures that need to be adopted if PTR-MS is to be used as a stand-alone technique for compound identification, that is, without the need to use other analytical instruments. Towards this goal Ambrose *et al.* have compared GC-FID and PTR-MS measurements of toluene based on ambient trace gas measurements at a forested site in New England, United States [25]. Toluene is quantified in PTR-MS studies by monitoring its protonated molecule parent ion at *m/z* 93. A potential problem is that other VOCs could contribute to the intensity of the peak at *m/z* 93, including fragments from protonated monoterpenes. Although an overall minor ($\sim$13%) positive bias associated with the PTR-MS measurements was observed, which was attributed to a variable toluene background, it was concluded that under the operating conditions used interferences associated with monoterpene sampling were not significant and therefore the ion signal intensity at *m/z* 93 can be used to reliably determine toluene concentrations in the atmosphere.

5.3.3 Plantations

Fares *et al.* suggested that cultivated *Citrus* plantations may play an important role in the chemistry of the local atmosphere in regions such as the Central Valley of California [99]. To investigate this, the researchers used both PTR-MS and GC–MS to measure biogenic VOC emissions (predominantly oxygenated compounds, monoterpenes and sesquiterpenes) for four major *Citrus* species that are grown in California (*Citrus sinensis* variety 'Parent Navel', *C. limon* variety 'Meyer', *C. reticulata* variety 'W. Murcott' and 'Clementine'). Methanol, acetone and acetaldehyde were found to be the dominant biogenic VOCs emitted from lemon and mandarin trees, while oxygenated monoterpenes, monoterpenes and sesquiterpenes were the main biogenic VOCs observed to be emitted from orange trees. It was also found that the VOC emissions from orange trees dramatically increased by about an order of magnitude after the trees had bloomed (produced flowers). From their study the researchers recommended that accurate emission models need to account for additional VOC emissions resulting from flowering.

Misztal *et al.* reported a series of studies investigating VOC emissions from an oil palm plantation in Malaysian Borneo [100, 101]. The first study investigated large estragole (methyl chavicol) emissions using a virtual disjunct eddy covariance technique [100]. The researchers comment that, in comparison to the oil palm plantation, no estragole emissions are detected from a nearby rainforest. The reason given for this is that oil palms emit estragole to attract the African oil palm weevil for pollination. In confirmation of this, the flowers rather than the leaves were determined to be the main source of estragole coming from oil palms. Based on their measurements, Misztal *et al.* estimated that ~500 Gg/year of estragole are emitted into the atmosphere from oil palm plantations. This may sound large but, to put this into some context, it is approximately three orders of magnitude less than the annual global isoprene emissions from plants. Although relatively small on the global scale, these emissions probably represent what is likely to be the highest single floral contribution of reactive carbon in the atmosphere. The second reported study provided the first above-canopy eddy-covariance measurements of VOC mixing ratios and fluxes from oil palms [101]. As observed for many other plants, isoprene was found to be the dominant VOC being emitted, with measurements at midday showing that isoprene made up 84% of all the VOC emissions. Isoprene was found to be predominantly emitted from the large, divided leaves (fronds). Other VOCs detected included methanol, acetaldehyde, acetone, methyl vinyl ketone/methacrolein (isoprene oxidation products), hydroxyacetone, hexanals, toluene and estragole. All of these VOCs were monitored at their protonated parent mass, with the exception of the hexanals (where the fragment ion peak at m/z 83 ($C_6H_{11}{}^+$) was monitored). Approximately 15% of the total biogenic VOC flux was attributed to floral emissions, with toluene being released at approximately the same rate as estragole.

5.3.4 Various Land Emissions

Grasslands, crops, woodland and shrubland cover a significant land area and are therefore potentially a major source of atmospheric VOCs. To determine their atmospheric influence, the VOC emission rates and concentrations need to be determined above these areas. Consequentially, in addition to the forest and plantation work reviewed above, other PTR-MS studies related to plant emissions into the atmosphere have included, for example, flux measurements associated with shrubland [12], an alfalfa field [102], grassland [13, 14, 103–105], woodland savannah [103] and an agricultural field in Europe [106], as detailed in the following pages.

5.3.4.1 Woodland and Grassland Savannahs

VOC emissions, associated with isoprene and its oxidation products, from woodland and grassland savannahs in Venezuela were reported by Holzinger *et al.* [103]. A diurnal cycle was observed which showed that lower concentrations of VOCs were emitted at night, as anticipated. Furthermore, emissions were found to be significantly reduced during the dry season. Daytime average volume mixing ratios from the woodland savannah site during the wet season were determined to be 1.62 ppbv for m/z 69 (isoprene), 0.98 ppbv for m/z 71 (methyl vinyl ketone and methacrolein), 0.12 ppbv for m/z 83 (3-methyl furan and unsaturated C_5-hydroxycarbonyls) and 0.16 ppbv for *m/z* 101 (isoprene hydroperoxides). In comparison, concentrations of all compounds for the grassland site were found to be approximately three times lower than found at the woodland site during the wet season. During the dry season the mixing ratios were found to be less by approximately a factor of two than found for the wet season. If the woodland savannah site is representative, then by extrapolation of their measurements, Holzinger *et al.* estimated that the global emission of isoprene into the atmosphere from tropical savannahs is between 53 and 79 Tg C per year [103].

5.3.4.2 Shrubland

Concentrations and emission rates of biogenic VOCs above a Mediterranean macchia ecosystem at a Mediterranean coastal site in western Italy have been measured using a disjunct eddy covariance method utilizing PTR-MS [12]. A diurnal cycle in mixing ratios was observed with daytime maxima being found for methanol, acetaldehyde, acetone and isoprene and a night-time maximum occurring for monoterpenes. An average basal emission rate[7] for the macchia vegetation was calculated by the researchers to be 430 $\mu g\ m^{-2}\ h^{-1}$ for isoprene and 1100 $\mu g\ m^{-2}\ h^{-1}$ for monoterpenes.

Owing to its vast distribution in North and South American deserts (the Mojave, the Sonoran and the Chihuahuan), Jardine *et al.* have investigated VOC emissions from the evergreen shrub *Larrea tridentata* (creosote bush) during field studies in southern Arizona [107]. In addition to PTR-MS, GC-PTR-MS and GS-MS were used to identify volatile compounds. During the North American monsoon season a strong diurnal pattern of branch emissions and ambient concentrations was found for a large number of VOCs, with maximum emissions being observed in the early afternoon. In addition to detecting VOCs that are typically observed in forest sites (i.e. oxygenated VOCs and volatile isoprenoids), other compounds, some of which had not been observed previously to have been emitted from plants, were recorded. These unusual compounds included 1-chloro-2-methoxy-benzene, isobutyronitrile, several aromatic compounds (which are usually considered to be from anthropogenic sources) such as benzene and a broad range of phenolics. That benzene is being observed raises questions about its use as a marker for anthropogenic emissions. High emissions of dimethyl sulfide (which is a volatile not commonly emitted by plants in significant quantities), nitriles and the oxidation products of isoprene (methyl vinyl ketone, methacrolein and 3-methyl furan) were also observed. In the case of nitrile emissions from creosote bush, it was suggested that this could be a significant but previously unaccounted for source of nitrogen loss from arid ecosystems.

5.3.4.3 Alfalfa and Grass Fields

The perennial plant alfalfa is an important forage crop that is cultivated in many countries. Warneke *et al.* measured the fluxes of selected oxygenated VOCs coming from an alfalfa field before, during and after harvesting, using disjunct eddy covariance methods combined

---

[7] The basal emission rate is defined as the emission rate measured under a standard set of light and temperature conditions. It is a fundamental parameter used in models to predict VOC emissions.

with PTR-MS [102]. Before cutting, the dominant VOC emitted was methanol. Interestingly, a peak flux of about 4 mg m^{-2} h^{-1} was observed for methanol at 8:00 am. It was proposed that a probable cause of this was the evaporation of dew which contained VOCs dissolved in it. VOCs such as hexenals, acetaldehyde and acetone were also observed to be emitted throughout the day, but in much smaller quantities than found for methanol. Prior to harvesting hardly any emissions of acetaldehyde and acetone were observed from the alfalfa field, and the small observed emission of hexenal was attributed to stress caused by factors such as wind or feeding insects. Following harvesting, enhancements in the emissions of methanol, acetaldehyde, acetone and hexenals were observed, with the most dramatic rises being for the latter three VOCs. These emissions remained enhanced during the time the alfalfa was drying (3 days). The researchers suggested their findings show that the emissions of methanol and acetone resulting from crop harvesting may not be important to atmospheric chemistry on the global scale. In comparison the overall methanol released from fields of alfalfa as the plant is growing was shown to be significant. Finally, it is suggested that the more atmospherically reactive acetaldehyde and hexenal compounds may have a local impact on atmospheric chemistry during harvesting.

Similarly, directly after cutting and during the hay drying phase, Davison *et al.* observed enhanced emissions of biogenic VOCs such as methanol, acetaldehyde, acetone, 2-butanone, *m*/*z* 81 (mainly from (*Z*)-3-hexenal) and *m*/*z* 83 (primarily the sum of (*Z*)-3-hexenol and hexenyl acetates) in a study of agricultural grasslands (VOC assignments were assisted by GC–MS and GC-FID-PTR-MS techniques) [105]. On the same experimental site, which is near Oensingen on the Swiss Central Plateau, Brunner *et al.* measured the concentrations and fluxes of methanol above two differently managed grassland fields [108]. One of these grassland sites is referred to as an intensive field, which means it is intensively managed (cut and fertilized four times a year), while the other is referred to as an extensive field (cut three times a year and not fertilized). The intensive field contained three main plant species (two graminoids (grasses) and one legume (white clover)), whereas the extensive field was composed of twelve species (six graminoids, four forbs (herbaceous flowering plants) and two legumes (the flowering plants Bird's-foot trefoil and white clover)). PTR-MS was used to take measurements of emitted VOCs to calculate concentrations and eddy covariance methods were then used to determine the fluxes. A distinct diurnal cycle was discovered for the methanol emissions, which showed a good correlation with global radiation and water vapour flux. Highest fluxes were recorded immediately after plants and grasses were cut, which was attributed to plant wounding. The extensively managed field, which had a greater biodiversity, was found to yield higher methanol emissions than released from the intensive field. This was expected given that the intensive field was dominated by graminoid species, which are known to be low methanol emitters, whereas the extensive field was dominated by non-graminoid plant species, which are stronger methanol emitters.

The eddy covariance method was used with PTR-MS to measure VOC fluxes emitted by mountain grassland in Stubai Valley, Austria, in a series of studies by Bamberger *et al.* [104], Müller *et al.* [13] and Ruuskanen *et al.* [14]. The study by Müller *et al.* was the first to use a high resolution time-of-flight mass spectrometer in PTR-MS for the investigation of biogenic emissions. The mass resolution was sufficiently high in their instrument that they could separate three ions with the same nominal mass of *m*/*z* 33, and only one of these at *m*/*z* 33.0335 (corresponding to protonated methanol) actually contributed to the eddy covariance fluxes. The other two isobaric ions were assigned to be $O_2H^+/O^{17}O^+$ (*m/z* 32.9957) and $DNOH^+$ (*m/z* 33.0204). Given the low natural abundance of deuterium (D) the apparent observation of $DNOH^+$ is unexpected. However, it must be appreciated that $DNOH^+$ has been assigned purely on the *m/z* value recorded and should therefore not be considered definitive. The point is that it is well separated from protonated methanol, independent of what the actual ion is at *m*/z 33.0204. Ruuskanen *et al.*, using the same high resolution PTR-TOF-MS instrument, observed 481 distinct mass peaks in their mass spectra in the mass range from *m/z* 0-315. During harvesting significant fluxes of 18 volatile compounds distributed over 43 ions were measured. The dominant biogenic VOCs observed were methanol, acetaldehyde, ethanol, hexenal and other C_6 leaf-wound compounds, acetone, acetic acid, monoterpenes and sesquiterpenes, with total VOC emission fluxes during cutting being estimated to being up to 200 nmol m^{-2} s^{-1}. Methanol accounted for approximately half of the oxygenated VOC emissions and a third of the carbon of all VOC emissions during harvesting.

By combining PTR-MS and relaxed eddy covariance methods, fluxes of methanol and acetone released from a bare agricultural field in northern Germany during one of the hottest weeks of a heat wave in Europe were determined by Schade and Custer [106]. Positive VOC fluxes from the bare ploughed soil for these oxygenated VOCs were found. Methanol fluxes were found to range from 0 to 0.20 mg carbon m^{-2} h^{-1}, while the acetone fluxes spanned -0.01 to 0.05 mg carbon m^{-2} h^{-1}. For acetone a significant increase in its mixing ratios was observed to occur at night, which is apparently consistent with a soil emission source. In the case of methanol emissions, it was shown that the mixing ratios correlated well with heat flux and peaked around noon. By assuming a biological production mechanism from organic matter in the topsoil, the researchers calculated an activation energy of 48 kJ mol^{-1} to liberate methanol from the topsoil. At the same well-characterized agricultural plot, the researchers also reported details of a field study which used PTR-MS in an eddy covariance setup to measure methanol (*m/z* 33) and acetaldehyde (*m/z* 45) exchange rates for Italian ryegrass (*Lolium multiflorum*) over a significant portion of its life cycle [109]. Mixing ratios of methanol and acetaldehyde were determined from measurements taken 2 m above the field and were found to range from 1 to 10 ppbv and 0.4 to 2.1 ppbv, respectively. Both the methanol and acetaldehyde mixing ratios were found to track ambient temperature. The methanol emissions rates were shown to be exponentially related to ambient temperatures and were found to be higher during rapid leaf area expansion and after it had rained. Acetaldehyde fluxes showed no obvious diurnal cycle, and therefore the researchers concluded that this VOC was not emitted from the ryegrass. Although a very slow net acetaldehyde deposition to the grass and soil surfaces was observed, it was also concluded that acetaldehyde was not deposited into ryegrass or into the soil in any significant amounts. This study confirmed that methanol emissions from grasslands contribute only a small percentage (~10%) to global green plant methanol emissions.

5.3.5 Oceans and Seas

5.3.5.1 Norwegian Fjord

Sinha *et al.* investigated air-sea fluxes of methanol, acetone, acetaldehyde, isoprene and dimethyl sulfide from a Norwegian fjord following a phytoplankton bloom in May/June

2005 [110]. A PTR-MS instrument was placed on a floating raft to which mesocosms[8] were attached. PTR-MS measurements were taken of air in the mesocosm enclosures anchored in the Rauenfjord in southern Norway. The use of mesocosms allows for a much more controlled environment for determining VOC uptake and emission than simply measuring the headspace air above an ocean's surface. Five ions were monitored at m/z values of 33, 45, 59, 63 and 69 and these were assigned to be protonated methanol, acetaldehyde, acetone, dimethyl sulfide and isoprene, respectively. The net flux of methanol was always into the ocean (with the estimated mean value being 0.26 ng m^{-2} s^{-1}) and this net flux was found to be strongest at night. In contrast, isoprene and acetaldehyde were found to have a net outward emission from the ocean and the associated fluxes were found to increase with light intensity and phytoplankton abundance. Dimethyl sulfide was also observed to be present in the headspace, but unlike isoprene and acetaldehyde its flux did not correlate significantly with light. The mean fluxes for acetone, acetaldehyde, isoprene and dimethyl sulfide were determined to be 0.21, 0.23, 0.12 and 0.3 ng m^{-2} s^{-1}, respectively. The main conclusion from this work is that the air-sea flux of each trace volatile compound behaves very differently according to the amount of light and biological activity. This therefore needs to be taken into account when estimating the influence of the global ocean on atmospheric VOC budgets, especially when one considers that on a global scale there will be large variations in light and biological conditions.

5.3.5.2 Coastal Regions

In clean marine air at Cape Grim in Tasmania, Galbally *et al.* monitored methanol, acetonitrile, acetaldehyde, acetone, isoprene, methyl vinyl ketone/methacrolein, methyl ethyl ketone and benzene at 104 m above sea level over a 14-day period in the late summer of February 2006 [111]. The concentrations of acetonitrile, acetaldehyde, acetone, methyl ethyl ketone and benzene in the marine air were found to be significantly less (by approximately a factor of four) than had been previously reported for a remote tropical ocean site in the northern hemisphere. The researchers attributed this to the fact that there is less land and therefore reduced biomass burning in the southern hemisphere.

In combination with a range of other measurements, Russell *et al.* investigated the growth of aerosol particles in a coastal region of the eastern United States [112]. Their data indicated that aerosol formation primarily occurs during mornings when there are peaks in the amounts of α- and β-pinene and ozone, which leads to the formation of condensable products from photochemical oxidation.

As part of a campaign to characterize the oxidizing environment of the atmospheric boundary layer along the coast of east Antarctica, the atmospheric concentrations of acetaldehyde and acetone were measured from the 30 December 2010 to 24 January 2011 [113], and the effects of penguin emissions on these two concentrations were also reported [114]. Background levels for acetaldehyde and acetone were found to be 80 ± 12 pptv and 128 ± 9 pptv, respectively, which are consistent with other marine boundary layer

[8] A mesocosm is an enclosure containing a well-defined amount of the environment being investigated (in this study it was seawater) which is maintained at close to natural conditions, but for which environmental factors can be controlled and manipulated. It therefore provides a bridge between laboratory-based and real-world measurements. In this study, the mesocosms were simply nine polyethylene enclosures (~20 m^3 water volume each, with ~4.3 m^3 headspace volume).

measurements (e.g. Galbally *et al.* [111] for acetone (118 ± 5 pptv) and Colomb *et al.* [115] for acetaldehyde (120 ± 40 pptv)). However, when the sampling site was impacted by emissions from nearby penguin colonies, the mixing ratios increased significantly to 270 ± 70 pptv for acetaldehyde and 230 ± 70 pptv for acetone.

5.3.5.3 Indian Ocean

PTR-MS has been used to measure atmospheric trace gases over the Indian Ocean [27, 65, 115]. Sprung *et al.* compared PTR-MS and Atmospheric Pressure Chemical Ionization Mass Spectrometric (APCI-MS) airborne measurements of acetone and acetonitrile during the Indian Ocean Experiment (INDOEX) in February and March 1999 [27]. Using backward trajectory analyses, it was found that the air being sampled contained air masses originating from India, the Bay of Bengal and remote southern hemispheric air. VOC concentrations from APCI-MS measurements were determined from in-flight and laboratory calibrations using permeation tubes (see Section 4.5.2). Although the VOC concentrations determined by PTR-MS measurements were found to be systematically higher than those obtained by APCI-MS, the values obtained were still within the combined error range for both instruments. Reported mixing ratios in the northern hemispheric boundary layer were up to 2.5 ppbv for acetone and up to 0.4 ppbv for acetonitrile. Measurements in the free troposphere gave values of 0.3–1.0 ppbv for acetone and 0.12–0.3 ppbv for acetonitrile. From a remote site over the southern Indian Ocean it was possible to sample pristine air, for which mixing ratios of between 0.3 and 0.8 ppbv for acetone and below 0.2 ppbv for acetonitrile were found.

In contrast to airborne measurements, Warneke and de Gouw reported ship-based marine boundary layer measurements of VOCs over the northwest Indian Ocean taken in April 2000 [65] while Colomb *et al.* [115] made ship-borne measurements of selected VOCs during the Measurement of Anthropogenic and Natural Compounds in the Southern Hemispheric Oceanic Troposphere (MANCHOT) campaign of December 2004. The Warneke and Gouw study included the on-line analysis of dimethyl sulfide, acetone and acetonitrile using PTR-MS and GC analyses of canister samples for C_2–C_7 hydrocarbons. From the observed diurnal variation of dimethyl sulfide the 24 h averaged OH concentration was estimated to be $(1.4 \pm 0.7) \times 10^6$ molecules cm^{-3}. The study by Colomb *et al.* included PTR-MS measurements of methanol, acetonitrile, acetaldehyde, acetone, dimethyl sulfide, isoprene, methacrolein, methyl vinyl ketone, benzene and terpenes. Dimethyl sulfide, which is biologically produced in the oceans, was found to be the most abundant VOC, although its mixing ratio showed considerable variation ranging from ~50 pptv up to 885 pptv.

5.3.5.4 Tropical Atlantic Ocean

A study in the tropical Atlantic by Williams *et al.* involved measuring acetone, methanol, acetonitrile and dimethyl sulfide within the atmospheric marine boundary layer and the upper ocean [116]. Measurements were taken on either side of the intertropical convergence zone. This zone, which sailors commonly refer to as the doldrums, is an area around the equator where the southeast and northeast trade winds meet. Consequently, it is a natural barrier to atmospheric mixing between the hemispheres. This results in hemispheric gradients for VOCs whose atmospheric lifetimes are shorter than interhemispheric mixing times (typically 1–2 years). Differences in certain VOC concentrations between the southern and northern hemispheres were observed. For examples, the methanol and acetone concentrations were found to be higher in the northern hemisphere, whereas the acetonitrile and dimethyl sulfide concentrations showed no significant inter-hemispheric gradient.

5.4 Anthropogenic VOCs

5.4.1 Background

In addition to biogenic sources, VOCs are added to the atmosphere as a result of human activities. These arise from many sources including motor vehicle exhaust, industrial activities, fossil fuel burning, oil refining, storage and distribution of liquid fossil fuels, landfills and agriculture practices. All of these processes can have negative effects on air quality. On a regional scale anthropogenic VOCs can damage ecosystems and considerably affect the air quality and hence, ultimately, the quality of life in that area. For example, anthropogenic VOCs play a significant role in the production of haze/smog and the formation of tropospheric ozone and aerosols in polluted air. It is therefore vital that we are able to detect and monitor these anthropogenic VOCs in the lower atmosphere so that we can determine their sources, concentrations, evolution and the role they play in and impact on the chemistry. Consequently, there have been many studies using PTR-MS to investigate anthropogenic VOC emissions.

Although the global release of anthropogenic VOCs into the atmosphere is significantly less than from biogenic sources, they are generally concentrated in densely populated and industrialized areas. The most densely populated areas are, of course, our cities, and it is with the applications of PTR-MS in cities that we shall begin this account of PTR-MS research on anthropogenic VOCs.

5.4.2 VOCs in Urban and Rural Sites

The use of PTR-MS to detect anthropogenic VOCs present in cities/megacities is a particularly fruitful area of investigation. Cities/megacities are regions of high pollution and hence have an impact on not only local atmospheric conditions but also on larger regional scales and thus on air-quality issues in general. Detailed studies have taken place in or close to a number of cities, including Caracas [117], Houston [118], Innsbruck [119–121], Tokyo [21, 122], Barcelona [123, 124], Manchester [125], London [126], Mexico City [15, 24, 127–133], Paris [134, 135] and Toronto [136], and each will be discussed below.

5.4.2.1 Innsbruck

The first city to be investigated using PTR-MS was the relatively small city of Innsbruck in Austria. In this study, Holzinger *et al.* continuously measured (with data points being taken every 25 s) various aromatic compounds coming predominantly from traffic emissions (benzene, toluene, xylene, C_9-benzenes and C_{10}-benzenes) and acetonitrile (which is considered to be predominantly biogenic in origin) for several days (21–26 May 1998) on a main road located on the western outskirts of Innsbruck, which at that time was being used by up to 20 000 vehicles/day [119]. In comparison to the global emission rate of acetonitrile from biomass burning, as measured also by Holzinger *et al.* (on the order of 0.8 Tg per year) [137], they determined that acetonitrile emissions resulting from fossil fuel burning (upper limit of 0.021 Tg per year (+ 150%, –50%)) would only contribute less than 6% to the total acetonitrile global budget. This figure was based on the assumption that their measurements are globally representative, which of course they may not be. Nevertheless, an important conclusion drawn from their work was the suggestion that acetonitrile can be used safely as a marker for biomass burning even in areas with considerable anthropogenic emissions. However, and as will be discussed later, other studies have shown that some care must be taken in adopting *m/z* 42 as a unique marker for acetonitrile in PTR-MS measurements.

In another series of measurements in Innsbruck, Beauchamp *et al.* used PTR-MS, together with traditional atmospheric composition measurements for CO, NO_x and PM_{10} particulates (particulate matter of a size less than 10 μm), to assess the impact on air quality of heavy-duty vehicles being driven on Austrian motorways [120]. A much longer continuous study, involving the above compounds and benzene and toluene, was undertaken by the same group whereby they conducted a year-long study of VOC levels at one site in the Inn valley (from 1 April 2004 until 31 March 2005). The aim of that study was to determine the local burden of increasing traffic levels in the Tyrol [121]. Changes in concentrations of monitored VOCs were observed to vary according to traffic type and frequency and meteorological conditions within the valley, with seasonal variations being observed for all compounds measured. The results that were obtained for benzene and toluene were compared to light and heavy-duty vehicle traffic frequency. As discussed by Schnitzhofer *et al.* [121], a strong increase in levels of these volatile compounds was observed in the morning following a night-time minimum. This strong increase in VOC concentrations was obviously caused by the start of the morning rush-hour traffic. Although traffic levels in fact remained relatively constant throughout the day, an unexpected drop in pollutant levels was observed later in the morning. The researchers attributed this to the air in the valley becoming well mixed (i.e. pollutants were being dispersed into a greater air volume). An observed rise of benzene and toluene levels in the afternoon was attributed to the return of more stable stratification of the valley air, until the decline in traffic volume led to an observed night-time minimum.

5.4.2.2 Caracas

A study in Venezuela measured concentrations of benzene, toluene, xylenes and C_9-benzenes, not only in Caracas but also in suburban (Altos de Pipe), rural (Calabozo) and remote (Parupa) sites [117]. The measurements taken in Caracas throughout the day were converted to average daytime mixing ratios for a number of VOCs, specifically benzene (1.1 ppbv), toluene (3.2 ppbv), xylenes (3.7 ppbv) and C_9-benzenes (2.7 ppbv). These concentrations are significantly lower than those that have been found in some cities in the United States and Europe.

5.4.2.3 Houston

Real-time continuous monitoring of VOCs at a site near the Houston Ship Channel during the Texas Air Quality Study 2000 was reported by Karl *et al.* [118]. The anthropogenic VOCs that were observed included aromatic compounds (e.g. benzene, toluene and xylenes), aliphatic hydrocarbons (e.g. propene and isoprene), oxygenated VOCs (e.g. formaldehyde, acetaldehyde, acetone, methanol and C_7 carbonyls), and three nitrogen-containing compounds (hydrogen cyanide, acetonitrile and acrylonitrile). This study found that biogenic VOCs made only minor contributions to the VOC profile at this site. The researchers used the fast time response capabilities of PTR-MS to great effect in order to monitor rapidly changing VOC plumes as they passed over the site of investigation. By combining the PTR-MS results with meteorological data, Karl *et al.* showed that it was possible to identify likely sources of the emissions, for example, they could differentiate industrial from residential urban emission sources.

5.4.2.4 *Tokyo*

The first investigation of VOC emissions in Tokyo combined both PTR-MS and GC-FID data [21] to measure isoprene (*m/z* 69), benzene (*m/z* 79), toluene (*m/z* 93), xylenes/ethylbenzene (*m/z* 107), trimethylbenzenes/propylybenzenes (*m/z* 121) and pinenes/limonene/camphene (*m/z* 137 and 81). Although the researchers obtained good correlations between the concentrations determined from the GC-FID and PTR-MS measurements, the actual values of the concentrations were not in good agreement and differed by a factor of 0.52–2.15. The actual value of this factor was found to be very much dependent on the type of hydrocarbon. Given that GC-FID concentrations are expected to be reasonably accurate, this discrepancy in concentrations illustrates yet again the need to characterize and calibrate a PTR-MS instrument for the task at hand. Oxygenated VOCs were shown to have a seasonal variation, with higher concentrations measured in the summer than in the autumn, which the researchers attributed to the enhanced photochemical reactivity that occurs in the summer. In contrast, aromatic hydrocarbons showed no seasonal variation, as would be expected given that they result mainly from automobile exhaust. A second study of Tokyo emissions used a calibrated PTR-MS instrument to measure ~30 oxygenated VOCs, aromatics and biogenic VOCs [122]. These compounds showed characteristic diurnal variations. For example, aldehydes, acetone and isoprene concentrations were found to peak during the daytime, whereas in comparison aromatics had their peak intensities in the evening. Methanol concentrations showed no diurnal variations and were found to be almost constant throughout the day.

5.4.2.5 *Barcelona*

Fiella and Peñuelas used PTR-MS to monitor the daily, weekly and seasonal patterns of air concentrations of various VOCs in the vicinity of a highway in a semi-urban site in Barcelona [123, 124] with the measurements being taken on a minute-by-minute basis. The actual site for the measurements was at Autónoma University, which is 10 miles northwest of Barcelona. The first reported study showed that the diurnal, weekly and seasonal fluctuations in measured VOC concentrations depended on various factors, including photochemical activity and meteorological conditions [123]. The VOCs that were found to have the greatest abundances were methanol, acetone and acetaldehyde. Importantly, it was found that none of the VOC concentrations exceeded any air quality control guidelines. In their second paper, Fiella and Peñuelas identified the most important VOC species involved in ozone production via photochemical processes [124], which is of interest to air quality control in the troposphere. Ozone is harmful to humans and can damage vegetation. In order of importance, the VOCs which were found to generate the highest ozone formation were formaldehyde, acetaldehyde, methanol, toluene, isoprene and acetone.

5.4.2.6 Manchester and London

Langford *et al.* used PTR-MS to determine VOC fluxes in the UK urban areas of Manchester [125] and London [126]. For the Manchester study, six VOCs were selected for investigation: methanol, acetaldehyde, acetone, isoprene, benzene and toluene. Both disjunct and virtual disjunct eddy covariance techniques were used by the researchers to calculate the volatile fluxes from the measured mixing ratios. For the London measurements eight VOCs were selected for investigation, namely methanol, acetonitrile, acetaldehyde, acetone, isoprene, benzene, toluene and C_2 benzenes (such as ethylbenzene, benzaldehyde and the xylene isomers). The fluxes of methanol, acetone, benzene and toluene were found to be larger in London than in Manchester, which is consistent with the larger organic aerosol emission fluxes found above London compared with those found in Manchester.

5.4.2.7 Mexico City

Mexico City is one of the world's largest megacities, being exceeded only by the metropolitan area of Tokyo, and hence has one of the worst air pollution problems in the world [138]. As a result of this the VOCs present in the ambient air in the Mexico City area have been extensively studied with PTR-MS. As part of the Mexico City Metropolitan Area (MCMA) field campaigns (2002–2003 and 2006) two PTR-MS instruments were used for measurements of ambient VOCs, either from fixed sites or from a mobile laboratory [15, 130–133]. Kolb *et al.* [139] and Herndon *et al.* [132] have provided details on the mobile laboratory, which has been used to obtain rapid real-time measurements of urban and regional trace gases (in addition to particulate distributions and emission source characteristics). The advantage of using a mobile laboratory is that either individual vehicle emissions can be investigated or 'fleet-averaged' exhaust emissions can be measured under real-world driving conditions. Herndon *et al.* found that fleet-averaged exhaust emissions of formaldehyde, acetaldehyde and benzene are substantial in Mexico City.

The mobile laboratory mentioned above was also used for on-road VOC measurements during the MCMA 2003 field campaign by Rogers *et al.* [15] and Zavala *et al.* [133]. Rogers *et al.* adapted a PTR-MS instrument to handle high impact shocks and the thermal loads that would be experienced within a mobile laboratory used as a chase vehicle for on-road operation. Fast time response (<2 s) measurements were made for the following volatiles: methanol, acetaldehyde, acetone, methyl tertiary butyl ether, benzene, toluene, C_2-benzenes (the sum of xylene isomers/ethylbenzene, and benzaldehyde) and C_3-benzenes (the sum of C_9H_{12} isomers with possible contributions from C_8H_8O isomers). This work found that there are significant sources of toluene which did not come from vehicles. Some of the PTR-MS results were compared with GC-FID analyses of air samples collected into air canisters. This showed that the VOC concentrations derived from PTR-MS agreed well with corresponding GC-FID measurements. Comparison of PTR-MS and GC-FID concentration measurements was used to show that ions found in the PTR-MS mass spectra at *m*/*z* 57, 79, 93, 107, and 121 were derived predominantly from methyl tertiary butyl ether/butenes, benzene, toluene, C_2-benzenes and C_3-benzenes, respectively. Although the contribution to the signal at *m*/*z* 107 was found to be dominated by C_2-benzenes, the correlation for the

C_2-benzenes was found to be not as good as that for the other compounds, with the PTR-MS measurements overestimating the concentration by ~30%. It was believed that in part this could have resulted from a contribution of benzaldehyde (C_7H_6O), whose protonated mass is at *m/z* 107. However, the researchers mention that at best this compound could have only contributed about 5% to the overall measurement and therefore it was proposed that other VOCs must have caused the larger and spurious PTR-MS measurements. Similarly, the concentrations derived for the VOCs contributing to *m/z* 121 ions were found to be 50% larger than those determined by GC-FID. Part of the discrepancy was associated with the omission of C_8H_8O isomers from the GC-FID data. However, this was insufficient to account for a 50% discrepancy. Zavala *et al.* used the mobile laboratory and the PTR-MS instrument adapted by Rogers *et al.*, for a complete analysis of motor vehicle exhaust emissions. From their estimates the exhaust emissions of benzene, toluene, formaldehyde and acetaldehyde were found to be higher than had been previously reported.

Velasco *et al.* reported a series of fixed-site studies relating to the measurement of VOCs during the MCMA field campaign 2002–2003 [131, 140] and the Megacity Initiative: Local and Global Research Observations (MILAGRO)/MCMA 2006 project [130]. The first of these studies was performed in a densely populated residential district of Mexico City [131]. This investigation demonstrated the feasibility of using PTR-MS combined with disjunct eddy covariance to calculate the fluxes of methanol, acetone, toluene and C_2-benzenes. The flux measurements were found to exhibit a diurnal pattern that was strongly related to the amount of traffic. This was corroborated by Velasco *et al.* during the MILAGRO/MCMA-2006 field campaign by a second set of flux measurements in a different district of Mexico City [130]. Using, once again, a disjunct eddy covariance method, PTR-MS was used to measure fluxes of methanol, benzene, toluene and C_2-benzenes and these were analysed by the researchers in terms of diurnal patterns and vehicular activity. After allowances were made for differences in local traffic between the two sites, the results from the MILAGRO/MCMA-2006 field campaign were found to be in good agreement with those obtained in the first field campaign. These studies demonstrated for the first time that eddy covariance techniques coupled with the fast time response measurements that are possible with PTR-MS provide a new and valuable tool for directly evaluating emission inventories, which the researchers proposed could be used to improve air quality management processes.

As part of the MCMA-2002 (7–24 February 2002) and MCMA 2003 (1 April–5 May 2003) field campaigns, Velasco *et al.* undertook a comprehensive study of VOCs from various fixed sites in the Valley of Mexico that included locations in the urban core, in a heavily industrial area and at boundary sites in rural landscapes [140]. PTR-MS measurements were combined with GC-FID, a technique known as a Fast Olefin Sensor,[9] and long path measurements using UV Differential Optical Absorption Spectrometry (DOAS). The use of these different and independent techniques allowed a wide range of VOCs to be monitored with different spatial and temporal scales. Generally, good agreement was found by comparing the results obtained from these various techniques for selected VOCs. PTR-MS was specifically used to monitor benzene, toluene, styrene, C_2-benzenes, C_3-benzenes, naphthalene, phenol, cresols, methanol, acetaldehyde, acetone and acetonitrile.

[9] As the name implies, this technique can measure olefins (alkenes). It is based on the use of a chemiluminescence reaction between the alkene and ozone to enable detection.

Elevated levels of toxic hydrocarbons, such as 1,3-butadiene, benzene, toluene and xylenes were found. This study also showed that vehicle exhaust is the major source of VOCs in Mexico City.

Jobson *et al.* reported and compared aromatic hydrocarbon measurements using several techniques, namely PTR-MS and DOAS along with GC-FID of canister samples taken from a roof-top site at the Mexican National Centre for Environmental Research and Training in the district of Iztapalapa [24]. The point sampling measurements employed by PTR-MS and GC-FID showed good correlations and were found to be in reasonable agreement for toluene, C_2-alkylbenzenes and C_3-alkylbenzenes. However, the benzene mixing ratio determined from PTR-MS was found to be greater than that obtained from the GC-FID measurements by on average a factor of 1.8 ± 0.6. The researchers suggested that the PTR-MS data for benzene suffered positive interference from fragment ions contributing to the ion signal at *m/z* 79, resulting from dissociative proton transfer to ethyl benzene and propyl benzene and thereby leading to an overestimation of the benzene concentration. Mixing ratios measured by DOAS were found to be on average a factor of 1.7 times greater than those obtained from the PTR-MS data for toluene, C2-alkylbenzenes, naphthalene and styrene. This again highlights the potential problems of using the intensities associated with a single *m/z* signal in a PTR-MS mass spectrum for quantitative measurements.

Bon *et al.* used a proton transfer reaction ion trap mass spectrometer (PTR-IT-MS) for the on-line quantification of 12 VOC species, including oxygenated VOCs and aromatics, at a suburban ground site during the MILAGRO campaign in March 2006 [128]. PTR-IT-MS has the advantage over quadrupole and time of flight mass spectrometers in that ions with a specific *m/z* can be isolated and trapped and then subjected to collisional processes to induce fragmentation, from which the resulting fragment pattern can help in the identification of the original VOC, as previously mentioned in Section 3.5.3.2. To further aid in the identification of the VOCs a GC separation system in conjunction with the PTR-IT-MS was also used. The diurnal cycles observed for the various VOC emissions were found to be significantly influenced by vehicle traffic and emissions from liquid propane gas during the night and photochemical processing during the afternoon. Also as part of the MILAGRO 2006 field campaign, Fortner *et al.* [129] used PTR-MS to detect VOCs on a rooftop in a mixed residential and industrial area northeast of downtown Mexico City. Thirty-eight individual ions were monitored, with many species being quantified including methanol, acetaldehyde, toluene, the sum of C_2 benzenes, the sum of C_3 benzenes, acetone, isoprene, benzene and ethyl acetate. This study resulted in an improved understanding of the type of VOCs present in the MCMA including their diurnal patterns and their origins.

5.4.2.8 *Toronto and Environs*

Over a 5-week period (May–June 2007) Vlasenko *et al.* used PTR-MS (in addition to GC-FID and GC–MS) to measure VOCs at a rural site in Ontario which consisted of farmland (mainly pasture, soybeans, alfalfa and corn) and mixed forest (primarily cedar and maple), some 70 km north of Toronto [45]. A southerly wind flow resulted in the detection of VOCs of anthropogenic origin (e.g. aromatic compounds were observed), whereas wind flows from the northwest resulted in VOCs that were substantially biogenic in origin (e.g. monoterpenes and isoprene were detected). Comparison between PTR-MS, GC-FID and GC–MS for toluene and benzene showed that the PTR-MS mixing ratios were 1.42 and 1.65 times higher, respectively, than those determined from the GC measurements. This discrepancy was attributed to the relatively high *E/N* used (135 Td), which resulted in fragment ions from higher aromatics contributing to the ion signals at *m/z* 79 where protonated benzene was being measured and *m/z* 93 where protonated toluene was being monitored. PTR-MS measurement of monoterpenes recorded at *m/z* 137 (protonated parent $C_{10}H_{16}H^+$) and at *m/z* 81 ($C_6H_9{}^+$, an ion fragment from protonated monoterpenes) resulted in a mixing ratio which was found to be approximately a factor of two higher than had been deduced from GC-FID, a discrepancy which was considered to have occurred because certain monoterpene or sesquiterpene compounds were not being detected by GC-FID. As expected for PTR-MS measurements, the mixing ratio for formaldehyde showed considerable variation with humidity for reasons discussed in Section 5.1.

Slowik *et al.* determined VOC concentrations in an urban area downtown of Toronto during the winter component of the Seasonal Particle Observations in the Region of Toronto (SPORT) field campaign [136]. PTR-MS data were combined with those obtained using an Aerodyne time-of-flight aerosol mass spectrometer and were then analyzed using positive matrix factorization (PMF). [10] This provided information on emission sources and atmospheric processing.

5.4.2.9 Paris

Another urban study was that by Gros and coworkers, who reported PTR-MS investigations of VOC sources in Paris during the spring of 2007 [134, 135]. The first paper provides a qualitative analysis of the detected VOCs, which demonstrates that the NMHCs were produced as a result of mainly local and regional traffic emissions, whereas the oxygenated VOCs predominantly came from air masses derived from the Atlantic Ocean and from north-eastern Europe. Mean mixing ratios for toluene and benzene from PTR-MS were found to be in excellent agreement with those determined using GC-FID. The second paper was a source apportionment study aided by PMF simulations using data recorded hourly by PTR-MS and GC-FID and focusing on 21 VOCs (consisting of alkanes, alkenes, alkynes, aromatics and oxygenated VOCs) [135]. The major VOC sources were reported to be associated with traffic-related emissions (vehicle exhaust (22%) and fuel evaporation (17%)), remote industrial sources (35%), natural gas and background (13%), local sources (7%), biogenic and fuel evaporation (5%) and wood burning (2%). Another PTR-MS study undertaken in Paris during the MEGAPOLI winter campaign of 2010 was a measurement of total OH reactivity using the Comparative Reactivity Method (see Section 5.6.2) [141]. It was found that when clean marine air masses entered the Paris atmosphere, the local OH reactivity of 20 s^{-1} could be explained by the observed VOC species. However, when there was a continental import of air masses, much higher levels of OH reactivity were observed (120–130 s^{-1}).

5.4.2.10 Boston, New York and Los Angeles

Warneke *et al.* undertook measurements downwind of urban areas in Boston, New York and Los Angeles to determine emission ratios of VOCs relative to CO [142]. The results obtained

[10] PMF is a statistical tool which can be applied to identify and quantify sources of air pollution by using concentration data.

were found to compare well with other measurements taken from urban areas and also to measurements of vehicle exhaust (at least, that is, to within a factor of two), showing that a large source of VOCs in urban areas is, as expected, from car exhaust. However, the measured emission ratios did not compare well with a frequently used anthropogenic emissions database (based on the EPA NEI-99 (Environmental Protection Agency National Emissions Inventory-99) database). The alkanes and oxygenated VOC measurements showed discrepancies of up to an order of magnitude.

5.4.3 Diesel Engine Emissions

Jobson *et al.* measured on-line VOC concentrations in the exhaust coming from diesel engines as a function of engine load [143]. in order to evaluate the analytical capabilities of PTR-MS for diesel engine emissions. A key aim of this work was to evaluate the analytical capabilities of PTR-MS for investigating diesel engine emissions and in particular to provide information that can and in particular to provide information that can be used to reduce VOC emissions. Although the diesel exhaust mass spectra obtained were found to be complex, they nevertheless displayed a pattern of strong ion signals at $m/z = 14n + 1$ ($n = 3 - 8$), with a relative ion abundance similar to that obtained from 70 eV electron impact ionization of alkanes. The researchers were able to assign about 75% of the organic ion signal and found that the relative abundances of benzene and toluene increased with engine load, which the researchers therefore attributed to be of pyrogenic origin.

5.4.4 Aircraft Emissions

Aircraft emissions at airports can have an impact on both regional and local air quality, and therefore it is important to have information about the types and concentrations of the VOCS emitted under all possible operational situations that occur at airports, for example, when an aircraft is idling, taxiing, taking off or landing. Together with tuneable infrared absorption spectroscopy, PTR-MS has been used by Herndon *et al.* to characterize VOC emissions from commercial aircraft during airport operations from downwind exhaust plumes [144]. The compounds monitored were methanol, acetaldehyde, acetone, benzene, toluene, C_2 benzenes and C_3 benzenes. The hydrocarbon emission index, which is a measure of the quantities of hydrocarbons emitted per kilogram of aircraft fuel burnt, was determined to be larger in idle and taxiway acceleration plumes relative to approach and take-off plumes. Direct measurements of hydrocarbon emissions from the engine exhausts of a DC-8 aircraft on the ground have also been made for different fuel types [145, 146]. These real-time measurements highlighted interesting variability with transient behaviour being observed on a time scale of several seconds.

5.4.5 VOC Emissions Associated with Farming

In this section, VOC emissions associated with certain farming activities will be considered. Although these emissions could be classified as biogenic in origin, they are a direct consequence of human activities and hence will be regarded as anthropogenic. There are plenty of sources of VOCs associated with human activity in a farming context including flooded rice fields, decomposition of organic materials at waste disposal sites and livestock husbandry [147]. One of the major VOCs emitted by livestock is methane. However, methane undergoes very little chemistry in the Earth's atmosphere. Furthermore, H_3O^+ does not react with methane. Therefore, methane emission will not be discussed any further in this section.

5.4.5.1 Cattle

Within the area of animal husbandry, Shaw *et al.* used PTR-MS to investigate emissions from lactating and dry (i.e. non-lactating) dairy cows and from their waste within an environmental chamber [148]. The aim of the research was to determine the potential impact of these emissions on photochemical ozone production. The data obtained showed that when cows and their waste were present, the VOCs with the highest fluxes were methanol, acetone/propanal (isomers), dimethyl sulfide and a compound giving a peak at *m/z* 109 in the mass spectrum, which the researchers suggested is protonated 4-methylphenol (cresol, C_7H_8O). The compounds with highest fluxes from fresh waste (urine and faeces) were found to be methanol, *m/z* 109 and *m/z* 60 (the latter ion was assigned to be protonated trimethylamine, C_3H_9N). Interestingly, several VOCs that were expected to be present, namely formaldehyde, methylamine and dimethylamine, were not detected. An important result from this work was that the sum of reactive VOC fluxes measured when cows were present was found to be a factor of 6–10 less than estimates that were being used for regulatory purposes. In addition, ozone formation potentials of the dominant VOCs were ~10% of those for typical combustion or biogenic VOCs. The major and perhaps unexpected finding from this work is that VOCs emitted by dairy cattle have a comparatively small impact on ozone formation.

Trimethylamine has also been detected by PTR-MS above a manured grass field by Twigg *et al.* [149]. In their study, the signals of ions at *m/z* 58 ($C_3H_8N^+$) and *m/z* 60 ($C_3H_{10}N^+$) were added to determine the concentrations of trimethylamine emitted from a managed grassland recently fertilized with 164 kg N (nitrogen) ha^{-1} of cattle slurry. The peak at *m/z* 60 is from the protonated parent molecule while *m/z* 58 represents a fragment ion corresponding to loss of H_2 from the protonated parent molecule, which the researchers claimed should have a branching ratio of between 10% and 34%. Twigg *et al.* estimated that in the first 31 h following the initial slurry application the average trimethylamine flux amounted to 0.38% of the ammonia/reactive nitrogen (NH_3/N) emissions.

5.4.5.2 Pigs and Sheep

PTR-MS has also been used to investigate the emissions from intensive pig production in Denmark [150–152], which is an obvious and often cited cause of foul air in populated rural areas. The highest emissions detected by PTR-MS were found to be associated with hydrogen sulfide and acetic acid. Also in Denmark, PTR-MS was used to quantify and study the temporal variation of gas emissions from a model pig house (scale 1:12.5 of a commercial pig house) under varying ventilation rates [153]. The aim of such model studies is to aid in the development of emission abatement technology. In order to do that, it is first necessary to identify the different factors that can affect the emission of VOCs. Concentrations of two inorganic volatiles (NH_3 and H_2S) and fourteen VOCs were measured coming from a slurry pit and at the outlet of the model pig house. The fourteen VOCs and their protonated *m/z* values were methanethiol (*m/z* 49), acetone (*m/z* 59), trimethylamine (*m/z* 60), acetic acid (*m/z* 61), dimethyl sulfide (*m/z* 63), C_4-carbonyls (e.g. 2-butanone) (*m/z*

73), propanoic acid (*m/z* 75), 2,3-butanedione (*m/z* 87), phenol/dimethyl disulfide (*m/z* 95), C_5 carboxylic acid (*m/z* 103), 4-methylphenol (*m/z* 109), indole (*m/z* 118), 4-ethylphenol (*m/z* 123) and 3-methyl-1*H*-indole (*m/z* 132). The researchers found that the emissions of H_2S and the 14 VOCs were lower compared to those found in real pig buildings, which indicated possible other sources of these compounds in the real pig building.

In related work, VOCs were recorded in a pigsty and a sheep shed at the Federal Agricultural Research Centre (FAL) in Mariensee, Germany, by Ngwabie *et al.* [154]. Some 17 major VOCs were measured in the sheep shed, while in the pigsty 15 main VOCs were identified. Of these the VOC emissions from both sheds were dominated by alcohols. From the sheep study the major VOC was found to be ethanol, whereas it was found to be methanol from the pigs. An important conclusion from this work is that the total annual VOC emissions from the animal husbandry sector in Germany results in ~150 Tg carbon, which is less than had been previously estimated.

5.4.6 Other Studies of Anthropogenic VOCs

5.4.6.1 Air Quality

Gas-phase PTR-MS VOC measurements have been combined with aerosol measurements of polluted and clean air masses off the coast of the north-eastern United States as part of a New England Air Quality Study that took place in 2002. The results were mainly used to determine the overall budget of organic carbon and to separate the sources of organic carbon in an attempt to categorize them quantitatively for primary anthropogenic emissions, biogenic sources and the photochemical age [155]. As part of the same study, a comparison of the daytime and night-time oxidation of biogenic and anthropogenic VOCs in summer was undertaken [156]. The night-time chemistry was found to have dominated the removal of biogenic VOCs from the air.

5.4.6.2 Firework Emissions

An unusual application of PTR-MS, but one which is concerned with anthropogenic VOCs released into the atmosphere, was an investigation of VOCs emitted during New Year firework festivities on 31 December 2004 at a site close to the Rhine River in Mainz, Germany [157]. From the mass spectra recorded it was found that the fireworks introduced a significant quantity of VOCs into the atmosphere consisting predominantly of acetone, acetonitrile and methanol. The VOCs showed a sharp increase in concentration observed just after midnight, which was then followed by a brief decline of several minutes. The trace gas concentrations were then found to increase again and ultimately peaked ~50 min after midnight. The observed reduction in emissions just after midnight was attributed by the researchers to have resulted from a '*champagne dip*', that is, there was a pause in setting off the fireworks whilst celebrating the New Year.

5.5 Biomass Burning

Biomass burning corresponds to the burning of living and/or dead vegetation, which can be anthropogenic (e.g. burning for the clearing of land, land management or change in land use, cooking, charcoal or brick making, heating, lighting etc.) or natural in origin (e.g. wildfires caused by, for example, lightning strikes or volcanic activity). Of these two major types of biomass burning, human-initiated burning of vegetation is the dominant source. Biomass burning represents a major and complex source of a wide range of trace VOCs in the global troposphere [158, 159], and hence it has a significant influence on atmospheric chemistry. For example, some of the VOCs released, such as the light aromatics, are precursors for aerosol production. A problem particularly associated with PTR-MS for investigating biomass burning is the complicated VOC mix. Specific compound identification can therefore be very challenging and thus GC separation is helpful for accurate speciation [160].

As mentioned earlier, Holzinger *et al.* suggested that acetonitrile, CH_3CN, is a useful marker for biomass burning even in areas with considerable traffic density [137]. This marker makes PTR-MS a powerful tool for identifying and monitoring biomass burning plumes [33, 44, 62, 161–164] and, in combination with CO measurements, it has been used to delineate biomass burning from fossil fuel emissions [163]. However, a word of warning with regards to the use of m/z 42 as a marker for acetonitrile is required. Sanhueza *et al.* reported significant concentrations of acetonitrile over Venezuelan woodland savannah during the wet season, which they attributed to its release from the warm waters of the Caribbean Sea [165]. Certainly, there is evidence from PTR-MS measurements that the ocean is an effective sink for acetonitrile [65, 161]. Although it is expected that acetonitrile concentrations should be much higher in biomass burning plumes, Sanhueza *et al.* commented that as a result of their observations the use of acetonitrile as a biomass burning marker should be treated with some caution. Furthermore, Dunne *et al.* have also shown that an ion signal at m/z 42 cannot be assumed to arise solely from protonated acetonitrile [166]. In this study, it was shown that the ^{13}C isotopologues of $C_3H_5{}^+$ and the product ion $C_3H_6{}^+$, which can be produced from reactions of NO^+ and $O_2{}^+$ with a number of VOCs (e.g. alkanes and alkenes), contributed to the ion signal at m/z 42. As mentioned earlier in this book NO^+ and $O_2{}^+$ are impurity reagent ions that are formed in PTR-MS as a result of air back streaming from the drift tube into the ion source and hence are always present to a greater or lesser extent in the drift tube. Although in low concentrations compared to H_3O^+ reagent ions (typically their combined intensities are ~2–3% of the H_3O^+ ion abundance), their reactions with VOCs cannot always be ignored. To quantify the contribution of reactions of NO^+ and $O_2{}^+$ to the protonated acetonitrile signal at m/z 42, Dunne *et al.* used a PTR-MS instrument with a switchable reagent ion source. This instrument was employed for a series of sequential measurements with H_3O^+, NO^+ and $O_2{}^+$ as reagent ions taken in urban air during 17 days of the summer (February–March 2011) in Sydney, Australia. Of some concern is that it was shown that the non-acetonitrile contributions to the signal at m/z 42 could be significant, with the percentage ranging from 5% to 41%, depending on pollution levels at the time of the measurements. The researchers therefore concluded that corrections to the PTR-MS data will be required that allows for the presence of alkenes and alkanes in urban air in order to determine correct acetonitrile concentrations.

Although some care must be taken when determining acetonitrile concentrations from the ion intensity at m/z 42, acetonitrile is still the dominant VOC contributing to that mass channel. From the PTR-MS airborne measurements of acetone, methanol, PAN and acetonitrile over the Mediterranean Sea during the Mediterranean Intensive Oxidants Study

(MINOS August 2001), Holzinger *et al.* [44] identified 10 biomass burning plumes over the eastern Mediterranean from strongly elevated acetonitrile mixing ratios. The characteristic biomass burning signatures obtained from these plumes showed higher concentrations of acetone and methanol than found in controlled experimental fires, suggesting that secondary production of these compounds must have taken place in the biomass burning plumes. The researchers commented that if this is ignored then the biomass burning contributions of these two VOCs would be significantly underestimated.

An important aspect of biomass burning is to determine the transport processes of the emissions and in particular how far they can travel in the atmosphere and hence influence air quality at a remote site. Another important aspect to quantify is the removal of VOCs by reactions in the atmosphere as they are being transported. Flying over New England in a NOAA WP-3 aircraft, de Gouw *et al.* intercepted aged forest fire plumes emanating from Alaska and western Canada during the New England Air Quality Study-Intercontinental Transport and Chemical Transformation (NEAQS-ITCT) 2k4 mission in 2004 [33]. Measurements of acetonitrile clearly proved that the air masses had been influenced by biomass burning. The findings showed that the emissions from forest fires were rapidly transported upward by fire-induced convection and were then efficiently transported over many thousands of kilometres without extensive atmospheric chemical removal. The researchers commented that if the fire plumes mix with the continental boundary layer, then these reactive VOCs can participate in the VOC-NO_x chemistry and thereby contribute to regional formation of ozone and organic aerosols.

Other PTR-MS studies have also shown that the effects of biomass burning can be transported over extremely long distances. Real-time continuous monitoring of VOCs using PTR-MS was performed at the Mauna Loa Baseline Station in Hawaii in March–April 2001 [162]. Evidence was obtained for the long-range transport of biomass plumes, but this time coming from Southeast Asia and the Indian subcontinent. Similarly, work by de Gouw *et al.*, who used a combination of aircraft-based PTR-MS measurements of the air over the Pacific coast of the United States together with air back-trajectory analysis, which permits the identification of a plume's origin, showed not only that long-range transport processes occur, but that contributions from air masses were coming from Southeast Asia and China, where biomass-burning emissions are known to be relatively high [164]. It was found that these air masses were transported across the Pacific Ocean at higher altitudes than air masses that were coming from Japan and Korea, which contain greater concentrations of those VOCs deriving from fossil fuel combustion. Long-range transport of biomass burning has also been observed in Nova Scotia from intense forest fires in Alaska and the Yukon Territory [167].

From ground-based measurements in Crete taken in 2001, during the MINOS campaign for the months July–August, Salisbury *et al.* measured the presence of acetonitrile together with benzene, toluene, methanol and acetone in the air during three time periods [62]. The first period involved air transported from Eastern Europe, which showed little influence from biomass burning. In the second period, the air investigated came from Western Europe. In the third period, the air was predominantly transported from Eastern Europe which had passed east and south of/over the Sea of Azov. From this third and last period of air sampling, the researchers determined a mean acetonitrile mixing ratio of $\sim$0.44 ppbv, which indicated that significant biomass burning had occurred. That high biomass burning had occurred in Eastern Europe was confirmed by the use of satellite images available from NASA.

In addition to airborne and ground-based measurements, for which VOCs were observed that had resulted from biomass burning, PTR-MS was used by Wisthaler *et al.* for fast-response VOC measurements on-board the NOAA research vessel Ronald H. Brown during leg 2 (4–23 March) of the INDOEX 1999 cruise [163]. By using an air flow trajectory analysis, their data showed that biomass burning coming from West India had a strong impact on the air being sampled. Mixed pollution sources from North-East India and the dominance of fossil fuel combustion in the Middle East were also observed. Elevated abundances of acetone and acetaldehyde were found which were attributed to unidentified sources in both biomass-burning impacted air masses and remote marine air.

From both airborne and ground-based field campaigns during the 2004 Amazon dry season as part of the Tropical Forest and Fire Emissions Experiment (TROFFEE), Yokelson *et al.* quantified the volatile emissions from a pristine tropical forest and several plantations [168]. In addition to this the emissions, fuel consumption and fire ecology of tropical deforestation fires were also quantified. About 80% of the mass of non-methane organic compounds emitted by the tropical deforestation fires were found to be associated with reactive, oxygenated VOCs. Recommended values for the emission factors for most of the major compounds released by deforestation fires globally were provided.

As part of the Megacity Initiative Local and Global Research Observations (MILAGRO) project, a comprehensive airborne study by Yokelson *et al.* reported the first detailed field measurements of biomass emissions in the Northern Hemisphere tropics [169]. Volatile emissions were measured from 20 deforestation and crop residue fires on the Yucatan peninsula. This included two trace gases which are often considered to be useful as indicators of biomass burning. One we have discussed before, namely acetonitrile, and the other is hydrogen cyanide. A variety of instrumentation was co-deployed for this investigation (FTIR spectroscopy, GD-FID, a GC-Trace Analytical Reduction Gas Detector, fluorescence and chemiluminescence instruments and various other spectrometers). PTR-MS was used to monitor methanol, acetonitrile, acetaldehyde, acetone, methyl ethyl ketone, methyl propanal, hydroxyacetone plus methyl acetate, benzene and 13 other volatile species.

Inomata *et al.* reported an intensive field campaign at Mount Tai, China, in June 2006 (MTX2006), using PTR-MS [26]. On one of the nights high concentrations of non-methane VOCs were observed, which were attributed to the burning of crop residues. In particular, the PTR-MS measurements showed an increase in acetonitrile. In addition to acetonitrile, it was found that oxygenated VOCs, rather than aromatics, also showed increased mixing ratios.

5.6 Applications of PTR-MS to Laboratory Studies of Atmospheric Chemistry

Although the use of PTR-MS in the environmental science area has been primarily used in field campaigns, it has also been successfully used in laboratory and enclosed environments for VOC emission investigations of importance to atmospheric chemistry. Laboratory studies, which are divorced from the complex and uncontrolled chemical environment of the atmosphere, can considerably improve our understanding of the reaction processes occurring within the drift tube of the PTR-MS. In particular, the investigation of individual chemical compounds of atmospheric relevance permits a study of ion production as a function of *E/N* in order to ascertain what fragment ions are produced and how they may contribute to signal intensities at a particular *m/z*. This is crucial information needed for field measurements of VOCs.

GC–MS has traditionally been used to distinguish isobaric and isomeric compounds in the atmosphere, i.e. compounds with the same nominal (integer) molecular mass (isobaric) or identical molecular mass (isomeric). A potential alternative approach is to use PTR-MS and exploit any *E/N* dependences in fragmentation in an attempt to distinguish isobaric/isomeric compounds via spectral fingerprinting. Tentative first steps towards the identification of structural isomers in mixtures via PTR-MS using this approach were performed for monoterpenes by Steeghs *et al.* [170] and Misztal *et al.* [171] and for sesquiterpenes by Kim *et al.* [172].

The first of these studies not only reported the potential use of *E/N* dependences, but also presented details on the use of collision-induced processes in a quadrupole ion trap to differentiate monoterpenes. Although it was possible to identify specific monoterpenes based on collision-induced dissociation, the identification of several monoterpenes in a complex mixture would still be difficult. Misztal *et al.* also investigated the effect of varying *E/N* over a limited range of approximately 100–140 Td in a bid to extend the selectivity of PTR-MS; this was specifically attempted in order to discriminate individual monoterpenes. By using a detailed analysis of fragmentation pathways the researchers have proposed that PTR-MS could be used to determine individual monoterpenes in a complex chemical mixture.

Kim *et al.* explored the ion fragmentation patterns for a series of sesquiterpenes (β-caryophyllene, α-humulene, aromadendrene, α-cubebene, neoclovene, isolongifolene, d-cadinene and triisopropylbenzene) as a function of *E/N* over the range of 76–141 Td and their dependence on the humidity in the drift tube [172]. All of these sesquiterpene compounds were found to produce the same product ions upon reaction with H_3O^+ but the relative product ion branching percentages were found to be compound specific. However, the differences in branching ratios are unlikely to be sufficient to provide a procedure that could be used to determine individual sesquiterpenes in a complex chemical matrix. Instead, Kim *et al.* were able to demonstrate that quantitative PTR-MS measurements are capable of determining total sesquiterpene concentrations and fluxes with an approximate 20% accuracy if two of the most abundant product ions at *m/z* 149 (fragment ion) and 205 (protonated parent molecule), out of the seven main product ions (*m/z* 81, 95, 109, 123, 135, 149 and 205), were monitored. The findings were applied to a field dataset collected above a deciduous forest to determine average daytime ecosystem scale mixing ratios of the sum of sesquiterpenes.

The studies by Steeghs *et al.* [170] and Misztal *et al.* [171] provided evidence that suggested PTR-MS could be used to discriminate certain isomeric VOCs by varying *E/N* and monitoring the corresponding mass spectral patterns. Further investigations of *E/N* fragmentation dependences may allow algorithms and measuring protocols to be developed, which could potentially be used to deconvolute contributions of various VOCs in a real-world atmospheric sample from PTR-MS measurements. At present such algorithms and protocols do not exist and therefore more fundamental research is required to determine if they are capable of providing sufficient discrimination between isomeric compounds present in the same chemical environment.

5.6.1 Laboratory Studies of Biomass Burning

It is often useful to study the burning of biomass in the controlled environment of a laboratory before attempting field investigations (see Section 5.5). Laboratory measurements prior to a field campaign are divorced from the complications of sampling and analyzing complex air samples and they permit validation of the use of PTR-MS in smoke-filled environments to be made. Furthermore, such studies help in planning the most appropriate sampling protocols and aid in determining product ion distributions and hence quantifying the concentrations of VOCs produced from biomass burning.

Using a combustion facility available at the U.S. Department of Agriculture Forest Service Fire Sciences Laboratory, Christian *et al.* measured trace gas and particle emissions from 47 laboratory fires burning 16 major fuel types (e.g. grass, leaves, twigs, peat, forest floor litter, shrubs, brush, hardwood log, shredded hardwood, soil and pine needles) collected predominantly from Africa and Indonesia [173]. A number of previously unquantified oxygenated compounds emitted from simulated African fires were identified including phenol, acetol, glycoaldehyde and furan. Oxygenated VOCs were found to account for ~70% of the non-methane organic compounds emitted by African savannah fires and 77% from Indonesian fires, which is consistent with the known domination of oxygenated VOCs in global atmospheric organic chemistry. This was confirmed by the same group in a companion paper [29], which provided for the first time a comparison of the quantitative performance of open-path Fourier transform infrared (OP-FTIR) spectroscopy, PTR-MS, GC–MS, GC-FID and GC-electron capture detection. The agreement between mixing ratios obtained using OP-FTIR and PTR-MS techniques was found to be excellent for methanol, good for phenol and acetol and for the combinations of furan/isoprene (isobaric) and acetic acid/glycolaldehyde (isomers). However, poor agreement was found for acetone and methyl vinyl ether, which are isomers. This led the researchers to suggest that one of these compounds must have been underestimated by OP-FTIR or that this technique was incapable of detecting other compounds that could have contributed to the PTR-MS ion signal at *m/z* 59 (which predominantly comes from protonated acetone and methyl vinyl ether). On the other hand good agreement between PTR-MS and GC measurements was found for benzene and toluene.

Other investigations using the Forest Service Fire Sciences Laboratory facility as part of the TROFFEE campaign have been reported by Karl *et al.* [17] and Yokelson *et al.* [174]. Karl *et al.* quantified VOC emissions from fires in tropical forest fuels using a combination of PTR-MS, FTIR spectroscopy and GC coupled to PTR-MS. From this combination of analytical techniques 72% of all emitted species over a mass range that went up to *m/z* 205 were identified. This investigation in fact involved both laboratory and field studies. The laboratory measurements consisted of 19 controlled laboratory fires while the field measurements involved an intensive airborne field campaign of 16 real forest fires during the peak of the burning season in Brazil. The results showed that a wide range of oxygenated species (OVOCs) were produced and that these dominated NMHC emissions from fires; the OVOC/NMHC ratio was determined to be approximately 4:1. The most abundant oxygenated volatile compounds were found to be formaldehyde, acetaldehyde, methanol, acetone and acetic and formic acid.

Yokelson *et al.* have also reported results from the burning of tropical forest fuels using various analytical methods, including FTIR spectroscopy, PTR-MS, GC, and GC/PTR-MS

[174]. Using all of the analytical methods, the researchers were able to identify ~50% of the mass associated with non-methane VOCs. The emission factors determined from the laboratory fires were fed into those obtained from their airborne and ground-based field campaigns. This revealed the dependence of emission factors on the ratio of flaming to smouldering combustion and fuel characteristics. This study has also provided recommended emission factors for all measured species for both primary deforestation fires and pasture maintenance fires.

Maleknia *et al.* have provided a detailed PTR-MS investigation of the fragmentation patterns for various individual biogenic VOCs as a function of *E*/*N* (81-155 Td), including alcohols (methanol, ethanol and butanol), carbonyl-containing compounds (acetic acid, acetone and benzaldehyde), isoprene, acetonitrile, tetrahydrofuran, pyrazine, toluene and xylene, a series of terpenes (*p*-cymene, camphene, 2-carene, limonene, β-myrcene, α-pinene, β-pinene, γ-terpinene and terpinolene) and two oxygen-containing terpenes (1,8-cineole and linalool) [175]. These reference results were then used to aid the identification of VOCs emitted from *Eucalyptus grandis* leaves when subjected to temperatures ranging from ambient (30°C) up to 100°C. It was found that the signals from ions detected at *m*/*z* 81 and 137 showed an approximate fivefold increase as the temperature increased from 35°C to 100°C. From their study of individual compounds the researchers assigned the *m*/*z* 81 ions to protonated pyrazine ($C_4H_5N_2^+$) or $C_6H_9^+$ fragment ions from terpenes. Even more dramatic increases in ion intensities were observed at *m*/*z* 153 (assigned to protonated oxygen-containing terpenes, $C_{10}H_{16}O$) and *m*/*z* 155 (considered to be protonated 1,8-cineole (eucalyptol)). Other ions detected as a result of emissions from the *E. grandis* leaves included signals at *m*/*z* 79 (protonated benzene and a fragment ion deriving from protonated xylene), *m*/*z* 69 (which is protonated isoprene ($C_5H_9^+$) and/or furan ($C_4H_5O^+$)), *m*/*z* 61 (protonated acetic acid), *m*/*z* 59 (protonated acetone), *m*/*z* 45 (protonated acetaldehyde) and *m*/*z* 43 (acetyl fragments).

Maleknia *et al.* have also reported further investigations of the temperature dependence of VOC emissions from eucalyptus [176]. The importance of their work stems from the fact that eucalyptus is a major biogenic source of VOCs in Australia (with the annual emission into the atmosphere being estimated to be in the region of a teragram (10^{12} g)) and that eucalyptus is often involved in wildfires. Wildfires, which are common in Australia, will result in sudden increases in the amount of VOCs released into the atmosphere. Using a combination of PTR-MS, GC–MS and DART-MS techniques the release of VOCs as a function of temperature ranging from ambient up to 200°C was investigated. Ion abundances (and therefore VOC emissions) could be correlated with boiling points and vapour pressures of specific compounds. The PTR-MS analysis of VOCs resulting from heating fresh leaf (*E. grandis*) material suggested that emissions of methanol (protonated at *m*/*z* 33) and acetaldehyde (protonated at *m*/*z* 45) were greatest at ~60°C, while ion intensities at *m*/*z* 137 and 153 (which were attributed to protonated terpenes) showed monotonic increases over a wide temperature range, from ambient values up to 200°C. The PTR-MS combustion studies of senescent leaves *(E. grandis)* identified two distinct temperature-dependent VOC compositions. The first corresponds to temperatures below which burning did not occur, which the researchers defined as before smoke is observed. At such temperatures the VOC composition was found to remain consistent and was found to correlate well with various naturally occurring isoprenoids, as observed in temperature profile studies. The second region corresponds to temperatures for which the eucalyptus plants started to burn. For this temperature regime product ions at *m/z* 75, 85, 87, 99, 111 and 125 were observed, which were associated with protonated aldehydes, ketones, furans and substituted benzenes. The researchers attributed the emission of these volatiles to the pyrolysis of polycarbohydrates (cellulose and lignin) commonly found in many types of wood.

In a similar but earlier study, Greenberg *et al.* investigated the volatile emissions of leaf and woody plant tissue over a temperature range of 30–300°C [177]. Oxygenated VOCs were found to dominate the volatile emissions, with the main compounds being assigned to be acetic acid, furylaldehyde (furfural), acetol, pyrazine, terpenes, 2,3-butadione, phenol and methanol. Less dominant were emissions assigned to furan, acetone, acetaldehyde, acetonitrile and benzaldehyde. It was suggested that VOC emissions under low turbulence conditions may produce flammable concentrations near leaves that could facilitate the propagation of wildfires.

A laboratory study by Grieshop *et al.* undertook experiments to investigate the effects of photochemical oxidation on organic aerosol emissions from flaming and smouldering hard- and softwood fires under plume-like conditions [178]. To detect the VOCs they co-deployed both PTR-MS and GC–MS analytical techniques, while aerosol size and composition were determined using an aerosol mass spectrometer and two scanning mobility particle sizers. Production of unexplained organic aerosols was attributed to the oxidation of low-volatility organic vapours.

Warneke *et al.* undertook a detailed series of laboratory biomass burning measurements at the Fire Science Laboratory using PTR-MS and PTR-IT-MS in order to help in the identification and quantification of VOCs generated by biomass burning occurring in the south-east and south-west United States [160]. A primary aim of that work was to improve our understanding of the impact of biomass burning emissions on air quality and climate resulting from ozone and secondary organic aerosol formation. To aid in the identification of the VOCs released, gas chromatography pre-separation methods were used for the PTR-IT-MS measurements and the results obtained were then compared with a comprehensive set of measurements using other instruments (GC–MS, FTIR and Negative Ion Proton Transfer Chemical Ionization Mass Spectrometry). The identified and unidentified mass fractions of the total PTR-MS signal were determined for the smouldering and the flaming phase of the fires. It was found that in some cases only half of the ions were successfully assigned to specific molecules. The majority of the unidentified ions had an *m/z* $\geq$100. This large unidentified mass fraction was considered by the researchers to be important for the production of secondary organic aerosol. Warneke *et al.* suggested that after only a few oxidation steps it is likely that a large fraction of this unidentified organic material ends up in the aerosol phase.

5.6.2 Reaction Products and Reactive Species

An important laboratory application of PTR-MS, in relation to elucidating atmospheric chemistry, is the measurement of reaction products that are difficult to observe or quantify by other analytical techniques. This includes products resulting from reactions involving the important tropospheric and highly reactive hydroxyl radical, OH, which significantly affects the lifetime of many VOCs in the atmosphere [10, 179, 180] and is the most common atmospheric radical involved in oxidation pathways.

In a chamber study, Wisthaler *et al.* measured acetone and pinonaldehyde yields from OH-initiated oxidation of α- and β-pinene in air in the presence of NO_x with the concentrations of monoterpenes, acetone and other gas-phase products being determined from PTR-MS measurements [10]. It was suggested that acetone can be formed via two processes: one is what is referred to as a prompt process, resulting from a reaction of OH with a terpene, while the second is a slower process, which results from the degradation of stable non-radical intermediates such as pinonaldehyde and nopinone. The product yields from α- and β-pinene were found to be in good agreement with values derived from other mass spectrometric or gas-chromatographic methods, but less so with spectroscopic methods. It was suggested that this could have resulted from an interfering aldehyde absorption band used to quantify pinonaldehyde.

Reactions of OH with isoprene have been investigated by Zhao *et al.* [179]. In this case, PTR-MS was used to detect C_4 ($C_4H_7O_2{}^+$, *m/z* 87) and C_5 ($C_5H_9O_2{}^+$, *m/z* 101) hydroxy-carbonyls resulting from OH-initiated oxidation of isoprene. From these measurements the hydroxycarbonyl yields (Y) were determined according to the equation:

$$Y(\%) = \frac{k_{\text{iso}}}{k_{\text{car}}} \times \frac{I_{\text{car}}}{I_{\text{iso}}} \times 100 \tag{5.3}$$

where k_{iso} and k_{car} are the rate coefficients for the reaction of H_3O^+ with isoprene and the hydroxycarbonyls, respectively, which were calculated by Zhao *et al.* using the average dipole orientation (ADO) theory (see Section 2.2.3.2). I_{car} and I_{iso} are the ion signal intensities of protonated hydroxycarbonyl and isoprene, respectively. By varying the isoprene concentration, a linear plot of the dependence of $I_{\text{car}} \times k_{\text{iso}}/k_{\text{car}}$ on I_{iso} was determined, the gradient of which provided the yield, Y. For the C_4- and C_5-hydroxycarbonyls the yields obtained for OH-initiated oxidation of isoprene in the presence of NO were (3.3 ± 1.6)% and (19.3 ± 6.1)%, respectively. In addition to the hydroxylcarbonyls, an ion peak was also observed at *m/z* 85, which was assigned to a protonated C_5-carbonyl ($C_5H_9O^+$) with an estimated yield of (8.4 ± 2.4)%. By taking the sum of the measured carbonyl yields and averaged yield values available in the literature for methyl vinyl ketone, methacrolein, 3-methyl furan and organic nitrates, the majority of the total reacted isoprene could be accounted for, with carbonyls making a significant contribution.

Products of the reaction of OH radicals with C_3–C_6 cycloalkanecarbaldehydes, which are second-generation oxidation products of monoterpene oxidation, in the presence of NO_x have been reported in a paper by D'Anna *et al.* [180], which combined PTR-MS with FTIR spectroscopy to determine carbon balances. FTIR spectroscopy was used to quantify CO, CO_2, and HCHO, while PTR-MS was used to identify a range of VOC products that included nitroperoxycarbonyl cycloalkanes, cycloketones, cycloalkyl nitrates and a number of other multifunctional compounds containing carbonyl, hydroxy and nitrooxy functional groups. Owing to the complication of ion fragmentation for some of these rather fragile compounds when using *E/N* values >100 Td [181], the *E/N* used was 80 Td. This resulted in $H_3O^+(H_2O)$ (*m/z* 37) being more abundant than H_3O^+, which complicated the analysis of the data. Nevertheless, it was shown that the dominant product resulting from the reaction of OH with cyclopropanecarbaldehyde is nitroperoxy carbonyl cyclopropane, with an estimated carbon yield of (35.9 ± 1.5)%. This product was monitored at *m/z* 103, *m/z* 69 and *m/z* 41, all of which correspond to fragment ions (no protonated parent molecule was observed). Another reaction product was formaldehyde (at *m/z* 31, the protonated parent molecule), with a carbon yield of (19.6 ± 2.0)%. For cyclobutanecarbaldehyde, the dominant product was found to be nitroperoxy carbonyl cyclobutane with a carbon yield of (30.3 ± 1.3)%, determined by monitoring fragment ions at *m/z* 55, 83 and 117. The next most abundant product was assigned to be 3-hydroxy-propanal, with a carbon yield of (18.8 ± 1.0)% and was identified via ions at *m/z* 75 (MH^+) and 57 ($[MH\text{-}H_2O]^+$). The carbon yield for formaldehyde was determined to be (10.5 ± 0.8)%. For cyclopentanecarbaldehyde, the dominant product was found to be nitrooxy carbonyl cyclopentane with a carbon yield of (33.4 ± 1.3)%, and this was detected at *m/z* 131 ($[MH + H_2O - HNO_3]^+$), 97 (cyclopentylcarbonyl fragment ion) and 69 (predominantly a cyclopentyl fragment ion). 4-hydroxy-butanal, which was assigned via ions at *m/z* 89 ($[MH]^+$) and 71 ($[MH - H_2O]^+$), has a measured carbon yield of (12.2 ± 0.9)%. Next comes 5-nitrooxy-pentanal, attributed to an ion signal at *m/z* 85 $[MH - HNO_3]^+$), with a carbon yield of (8.7 ± 0.6)%. Nitroperoxy carbonyl cyclohexane was found to be the major reaction product resulting from the reaction with cyclohexanecarbaldehyde. In order to determine the carbon yield (33.2 ± 1.1)% associated with this product, the ion signals at *m/z* 145 ($[MH + H_2O - HNO_3]^+$), 111 (cyclohexylcarbonyl fragment ion), and 83 (cyclohexyl fragment ion) were recorded. Another prominent compound observed was cyclohexanone (*m/z* 99 (MH^+)), with a carbon yield of (22.2 ± 0.5)%, followed by cyclohexyl nitrate, *m/z* 101 ($[MH + H_2O - HNO_3]^+$) and 83 ($[MH - HNO_3]^+$), with a carbon yield of (13.2 ± 1.0)%.

Given that OH radicals are the main oxidants in the troposphere and the principal sink of important trace gases such as CO and VOCs, an important parameter to know in atmospheric chemistry is the total OH reactivity. Sinha *et al.* have shown that a '*comparative reactivity method*' employing PTR-MS may provide a viable methodology for exploring this, and therefore it has the potential of providing an important alternative to the currently used expensive laser systems to determine OH concentration measurements [182]. In the PTR-MS approach, hydroxyl is introduced at a constant rate into a glass reactor that contains a reactive compound X, which is not normally present in air, either in zero air or in ambient air containing VOCs and other OH reactive species. PTR-MS is used to determine the concentration of X leaving the reactor in zero air with no OH (C_1), in zero air with OH (C_2) ([OH] < [X]), which gives the initial concentration of OH (C_1–C_2), and finally in air containing reactive VOCs with OH (C_3). The introduced VOCs compete with X for reaction with OH, and thus $C_2 < C_3 < C_1$. From the measured C_1, C_2 and C_3 values the total OH reactivity of an air sample (R_{air}) can be determined from the equation:

$$R_{air} = \frac{C_3 - C_2}{C_1 - C_3} \times k_X C_1 \qquad (5.4)$$

where k_X is the rate coefficient for the reaction of OH with X. In their proof-of-principle study, Sinha *et al.* used pyrrole (C_4H_4NH) as X ($k_X = (1.20 \pm 0.16) \times 10^{-10}$ cm^3 molecule^{-1} s^{-1}) and they tested their system with different single and mixed hydrocarbon standards. Field tests in the tropical rainforest of Surinam and the urban atmosphere of Mainz, Germany, gave realistic average values for the total OH reactivity to be ~53 s^{-1} and ~10 s^{-1}, respectively, thereby providing confidence in the methodology employed. Importantly, in comparison to calculated total OH reactivity, the results from the tropical rainforest investigation indicated that a significant fraction of OH reactive species was being missed in measurements at forest sites.

In order to explore possible causes of missing OH reactivity, Kim *et al.* used the comparative reactivity method to investigate OH reactivity within a branch enclosure for four different tree species (red oak, white pine, beech and red maple) [183]. Their results showed that isoprene and monoterpenes, and in one case (red maple) a sesquiterpene (α-farnesene, detected using GC–MS), could account for the measured OH reactivity. Thus no unknown biogenic VOCs could have been contributing to the OH reactivity in any significant way, and therefore no noteworthy difference between measured and calculated OH reactivity was observed. This is in contrast to the tropical forest study by Sinha *et al.* mentioned above [182].

An interesting instrumental development has been described by Hanson *et al.* [184], who demonstrated the ability of PTR-MS to detect peroxy radicals. These radicals were produced in a laminar flow reactor via an initial reaction of atomic chlorine with an alkane to produce an alkyl radical, which was followed by reaction with O_2 to generate the peroxy radical. The PTR-MS instrument was operated under extremely low *E/N* conditions (18–34 Td), which ensured that $H_3O^+(H_2O)_n$ cluster ions ($n > 0$) dominated rather than bare H_3O^+ reagent ions, because alkanes do not react with protonated water clusters. The detection of methyl and ethyl peroxy radical species was found to be adversely affected by water vapour, whereas that for the cylcohexyl peroxy radical was found to be much less affected. A particularly powerful feature of this work was the ability to map the formation of product distributions in organic reaction systems involving peroxy radicals.

5.6.3 Simulation Chamber and Container Measurements

Atmospheric simulation chambers provide a controlled environment for exploring aspects of atmospheric chemistry, bridging the divide between the study of elementary reactions in the laboratory and those occurring in the complex chemical environment of the atmosphere. The measurements performed in such chambers can provide important insights into the behaviour of atmospheric trace gases and allow the investigation of tropospheric chemistry under pre-determined and controlled conditions (e.g. temperature, pressure, humidity and irradiation). Individual VOCs or a group of VOCs can be investigated in addition to the chemical roles of oxidants and water.

Using a simulation chamber PTR-MS and GC–MS studies of the biogenic VOC emissions from a 3-year-old potted European Beech (*Fagus sylvatica* L.) were investigated by Joó *et al.* [185]. This study revealed high emissions of linalool, an oxygenated monoterpene ($C_{10}H_{18}O$). This compound has a mass spectral signature at *m/z* 137 ($C_{10}H_{17}{}^+$) in PTR-MS corresponding to the elimination of H_2O from the protonated parent molecule but, and as we have already mentioned, an ion signal at *m/z* 137 is generally used as a measure of total monoterpene levels. Joó *et al.* showed that if linalool is present in the emissions of plants, then that will artificially elevate emission rates of monoterpenes measured by PTR-MS if its contribution is not taken into account. The researchers recommended that care must therefore be taken when quantifying monoterpene emissions from vegetation using PTR-MS measurements.

VOC emissions from three Norwegian spruce trees (*P. abies* L. Karst) kept in 50 L containers for 6 months prior to laboratory measurements were investigated in response to changes in temperature over a small range of values from 23°C to 35°C in approximate steps of 4°C at a fixed light intensity of 200 μmol photons m^{-2} s^{-1} by Filella *et al.* [186]. It was found that the emission rates of monoterpenes (recorded at *m/z* 137), acetone (*m/z* 59), methanol (*m/z* 33), hexanal (recorded at *m/z* 83), hexanol (recorded at *m/z* 85) and methyl salicylate (*m/z* 153) all increased exponentially with increasing temperature. However, in contrast to this temperature dependence, the acetic acid and acetaldehyde emission rates were only found to increase with temperature until saturation levels were reached at $\sim$30°C. The research team suggested that calculated VOC emission rates, and particularly those for the short-chained carboxylic acids and aldehydes, may be inadequate if only incident irradiance and leaf temperature are used in empirical algorithms to predict VOC emission rates.

Using a controlled environment cuvette,[11] on-line PTR-MS analysis of VOCs emitted from the commercially important coniferous species Sitka spruce (*Picea sitchensis* Bong.) were undertaken by Hayward *et al.* [187]. By taking measurements for a range of controlled light and temperature regimes, isoprene was found to be the most abundant VOC emitted, with the emission rate being found to be strongly correlated with both light level and temperature. The emissions of other VOCs (i.e. monoterpenes, methanol and acetaldehyde), which represented less than one-third of the total observed VOC emissions, were found to be dependent on the temperature but not the light intensity. However, in response to a sudden light to dark transition, a rapid fall in isoprene level was observed which was mirrored by a burst of acetaldehyde production, with the acetaldehyde maximum coinciding with the isoprene minimum (at $\sim$7 min post illumination). Another study involving a cuvette system investigated isoprene emissions from shoot cultures of Grey poplar trees. The results showed that under continuous irradiation the emissions were coupled to circadian rhythms [188].

Gómez Alvarez *et al.* coupled PTR-MS with the European PHOtoREactor (EUPHORE) atmospheric simulation chamber (Valencia, Spain) to provide experimental confirmation of the dicarbonylic mechanism in the photooxidation of toluene and benzene [189]. The particular benefit of PTR-MS in this context is its relatively fast response time, which provides data that can be tested against a Master Chemical Mechanism (MCMv3.1) model. Differences in mass spectral fragmentation patterns also allowed PTR-MS to distinguish between *cis*- and *trans*-butenedial, which are two of the products resulting from the photooxidation process.

Using the Paul Scherrer Institute (PSI) environmental chamber (Villigen, Switzerland) under various NO_x conditions, Müller *et al.* employed a high resolution PTR-TOF-MS instrument to investigate the photooxidation of 1,3,5-trimethylbenzene (C_9H_{12}) [190]. Approximately 200 distinct ion signals were observed. These ion signals were grouped according to the number of C, N and O atoms. This made it possible to determine the time evolution of the O:C ratio and the average carbon oxidation state of the reaction mixture. The measured amount of carbon was found to be only $\sim$15% lower than the total amount of carbon available from the degradation of 1,3,5-trimethylbenzene.

Laboratory chamber studies utilizing PTR-MS for VOC measurements have also been used to elucidate the chemical routes for the formation of secondary organic aerosols [191–197]. An example is aerosol formation from the 1,3,5-trimethylbenzene-propene-NO_x-water vapour system, which was investigated at the PSI smog chamber [192,196]. This work classified the VOCs according to their temporal profile and identified oligimerization as a key secondary organic aerosol formation mechanism.

[11] A cuvette system is a sealed chamber, for example, made from Teflon or glass, within which air circulates with incoming and outgoing gas streams.

Chamber (Caltech Indoor Chamber Facility) experiments have also been carried out by Lee *et al.* to monitor the time evolution and yields of gas-phase VOCs produced by the ozonolysis of ten different terpenes [197]: six monoterpenes (α-pinene, β-pinene, 3-carene, terpinolene, α-terpinene and myrcene), two sesquiterpenes (α-humulene and β-caryophyllene) and two oxygenated terpenes (methyl chavicol and linalool). Several gas-phase oxidation products were identified including formaldehyde, acetaldehyde, formic acid, acetone, acetic acid, nopinone and pinonaldehyde.

In addition to ozonolysis, photooxidation is another important terpene loss process. Lee *et al.* have therefore also investigated the photooxidation of 16 different terpenes, namely isoprene, eight monoterpenes, three oxygenated monoterpenes, and four sesquiterpenes, at the Caltech Indoor Chamber Facility under atmospherically relevant hydrocarbon/NO_x ratios [193]. PTR-MS was used to monitor the time evolution and yields of first- and second-generation oxidation products in real-time. This type of research aids in the elucidation of reaction pathways leading to secondary organic aerosols. Both PTR-MS and GC-FID were used to determine the gas-phase concentrations of the parent terpene. Pure commercial standards were employed to calibrate the PTR-MS instrument for various compounds, including formaldehyde, acetaldehyde, formic acid, acetone, acetic acid, nopinone and methacrolein/methyl vinyl ketone. Concentrations of other oxidation products for which calibration standards were not available, such as pinonaldehyde, were inferred from using equation (4.1), which relates the concentration of a volatile to the ion signal intensities (reagent and product ions), the reaction rate coefficient and the reaction time.

Ng *et al.* have provided a more detailed analysis of the first and second-generation oxidation products in the formation of secondary organic aerosols [194]. By using Differential Mobility Analyzers, aerosol growth was determined and this growth was then compared with concentrations of VOCs in the gas-phase determined from PTR-MS measurements. From this the general mechanisms of secondary organic aerosols was studied. As an illustration of this, Figure 5.1 shows the photooxidation decay of α-pinene and the corresponding

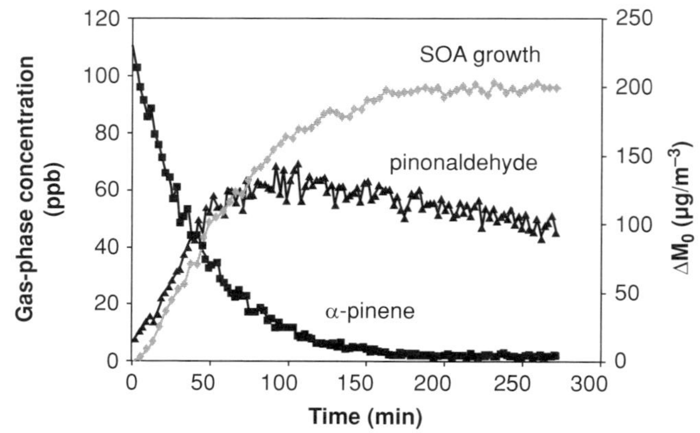

Figure 5.1 *Time evolution of α-pinene and pinonaldehyde (resulting from the photooxidation of α-pinene) as measured by PTR-MS in a reaction chamber experiment. Also shown is the corresponding secondary organic aerosol (SOA) growth curve, as measured by aerosol mass spectrometry. Reproduced with permission from [194]. Copyright © 2006, American Chemical Society.*

growth of pinonaldehyde as a function of time. Pinonaldehyde was found to decrease in concentration once α-pinene had been almost consumed. This loss was attributed to reaction with OH and photodissociation. Secondary organic aerosol growth is seen to occur ~10 min after reaction initiation.

Presto and Donahue have developed a method which uses VOC measurements from PTR-MS coupled to particle measurements to determine secondary organic aerosol production at low total organic mass density ($<10\ \mu g\ m^{-3}$) [195]. This work showed that extrapolations of the partitioning models used at the time of their study to conditions of low organic mass concentration significantly underestimated secondary organic aerosol production under dark and low-NO_x conditions. In contrast, secondary organic aerosol production under illuminated and higher NO_x conditions, which are more typical of polluted regional air masses, was found to be overestimated.

It has already been mentioned that many oxygenated VOCs result from the photooxidation of VOCs and that they contribute to secondary organic aerosol formation. Given the atmospheric importance of secondary organic aerosols and the fact that their concentrations can change rapidly, an analytical technique such as PTR-MS would be useful for investigations of the molecular constituents of particles in an aerosol, providing this material could be evaporated into the gas phase. Hellén *et al.* have succeeded in doing this by modifying the sampling inlet system of a PTR-MS instrument [198]. This new sampling system consists of two inlet lines, one of which allows passage of both gases and particles. The other inlet line has two possible pathways; one route contains a denuder to strip off the gas phase so that only particles pass through and the other route contains a Teflon filter to remove particles so that any contributions to ion signals from the gas phase to the particle measurements can be determined. The outlets for both inlets are coupled through a common heater, which is used to vaporise the aerosol particles to allow the products to enter into the drift tube of the PTR-MS instrument. The capabilities of this technique were demonstrated by generating *cis*-pinonic acid particles, as a test aerosol generated with a nebuliser. This was followed by investigating the application of this technique to the detection of secondary organic aerosol particles formed during the photooxidation of 1,3,5-trimethylbenzene and α-pinene with NO_x in a smog chamber. For the pinonic acid study, an aerodynamic particle sizer (APS) was used to determine the aerosol mass concentration (the mass of the total suspended particulates per unit volume). This was then compared with the mass concentrations derived from the PTR-MS measurements. Although a good correlation was found between the results from the two techniques for the concentration measurements, the mass concentration determined by PTR-MS was 69% of that determined by APS. By correcting for losses in the sampling system the researchers reduced this difference to 16%. To illustrate the performance of this technique, Figure 5.2(a) summarizes the results for the photooxidation of α-pinene for a typical cycle of measurements whereby ions were recorded at *m/z* 59 (with acetone being considered to be making the highest contribution in the gas phase while glyoxal dominates the particle phase), *m/z* 151 (a fragment ion corresponding to the loss of H_2O from protonated pinonaldehyde), *m/z* 167 (the molecule responsible was not identified) and *m/z* 171 (not identified), with each cycling mode (gas + particle, particle and background) shown. Figure 5.2(b) shows the temporal evolution of secondary organic aerosol products (measured using a scanning mobility particle sizer) and ion signal intensities for both the α-pinene and 1,3,5-trimethylbenzene measurements. From Figure 5.2(b) it can be seen that the starting compounds steadily reduce as a result of

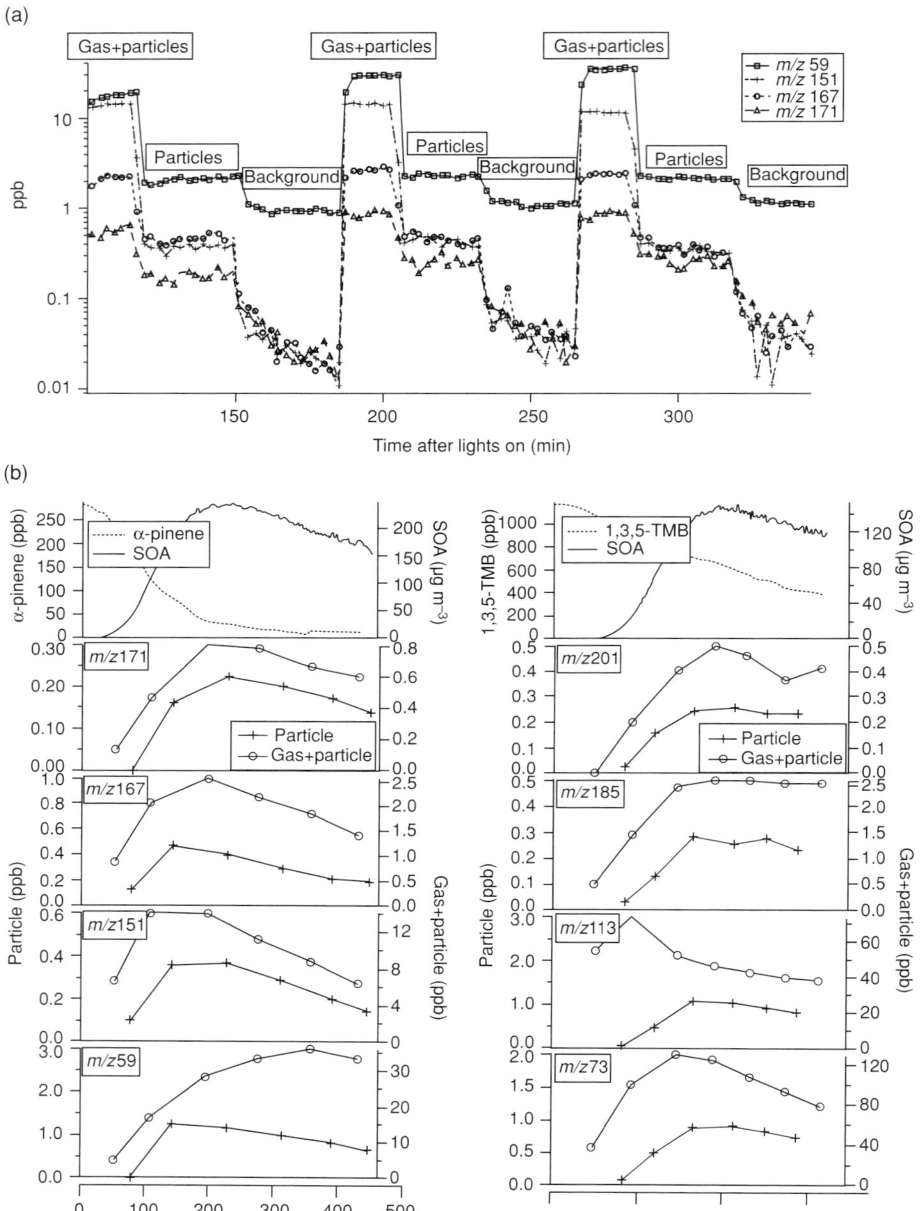

Figure 5.2 *(a) Measured PTR-MS signals for m/z 59, 151, 167 and 171 from a smog chamber experiment involving α-pinene. (b) Temporal evolution of aerosol mass, α-pinene and 1,3,5-trimethylbenzene concentrations in the smog chamber and the ions for the particle and gas + particle at m/z values of 59, 151, 167 and 171 in the α-pinene experiment and m/z 73, 113, 185 and 201 in the 1,3,5-trimethylbenzene experiment. Reproduced with permission from [198]. Copyright © 2008, American Chemical Society.*

reactions (with OH for 1,3,5-trimethylbenzene and with OH and ozone for α-pinene). The loss of α-pinene is faster than observed for 1,3,5-trimethylbenzene, having decreased by a factor of two within 90 minutes of irradiation (whereas 1,3,5-trimethylbenzene reduces by a factor of two within about 400 minutes). For the 1,3,5-trimethylbenzene experiment, ion signals at *m/z* 73, 113, 185 (not identified) and 201 (not identified) were monitored. *m/z* 73 was assigned to be protonated methylglyoxal, and *m/z* 113 was tentatively assigned by the researchers to protonated 2-methyl-4-oxo-2-pentenal and 3,5-dimethyl-5(2H)-2-furanone.

Other aerosol work with PTR-MS has been undertaken by Liggio *et al.* to investigate the direct polymerization of isoprene and α-pinene on acidic sulfate aerosols [199]. For this study they utilized an aerosol mass spectrometer together with a Teflon reaction chamber of volume 2 m^3, while PTR-MS was employed to monitor the gas-phase concentrations of isoprene and α-pinene. It was found that isoprene and α-pinene were taken up directly by acidic aerosols, with polymers being formed that contain at least four isoprene and two α-pinene repeating units.

5.7 Plant Studies

Owing to their immobility plants have evolved to produce and emit volatile compounds so that they can communicate and interact with their surroundings. Furthermore, in response to changes in environmental parameters such as light, temperature, nutrients and water, plants will adjust the emission rates of many different VOCs. Volatiles are released by the flowers, leaves and fruits of plants, but in addition it is important to appreciate that volatiles are also released from their roots into the soil [200].

Volatiles emitted by plants serve many purposes: for example, they attract pollinators, seed disseminators and natural enemies of attacking insects [201]. They are also used as signals of stress, they can have antimicrobial properties, and they can serve to warn neighbouring plants of potential attack [202–208]. Using various analytical measurements, ~1700 VOCs have been identified from more than ninety plant families [200]. Many of these volatiles are found to be oxygenated (e.g. formic and acetic acids, acetone, formaldehyde, acetaldehyde, methanol and ethanol). However, in terms of emission rates from plants, isoprene and monoterpenes are quantitatively the largest [88].

Given the quantity and diversity of volatile compounds being released by plants, it is not surprising that PTR-MS has been used to great effect in investigations of such emissions, the results of which are improving our understanding of the biosynthesis and ecology of plant VOCs. Although clearly linked to Section 5.3 in this chapter, which dealt with biogenic emissions over large length scales, the focus on the PTR-MS work in this section will be far more local, that is, in the immediate vicinity of the plants. PTR-MS is particularly beneficial for applications in botanical studies owing to its short response time so that rapid changes in VOC concentrations being emitted from plants can be monitored. Furthermore, little or no sample preparation (e.g. no pre-concentration) is required. Of course, there are other techniques that can also be used in the analysis of plant VOCs and Tholl *et al.* have provided a useful overview of the various methods [208], including PTR-MS, in which they have described the advantages and disadvantages of the various instrumental methods and what practical approaches are required.

5.7.1 Isoprene Emissions

Given that isoprene (C_5H_8) is a major biogenic VOC emitted globally and a substantial contributor to ozone and secondary organic aerosol production, it is important to have a thorough understanding of the production and consumption of isoprene. Accurate measurements of isoprene concentrations in the atmosphere are also critical to achieve a quantitative understanding of atmospheric chemical processing. A description of PTR-MS investigations of isoprene emissions by plants therefore merits its own separate sub-section.

A major problem when using low mass resolution PTR-MS (e.g. an instrument based on a quadrupole mass spectrometer, with a typical mass resolution of about 1 Da) is that the ion peak at a nominal value of *m/z* 69 used to identify isoprene is not necessarily unique. Potential interfering compounds producing ions with the same nominal *m/z* are often present in the atmosphere, notably furan (C_4H_4O), a fragmentation product of MBO (which is co-emitted with isoprene by some tree species and in the case of coniferous ecosystems is far more dominant than isoprene) and 2- and 3-methyl butanal [2]. As we have seen elsewhere in this chapter there are also problems in determining the individual concentrations of two of the degradation products of isoprene, namely methyl vinyl ketone and methacrolein, because they are structural isomers and therefore have identical masses (protonated mass at *m/z* 71.050). The protonated ion signal intensity at a nominal *m/z* 71 is typically attributed to the sum of these two compounds. However, this is far from satisfactory because both species can be formed by different routes and they contribute differently to the production of ozone and secondary organic aerosols. Furthermore, *m/z* 71 may not be unique to methyl vinyl ketone and methacrolein.

Although there has been considerable research devoted to investigating biogenic isoprene emission, there is still much debate and uncertainty about the role it plays in plants. When you consider it, any isoprene emission results in a loss of both energy and carbon from a plant. Thus the fact that isoprene emission is ubiquitous suggests that there must be some benefit for the plant which more than compensates for these losses. However, the precise nature of these benefits still remains to be determined. One possible explanation comes from the observation that the actual amount of isoprene (and terpenes) that is emitted by trees, plants, agricultural crops and grass is found to be dependent on light intensity and temperature (in addition to the stage of a plant's life cycle). It has therefore been proposed that the emission of isoprenoid compounds, and indeed other VOCs, is a mechanism that plants have evolved for removing excess energy resulting from, for example, high temperatures [209, 210] and high light intensities [211] and as a protection against oxidative stress [212, 213]. Indeed some researchers have described the synthesis of isoprene as a plant's metabolic '*safety valve*' [214].

Loreto *et al.* investigated the emission of isoprene and other VOCs from reed leaves either exposed to high temperatures (45°C) for a period of 150 min or to high light intensities for 40 min (increasing from light levels where photosynthesis was close to light saturation up to four times that amount) [215]. Following the exposure of the leaves to 45°C a very rapid increase of isoprene emission resulted (see Figure 5.3(a)). The maximum was reached after about an hour and then a drop in intensity occurred going down to below preheating levels. This was then followed by a slow increase back to the level originally measured at 30°C (marked as NS on Figure 5.3(a)). The rapid rise and decay of isoprene was expected by the researchers, because it is related to the optimal temperature for isoprene emission (the

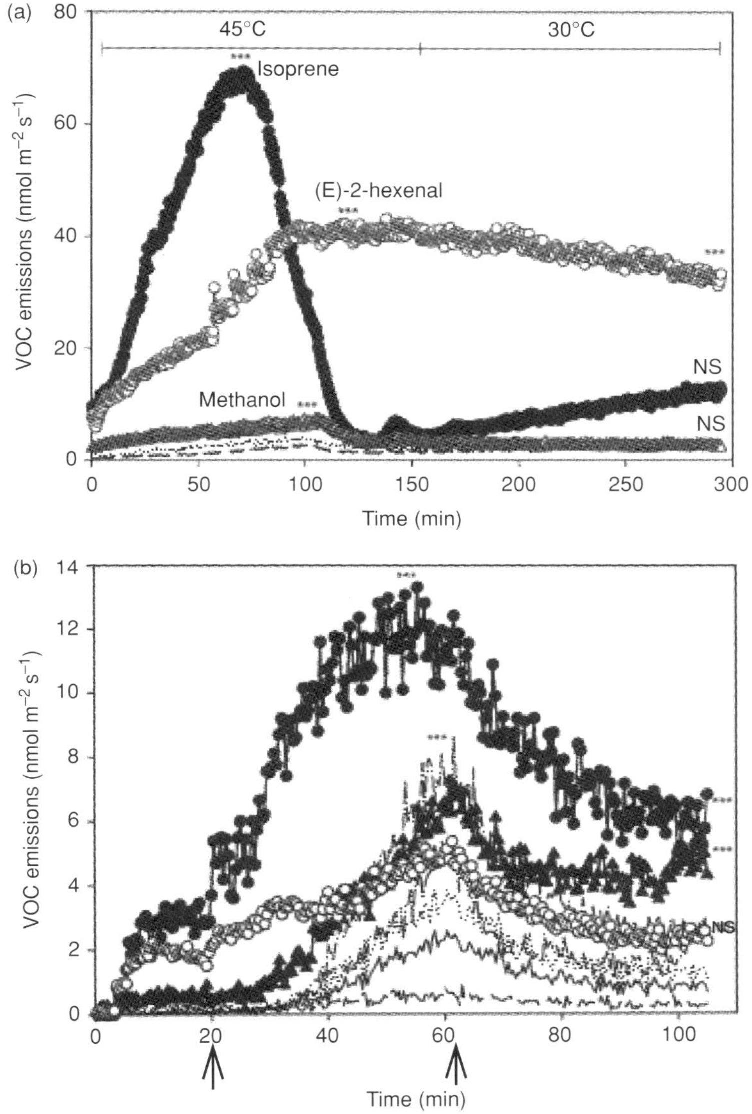

Figure 5.3 *(a) The time evolution of isoprene, (E)-2-hexanal and methanol following exposure of reed leaves to a temperature of 45°C (first 150 min) and then to 30°C (last 150 min). Other C_6 compounds are identified by lines that represent (Z)-3-hexenal (m/z 81 + m/z 99) and the sum of (Z)-3-hexenol and (E)-2-hexenol (m/z 83 + m/z 101). (b) Time evolution of various VOCs emitted during high-light treatment (initiated at the time shown by the first arrow and stopped at the time shown by the second arrow). Isoprene, acetaldehyde and (E)-2-hexenal are identified by symbols (solid circles, solid triangles and hollow circles, respectively). Other VOCs (C_6 compounds) are identified by the lines. Reproduced with permission from [215]. Copyright © 2006 John Wiley & Sons, Ltd.*

rise) and the regulation of isoprene synthesis (the decay). An interesting observation was that the level of a VOC that indicates plant damage, namely (*E*)-2-hexenal (measured at *m/z* 57), also increased with the raised temperature. In comparison with isoprene, the rise of (*E*)-2-hexenal was slower (see Figure 5.3(b)) and a maximum was reached at ~100 min (with a final concentration approximately half that reached by isoprene). However, unlike isoprene, once the maximum was reached, the (*E*)-2-hexenal intensity did not decline. Instead it maintained this level for some time, even 150 min after the temperature was decreased from 45°C down to 30°C. The research team found that exposure to high light levels (2000 μmol photons m^{-2} s^{-1}) also resulted in a rapid rise in isoprene levels and a more gradual rise in concentrations of other VOCs, namely acetaldehyde and C_6 compounds (see Figure 5.3(b)). Upon reducing the light intensity, the VOCs quickly returned to their original levels. The C_6 compounds observed as a result of both the high temperature and high light conditions are known markers commonly associated with plants that are under stress or are being damaged.

Pegoraro *et al.* undertook experiments in the tropical rainforest ecosystem of Biosphere 2, which is an artificial closed biological system, to assess the effects on isoprene sources and sinks (soil) for elevated CO_2 and drought conditions and during mild water stress [216, 217]. In addition, they assessed the relationship of isoprene levels to changes in light and temperature. It was found that elevated CO_2 had no effect on isoprene uptake for litter-free soil (i.e. soil that has no organic material on its surface), but did strongly depress isoprene production. An important conclusion from their first study was that whilst the soil uptake of atmospheric isoprene in the mesocosm is almost certainly modest, it is still at a significantly high enough level for it to have to be taken into account if a good estimate of the global isoprene budget is to be determined [216]. In the second study, it was found that overall isoprene production was hardly affected by an imposed mild water stress [217]. The reason given for this was that the isoprene emitters were mainly distributed among the large trees of the canopy layer, and these trees have well developed deep roots in the lower soil profile for which the water content would have decreased much less than in the top 30 cm of the soil.

A number of groups have used isotopic labelling in PTR-MS experiments to investigate the biosynthesis pathways of isoprene production in plants [218–220]. By feeding [U-^{13}C]-labelled[12] α-D-glucose ($^{13}C_6H_{12}O_6$) via cut ends of stems into the xylem of *Quercus robur* seedlings, Kreuzwieser *et al.* were able to observe the emission of ^{13}C-labelled isoprene in real time [218]. Protonated isoprene was monitored at all of the possible isotopic combinations, that is, *m/z* 69 ($^{12}C_5H_9{}^+$), *m/z* 70 ($^{13}C^{12}C_4H_9{}^+$), *m/z* 71 ($^{13}C_2{}^{12}C_3H_9{}^+$), *m/z* 72 ($^{13}C_3{}^{12}C_2H_9{}^+$), *m/z* 73 ($^{13}C_4{}^{12}C_1H_9{}^+$) and *m/z* 74 ($^{13}C_5H_9{}^+$). The researchers demonstrated that xylem-transported glucose (or its degradation products) has the potential for use as an additional precursor in the biosynthesis of isoprene and that this carbon source becomes more important when photosynthesis is restricted. At about the same time as these studies by Kreuzwieser *et al.*, Karl *et al.* used PTR-MS to investigate the kinetics of the formation of ^{13}C-isoprene following the exposure of oak (*Quercus agrifolia Nee*) and cottonwood (*Populus deltoides Barr.*) leaves to $^{13}CO_2$ [219]. The loss of ^{13}C after the atmosphere was returned to a normal CO_2 isotope composition was followed. This seminal

[12] The U refers to uniformly ^{13}C labelled, i.e. all of the carbon atoms in the glucose are ^{13}C.

work demonstrated a link between photosynthesis and isoprene production and emission, providing new insights into the subcellular origins of isoprene. ^{13}C-labelled experiments have also been used to probe different carbon pools within young poplar leaves, demonstrating that isoprene comes from a combination of CO_2 biosynthesis, xylem-transported carbon and internal stores such as starch [220].

Finally, we comment on a study by Brilli *et al.* that found that light-dependent isoprene emission was continued and even enhanced for hours after inhibiting photosynthesis as a result of leaf cutting of the strong isoprene-emitter *Populus alba* [221]. It was concluded that isoprene emission does not have to be directly coupled to photosynthesis because it may occur even after leaves or branches have been cut, for example, by agricultural practices or because of abiotic stress (i.e. stress caused by non-biological factors such as drought, overwatering, high and low temperatures and wind damage) and biotic stress (i.e. stress caused by living organisms such as bacteria, fungi and herbivores).

5.7.2 Acetaldehyde Emissions

In addition to detailed investigations of isoprene emission, there have been a number of studies focused on the association of plants and the emission of acetaldehyde, which also showed interesting behaviour with regard to light levels.

Holzinger *et al.* investigated VOC emissions from Mediterranean Holm oak (*Quercus ilex* L.) in a controlled chamber and observed bursts of acetaldehyde under fast light/dark changes [222]. Further investigations of the transient release of acetaldehyde were reported by Karl *et al.* [223]. Confirmation that the ion observed at m/z 45 in the PTR-MS mass spectrum is indeed acetaldehyde was obtained by analyzing the leaf emissions using high-performance liquid chromatography (HPLC). As mentioned before in this book, techniques that use separation methods, such as HPLC and GC–MS, are useful for an accurate identification of VOCs. However, the trade-off is that rapid changes in VOC emissions cannot be measured and this is illustrated in the case of acetaldehyde by the fact that the emissions had to be concentrated over a period of 15–30 min before an HPLC measurement could be taken.

In confirmation of the findings by Holzinger *et al.* with oak leaves [222], Karl *et al.* observed rapid bursts of acetaldehyde when the leaves of various trees (sycamore, aspen, cottonwood and maple), which were separately placed in a cuvette constructed from a Teflon cylinder, were exposed to light–dark transitions [223]. This is illustrated in Figure 5.4 for sycamore leaves (*Plantanus acerifolia*). This diagram also presents the observed changes in isoprene emissions, which show an opposite effect to that observed for acetaldehyde. It was found that when the leaf was darkened, isoprene levels rapidly decreased whereas acetaldehyde levels rapidly increased. Notably, when undergoing these light–dark transitions, no wound VOCs (e.g. (*E*)-2-hexenal and (*Z*)-3-hexenal recorded at m/z 57 (fragment ion), 81 (fragment ion) and 99 (protonated parent molecule) [224]) were observed to accompany the acetaldehyde emission. This suggested that the light–dark transitions were not causing any damage to the leaves and that acetaldehyde is being released as a result of a normal metabolic process. It was suggested from a ^{13}C isotope labelling experiment that the production of acetaldehyde could have resulted from a pyruvate overflow mechanism controlled by cytosolic pyruvate levels and pyruvate decarboxylase activity. In this proposed mechanism, pyruvate decarboxylase acts as a safety valve to convert excess cytosolic

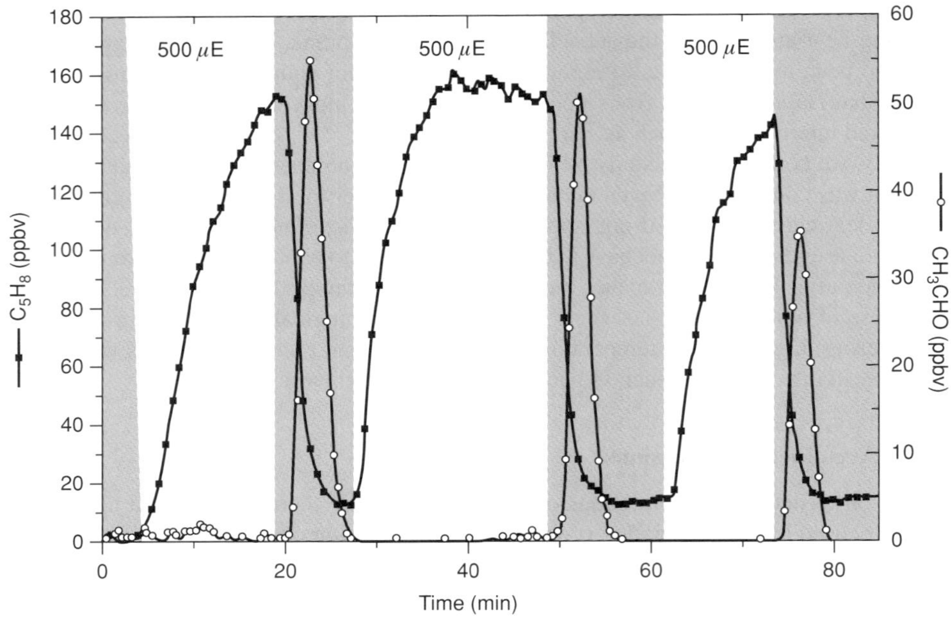

Figure 5.4 *Changes in acetaldehyde and isoprene levels following light–dark transitions subjected to sycamore leaves. Note that acetaldehyde shows a rapid burst and decline during the dark period. The ability to measure such rapid changes in volatiles emitted from leaves illustrates one of the key benefits of PTR-MS. These results were replicated in several leaf varieties. Light intensity is given in terms of irradiance (E has units of W m^{-2}). Reproduced with permission from [223]. Copyright © 2002 John Wiley & Sons, Ltd.*

pyruvate into acetaldehyde. In addition to trying to understand the metabolic pathways resulting in the release of acetaldehyde (and other VOCs), the interest in acetaldehyde emissions stems from the fact that acetaldehyde is converted in the atmosphere to peroxy acetic nitric anhydride (known as PAN and already encountered several times earlier in this chapter), which has a longer atmospheric lifetime than acetaldehyde.

The prompt and transient release of acetaldehyde (and other oxygenated VOCs) from leaves during light–dark transitions have also been reported by Graus *et al.* [225]. In their study, they investigated grey poplar trees with an objective to characterize, in more detail, the acetaldehyde emissions after darkening. The PTR-MS mass scans revealed that, following darkening, isoprene emissions rapidly decreased whereas those associated with oxygenated VOCs rapidly increased. The first observed oxygenated VOC to be emitted was hexenal, followed by acetaldehyde and other C_6-VOCs. Under oxygen-free (anoxic) conditions only acetaldehyde was found to be released after switching off the light. This provided sufficient evidence for the researchers to suggest that hexenal and other C_6-VOCs are released from the lipoxygenase reaction during light–dark transitions under aerobic conditions. Additional experiments undertaken by these researchers using enzyme inhibitors to artificially increase cytosolic pyruvate suggested that the acetaldehyde burst after a light–dark transition could not be explained by the pyruvate overflow mechanism proposed by Karl *et al.* [223]. Thus the metabolic pathway leading to acetaldehyde emissions following rapid darkening is still a matter of debate.

Brilli *et al.* used a more sensitive and higher resolution PTR-MS instrument than had been previously used in other plant studies to further characterize the burst of acetaldehyde when exposing various plant species to rapid light-to-dark fluctuations (in addition to investigating plant volatiles resulting from wounding – see Section 5.7.7) [221]. A PTR-TOF-MS instrument allows complete and almost instantaneous detection of whole spectra (up to *m/z* 315 in their studies) with a time resolution of less than 1 s. For the plant species *Dactylis glomerata*, light–dark transitions resulted in a burst of green leaf volatiles (first hexenals (*m/z* 81.070), and then sequentially hexenols and hexanal (*m/z* 83.085) and hexenyl acetate (*m/z* 143.107)), but no acetaldehyde. For *Populus alba* light–dark transitions also triggered a burst of green leaf VOCs, but for this plant these were found to be accompanied by a strong release of acetaldehyde (*m/z* 45.054) and a drop in the isoprene (*m/z* 69.070) emission. The plant species *Q. ilex* showed intense monoterpene emission (as measured at the fragment ion *m/z* 81.070 and the protonated parent *m/z* 137.133) in illuminated leaves that was found to decline rapidly as soon as the light was switched off. At the point when the light was switched off a burst of hexenyl acetate and acetaldehyde was detected. These are fascinating processes in plant biochemistry which have been detected by PTR-MS measurements and which still await a full explanation.

Jardine *et al.* investigated the physiological and environmental factors that control the exchange of acetaldehyde between forest canopies and the atmosphere [226]. During the day, they discovered that the upper canopy behaved as a net source of acetaldehyde while, in comparison, at lower heights reduced emission rates or net uptake occurred. During the night it was found that uptake dominated throughout the whole of the canopy. This study confirmed that only living leaves are involved in the exchange of acetaldehyde. They found that acetaldehyde exchange rates depended on a number of factors, including ambient acetaldehyde concentrations, stomatal resistance to acetaldehyde, light levels and temperature.

5.7.3 Pollination

The attraction of pollinating insects, such as bees, to plants is a complex process involving both chemical and visual cues through processes which are still not fully understood. Although many bee varieties are attracted to many plant species, it so happens that other bee species will only collect pollen from a single genus of flowering plants. One such bee species is *Hoplitis adunca*, which is a Eurasian cavity-nesting solitary bee that collects pollen exclusively from *Echium* spp.[13] (Boraginaceae). Using a combination of PTR-MS, GC/GC–MS and olfactometer bioassays, Filella *et al.* investigated whether the floral scent from *Echium vulgare* plays a major role in the attraction of *H. adunca* females and aimed to identify the components of the scent that are involved [227]. Towards this goal the scents of fresh flowers, nectar, pollen and whole plants of *E. vulgare* and another Boraginaceae (*Anchusa officinalis*) which is known not to attract the bees, were compared. The olfactometer bioassay measurements showed that *H. adunca* bees were strongly attracted to

[13] The abbreviation spp. means something that applies to many species within a genus.

E. vulgare scent but not to the scent of *A. officinalis*. For the VOC measurements, samples of flowers, pollen and nectar of *E. vulgare* and *A. officinalis* were enclosed in a leaf cuvette. The air leaving the cuvette was then analyzed using PTR-MS. GC–MS was also used to identify the VOCs and in particular the different monoterpenes so that they could then be quantified by PTR-MS. The chemical analysis of the VOC emissions of the two plants showed significant differences. In comparison to *A. officinalis*, the pollen of *E. vulgare* emitted far higher concentrations of VOCs yielding ion signals at *m/z* 55 (protonated butanal, 1,3-butadiene or potentially other VOCs with an unprotonated nominal mass of *m/z* 54) and *m/z* 83 (possibly fragment ions from protonated hexanal, hexenols, hexenyl acetate). Signals at *m/z* 109 (tentatively assigned to protonated 1,4-benzoquinone) and *m/z* 123 were characteristic ions resulting from volatiles emitted from the nectar of *E. vulgare*. In terms of the VOC emission from the flowers, *E. vulgare* was found to emit limonene, longicyclene, junipene, *trans*-caryophyllene and α-humulene, all of which were not emitted by *A. officinalis*. The most dominant monoterpenes emitted by *E. vulgare* were found to be α-pinene, junipene and limonene, whereas in comparison the most emitted terpenoid by *A. officinalis* was found to be α-pinene. This study confirmed that floral scent plays a key role in attracting bees to flowers, although it is still unclear whether these insect pollinators use only a few or all of the identified scent compounds for finding the correct plant.

5.7.4 Roots and Soil

It is not only those parts of a plant exposed to the air that can emit VOCs. The roots of plants and trees are also known to exchange VOCs with the surrounding soil. However, only a limited number of PTR-MS studies on this topic have so far been reported in the research literature. Two such studies, which were reported by Pegoraro *et al.*, have been briefly described previously in the section dealing with isoprene emissions (see Section 5.7.1) [216, 217].

The soil is also involved with the exchange of VOCs from and to the atmosphere. Asensio and coworkers used PTR-MS to try and quantify the rate of exchange of VOCs between soil in a Mediterranean forest and the atmosphere, and how this exchange is influenced by soil moisture, temperature and the presence of plant roots [228]. For this particular study the researchers used soil samples collected from the top soil layer in a Holm oak forest in south-eastern Catalonia. This soil was used to fill six pots, into which *Q. ilex* plants were placed and another six pots without any plants. All of these pots were maintained in Mediterranean winter-like conditions for 3 months covering the months November–February, before any measurements were undertaken. During this period half of these pots (three 'soil + plant' and three 'soil without plant') were treated to a low irrigation regime, which corresponded to a 5% volume soil moisture. The other six pots were well watered and had 21% soil moisture. For the pots that had undergone a low irrigation regime, the measurements in February involved monitoring VOCs released while maintaining the soil temperature at 20°C and gradually increasing the soil water gradient (5–27.5% in steps of 7.5%). The emissions from the other six pots were monitored while changing the temperature from 10°C to 35°C. To ensure that soil emissions were being measured a metal collar was inserted 10 cm into the soil 1 h before any measurements were taken and hermetically sealed with a 'soil hood chamber', through which ambient air could be introduced and headspace air sampled by PTR-MS for VOC analysis. VOC samples were also trapped on carbon-trap adsorption tubes for GC–MS analysis so that the different monoterpenes could be identified. Most of the compounds detected by PTR-MS were found to be oxygenated VOCs such as alcohols, aldehydes and ketones, as well as aromatic hydrocarbons. The soil was found to act as a sink rather than a source of VOCs in both soil moisture and temperature treatments. Furthermore, this work showed that roots decreased the emission of many compounds when the moisture content of the soil increased and under conditions of increasing soil temperature. Only low root/soil exchange rates were recorded, which led the researchers to suggest that soil-atmosphere VOC exchange in this system does not play an important role in atmospheric chemistry.

Asensio and colleagues extended their research into soil emissions to determine the response of interannual and interseasonal fluxes of VOCs (and CO_2) to drought in a Holm oak forest during two periods from April 2003 to April 2004 and from November 2004 to July 2005 [229]. Twenty-four forest plots, each of area 10 m^2, were randomly selected for the study, with all being at the same altitude. Half of the plots were subjected to an artificial drought. The measurements were carried out using an infrared gas analyzer (for CO_2 measurements) and GC–MS and PTR-MS (for VOC measurements). Although only low exchange rates were observed for soil monoterpenes, and therefore their contribution to the total monoterpene atmospheric budget is considered to be negligible, they were nevertheless found to be affected by variations in soil moisture and temperature. From measurements taken at the control plots, the researchers found that high soil temperatures increased the monoterpene uptake, but that individual monoterpenes (as determined using GC–MS) responded differently to drought treatment and to seasonal variations. The high soil temperatures that occurred in the summer of 2005 resulted in an increase in VOC exchange rates.

5.7.5 Other Plant Studies

5.7.5.1 Root-secreted VOCs

Steeghs *et al.* have described the use of PTR-MS, in conjunction with GC–MS, to explore VOCs in the rhizosphere[14] and their generation resulting from biotic stresses [230]. This area of research is important because plant roots are believed to release between 5% and 20% of all photosynthetically fixed carbon and therefore these roots create and maintain a carbon-rich environment for rhizosphere organisms, such as plant pathogens and symbiotic microbes from which VOCs can also be emitted. The research team focused on root-secreted VOCs from *Arabidopsis*. Simple volatile metabolites were mainly observed, such as ethanol, acetaldehyde, acetic acid, ethyl acetate, 2-butanone, 2,3,-butanedione and acetone, together with the monoterpene 1,8-cineole. Some VOCs were produced regardless of how the plants were treated (e.g. with compatible or incompatible bacteria), whereas other VOCs were induced specifically as a result of different compatible and incompatible interactions between microbes and insects and *Arabidopsis* roots.

[14] This is defined to be the region of soil that surrounds a plant root and is directly influenced by root secretions and soil microorganisms.

5.7.5.2 Methanol Release and Bacterial Growth: Plant-Methylobacterium Association

Abanda-Nkpwatt *et al.* have made use of PTR-MS and other analytical techniques to elucidate the effect of pink-pigmented facultative methylotrophs, which are a surface bacteria found on plants [231]. The aim of the study was to investigate the potential benefits of methylotrophic bacteria to promote seedling growth and germination of a number of plants, namely cultivated tobacco, tomato, white mustard and strawberry. During this study the levels of methanol emitted through the stomata of plants, which can be used as a carbon and energy source for certain types of bacteria, along with the consumption of methanol by the methylobacteria, were determined using PTR-MS. For example, it was found that the cultivated tobacco seedlings emitted methanol with a mixing ratio of 0.4–0.7 ppbv. However, when the seedlings were co-cultivated with a particular strain of *Methylobacterium extorquens* it was found that the methanol concentration fell owing to the consumption of methanol by the bacteria. By nourishing themselves with the methanol released from a plant the bacteria in turn release beneficial metabolite compounds which can be used to promote growth. However, not all plant–methylobacterium associations are mutually beneficial. The researchers found, for instance, that the germination of six other seed varieties (bean, barley, maize, pea, carrot and wheat) was not affected by the bacteria.

5.7.5.3 Comparison of VOC Emissions from Young and Mature Leaves

Bracho-Nunez *et al.* have investigated VOC emissions from young and mature leaves of several Mediterranean tree and shrub species in terms of their physiological and developmental changes during the growing period within a controlled chamber environment [232]. As part of this study the researchers estimated the standard emission rates (defined by the researchers as the VOC emission rate at standard light and leaf temperature conditions of 1000 μmol photons m^{-2} s^{-1} Photosynthetic Active Radiation and 30°C, respectively). Isoprenoid and oxygenated VOC emissions were measured online by use of PTR-MS and offline measurements were made with GC–MS and GC-FID. The PTR-MS measurements monitored 30 ions, with the main VOCs detected being methanol (at m/z 33), acetone (at m/z 59), isoprene (at m/z 69), monoterpenes (at m/z 137 and m/z 81 (fragment)) and sesquiterpenes (at m/z 205). The results suggested that VOC emission is a very much dependent on the stage of a plant's development and that the type of VOC emitted and its concentration are specific to a particular plant species.

5.7.6 Stress-Related Emissions

VOCs are also released by plants as a result of stress for example see references 35, 36, 78, 204, 209, 211–217, 221, 222, 224, 230, 233–244. Stress can be brought about, for example, by high temperature, high irradiance by light, drought, exposure to chemicals, flooding or pollution. Stress on plants will also be caused by wounding which can either be through natural causes (e.g. weather damage or feeding herbivores) or can be deliberate (e.g. harvesting).

An example of a chemically induced stress-related PTR-MS study is one by Beauchamp *et al.* [235]. In conjunction with GC–MS (used for compound identification), PTR-MS was used to investigate the emission kinetics and thereby show that tobacco plants (*Nicotiana tabacum* L. var. *Bel W3*) were stressed as a result of artificial exposure to ozone. VOCs were found to have been produced within the octadecanoid pathway (sum of (*E*)-2-hexanal, 1-hexanol, hexanal isomers and (*E*)-3-hexenol and (*Z*)-3-hexenylacetate). The researchers related an observed delay in emission directly to ozone flux density into the plants. Quantitative measurements showed that approximately one C_6 product was emitted per five O_3 molecules taken up by the plant. Furthermore, it was discovered that the O_3 flux density had to be greater than $(1.6 \pm 0.7) \times 10^{-8}$ mol m^{-2} s^{-1} for C_6 emission to occur.

Holzinger *et al.* investigated the effects of flooding on biogenic emissions [222]. It was found that when the roots of the Mediterranean holm oak were flooded with water there was a massive increase in acetaldehyde and ethanol production from the leaves of the trees. The effects of flooding on VOC emissions from four Amazonian floodplain tree species have also been reported by Rottenberger *et al.* [236]. The exchange of volatile C_2-compounds (ethanol, acetaldehyde and acetic acid) was investigated over several days of measurements. Both PTR-MS and HPLC ion chromatography were used in this study. It was observed that none of the four species of tree emitted any of the compounds being monitored under non-flooded conditions: instead, low levels of deposition of acetaldehyde and acetic acid were observed. After the roots of the tree species were flooded, ethanol and acetaldehyde were released. Only for those species of tree emitting the highest rates of ethanol and acetaldehyde was acetic acid observed. These three compounds were found to follow a similar diurnal emission profile, with an emission burst in the morning followed by a decline in the evening. The researchers proposed that this behaviour is due to all three compounds being derived from fermentation, with ethanol being transported to the leaves through the transpiratory stream and then being partly converted to acetaldehyde and acetic acid by enzymatic processes.

There are a number of biogenic VOCs which can be released when a plant is under stress and one of these is methyl salicylate, which is a semivolatile plant hormone. A compound associated with regulating stress in plants is jasmonic acid. It has been reported that this acid can elicit the emission of methyl salicylate from plants in addition to other VOCs. Using a leaf cuvette system, Filella *et al.* investigated this by spraying jasmonic acid onto the leaves of the Mediterranean tree species *Q. ilex* and subsequently studied the dynamics of biogenic emissions using both PTR-MS and GC–MS following a dark–light transition [237]. The GC–MS technique was used to determine the relative composition of the monoterpenes. The researchers found that jasmonic acid significantly changed the emissions of three VOCs: methyl salicylate (*m/z* 153), the monoterpenes (*m/z* 137, 81 and 67) and formaldehyde (*m/z* 31). Monoterpene emissions increased following a dark–light transition and were enhanced by ca. 20–30% after jasmonic acid spraying, as was methyl salicylate emission (more than twofold) 1 hour after treatment, but in the case of methyl salicylate this effect had disappeared after 24 h. In contrast, it was found that the leaf cuvette concentration of formaldehyde decreased after dark–light transitions in both control leaves and for those leaves sprayed with jasmonic acid, which indicated a leaf uptake of formaldehyde in light. However, it was found that this uptake was lower (approximately half the rate) for leaves treated with jasmonic acid.

Using GC–MS and PTR-MS, Peñuelas *et al.* assessed the potential effects of fumigating Holm oak (*Q. ilex L.*) with methyl salicylate on the production and emissions of monoterpenes, both at whole plant-scales and leaf-scales levels, whilst exposed to temperatures ranging from 25°C to 50°C [238]. Comparisons were made using control plants, that is, plants that were not fumigated. Fumigated plants (at levels of ~60 mm^3 m^{-3} in air) showed an approximately threefold to fourfold increase in leaf monoterpene concentrations and emission rates compared to those measured from control (unfumigated) plants. Given that oak species do not store monoterpenes, it was concluded that the monoterpenes were produced biochemically immediately prior to emission.

In another study involving *Q. ilex* by Peñuelas and colleagues, isoprene and monoterpene emissions from seedlings under drought conditions and after re-watering were investigated [239]. Following re-watering of *Q. ilex* seedlings that had been previously subjected to drought conditions, the emission rates for these compounds were found to dramatically increase. The same group also investigated the emission of short-chained oxygenated VOCs from *P. halapensis* saplings in response to changes in water availability [240]. In that study, it was found that the oxygenated VOC exchanges, with the exception of acetone, were dependent on stomatal conductance[15] and transpiration. The conclusion was that the emission of oxygenated VOCs as a result of a water deficit is mainly driven by water stress on stomatal closure together with oxygenated VOC concentrations in the ambient air.

Karl *et al.* have investigated the emission of methyl salicylate from trees [241]. In their study, they examined its concentration above a walnut (Paradox Hybrid; *Juglans californica* × *Juglans regia*) agroforest. Air concentrations at the canopy level were measured using PTR-MS. These revealed that the highest methyl salicylate fluxes occurred after plants had experienced low night time temperatures (~7.5°C) followed by a large daytime temperature increase (22°C). The observations showed that methyl salicylate emissions can be used as an ecosystem-scale stress marker before visible damage to a plant becomes apparent.

A link between stress resulting from drought and biogenic VOCs (mainly monoterpenes) in a young (4-year-old) potted beech tree *Fagus sylvatica L.* was investigated by Šimpraga *et al.* [242], who used PTR-MS over a 21 day period for continuous measurements of the biogenic VOCs in the growth room of the Laboratory of Plant Ecology at Ghent University in Belgium. The tree was watered on day 1 and then was not watered again until day 18 of the study. In order to determine which VOC contributed to a given *m/z* in the PTR-MS mass spectrum, thermal desorption GC–MS analysis was also undertaken. This revealed that the isoprenoid emissions from beech trees consisted of monoterpenes, which were recorded in PTR-MS at *m/z* 137 (protonated parent, $C_{10}H_{17}^+$) and at *m/z* 81 (fragment ion). The experiments were performed under carefully controlled conditions of radiation and air temperature. It was found that the monoterpene emissions first increased when drought conditions were imposed (from day 1), reached a maximum at approximately day 9, after which they were observed to decrease. However, *m/z* 81 did not follow this trend near to the end of the drought stress period, since the signal intensity associated with this ion channel was suddenly observed to have increased on day 18. This was found to happen when the plant was suffering from imposed drought stress, as indicated by pronounced negative stem growth. Upon watering, the monoterpene peak at *m/z* 137 decreased further in intensity, but the *m/z* 81 peak showed a sudden decrease in intensity. The conclusion was that the *m/z* 81 signal had a contribution from an unidentified VOC which was being emitted when the plant was under severe stress and it was speculated to result from the green leaf volatile Z-3-hexenal.

[15] Stomatal conductance is a measure of the rate of either CO_2 entering or water vapour exiting the stomata (pores) of a leaf. It depends on factors such as humidity and light intensity.

5.7.7 VOC Emissions Linked to Plant Damage

So far we have reviewed VOC emissions predominantly being released from non-stressed or stressed but intact plants (e.g. in response to a plant's exposure to ozone, water conditions or intense light). We have seen that VOC emissions from the intact plants change as a result of stress, both in terms of composition and intensities, and that they can provide important and useful information as to how a particular plant responds to environmental changes. In the sections that follow, we describe PTR-MS studies which have investigated VOC emissions derived from damaged (wounded) of plants, which can be through natural causes (e.g. weather conditions or insect, bacterial or fungal attack), or through human activity (e.g. harvesting of crops or the mowing of grass). We shall see that nearly all plants behave in a similar way to damage by releasing aldehydes, alcohols and their derivatives (e.g. hexanals). These VOC emissions must somehow be beneficial to a damaged plant. One possibility is that the wound compounds are produced and released because of their antibacterial and antifungal properties [243], which aid in the healing of the plant and inhibit the spread of bacteria and other microorganisms into damaged tissue. Other uses for these emissions are the attraction of predators and parasitoids of herbivores and for inducing plant defence mechanisms [244].

5.7.7.1 Mechanical Wounding

The first PTR-MS study that investigated the mechanical (deliberate) wounding of plants was by Fall *et al.* [224], who demonstrated that the formation and emission of hexenal family compounds after the intentional wounding of aspen (*Populus tremula*) and beech (*Fagus sylvatica*) leaves and white clover (*Trifolium repens*) can be monitored on-line. The quantity of hexenal emission was found to be proportional to the degree of wounding and not dependent on light conditions. It was also observed that the emissions were enhanced for detached leaves as they dried. Such studies are complicated because many of the VOCs associated with wounding (or stress) react dissociatively with H_3O^+. For example, *m/z* 83 ($C_6H_{11}{}^+$) is the dominant fragment ion observed for hexanals, hexenols and hexenyl acetates. It is therefore usual to take the intensity of the peak at *m/z* 83 in the mass spectrum to represent the sum of hexenols plus hexanals present in the headspace. However, occasionally a fragment ion can be used as a marker for a specific compound. For example, Fall *et al.* showed that *m/z* 81 ($C_6H_9{}^+$) is the major fragment ion (69%) associated with (*Z*)-3-hexenal, which results from the loss of H_2O from the protonated parent molecule, whereas the (*E*)-2-hexenal branching ratio for *m/z 81* is small (13%). Dissociative proton transfer is a dominant reaction pathway for other compounds (e.g. saturated alcohols [245]), with the amount of fragmentation being dependent on the value of *E*/*N* being applied to the drift tube. When dealing with compounds which show ion fragmentation, considerable care has to be taken with their analysis, especially when using ion intensities to determine their atmospheric concentrations.

The emission of volatiles from wounded plants is rapid and they follow interesting time evolutions. It is therefore of considerable interest to track these time dependences. PTR-MS is particularly well suited to this area of plant science because of its fast response capabilities. This was nicely demonstrated in the early work of Fall *et al.* [224], mentioned above, where they were able to quantify the emission of (*Z*)-3-hexenal (based on the ion signal at *m/z* 81) within just 1 or 2 s after the wounding of aspen leaves. The disappearance of (*Z*)-3-hexenal was then monitored as the appearance and rise of its metabolites ((*E*)-2-hexenal (*m/z* 57), hexenols (*m/z* 83) and hexenyl acetates (*m/z* 143) were being observed. (*Z*)-3-hexenal is derived from oxidative cleavage of α-linolenic acid and therefore this work provided unique insight into the underlying biochemistry associated with leaf damage. Fall *et al.* found that leaf wound VOCs were not emitted from wounded leaves when placed in a nitrogen atmosphere, which suggests that the hexenyl compounds are not stored in the leaves, otherwise their emissions would be seen in an oxygen-depleted atmosphere. Instead Fall *et al.* showed that oxygen is clearly required for the formation of hexanal and hexenal families of VOCs when a plant is wounded.

Another example of a plant wounding study that illustrates the advantages of using the fast time response available with PTR-MS was by D'Auria *et al.* [244]. The green leaf volatiles (*Z*)-3-hexenal, (*Z*)-3-hexen-1-ol and (*Z*)-3-hexen-1-yl acetate emitted after mechanically wounding *Arabidopsis thaliana Ler* plants were detected. (*Z*)-3-hexenal was the first VOC to be emitted and its concentration was found to quickly reach a maximum just 30–45 s after the leaf had been damaged. The peak emission of (*Z*)-3-hexen-1-ol occurred later, at ~2.5 min. It was found that (*Z*)-3-hexen-1-yl acetate was the dominant green-leaf volatile released after leaves were damaged, peaking after about 5 min and then declining to pre-damaged levels after about 90 min. As observed by Fall *et al.* [224], the rise of (*Z*)-3-hexen-1-yl acetate is mirrored by the fall in (*Z*)-3-hexenal and Z-3-hexen-1-ol because they are involved in its formation.

In a more recent study and the first application of high resolution PTR-TOF-MS to investigate VOC emissions resulting from leaf wounding (and darkening), Brilli *et al.* measured the peak positions of the mass spectral features to a very high mass accuracy [221]. In this way, following the wounding of *D. glomerata* leaves, specific leaf-wound VOCs could be assigned with high confidence without the need for GC–MS analysis. The researchers observed emissions of methanol, C_5 compounds and isoprene from plant species that do not actively emit isoprene. From the time profiles of the recorded VOC intensities, it was possible to determine when enzymatic conversions in the lipoxygenases pathway occurred in the mechanically wounded plants. In this study, the conversion process was followed from the initial membrane breakdown, which was found to produce rapid emissions of C_6 aldehydes, through to the slower conversion of the intermediate C_6 alcohols into hexyl and hexenyl acetates.

5.7.7.2 Weather Damage

As mentioned in the previous section, the study by Fall *et al.* showed that the quantity of wound VOCs released following plant injury is directly proportional to the degree of damage [224]. Thus the increase in VOC emissions as a result of plant wounding can be significant if damage is extensive. An example of this occurred inadvertently during a period of VOC emission observations at Bago State Forest during November 2006 (which is the late spring season in the southern hemisphere) [176]. Measurements from an eucalyptus forest, with average tree height of 40 m and dominant species *E. delegatensis* (alpine ash) and *E. dalympleana* (mountain gum), were interrupted by a storm which resulted in substantial damage to the trees. As a result of this it was found that the concentrations of monoterpenes increased by about a factor of four (to approximately 2 ppbv) during the storm, a finding that was attributed to extensive leaf damage (see Figure 5.5). In contrast, it was observed

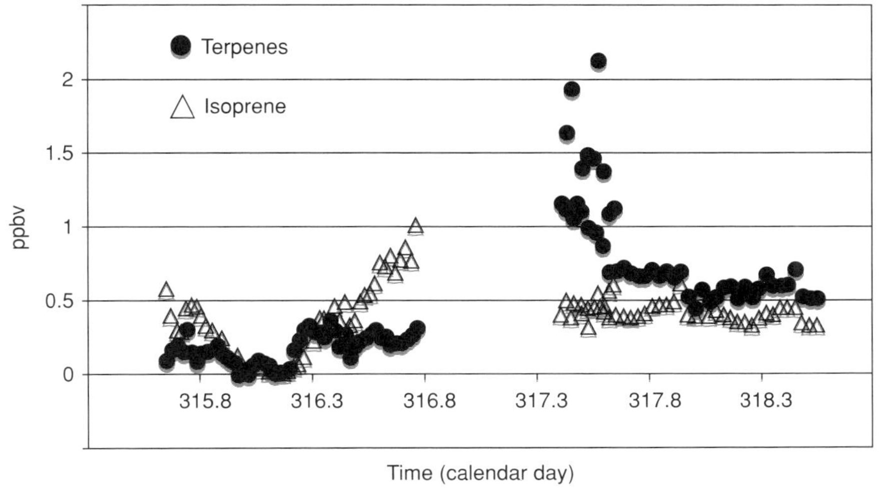

Figure 5.5 *Mixing ratios of isoprene (m/z 69) and total terpenes measured at Bago State Forest in November 2006. The sharp rise in the terpene concentration at day 317.3 is a result of leaf damage caused by a thunderstorm. Reprinted from [176], with permission from © copyright 2009, Elsevier.*

that the isoprene levels remained at a low (0.5 ppbv) concentration, which was presumed to be in response to the cold temperatures (~8°C) that occurred after the storm.

With regard to weather conditions, it is not only storms that can lead to leaf damage. As mentioned earlier in this chapter (Section 5.3.2.3), Fall *et al.* recorded VOC abundances after a hard freeze in central Austria [35,78]. Substantial amounts of C_5 VOCs, ranging from 300 pptv to 6 pptv, and including 1-penten-3-ol, methylbutanals and probably pentenone, were detected at the site for several days. The conclusion made by Fall *et al.* was that leaf-freezing events in forests cause leaf damage which gives rise to the release of substantial levels of reactive C_5 and C_6 compounds. It was suggested that these VOCs will contribute to regional tropospheric chemistry.

5.7.7.3 Harvesting and Mowing

It is to be expected that any damage to plants will release VOCs and none is more apparent to most people than when mowing grass or harvesting of crops, which leads to characteristic smells derived from the emission of a range of oxygenated VOCs [36, 80, 102, 246–248].

Karl *et al.* undertook eddy covariance measurements using PTR-MS of oxygenated VOC fluxes during harvesting for hay in the eastern part of the Austrian Tirol [80, 247]. Significant fluxes of methanol (*m/z* 33), acetaldehyde (*m/z* 45), acetone (*m/z* 59), pentenols plus 2-methylbutanal (*m/z* 69), hexenals (*m/z* 57, 81 and 99), and hexanal plus hexenols (*m/z* 83) were detected resulting from the harvesting of grass. From their measurements values for the high fluxes that occurred for about 2 days after cutting were determined to be 1–8.4 mg m^{-2} h^{-1}, 0.5–3 mg m^{-2} h^{-1}, 0.1–1.5 mg m^{-2} h^{-1} and 0.1–1.5 mg m^{-2} h^{-1} for methanol, acetaldehyde, hexenals and acetone, respectively.

As discussed earlier in Section 5.3.4.3, oxygenated VOCs have also been observed to be emitted from alfalfa fields during harvesting [102]. In addition to the standard burst of wound compounds such as hexenals being released immediately after cutting, it was found that the emissions of methanol, acetaldehyde and acetone were also enhanced following cutting. As the alfalfa dried, which results in further cell damage, the concentrations of these VOCs were observed to remain high for the three days of drying. Similar results were obtained from an earlier study dealing with the release of VOCs from the cutting and drying of grass and clover leaves and stems by de Gouw *et al.* [248]. After the initial burst of VOCs following the wounding that resulted from the cutting of leaves and stems, (including the normal wound compounds (Z)-3-hexenal, (Z)-3-hexenol and hexenyl acetate), emissions of a wide range of oxygenated VOCs remained enhanced during the drying phase, including methanol, acetone, acetaldehyde, butanone and formaldehyde. When integrated over time, the drying phase of the cut plant material accounts for much higher emissions than result from the initial wounding of plants.

The release of C_6 compounds, including (*Z*)-3-hexenal, (*E*)-2-hexenal, hexenol and hexanal, were observed by Karl *et al.* as a result of urban lawn mowing on the outskirts of Innsbruck during continuous ambient air measurements in July and August 1998 [246]. It was noted that the peak levels of biogenic VOCs following mowing were found to be in the same concentration range (20–60 ppbv) as those that come from the internal combustion engines of lawn mowers and that the integrated biogenic emissions were much greater in the drying grass clippings. Following rain, additional emissions of acetone and other VOCs from the wet clippings were also observed.

In a series of laboratory experiments using PTR-MS, PTR-IT-MS and GC/PTR-MS, Karl *et al.* monitored the release of biogenic VOCs from grass crop species, rice and sorghum in simulated leaf drying/senescence experiments [249]. The grass crops were found to release a large variety of oxygenated VOCs upon drying, primarily from the leaves rather than from the stems. The VOC release from paddy rice varieties was observed to be much greater than from sorghum. The major VOCs were unambiguously identified by the use of GC-PTR-MS and included methanol, acetaldehyde, acetone, pentanal, methyl propanal, hexenol, hexanal, *cis*-3-hexenal and *trans*-2-hexenal. The results obtained with the PTR-IT-MS were found to be comparable to those obtained with standard PTR-MS. However, the application of collision-induced dissociation of trapped ions within the ion trap permitted the ratios of the *cis*- and *trans*-hexenals during different phases of drying to be unambiguously determined.

5.7.7.4 Biofuel Crops

Eller *et al.* [250] and Crespo *et al.* [251] have investigated VOC emissions from plant species which are currently being considered for use in the production of biofuel. The interest in the volatile emissions from such plants results from the fact that, if selected as a biofuel crop, they would be grown on such a large scale that substantial increases in local VOC emissions would result. Ultimately this could lead to changes in the local atmospheric chemistry where these plants are being grown.

Using laboratory chamber experiments, Eller *et al.* determined VOC emission rates during the growth and simulated harvest phases for switchgrass (*Panicum virgatum*), which is a grass native to North America and which is being considered for use in second-generation biofuel production [250]. These measurements suggest the following emission levels over 1 year: 3 kg carbon ha^{-1} (i.e. 3 kg of carbon per hectare) for methanol (*m/z* 33), 1 kg carbon ha^{-1} for acetaldehyde (*m/z* 45), 1 kg carbon ha^{-1} for acetone (*m/z* 59), 0.9 kg carbon ha^{-1} for the monoterpenes (*m/z* 137), 0.5 kg carbon ha^{-1} for isoprene (*m/z* 69) and another compound (considered to be a fragment associated with 1-penten-3-ol: GC–MS measurements suggested that the majority of the pre-cut emissions being measured at *m/z* 69 were in fact associated with isoprene and that the post-cut emissions were mainly1-penten-3-ol), 0.2 kg carbon ha^{-1} for hexenals (*m/z* 81 and 99) and 0.1 kg carbon ha^{-1} for hexenols (*m/z* 83 and 101). Eller *et al.* claimed that the total annual emissions obtained are less than those expected from *Eucalyptus* or poplar plantations, which are also under consideration as sources of biofuels. It was therefore suggested that in terms of environmental factors switchgrass would be a better choice for biofuel because it would produce less tropospheric ozone and secondary organic aerosols.

Crespo *et al.* studied the VOC emissions from two plant species, elephant grass (*Miscanthus gigantus*) and black bamboo (*Phyllostachys nigra*), during growth, harvesting and post harvesting [251]. For these studies both standard quadrupole PTR-MS and PTR-IT-MS instruments were employed. The advantage of using a quadrupole mass spectrometer is the greater sensitivity that can be achieved in comparison to an ion trap instrument, while the advantage of an ion trap mass spectrometer compared with a quadrupole mass spectrometer is its greater selectivity through its MS/MS capability. Bamboo was found to generate more VOCs than elephant grass, especially the wound compounds (e.g. hexanal). Furthermore, the VOC emissions from harvested elephant grass were found to be strongly dependent on the season. An important conclusion from this work is that an overestimation of the impact on local air quality in dry periods will occur if the seasonal variations in VOC emissions are not properly taken into account, and therefore the researchers suggested that post-harvest emissions should be measured for improved emission estimates. Furthermore, this work confirmed the conclusion made in the work by Eller *et al.* mentioned above [250]; that is, owing to their lower VOC emissions, the use of perennial grasses for biofuel production is a better choice than woody species in those regions where regional atmospheric chemistry could be significantly affected.

5.7.7.5 Herbivore Attack by Small Predators

Herbivore wounding by small predators (e.g. spider mites) induces plants to emit VOCs that, for example, can act as a chemical communicator to attract natural predators that either kill or become a parasite of the herbivore causing the damage. The emitted VOCs can also warn other parts of the plant or surrounding plants of a possible attack. These VOC emissions are highly complex and change rapidly. Again owing to its fast response, which provides the opportunity to follow chemical and biological processes in real-time, PTR-MS has been used to investigate the VOCs being emitted by plants in response to wounding by small herbivores, and not only from the site of injury [252–254].

The first study was by Peñuelas *et al.* [203]. This revealed that *Euphydryas aurinia* caterpillars feeding on *Succisa pratensis* leaves induced emissions of large amounts of methanol and other VOCs (including monoterpenes, sesquiterpenes and lipoxygenase-derived volatile compounds (e.g. (Z)-3hexenal and (Z)-3-hexenyl acetate)). VOC release was monitored immediately after leaf attack (within the first hour) and after 1 day of continuous herbivorous feeding. Some of the VOCs observed at the start of the measurements were still found to be increasing 24 h later, with methanol being the most significant. Similarly, von Dahl *et al.* showed that when *Manduca sexta* larvae attacked *Nicotiana attenuata* plants methanol emission increased dramatically [252]. Twenty-four hours after the herbivores had started to feed on the plants there was sustained methanol emission at substantially greater levels than that of (*E*)-2-hexenal, which is a standard VOC marker for leaf damage. The research team suggested that the enzymatic demethylation of pectin by pectin methylesterase is a likely source of this methanol. However, we comment that methanol production in plants is not well understood and a more recent study by Oikawa *et al.* showed that pectin methylesterase activity is not necessarily a good predictor of methanol emission [255].

Brilli *et al.* reported details on monoterpene emissions during a 7 day period when *Chrysomela populi* beetles were feeding on *Populus* × *euroamericana* leaves [256]. Poplar leaves are among the strongest isoprene-emitting species and do not strongly emit monoterpenes. The observations showed that inexperienced *Chrysomela* adults (i.e. newly emerged adult beetles from the pupal stage that had not been fed with poplar leaves) have a preference to land and feed on uninfected young leaves, but are more attracted to mature attacked leaves than to young attacked ones. The researchers investigated whether there are differences in the volatiles emitted by young and adult leaves that can provide olfactory clues for the beetles. For the study involving the gas exchange measurements, a single leaf was enclosed in a 0.5 L cuvette without detaching it from the plant. PTR-MS was used to monitor isoprene (at *m*/*z* 69), monoterpenes (at *m*/*z* 137) and linalool (at *m*/*z* 155) every 10 s in the gas sampled from the cuvette. In addition to PTR-MS, GC–MS was used to analyze concentrated air samples to separate the isobaric monoterpenes and to detect other biogenic VOCs and, in particular, salicylaldehyde. Measurements were made before insect feeding (day 0), and for days 1, 4 and 7 after the beetles started feeding. For day 0, the PTR-MS measurements showed that young leaves emitted far higher monoterpene concentrations than was being emitted from mature leaves, but that these were still in relatively low concentrations. The GC–MS work found that in both mature and young leaves the dominant VOCs emitted were the monoterpene (*E*)-β-ocimene ($C_{10}H_{16}$) and a monoterpene derivative, linalool ($C_{10}H_{18}O$). However, for young leaves these volatiles were emitted in approximately equal amounts, whereas for mature leaves it was found that (*E*)-β-ocimene was three times more abundant than linalool. Furthermore, it was observed that isoprene emissions were higher (by approximately a factor of four) for mature leaves than for young leaves.

Following attack by beetles the emission profiles were found to change dramatically. It was found that the emission of monoterpenes from mature leaves increased and peaked after 4 days, reaching rates (2.0 nmol m^{-2} s^{-1}) that were nearly two orders of magnitude higher than measured from the same leaves before insect attack. By day 7 the monoterpene emission rate had decreased. Although similar changes in monoterpene emissions were observed from young leaves being attacked by insects, the increase in monoterpene emissions observed on day 4 were only about a factor of two of that determined on day 0. During the period for which monoterpene emission rates were increasing for the mature leaves, it was found that the isoprene emission rates were decreasing and that these only started to increase again on day 7, when monoterpene emissions started to decrease, that is, the emissions of monoterpenes and isoprene were inversely correlated after attack. Figure 5.6 provides a summary of the key results that were found. Interestingly, the researchers also found that monoterpene emissions were induced from adjacent leaves that had not

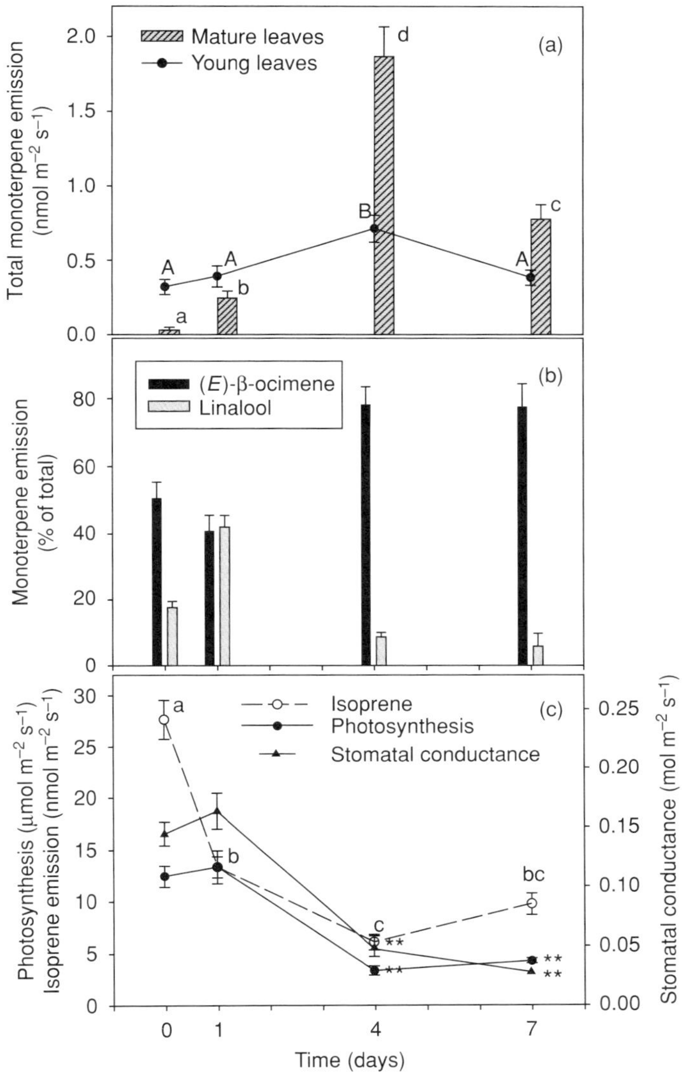

Figure 5.6 *(a) The total emission of monoterpenes from mature (bars) and young Poplar leaves (filled circles) determined from PTR-MS measurements. (b) The percentage of the two principal monoterpenes, (E)-β-ocimene and linalool (determined using GC–MS) emitted for day 0 (no insect attack) and for days 1, 4 and 7 after beetle attack. In (c), the rates of isoprene emission and photosynthesis and the stomatal conductance are provided for mature leaves. Reproduced with permission from [256]. Copyright © 2009, Blackwell Publishing Ltd.*

been attacked by the beetle and even in those that had no vascular connection to the wounded leaves.

In contrast to the laboratory study by Brilli *et al.* [256], in a field study using an enclosed branch method Schaub *et al.* investigated green leaf volatiles being emitted as *Epirrita autumnata* (an autumnal moth) larvae were feeding on *Populus tremula* L. × *P. tremuloides* Michx. [257]. Measurements were taken over a period of 20 h in order to determine the timing and pattern of biogenic VOC emissions during the early stages of herbivore attack. In addition to the real-time PTR-MS studies, headspace air samples were also collected for later GC–MS analysis to aid in the identification of the VOCs being emitted. It was observed that green leaf VOC emission peaks occurred intermittently from the outset, which was taken to indicate herbivore activity, while terpene emissions were found to be induced within 16 h. Of note is that the green leaf emissions were found to be quantitatively and qualitatively different from those resulting from accidental mechanical damage: specifically, herbivore-induced emissions were found to have sporadic short-lived increases.

In addition to investigating herbivore attack by small predators above ground, PTR-MS has also been used to detect herbivore activity below ground, which is far more serious for a plant's survival because herbivore attack of the roots can inflict far more lasting damage on a plant than herbivore attack above ground. The first study was by Steeghs *et al.* [230], who analyzed VOC emissions from *Arabidopsis* roots cultured *in vitro*. Infection with a pathogen, *Pseudomonas syringae*, and the aphid *Diuraphis noxia*, was found to induce the release of compounds such as acetic acid, acetone and ethanol. In addition to these simple metabolites, a rapid release of a single monoterpene, 1,8-cineole, was observed. It is noteworthy that this compound does not occur when the roots of plants are mechanically injured, that is, Steeghs *et al.* found that green leaf volatiles are not emitted by damaged roots, and therefore roots respond differently to above-ground plant damage. Although this seminal study opens the way for root VOC emission analysis with PTR-MS, it is not clear from this study how representative *in vitro* measurements are for plants growing in soil.

To explore this, Danner *et al.* used PTR-MS to measure root-herbivore-induced VOC from potted plants. They clearly demonstrated that PTR-MS can be applied *in vivo* to detect specific root-herbivore-induced VOCs from *Brassica* plants [258], the genus of which includes many important crops, such as cabbage, broccoli and oil seed rape. The VOCs released from the roots of non-infested potted turnip plants (*Brassica rapa* subsp. *rapa* var. Nancy) were compared with those resulting from infestation with the larvae of the cabbage root fly (*Delia radicum*), which is a major pest of *Brassica* crops. As found for other herbivore attacks, signals at m/z 63 and m/z 95, which were assigned to protonated dimethyl sulfide and dimethyl disulfide, respectively, substantially increased following herbivore attack. As well as these compounds, an increase in other VOC emissions was found that had not previously been observed from infested plants. The main species identified were methanethiol (CH_4S), which produces an ion at m/z 49 corresponding to the protonated parent molecule, and a sulfur-containing marker compound at m/z 60 (which the researchers assigned to a glucosinolate[16] break-down product, but a specific compound assignment was not made). In addition to the turnip plants, Danner *et al.* reported VOC

[16] Glucosinolates are plant-produced organic compounds that contain sulfur and nitrogen and are well-known defence compounds used by a plant.

emissions from *Brassica juncea* (a species of mustard plant) after infestation with *Delia radicum*. In particular, and based on the results from the turnip plant, the intensities of the ion signals at *m/z* 49, 60, 63 and 95 were recorded as a function of time. Only low level emissions (several ppbv) were observed initially but then the emission levels increased steadily with time.

In a follow-up study the same research group investigated the VOC emission associated with the ion measured at *m/z* 60 from other species of *Brassica* that had been damaged, in this case either through artificial means or after infestation with *D. Radicum* larvae [259]. An investigation was also carried out to determine if methanethiol, dimethyl sulfide and dimethyl disulfide are only emitted after larval infestation. Focussing on these VOCs, the damaged roots of six *Brassica* species were investigated (*B. nigra*, *B. juncea*, *B. napus*, *B. rapa*, *B. oleracea* and *B. carinata*). The intensity of the *m/z* 60 ion was employed to determine if there was any correlation with the glucosinolate composition of the roots. The measurements revealed that the species *B. nigra*, *B. juncea* and *B. napus* emitted *m/z* 60 directly after artificial or larval damage, while *B. oleracea* and *B. carinata* showed an increase in *m/z* 60 emission only after larval damage. Sulfides and methanethiol emissions were also found to increase as a result of larval damage for *B. nigra* and *B. juncea*, but only 6–12 h after the damage had occurred. It was found that *B. rapa*, *B. oleracea* and *B. carinata* principally emitted methanethiol after artificial or larval damage. An important conclusion from this work is that it appears that the species of plant determines the principal VOCs that are emitted, and not the type of damage inflicted or root glucosinolate composition.

Additional work dealing with the on-line detection of root-induced volatiles in *B. nigra* plants infested with the root fly larvae *D. radicum* has been reported by Crespo and colleagues [260, 261]. This study used both PTR-MS and GC–MS for analysis. The *m/z* 60 ion mentioned in the previous two paragraphs was assigned to protonated thiocyanic acid ($HSCNH^+$). The PTR-MS measurements revealed that the emission of thiocyanic acid stopped as soon as the larvae had pupated. It was therefore suggested that the *m/z* 60 ion could be a useful marker for larvae that are actively feeding. In comparison, the concentrations of methanethiol, dimethyl sulfide and dimethyl disulfide were found to increase much later from the infested roots, which the researchers have suggested indicates that enzymes or genes need to be activated for the production of these compounds. The researchers also suggested that identification of the root-induced responses should aid in designing non-invasive analytical procedures to assess root infestations.

Finally, we comment on the work of Schausberger *et al.* who have reported the effects of the mycorrhizal fungus *Glomus mosseae* on the VOC composition of plant volatiles emitted above ground by bean plants (*Phaseolus vulgaris*) that were under attack by spider mites, *Tetranychus urticae* [254]. The aim of the study was to investigate how the VOCs emitted influence the attraction of the natural enemy of the spider mite, the predatory mite *Phytoseiulus persimilis*. It was observed that the fungus changed the herbivore-induced plant volatile composition by predominantly increasing the emission of two compounds, β-ocimene and β-caryophyllene, immediately upon spider mite attack, with the predators preferring herbivore-induced plant volatiles of mycorrhizal plants to those from non-mycorrhizal plants. The researchers commented that their study has provided a key example of where a below-ground microorganism has affected the volatile emissions from a plant resulting from an above-ground herbivore attack to attract an above-ground carnivorous predator.

5.7.7.6 *Large Herbivore Attack*

In comparison to an insect attack, there has been much less research associated with VOC emissions resulting from the attack on plants by large herbivores such as mammals. However, there has been one such study which has involved PTR-MS. In this, Bartolome *et al.* monitored leaf volatiles in the pptv range emitted as a result of Mediterranean shrubland being browsed by horses [253]. The shrubland was dominated by *Rosmarinus officinalis*, *Erica multiflora*, *Ulex parviflorus* and *Globularia alypum*. In comparison to VOCs being emitted from a control plot, significant enhancement of hexanals, acetic acid and acetone from the plants was observed following the feeding process.

5.7.8 VOC Uptake by Plants

As well as plants emitting VOCs into the atmosphere, plants can also absorb VOCs from the atmosphere and this can represent a major and useful route for the removal of atmospheric VOCs. PTR-MS has been used to determine the uptake of VOCs by plants by Tani *et al.* [262] and Seco *et al.* [263]. In the first study, an exposure measurement system was developed that allowed an accurate determination of VOC uptake rates by plants under controlled environmental conditions. This device consisted of a PTR-MS instrument, a CO_2 analyzer, a diffusion device and a leaf enclosure. The VOC uptake was determined using a flow-through chamber by measuring the change in concentration between the incoming and outgoing air. Using this approach, Tani *et al.* measured the uptake of VOCs by the species golden pothos (*Epipremnum aureum*). The research team particularly focused on methyl isobutyl ketone, because it is ubiquitous in the atmosphere. It was found that even at the lowest methyl isobutyl ketone levels (70 ppbv) more than 2.2% of the incoming VOC was lost to the leaf. In the study by Seco *et al.*, the emission and uptake of formaldehyde (m/z 31) by the Mediterranean trees *Q. ilex* and *Pinus halepensis* was investigated, with experiments conducted at leaf level under controlled laboratory conditions. Formaldehyde is a common volatile found in the atmosphere and it can be present in high concentrations in polluted areas which can have a serious impact on human health. Following corrections for the effect of humidity on the signal at m/z 31 (see Sections 4.6 and 5.1), formaldehyde exchange was found to be dependent on both the atmospheric mixing ratio and stomatal conductance. Formaldehyde exchange was at its greatest when stomatal conductance was at its maximum, which results from the high solubility of formaldehyde in water.

Using the flow-through chamber method developed by Tani *et al.* mentioned above, the uptake of VOCs by house plants, particularly of aldehydes and ketones, which are well-known pollutants that are emitted from many materials found in indoors, has been investigated by Tani and Hewitt [264]. The ketones used in their study were acetone, methyl ethyl ketone, diethyl ketone, methyl *n*-propyl ketone, methyl isopropyl ketone and methyl isobutyl ketone. The aldehydes investigated were propionaldehyde, *n*-butylaldehyde, iso-butylaldehyde, crotonaldehyde, methacrolein, *n*-valeraldehyde, iso-valeraldehyde and benzaldehyde. The interest in these types of investigations lies in the efficacy of common house plants to improve the quality of indoor air, which is achieved by removing VOCs through stomatal aperture absorption and by adsorption on the cuticle or outer surface of a plant's leaf. Two house plants were used in this study; the peace lily (*Spathiphyllum clevelandii*) and the golden pothos. The researchers found that the C_3–C_6 aldehydes and C_4–C_6 ketones were taken up by the plant leaves, but that the C_3 ketone, acetone, was not. Nevertheless, the results showed that the two house plants used in the study are efficient at scavenging aldehydes and ketones present in ambient air.

5.8 Outlook for Atmospheric and Environmental Applications of PTR-MS

A major problem in applying PTR-MS to the atmospheric and environmental sciences is the lack of knowledge concerning the identities and quantities of reactive VOCs being emitted by vegetation and their photochemical oxidation pathways. Measurements of biogenic VOCs and their photochemical oxidation products therefore remain a challenge. In addition, new fields of environmental applications are emerging which require further development, exploration and validation. For example, OH radicals are the main oxidants in the troposphere and are the principal sink of important trace gases such as CO and VOCs. Sinha *et al.* have recently shown that the PTR-MS based '*comparative reactivity method*' may provide a viable methodology for exploring total OH reactivity, an important parameter in atmospheric chemistry [182]. This requires further testing and investigation.

Scientific progress on particle-bound organic carbon has been slowed by analytical problems. Current research on particulate organic carbon is focused on understanding its role in affecting the Earth's climate and its impact on human health. Many of the current uncertainties in this field are associated with the poorly characterized organic fraction of the particulates. New techniques are needed for the time-resolved and *in situ* characterization of the organic composition of atmospheric particulates. Particle collection on a substrate followed by thermal desorption and PTR-MS analysis may provide a new analytical approach for the speciation of particulate organic carbon, but this is far from trivial because of problems in collecting micrometre-sized particles.

The groundbreaking investigations of VOC emissions from damaged plants, both above and below ground, using the non-destructive PTR-MS technique, have provided new and unique insights into plant VOC emissions and their interactions with the environment. Further studies will undoubtedly lead to an enhanced understanding of the role these VOCs play in the defence and communication of plants, their interactions with the local environment and how they interact with natural predators.

References

1. Hewitt, C. N., Hayward, S., Tani, A. (2003) The application of proton transfer reaction-mass spectrometry (PTR-MS) to the monitoring and analysis of volatile organic compounds in the atmosphere. *J. Environ. Monit.* **5**, 1.
2. de Gouw, J. A., Warneke, C. (2007) Measurements of volatile organic compounds in the Earth's atmosphere using proton-transfer-reaction mass spectrometry. *Mass Spectrom. Rev.* **26**, 223.
3. Blake, R. S., Monks, P. S., Ellis, A. M. (2009) Proton Transfer Reaction Mass Spectrometry. *Chem. Rev.* **109**, 861.
4. Lindinger, W., Hansel, A., Jordan, A. (1998) On-line monitoring of volatile organic compounds at pptv levels by means of proton-transfer-reaction mass spectrometry

(PTR-MS): medical applications, food control and environmental research. *Int. J. Mass Spectrom. Ion Processes* **173**, 191.

5. Hansel, A., Jordan, A., Warneke, C. *et al.* (1998) Improved detection limit of the proton-transfer reaction mass spectrometer: on-line monitoring of volatile organic compounds at mixing ratios of a few pptv. *Rapid Commun. Mass Spectrom.* **12**, 871.
6. Crutzen, P. J., Williams, J., Poschl, U. *et al.* (2000) High spatial and temporal resolution measurements of primary organics and their oxidation products over the tropical forests of Surinam. *Atmos. Environ.* **34**, 1161.
7. Williams, J., Poschl, U., Crutzen, P. J. *et al.* (2001) An atmospheric chemistry interpretation of mass scans obtained from a Proton Transfer Mass Spectrometer flown over the tropical rainforest of Surinam. *J. Atmos. Chem.* **38**, 133.
8. Warneke, C., Holzinger, R., Hansel, A. *et al.* (2001) Isoprene and its oxidation products methyl vinyl ketone, methacrolein, and isoprene related peroxides measured online over the tropical rain forest of Surinam in March 1998. *J. Atmos. Chem.* **38**, 167.
9. de Gouw, J. A., Middlebrook, A. M., Warneke, C. *et al.* (2011) Organic aerosol formation downwind from the Deepwater Horizon oil spill. *Science* **331**, 1295.
10. Wisthaler, A., Jensen, N. R., Winterhalter, R. *et al.* (2001) Measurements of acetone and other gas phase product yields from the OH-initiated oxidation of terpenes by proton-transfer-reaction mass spectrometry (PTR-MS). *Atmos. Environ.* **35**, 6181.
11. Northway, M. J., de Gouw, J. A., Fahey, D. W. *et al.* (2004) Evaluation of the role of heterogeneous oxidation of alkenes in the detection of atmospheric acetaldehyde. *Atmos. Environ.* **38**, 6017.
12. Davison, B., Taipale, R., Langford, B. *et al.* (2009) Concentrations and fluxes of biogenic volatile organic compounds above a Mediterranean macchia ecosystem in western Italy. *Biogeosciences* **6**, 1655.
13. Müller, M., Graus, M., Ruuskanen, T. M. *et al.* (2010) First eddy covariance flux measurements by PTR-TOF. *Atmos. Meas. Tech.* **3**, 387.
14. Ruuskanen, T. M., Müller, M., Schnitzhofer, R. *et al.* (2011) Eddy covariance VOC emission and deposition fluxes above grassland using PTR-TOF. *Atmos. Chem. Phys.* **11**, 611.
15. Rogers, T. M., Grimsrud, E. P., Herndon, S. C. *et al.* (2006) On-road measurements of volatile organic compounds in the Mexico City metropolitan area using proton transfer reaction mass spectrometry. *Int. J. Mass Spectrom.* **252**, 26.
16. de Gouw, J. A., Goldan, P. D., Warneke, C. *et al.* (2003) Validation of proton transfer reaction-mass spectrometry (PTR-MS) measurements of gas-phase organic compounds in the atmosphere during the New England Air Quality Study (NEAQS) in 2002. *J. Geophys. Res.* **108**, D4682.
17. Karl, T. G., Christian, T. J., Yokelson, R. J. *et al.* (2007) The Tropical Forest and Fire Emissions Experiment: method evaluation of volatile organic compound emissions measured by PTR-MS, FTIR, and GC from tropical biomass burning. *Atmos. Chem. Phys.* **7** 5883.
18. Brenninkmeijer, C., Crutzen, P. J., Boumard, F. *et al.* (2007) Civil Aircraft for the regular investigation of the atmosphere based on an instrumented container: The new CARIBIC system. *Atmos. Chem. Phys.* **7**, 4953.

19. Karl, T., Apel, E., Hodzic, A. *et al.* (2009) Emissions of volatile organic compounds inferred from airborne flux measurements over a megacity. *Atmos. Chem. Phys.* **9**, 271.
20. Warneke, C., Veres, P., Holloway, J. S. *et al.* (2011) Airborne formaldehyde measurements using PTR-MS: calibration, humidity dependence, inter-comparison and initial results. *Atmos. Meas. Tech.* **4**, 2345.
21. Kato, S., Miyakawa, Y., Kaneko, T., Kajii, Y. (2004) Urban air measurements using PTR-MS in Tokyo area and comparison with GC-FID measurements. *Int. J. Mass Spectrom.* **235**, 103.
22. Kuster, W. C., Jobson, B. T., Karl, T. *et al.* (2004) Intercomparison of volatile organic carbon measurement techniques and data at La Porte during the TexAQS2000 Air Quality Study. *Environ. Sci. Technol.* **38**, 221.
23. Ammann, C., Spirig, C., Neftel, A. *et al.* (2004) Application of PTR-MS for measurements of biogenic VOC in a deciduous forest. *Int. J. Mass Spectrom.* **239**, 87.
24. Jobson, B. T., Volkamer, R. A., Velasco, E. *et al.* (2010) Comparison of aromatic hydrocarbon measurements made by PTR-MS, DOAS and GC-FID during the MCMA 2003 Field Experiment. *Atmos. Chem. Phys.* **10**, 1989.
25. Ambrose, J. L., Haase, K., Russo, R. S. *et al.* (2010) A comparison of GC-FID and PTR-MS toluene measurements in ambient air under conditions of enhanced monoterpene loading. *Atmos. Meas. Tech.* **3**, 959.
26. Inomata, S., Tanimoto, H., Kato, S. *et al.* (2010) PTR-MS measurements of non-methane volatile organic compounds during an intensive field campaign at the summit of Mount Tai, China, in June 2006. *Atmos. Chem. Phys.* **10**, 7085.
27. Sprung, D., Jost, C., Reiner, T. *et al.* (2001) Acetone and acetonitrile in the tropical Indian Ocean boundary layer and free troposphere: aircraft-based intercomparison of AP-CIMS and PTR-MS measurements. *J. Geophys. Res.* **106**, D28511.
28. Inomata, S., Tanimoto, H., Kameyama, S. *et al.* (2008) Technical Note: Determination of formaldehyde mixing ratios in air with PTR-MS: laboratory experiments and field measurements. *Atmos. Chem. Phys.* **8**, 273.
29. Christian, T. J., Kleiss, B., Yokelson, R. J. *et al.* (2004) Comprehensive laboratory measurements of biomass-burning emissions: 2. First intercomparison of open-path FTIR, PTR-MS, and GC-MS/FID/ECD. *J. Geophys. Res.* **109**, D02311.
30. Warneke, C., van der Veen, C., Luxembourg, S. *et al.* (2001) Measurements of benzene and toluene in ambient air using proton-transfer-reaction mass spectrometry: calibration, humidity dependence, and field intercomparison. *Int. J. Mass Spectrom.* **207**, 167.
31. Hayward, S., Hewitt, C. N., Sartin, J. H., Owen, S. M. (2002) Performance characteristics and applications of a proton transfer reaction-mass spectrometer for measuring volatile organic compounds in ambient air. *Environ. Sci. Technol.* **36**, 1554.
32. de Gouw, J., Warneke, C., Holzinger, R. *et al.* (2004) Inter-comparison between airborne measurements of methanol, acetonitrile and acetone using two differently configured PTR-MS instruments. *Int. J. Mass Spectrom.* **239**, 129.
33. de Gouw, J. A., Warneke, C., Stohl, A. *et al.* (2006) Volatile organic compounds composition of merged and aged forest fire plumes from Alaska and western Canada. *J. Geophys. Res.* **111**, D10303.

34. Lee, A., Schade, G. W., Holzinger, R., Goldstein, A. H. (2005) A comparison of new measurements of total monoterpene flux with improved measurements of speciated monoterpene flux. *Atmos. Chem. Phys.* **5**, 505.
35. Fall, R., Karl, T., Jordan, A., Lindinger, W. (2001) Biogenic C5 VOCs: release from leaves after freeze–thaw wounding and occurrence in air at a high mountain observatory. *Atmos. Environ.* **35**, 3905.
36. de Gouw, J. A., Howard, C. J., Custer, T. G. *et al.* (2000) Proton-transfer chemical-ionization mass spectrometry allows real-time analysis of volatile organic compounds released from cutting and drying of crops. *Environ. Sci. Technol.* **34**, 2640.
37. Monks, P. S. (2005) Gas-phase radical chemistry in the troposphere. *Chem. Soc. Rev.* **34**, 376.
38. Singh, H. B., Kanakidou, M., Crutzen, P. J., Jacob, D. J. (1995) High concentrations and photochemical fate of oxygenated hydrocarbons in the global troposphere. *Nature.* **378**, 50.
39. Wyche, K. P., Blake, R. S., Ellis, A. M. *et al.* (2007) Technical Note: Performance of Chemical Ionization Reaction Time-of-Flight Mass Spectrometry (CIR-TOF-MS) for the measurement of atmospherically significant oxygenated volatile organic compounds. *Atmos. Chem. Phys.* **7**, 609.
40. Warneke, C., de Gouw, J. A., Kuster, W. C. *et al.* (2003) Validation of atmospheric VOC measurements by Proton-Transfer-Reaction Mass Spectrometry using a Gas-Chromatographic Preseparation Method. *Environ. Sci. Technol.* **37**, 2494.
41. Holzinger, R., Jordan, A., Hansel, A., Lindinger, W. (2001) Methanol measurements in the lower troposphere near Innsbruck (047316°N; 011324°E), Austria. *Atmos. Environ.* **35**, 2525.
42. Holzinger, R., Lee, A., Paw, U. K. T., Goldstein, A. H. (2005) Observations of oxidation products above a forest imply biogenic emissions of very reactive compounds. *Atmos. Chem. Phys.* **5**, 67.
43. Pöschl, U., Williams, J., Hoor, P. *et al.* (2001) High acetone concentrations throughout the 0–12 km altitude range over the tropical rainforest in Surinam. *J. Atmos. Chem.* **38**, 115.
44. Holzinger, R., Williams, J., Salisbury, G. *et al.* (2005) Oxygenated compounds in aged biomass burning plumes over the Eastern Mediterranean: evidence for strong secondary production of methanol and acetone. *Atmos. Chem. Phys.* **5**, 39.
45. Vlasenko, A., Slowik, J. G., Bottenheim, J. W. *et al.* (2009) Measurements of VOCs by proton transfer reaction mass spectrometry at a rural Ontario site: sources and correlation to aerosol composition. *J. Geophys. Res.* **114**, D21305.
46. Hunter E. P., Lias, S. G. (2011) Proton affinity evaluation. In: *NIST Chemistry WebBook*, NIST Standard Reference Database Number 69 (Eds P. J. Linstrom, W. G. Mallard), National Institute of Standards and Technology, Gaithersburg MD, 20899, http://webbook.nist.gov.
47. Ammann, C., Brunner, A., Spirig, C., Neftel, A. (2006) Technical note: Water vapour concentration and flux measurements with PTR-MS. *Atmos. Chem. Phys.* **6**, 4643.
48. Steinbacher, M., Dommen, J., Ammann, C. *et al.* (2004) Performance characteristics of a proton-transfer-reaction mass spectrometer (PTR-MS) derived from laboratory and field measurements. *Int. J. Mass Spectrom.* **239**, 117.

49. Knighton, W. B., Fortner, E. C., Midey *et al.* (2009) HCN detection with a proton transfer reaction mass spectrometer. *Int. J. Mass Spectrom.* **283**, 112.
50. Hansel, A., Wisthaler, A. (2000) A method for real-time detection of PAN, PPN and MPAN in ambient air. *Geophys. Res. Lett.* **27**, 895.
51. Ikezoe, Y., Matsuoka, S., Takebe, M., Viggiano, A. (Maruzen Comp, Ltd., Tokyo, 1987) "Gas Phase Ion-Molecule Reactions Rate Constants Through 1986".
52. Anicich, V. G. JPL Publication 03-19. Available as a pdf download on http://trs-new.jpl.nasa.gov/dspace/bitstream/2014/7981/1/03-2964.pdf. "An Index of the Literature for Bimolecular Gas Phase Cation-Molecule Reaction Kinetics".
53. Norman, M., Hansel, A., Wisthaler, A. (2007) $O_2{}^+$ as reagent ion in the PTR-MS instrument: detection of gas-phase ammonia. *Int. J. Mass Spectrom.* **265**, 382.
54. Wyche, K. P., Blake, R. S., Willis, K. A. *et al.* (2005) Differentiation of isobaric compounds using chemical ionization reaction mass spectrometry. *Rapid. Commun. Mass Spectrom.* **19**, 3356.
55. Knighton, W. B., Fortner, E. C., Herndon, S. C. *et al.* (2009) Adaptation of a proton transfer reaction mass spectrometer instrument to employ NO^+ as reagent ion for the detection of 1,3-butadiene in the ambient atmosphere. *Rapid Commun. Mass Spectrom.* **23**, 3301.
56. Karl, T., Hansel, A., Cappellin, L. *et al.* (2012) Selective measurements of isoprene and 2-methyl-3-buten-2-ol based on NO^+ ionization mass spectrometry. *Atmos. Chem. Phys.* **12**, 11877.
57. Sulzer, P., Edtbauer, A., Hartungen, E. *et al.* (2012) From conventional proton-transfer-reaction mass spectrometry (PTR-MS) to universal trace gas analysis. *Int. J. Mass Spectrom.* **321**, 66.
58. Lindinger, W., Hansel, A., Jordan, A. (1998) Proton-transfer-reaction mass spectrometry (PTR-MS): on-line monitoring of volatile organic compounds at pptv levels. *Chem. Soc. Rev.* **27**, 347.
59. Hansel, A., Jordan, A., Warneke, C. *et al.* (1999) Proton-transfer-reaction mass spectrometry (PTR-MS): on-line monitoring of volatile organic compounds at volume mixing ratios of a few pptv. *Plasma Sources Sci. Technol.* **8**, 332.
60. Warneke, C., Karl, T., Judmaier, H. *et al.* (1999) Acetone, methanol, and other partially oxidized volatile organic emissions from dead plant matter by abiological processed: significance for Atmospheric HO_x chemistry. *Global Biogeochem. Cycles.* **13**, 9.
61. Murphy, J. G., Oram, D. E., Reeves, C. E. (2010) Measurements of volatile organic compounds over West Africa. *Atmos. Chem. Phys.* **10**, 5281.
62. Salisbury, G., Williams, J., Holzinger, R. *et al.* (2003) Ground-based PTR-MS measurements of reactive organic compounds during the MINOS campaign in Crete, July–August 2001. *Atmos. Chem. Phys.* **3**, 925.
63. Williams, J., Fischer, H., Harris, G. W. *et al.* (2000) Variability-lifetime relationship for organic trace gases: a novel aid to compound identification and estimation of HO concentrations. *J. Geophys. Res.* **105**, D20473.
64. Karl, T., Crutzen, P. J., Mandl, M. *et al.* (2001) Variability-lifetime relationship of VOCs observed at the Sonnblick Observatory 1999 – estimation of HO-densities. *Atmos. Environ.* **35**, 5287.

65. Warneke, C., de Gouw, J. A. (2001) Organic trace gas composition of the marine boundary layer over the northwest Indian Ocean in April 2000. *Atmos. Environ.* **35**, 5923.
66. Eerdekens, G., Ganzeveld, L., Vilà-Guerau de Arellano, J. *et al.* (2009) Flux estimates of isoprene, methanol and acetone from airborne PTR-MS measurements over the tropical rainforest during the GABRIEL 2005 campaign. *Atmos. Chem. Phys.* **9**, 4207.
67. Müller, K., Haferkorn, S., Grabmer, W. *et al.* (2006) Biogenic carbonyl compounds within and above a coniferous forest in Germany. *Atmos. Environ.* **40**, S81.
68. Rinne, J., Ruuskanen, T. M., Reissell, A. *et al.* (2005) On-line PTR-MS measurements of atmospheric concentrations of volatile organic compounds in European boreal forest ecosystem. *Boreal Environ. Res.* **10**, 425.
69. Taipale, R., Ruuskanen, T. M., Rinne, J. *et al.* (2008) Technical Note: Quantitative long-term measurements of VOC concentrations by PTR-MS – measurement, calibration, and volume mixing ratio calculation methods. *Atmos. Chem. Phys.* **8**, 6681.
70. Lappalainen, H. K., Sevanto, S., Bäck, J. *et al.* (2009) Day-time concentrations of biogenic volatile organic compounds in a boreal forest canopy and their relation to environmental and biological factors. *Atmos. Chem. Phys.* **9**, 5447.
71. Williams, J., Crowley, J., Fischer, H. *et al.* (2011) The summertime Boreal forest field measurement intensive (HUMPPA-COPEC-2010): an overview of meteorological and chemical influences. *Atmos. Chem. Phys.* **11**, 10599.
72. Nölscher, A. C., Williams, J., Sinha, V. *et al.* (2012) Summertime total OH reactivity measurements from boreal forest during HUMPPA-COPEC 2010. *Atmos. Chem. Phys.* **12**, 8257.
73. Bouvier-Brown, N. C., Holzinger, R., Palitzsch, K., Goldstein, A. H. (2009) Large emissions of sesquiterpenes and methyl chavicol quantified from branch enclosure measurements. *Atmos. Environ.* **43**, 389.
74. Bouvier-Brown, N. C., Goldstein, A. H., Worton, D. R. *et al.* (2009) Methyl chavicol: characterization of its biogenic emission rate, abundance, and oxidation products in the atmosphere. *Atmos. Chem. Phys.* **9**, 2061.
75. Holzinger, R., Lee, A., McKay, M., Goldstein, A. H. (2006) Seasonal variability of monoterpene emission factors for a Ponderosa pine plantation in California. *Atmos. Chem. Phys.* **6**, 1267.
76. Karl, T., Harley, P., Guenther, A. *et al.* (2005) The bi-directional exchange of oxygenated VOCs between a loblolly pine (*Pinus taeda*) plantation and the atmosphere. *Atmos. Chem. Phys.* **5**, 3015.
77. Kim, S., Karl, T., Guenther, A. *et al.* (2010) Emissions and ambient distributions of Biogenic Volatile Organic Compounds (BVOC) in a ponderosa pine ecosystem: interpretation of PTR-MS mass spectra. *Atmos. Chem. Phys.* **10**, 1759.
78. Karl, T., Fall, R., Crutzen, P. J. *et al.* (2001) High concentrations of reactive biogenic VOCs at a high altitude site in late autumn. *Geophys. Res. Lett.* **28**, 507.
79. Karl, T., Potosnak, M., Guenther, A. *et al.* (2004) Exchange processes of volatile organic compounds above a tropical rain forest: implications for modeling tropospheric chemistry above dense vegetation. *J. Geophys. Res.* **109**, D18306.
80. Karl, T., Guenther, A., Lindinger, C. *et al.* (2001) Eddy covariance measurements of oxygenated volatile organic compound fluxes from crop harvesting using a redesigned proton-transfer-reaction mass spectrometry. *J. Geophys. Res.* **106**, D24157.

81. Rinne, H. J. I., Guenther, A., Warneke, C. *et al.* (2001) Disjunct eddy covariance technique for trace gas flux measurements. *Geophys. Res. Lett.* **28**, 3139.
82. Spirig, C., Neftel, A., Ammann, C. *et al.* (2005) Eddy covariance flux measurements of biogenic VOCs during ECHO 2003 using proton transfer reaction mass spectrometry. *Atmos. Chem. Phys.* **5**, 465.
83. Karl, T., Guenther, A. (2004) Atmospheric variability of biogenic VOCs in the surface layer measured by proton-transfer-reaction mass spectrometry. *Int. J. Mass Spectrom.* **239**, 77.
84. Karl, T. G., Spirig, C., Rinne, J. *et al.* (2002) Virtual disjunct eddy covariance measurements of organic compound fluxes from a subalpine forest using proton transfer reaction mass spectrometry. *Atmos. Chem. Phys.* **2**, 279.
85. Graus, M., Hansel, A., Wisthaler, A. *et al.* (2006) A relaxed-eddy-accumulation method for the measurement of isoprenoid canopy-fluxes using an online gas-chromatographic technique and PTR-MS simultaneously. *Atmos. Environ.* **40**, S43.
86. Grabmer, W., Kreuzwieser, J., Wisthaler, A. *et al.* (2006) VOC emissions from Norway spruce (*Picea abies* L. [Karst]) twigs in the field – results of a dynamic enclosure study. *Atmos. Environ.* **40**, S128.
87. Grabmer, W., Graus, M., Lindinger, C. *et al.* (2004) Disjunct eddy covariance measurements of monoterpene fluxes from a Norway spruce forest using PTR-MS. *Int. J. Mass Spectrom.* **239**, 111.
88. Guenther, A., Hewitt, C. N., Erickson, D. *et al.* (1995) A global model of natural organic compound emissions. *J. Geophys. Res.* **100**, D8873.
89. Schade, G. W., Solomon, S. J., Dellwik, E. *et al.* (2011) Methanol and other VOC fluxes from a Danish beech forest during late springtime. *Biogeochem.* **106**, 337.
90. Rinne, J., Taipale, R., Markkanen, T. *et al.* (2007) Hydrocarbon fluxes above a Scots pine forest canopy: measurements and modeling. *Atmos. Chem. Phys.* **7**, 3361.
91. Ruuskanen, T. M., Kolari, P., Back, J. *et al.* (2005) On-line field measurements of monoterpene emissions from Scots pine by proton-transfer-reaction mass spectrometry. *Boreal Environ. Res.* **10**, 553.
92. Karl, T., Guenther, A., Yokelson, R. J. *et al.* (2007) The tropical forest and fire emissions experiment: emission, chemistry, and transport of biogenic volatile organic compounds in the lower atmosphere over Amazonia. *J. Geophys. Res.* **112**, D18302.
93. Karl, T., Guenther, A., Turnipseed, A. *et al.* (2009) Rapid formation of isoprene photo-oxidation products observed in Amazonia. *Atmos. Chem. Phys.* **9**, 7753.
94. Langford, B., Misztal, P. K., Nemitz, E. *et al.* (2010) Fluxes and concentrations of volatile organic compounds from a South-East Asian tropical rainforest. *Atmos. Chem. Phys.* **10**, 8391.
95. Pugh, T. A. M., MacKenzie, A. R., Hewitt, C. N. *et al.* (2010) Simulating atmospheric composition over a South-East Asian tropical rainforest: performance of a chemistry box model. *Atmos. Chem. Phys.* **10**, 279.
96. Pugh, T. A. M., MacKenzie, A. R., Langford, B. *et al.* (2011) The influence of small-scale variations in isoprene concentrations on atmospheric chemistry over a tropical rainforest. *Atmos. Chem. Phys.* **11**, 4121.
97. Robinson, N. H., Hamilton, J. F., Allan, J. D. *et al.* (2011) Evidence for a significant proportion of Secondary Organic Aerosol from isoprene above a maritime tropical forest. *Atmos. Chem. Phys.* **11**, 1039.

98. Laffineur, Q., Aubinet, M., Schoon, N. *et al.* (2011) Isoprene and monoterpene emissions from a mixed temperate forest. *Atmos. Environ.* **45**, 3157.
99. Fares, S., Gentner, D. R., Park, J.-H. *et al.* (2011) Biogenic emissions from Citrus species in California. *Atmos. Environ.* **45**, 4557.
100. Misztal, P. K., Owen, S. M., Guenther, A. B. *et al.* (2010) Large estragole fluxes from oil palms in Borneo. *Atmos. Chem. Phys.* **10**, 4343.
101. Misztal, P. K., Nemitz, E., Langford, B. *et al.* (2011) Direct ecosystem fluxes of volatile organic compounds from oil palms in South-East Asia. *Atmos. Chem. Phys.* **11**, 8995.
102. Warneke, C., Luxembourg, S. L., de Gouw, J. A. *et al.* (2002) Disjunct eddy covariance measurements of oxygenated volatile organic compounds fluxes from an alfalfa field before and after cutting. *J. Geophys. Res.* **197**, D4067.
103. Holzinger, R., Sanhueza, E., von Kuhlmann, R. *et al.* (2002) Diurnal cycles and seasonal variation of isoprene and its oxidation products in the tropical savanna atmosphere. *Global Biogeochem. Cycles.* **16**, 1074.
104. Bamberger, I., Hörtnagl, L., Schnitzhofer, R. *et al.* (2010) BVOC fluxes above mountain grassland. *Biogeosciences* **7**, 1413.
105. Davison, B., Brunner, A., Ammann, C. *et al.* (2008) Cut-induced VOC emissions from agricultural grasslands. *Plant Biol.* **10**, 76.
106. Schade, G. W., Custer, T. G. (2004) OVOC emissions from agricultural soil in northern Germany during the 2003 European heat wave. *Atmos. Environ.* **38**, 6105.
107. Jardine, K., Abrell, L., Kurc, S. A. *et al.* (2010) Volatile organic compound emissions from *Larrea tridentata* (creosotebush). *Atmos. Chem. Phys.* **10**, 12191.
108. Brunner, A., Ammann, C., Neftel, A., Spirig, C. (2007) Methanol exchange between grassland and the atmosphere. *Biogeosciences* **4**, 395.
109. Custer, T., Schade, G. (2007) Methanol and acetaldehyde fluxes over ryegrass. *Tellus* **59**, 673.
110. Sinha, V., Williams, J., Meyerhofer, M. *et al.* (2007) Air–sea fluxes of methanol, acetone, acetaldehyde, isoprene and DMS from a Norwegian fjord following a phytoplankton bloom in a mesocosm experiment. *Atmos. Chem. Phys.* **7**, 739.
111. Galbally, I. E., Lawson, S. J., Weeks, I. A. *et al.* (2007) Volatile organic compounds in marine air at Cape Grim, Australia. *Environ. Chem.* **4**, 178.
112. Russell, L. M., Mensah, A. A., Fischer, E. V. *et al.* (2007) Nanoparticle growth following photochemical α- and β-pinene oxidation at Appledore Island during International Consortium for Research on Transport and Transformation/Chemistry of Halogens at the Isles of Shoals 2004. *J. Geophys. Res.* **112**, D10S21.
113. Preunkert, S., Ancellet, G., Legrand, M. *et al.* (2012) Oxidant Production over Antarctic Land and its Export (OPALE) project: an overview of the 2010–2011 summer campaign. *J. Geophys. Res.* **117**, D15307.
114. Legrand, M., Gros, V., Preunkert, S. *et al.* (2011) A reassessment of the budget of formic and acetic acids in the boundary layer at Dumont d'Urville (coastal Antarctica): the role of penguin emissions on the budget of several oxygenated volatile organic compounds. *J. Geophys. Res.* **117**, D06308.
115. Colomb, A., Gros, V., Alvain, S. *et al.* (2009) Variation of atmospheric volatile organic compounds over the Southern Indian Ocean (30–49°S). *Environ. Chem.* **6**, 70.

116. Williams, J., Holzinger, R., Gros, V. *et al.* (2004) Measurements of organic species in air and seawater from the tropical Atlantic. *Geophys. Rev. Lett.* **31**, L23S06.
117. Holzinger, R., Kleiss, B., Donoso, L., Sanhueza, E. (2001) Aromatic hydrocarbons at urban, sub-urban, rural (8°52′N; 67°19′W) and remote sites in Venezuela. *Atmos. Environ.* **35**, 4917.
118. Karl, T., Jobson, T., Kuster, W. C. *et al.* (2003) Use of proton-transfer-reaction mass spectrometry to characterize volatile organic compound sources at the La Porte super site during the Texas Air Quality Study 2000. *J. Geophys. Res.* **108**, D4508.
119. Holzinger, R., Jordan, A., Hansel, A., Lindinger, W. (2001) Automobile emissions of acetonitrile: assessment of its contribution to the global source. *J. Atmos. Chem.* **38**, 187.
120. Beauchamp, J., Wisthaler, A., Grabmer, W. *et al.* (2004) Short-term measurements of CO, NO, NO_2, organic compounds and PM_{10} at a motorway location in an Austrian valley. *Atmos. Environ.* **38**, 2511.
121. Schnitzhofer, R., Beauchamp, J., Dunkl, J. *et al.* (2008) Long-term measurements of CO, NO, NO_2, benzene, toluene and PM_{10} at a motorway location in an Austrian valley. *Atmos. Environ.* **42**, 1012.
122. Miyakawa, Y., Kato, S., Kajii, Y. (2005) Calibration of the proton transfer reaction mass spectrometry (PTR-MS) instrument for oxygenated volatile organic compounds (OVOCs) and the measurement of ambient air in Tokyo. *Taiki Kankyo Gakkaishi (Journal of Japan Society for Atmospheric Environment)* **40**, 209.
123. Fiella, I., Peñuelas, J. (2006) Daily, weekly, and seasonal time courses of VOC concentrations in a semi-urban area near Barcelona. *Atmos. Environ.* **40**, 7752.
124. Fiella, I., Peñuelas, J. (2006) Daily, weekly and seasonal relationships among VOCs, NO_x and O_3 in a semi-urban area near Barcelona. *J. Atmos. Chem.* **54**, 189.
125. Langford, B., Davison, B., Nemitz, E., Hewitt, C. N. (2009) Mixing ratios and eddy covariance flux measurements of volatile organic compounds from an urban canopy (Manchester, UK). *Atmos. Chem. Phys.* **9**, 1971.
126. Langford, B., Nemitz, E., House, E. *et al.* (2010) Fluxes and concentrations of volatile organic compounds above central London, UK. *Atmos. Chem. Phys.* **10**, 627.
127. Molina, L. T., Kolb, C. E., de Foy, B. *et al.* (2007) Air quality in North America's most populous city – overview of the MCMA-2003 campaign. *Atmos. Chem. Phys.* **7**, 2447.
128. Bon, D. M., Ulbrich, I. M., de Gouw, J. A. *et al.* (2011) Measurements of volatile organic compounds at a suburban ground site (T1) in Mexico City during the MILAGRO 2006 campaign: measurement comparison, emission ratios, and source attribution. *Atmos. Chem. Phys.* **11**, 2399.
129. Fortner, E. C., Zheng, J., Zhang, R. *et al.* (2009) Measurements of volatile organic compounds using proton transfer reaction–mass spectrometry during the MILAGRO 2006 Campaign. *Atmos. Chem. Phys.* **9**, 467.
130. Velasco, E., Pressley, S., Grivicke, R. *et al.* (2009) Eddy covariance flux measurements of pollutant gases in urban Mexico City. *Atmos. Chem. Phys.* **9**, 7325.
131. Velasco, E., Lamb, B., Pressley, S. *et al.* (2005) Flux measurements of volatile organic compounds from an urban landscape. *Geo. Phys. Res. Lett.* **32**, L20802.

132. Herndon, S. C., Jayne, J. T., Zahniser, M. S. *et al.* (2005) Characterization of urban pollutant emission fluxes and ambient concentration distributions using a mobile laboratory with rapid response instrumentation. *Faraday Discuss.* **130**, 327.
133. Zavala, M., Herndon, S. C., Slott, R. S. *et al.* (2006) Characterization of on-road vehicle emissions in the Mexico City Metropolitan Area using a mobile laboratory in chase and fleet average measurement modes during the MCMA-2003 Field Campaign. *Atmos.Chem. Phys.* **6**, 5129.
134. Gros, V., Gaimoz, C., Herrmann, F. *et al.* (2011) Volatile organic compounds sources in Paris in spring 2007. Part I: qualitative analysis. *Environ. Chem.* **8**, 74.
135. Gaimoz, C., Sauvage, S., Gros, V. *et al.* (2011) Volatile organic compound sources in Paris in spring 2007. Part II: source apportionment using positive matrix factorisation. *Environ. Chem.* **8**, 91.
136. Slowik, J. G., Vlasenko, A., McGuire, M. *et al.* (2010) Simultaneous factor analysis of organic particle and gas mass spectra: AMS and PTR-MS measurements at an urban site. *Atmos. Chem. Phys.* **10**, 1969.
137. Holzinger, R., Warneke, C., Hansel, A. *et al.* (1999) Biomass burning as a source of formaldehyde, acetaldehyde, methanol, acetone, acetonitrile, and hydrogen cyanide. *Geophys. Res. Lett.* **26**, 1161.
138. Molina, L. T., Molina, M. J. (Eds) (2002) *Air Quality in the Mexico Megacity: An Integrated Assessment*, Kluwer Academic, Dordrecht.
139. Kolb, C. E., Herndon, S. C., McManus, J. B. *et al.* (2004) Mobile laboratory with rapid response instruments for real-time measurements of urban and regional trace gas and particulate distributions and emission source characteristics. *Environ. Sci. Technol.* **38**, 5694.
140. Velasco, E., Lamb, B., Westberg, H. *et al.* (2007) Distribution, magnitudes, reactivities, ratios and diurnal patterns of volatile organic compounds in the Valley of Mexico during the MCMA 2002 & 2003 field campaigns. *Atmos. Chem. Phys.* **7**, 329.
141. Dolgorouky, C., Gros, V., Sarda-Esteve, R. *et al.* (2012) Total OH reactivity measurements in Paris during the 2010 MEGAPOLI winter campaign. *Atmos. Chem. Phys.* **12**, 9593.
142. Warneke, C., McKeen, S. A., de Gouw, J. A. *et al.* (2007) Determination of urban volatile organic compound emission ratios and comparison with an emissions database. *J. Geophys. Res.* **112**, D10S47.
143. Jobson, B.T., Alexander, M.L., Maupin, G.D., Muntean, G.G. (2005) On-line analysis of organic compounds in diesel exhaust using a proton transfer reaction mass spectrometer (PTR-MS). *Int. J. Mass Spectrom.* **245**, 78.
144. Herndon, S. C., Rogers, T., Dunlea, E. J. *et al.* (2006) Hydrocarbon emissions from in-use commercial aircraft during airport operations. *Environ. Sci. Technol.* **40**, 4406.
145. Yelvington, P. E., Herndon, S. C., Wormhoudt, J. C. *et al.* (2007) Chemical speciation of hydrocarbon emissions from a commercial aircraft engine. *J. Propul. Power.* **23**, 912.
146. Knighton, W. B., Rogers, T. M., Anderson, B. E. *et al.* (2007) Quantification of aircraft engine hydrocarbon emissions using Proton Transfer Reaction Mass Spectrometry. *J. Propul. Power.* **23**, 949.
147. Hewitt, C. N. (Ed.) (1999) *Reactive Hydrocarbons in the Atmosphere*, Academic Press, San Diego.

148. Shaw, S. L., Mitloehner, F. M., Jackson, W. *et al.* (2007) Volatile organic compound emissions from dairy cows and their waste as measured by Proton-Transfer-Reaction Mass Spectrometry. *Environ. Sci. Technol.* **41**, 1310.
149. Twigg, M. M., House, E., Thomas, R. *et al.* (2011) Surface/atmosphere exchange and chemical interactions of reactive nitrogen compounds above a manured grassland. *Agr. Forest Meteorol.* **151**, 1488.
150. Feilberg, A., Liu, D., Adamsen, A.P.S., Hansen, M.J., Jonassen, K. E. N. (2010) Odorant emissions from intensive pig production measured by online Proton-Transfer-Reaction Mass Spectrometry. *Environ. Sci. Technol.* **44**, 5894.
151. Hansen, M. J., Liu, D., Guldberg, L. B., Feilberg, A. (2012) Application of Proton-Transfer-Reaction Mass Spectrometry to the assessment of odorant removal in a biological air cleaner for pig production. *J. Agric. Food Chem.* **60**, 2599.
152. Hansen, M. J., Adamsen, A. P. S., Pedersen, P., Feilberg, A. (2012) Prediction of odor from pig production based on chemical odorants. *J. Environ. Qual.* **41**, 436.
153. Saha, C. K., Feilberg, A., Zhang, G., Adamsen, A. P. S. (2011) Effects of airflow on odorants' emissions in a model pig house – a laboratory study using Proton-Transfer-Reaction Mass Spectrometry (PTR-MS). *Sci. Total Environ.* **410–411**, 161.
154. Ngwabie, N. M., Schade, G. W., Custer, T. G., Linke, S., Hinz, T. (2007) Volatile organic compound emission and other trace gases from selected animal buildings. *Landbauforschung Völkenrode*, **57**, 273.
155. de Gouw, J. A., Middlebrook, A. M., Warneke, C. *et al.* (2005) Budget of organic carbon in a polluted atmosphere: results from the New England Air Quality Study in 2002. *J. Geophys. Res.* **110**, D16305.
156. Warneke, C., de Gouw, J. A., Goldan, P. D. *et al.* (2004) Comparison of daytime and nighttime oxidation of biogenic and anthropogenic VOCs along the New England coast in summer during New England Air Quality Study 2002. *J. Geophys. Res.* **109**, D10309.
157. Drewnick, F., Hings, S. S., Curtius, J. *et al.* (2006) Measurement of fine particulate and gas-phase species during the New Year's fireworks 2005 in Mainz, Germany. *Atmos. Environ.* **40**, 4316.
158. Lobert, J. M., Scharffe, D. H., Hao, W. M., Crutzen, P. J. (1990) Importance of biomass burning in the atmospheric budgets of nitrogen-containg gases. *Nature* **346**, 552.
159. Akagi, S. K., Yokelson, R. J., Wiedinmyer, C. *et al.* (2011) Emission factors for open and domestic biomass burning for use in atmospheric models. *Atmos. Chem. Phys.* **11**, 4039.
160. Warneke, C., Roberts, J. M., Veres, P. *et al.* (2011) VOC identification and intercomparison from laboratory biomass burning using PTR-MS and PIT-MS. *Int. J. Mass Spectrom.* **303**, 6.
161. de Gouw, J. A., Warneke, C., Parrish, D. *et al.* (2003) Emission sources and ocean uptake of acetonitrile (CH_3CN) in the atmosphere. *J. Geophys. Res.* **108**, D4329.
162. Karl, T., Hansel, A., Märk, T. *et al.* (2003) Trace gas monitoring at the Mauna Loa Baseline Observatory using proton-transfer reaction mass spectrometry. *Int. J. Mass Spectrom.* **223–224**, 527.
163. Wisthaler, A., Hansel, A., Dickerson, R. R., Crutzen, P. J. (2002) Organic trace gas measurements by PTR-MS during INDOEX 1999. *J. Geophys. Res.* **107**, D8024.

164. de Gouw, J. A., Cooper, O. R., Warneke, C. *et al.* (2004) Chemical composition of air masses transported from Asia to the U.S. West Coast during ITCT 2K2: fossil fuel combustion versus biomass-burning signatures. *J. Geophys. Res.* **109**, D23S20.
165. Sanhueza, E., Holzinger, R., Kleiss, B. *et al.* (2004) New insights in the global cycle of acetonitrile: release from the ocean and dry deposition in the tropical savanna of Venezuela. *Atmos. Chem. Phys.* **4**, 275.
166. Dunne, A. E., Galbally, I. E., Lawson, S., Patti, A. (2012) Interference in the PTR-MS measurement of acetonitrile at *m/z* 42 in polluted urban air – a study using switchable reagent ion PTR-MS. *Int. J. Mass Spectrom.* **319–320**, 40.
167. Duck, T. J., Firanski, B. J., Millet, D. B. *et al.* (2007) Transport of forest fire emissions from Alaska and the Yukon Territory to Nova Scotia during summer 2004. *J. Geophys. Res.* **112**, D10S44.
168. Yokelson, R. J., Karl, T., Artaxo, P. *et al.* (2007) The Tropical Forest and Fire Emissions Experiment: overview and airborne fire emission factor measurements. *Atmos. Chem. Phys.* **7**, 5175.
169. Yokelson, R. J., Crounse, J. D., DeCarlo, P. F. *et al.* (2009) Emissions from biomass burning in the Yucatan. *Atmos. Chem. Phys.* **9**, 5785.
170. Steeghs, M. M. L., Crespo, E., Harren, F. J. M. (2007) Collision induced dissociation study of 10 monoterpenes for identification in trace gas measurements using the newly developed proton-transfer reaction ion trap mass spectrometer. *Int. J. Mass Spectrom.* **263**, 204.
171. Misztal, P. K., Heal, M. R., Nemitz, E., Cape, J. N. (2012) Development of PTR-MS selectivity for structural isomers: Monoterpenes as a case study. *Int. J. Mass Spectrom.* **310**, 10.
172. Kim, S., Karl, T., Helmig, D. *et al.* (2009) Measurement of atmospheric sesquiterpenes by proton transfer reaction-mass spectrometry (PTR-MS). *Atmos. Meas. Tech.* **2**, 99.
173. Christian, T. J., Kleiss, B., Yokelson, R. J. *et al.* (2003) Comprehensive laboratory measurements of biomass-burning emissions: 1. Emissions from Indonesian, African, and other fuels. *J. Geophys. Res.* **108**, D4719.
174. Yokelson, R. J., Christian, T. J., Karl, T. G., Guenther, A. (2008) The tropical forest and fire emissions experiment: laboratory fire measurements and synthesis of campaign data. *Atmos. Chem. Phys.* **8**, 3509.
175. Maleknia, S. D., Bell, T. L., Adams, M. A. (2007) PTR-MS analysis of reference and plant-emitted volatile organic compounds. *Int. J. Mass Spectrom.* **262**, 203.
176. Maleknia, S. D., Bell, T. L., Adams, M. A. (2009) Eucalypt smoke and wildfires: Temperature dependent emissions of biogenic volatile organic compounds. *Int. J. Mass Spectrom.* **279**, 126.
177. Greenberg, J. P., Friedli, H., Guenther, A. B. *et al.* (2006) Volatile organic emissions from the distillation and pyrolysis of vegetation. *Atmos. Chem. Phys.* **6**, 81.
178. Grieshop, A. P., Logue, J. M., Donahue, N. M., Robinson, A. L. (2009) Laboratory investigation of photochemical oxidation of organic aerosol from wood fires 1: measurement and simulation of organic aerosol evolution. *Atmos. Chem. Phys.* **9**, 1263.
179. Zhao, J., Zhang, R., Fortner, E. C., North, S. W. (2004) Quantification of Hydroxycarbonyls from OH-Isoprene Reaction. *J. Am. Chem. Soc.* **126**, 2686.

180. D'Anna, B., Wisthaler, A., Andreasen, Ø. *et al.* (2005) Atmospheric chemistry of C3-C6 cycloalkanecarbaldehydes. *J. Phys. Chem. A.* **109**, 5104.
181. Jensen, N. R. V., Karl, T., Prazeller, P. *et al.* (2003) Contribution: 1st International Conference on Proton-Transfer Reaction Mass Spectrometry and Its Applications.
182. Sinha, V., Williams, J., Crowley, J. N., Lelieveld, J. (2008) The Comparative Reactivity Method – a new tool to measure total OH reactivity in ambient air. *Atmos. Chem. Phys.* **8**, 2213.
183. Kim, S., Guenther, A., Karl, T., Greenberg, J. (2011) Contributions of primary and secondary biogenic VOC to total OH reactivity during the CABINEX (Community Atmosphere-Biosphere INteractions Experiments)-09 field campaign. *Atmos. Chem. Phys.* **11**, 8613.
184. Hanson, D., Orlando, J., Noziere, B., Kosciuch, E. (2004) Proton transfer mass spectrometry studies of peroxy radicals. *Int. J. Mass Spectrom.* **239**, 147.
185. Joó, É., Dewulf, J., Demarcke, M. *et al.* (2010) Quantification of interferences in PTRMS measurements of monoterpene emissions from *Fagus sylvatica* L. using simultaneous TD-GC-MS measurements. *Int. J. Mass Spectrom.* **291**, 90.
186. Filella, I., Wilkinson, M. J., Llusià, J. *et al.* (2007) Volatile organic compounds emissions in Norway spruce (*Picea abies*) in response to temperature changes. *Physiologia Plantarum.* **130**, 58.
187. Hayward, S., Tani, A., Owen, S. M., Hewitt, C. N. (2004) Online analysis of volatile organic compound emissions from Sitka spruce (*Picea sitchensis*). *Tree Physiol.* **24**, 721.
188. Loivamäki, M., Louis, S., Cinege, G. *et al.* (2007) Circadian rhythms of isoprene biosynthesis in grey poplar leaves. *Plant Physiol.* **143**, 540.
189. Gómez Alvarez, E., Viidanoja, J., Muñoz, A. *et al.* (2007) Experimental confirmation of the dicarbonyl route in the photo-oxidation of toluene and benzene. *Environ. Sci. Technol.* **41**, 8362.
190. Müller, M., Graus, M., Wisthaler, A. *et al.* (2012) Analysis of high mass resolution PTR-TOF mass spectra from 1,3,5-trimethylbenzene (TMB) environmental chamber experiments. *Atmos. Chem. Phys.* **12**, 829.
191. Paulsen, D., Dommen, J., Kalberer, M. *et al.* (2005) Secondary organic aerosol formation by irradiation of 1,3,5-trimethylbenzene-NO_x-H_2O in a new reaction chamber for atmospheric chemistry and physics. *Environ. Sci. Technol.* **39**, 2668.
192. Baltensperger, U., Kalberer, M., Dommen, J. *et al.* (2005) Secondary organic aerosols from anthropogenic and biogenic precursors. *Faraday Discuss.* **130**, 265.
193. Lee, A., Goldstein, A., Kroll, J. *et al.* (2006) Gas-phase products and secondary aerosol yields from the photooxidation of 16 different terpenes. *Geophys. Res.* **111**, D17305.
194. Ng, N. L., Kroll, J. H., Keywood, M. D. *et al.* (2006) Contribution of first- versus second-generation products to secondary organic aerosols formed in the oxidation of biogenic hydrocarbons. *Environ. Sci. Technol.* **40**, 2283.
195. Presto, A. A., Donahue, N. M. (2006) Investigation of α-pinene + ozone secondary organic aerosol formation at low total aerosol mass. *Environ. Sci. Technol.* **40**, 3536.
196. Kalberer, M., Paulsen, D., Sax, M. *et al.* (2004) Identification of polymers as major components of atmospheric organic aerosols. *Science* **303**, 1659.

197. Lee, A., Goldstein, A., Keywood, M. *et al.* (2006) Gas-phase products and secondary aerosol yields from the ozonolysis of ten different terpenes. *J. Geophys. Res.* **111**, D07302.
198. Hellén, H., Dommen, J., Metzger, A. *et al.* (2008) Using proton transfer reaction mass spectrometry for online analysis of secondary organic aerosols. *Environ. Sci. Technol.* **42**, 7347.
199. Liggio, J., Li, S. M., Brook, J. R., Mihele, C. (2007) Direct polymerization of isoprene and α-pinene on acidic aerosols. *Geophys. Res. Lett.* **34**, L05814.
200. Dudareva, N., Negre, F., Nagegowda, D. A., Orlova, I. (2006) Plant volatiles: recent advances and future perspectives. *Crit. Rev. Plant Sci.* **25**, 417.
201. Paré, P. W., Tumlinson, J. H. (1999) Plant volatiles as a defense against insect herbivores. *Plant Physiol.* **121**, 325.
202. Pichersky, E., Gershenzon, J. (2002) The formation and function of plant volatiles: perfumes for pollinator attraction and defense. *Current Opinion in Plant Biology*. **5**, 237.
203. Peñuelas, J., Filella, I., Stefanescu, C., Llusià, J. (2005) Caterpillars of *Euphydryas aurinia* (Lepidoptera : Nymphalidae) feeding on *Succisa pratensis* leaves induce large foliar emissions of methanol. *New Phytol.* **167**, 851.
204. Vickers, C. E., Gershenzon, J., Lerdau, M. T., Loreto, F. (2009) A unified mechanism of action for volatile isoprenoids in plant abiotic stress. *Nature Chem. Biol.* **5**, 283.
205. Peñuelas, J., Llusià, J., Estiarte, M. (1995) Terpenoids: a plant language. *Trends Ecol. E.* **10**, 289.
206. Baldwin, I. T., Halitschke, R., Paschold, A. *et al.* (2006) Volatile signaling in plant-plant interactions: "Talking Trees" in the genomics era. *Science* **311**, 812.
207. Knudsen, J. T., Gershenzon, J. (2006) The chemistry diversity of floral scent. In: *Biology of Floral Scent* (Eds N. Dudareva and E. Pichersky), CRC Press, Boca Raton, FL, pp. 27–52.
208. Tholl, D., Boland, W., Hansel, A. *et al.* (2006) Practical approaches to plant volatile analysis. *Plant J.* **45**, 540.
209. Singsass, E. L., Sharkey, T. D. (1998) The regulation of isoprene emission responses to rapid leaf temperature fluctuations. *Plant, Cell Environ.* **21**, 1181.
210. Copolovici, L. O., Filella, I., Llusià, J. *et al.* (2005) The capacity for thermal protection of photosynthetic electron transport varies for different monoterpenes in *Quercus ilex*. *Plant Physiol.* **139**, 485.
211. Peñuelas, J., Munné-Bosch, S. (2005) Isoprenoids: an evolutionary pool for photoprotection. *Trends Plant Sci.* **10**, 166.
212. Velikova, V., Pinelli, P., Pasqualini, S. *et al.* (2005) Isoprene decreases the concentration of nitric oxide in leaves exposed to elevated ozone. *New Phytol.* **166**, 419.
213. Vickers, C. E., Possell, M., Cojocariu, C. I. *et al.* (2009) Isoprene synthesis protects transgenic tobacco plants from oxidative stress. *Plant, Cell Environ.* **32**, 520.
214. Rosenstiel, T. N., Ebbets, A. L., Khatri, W. C. *et al.* (2004) Induction of poplar leaf nitrate reductase: a test of extrachloroplastic control of isoprene emission rate. *Plant Biol.* **6**, 12.
215. Loreto, F., Barta, C., Brilli, F., Nogues, I. (2006) On the induction of volatile organic compound emissions by plants as consequence of wounding or fluctuations of light and temperature. *Plant Cell Environ.* **29**, 1820.

216. Pegoraro, E., Abrell, L., Van Haren, J. *et al.* (2005) The effect of elevated atmospheric CO_2 and drought on sources and sinks of isoprene in a temperate and tropical rainforest mesocosm. *Global Change Biol.* **11**, 1234.
217. Pegoraro, E., Rey, A., Abrell, L. *et al.* (2006) Drought effect on isoprene production and consumption in Biosphere 2 tropical rainforest. *Global Change Biol.* **12**, 456.
218. Kreuzwieser, J., Graus, M., Wisthaler, A. *et al.* (2002) Xylem-transported glucose as an additional carbon source for leaf isoprene formation in *Quercus robur*. *New Phytol.* **156**, 171.
219. Karl, T., Fall, R., Rosenstiel, T. N. *et al.* (2002) On-line analysis of the $^{13}CO_2$ labeling of leaf isoprene suggests multiple subcellular origins of isoprene precursors. *Planta* **215**, 894.
220. Schnitzler, J. P., Graus, M., Kreuzwieser, J. *et al.* (2004) Contribution of different carbon sources to isoprene biosynthesis in poplar leaves. *Plant Physiol.* **135**, 152.
221. Brilli, F., Ruuskanen, T. M., Schnitzhofer, R. *et al.* (2011) Detection of plant volatiles after leaf wounding and darkening by Proton Transfer Reaction "Time-of-Flight" Mass Spectrometry (PTR-TOF). *PLoS ONE.* **6**, 1.
222. Holzinger, R., Sandoval-Soto, L., Rottenberger, S. *et al.* (2000) Emissions of volatile organic compounds from *Quercus ilex.* L. measured by proton transfer reaction mass spectrometry under different environmental conditions. *J. Geophys. Res.* **105**, D20573.
223. Karl, T., Curtis, A. J., Rosenstiel, T. N. *et al.* (2002) Transient releases of acetaldehyde from tree leaves – products of a pyruvate overflow mechanism? *Plant Cell Environ.* **25**, 1121.
224. Fall, R., Karl, T., Hansel, A. *et al.* (1999) Volatile organic compounds emitted after leaf wounding: on-line analysis by proton-transfer-reaction mass spectrometry. *J. Geophys. Res.* **104**, D15963.
225. Graus, M., Schnitzler, J. P., Hansel, A. *et al.* (2004) Transient release of oxygenated volatile organic compounds during light-dark transitions in grey poplar leaves. *Plant Physiol.* **135**, 1967.
226. Jardine, K., Harley, P., Karl, T. *et al.* (2008) Plant physiological and environmental controls over the exchange of acetaldehyde between forest canopies and the atmosphere. *Biogeosciences* **5**, 1559.
227. Filella, I., Bosch, J., Llusià, J. *et al.* (2011) Chemical cues involved in the attraction of the oligolectic bee *Hoplitis adunca* to its host plant *Echium vulgare*. *Biochem. System. Ecol.* **39**, 498.
228. Asensio, D., Peñuelas, J.; Filella, I., Llusià, J. (2007) On-line screening of soil VOCs exchange responses to moisture, temperature and root presence. *Plant Soil* **291**, 249.
229. Asensio, D., Peñuelas, J., Llusià, J. *et al.* (2007) Interannual and interseasonal soil CO_2 efflux and VOC exchange rates in a Mediterranean holm oak forest in response to experimental drought. *Soil Biol. Biochem.* **39**, 2471.
230. Steeghs, M., Bais, H. P., deGouw, J. *et al.* (2004) Proton-Transfer-Reaction Mass Spectrometry as a new tool for real time analysis of root-secreted volatile organic compounds in arabidopsis. *Plant Physiol.* **135**, 47.
231. Abanda-Nkpwatt, D., Müsch, M., Tschiersch, J. *et al.* (2006) Molecular interaction between Methylobacterium extorquens and seedlings: growth promotion, methanol consumption, and localization of the methanol emission site. *J. Exp. Botany.* **57**, 4025.

232. Bracho-Nunez, A., Welter, S., Staudt, M., Kesselmeier, J. (2011) Plant-specific volatile organic compound emission rates from young and mature leaves of Mediterranean vegetation. *J. Geophys. Res.* **116**, D16304.
233. Fukui, Y., Doskey, P. V. (1998) Air-surface exchange of nonmethane organic compounds at a grassland site: seasonal variations and stressed emissions. *J. Geophys. Res.* **1103**, D13153.
234. Heiden, A. C., Kobel, K., Langebartels, C. *et al.* (2003) Emissions of oxygenated volatile organic compounds from plants part i: emissions from lipoxygenase activity. *J. Atmos. Chem.* **45**, 143.
235. Beauchamp, J., Wisthaler, A., Hansel, A. *et al.* (2005) Ozone induced emissions of biogenic VOC from tobacco: relationships between ozone uptake and emission of LOX products. *Plant Cell Environ.* **28**, 1334.
236. Rottenberger, S., Kleiss, B., Kuhn, U. *et al.* (2008) The effect of flooding on the exchange of the volatile C2-compounds ethanol, acetaldehyde and acetic acid between leaves of Amazonian floodplain tree species and the atmosphere. *Biogeosciences* **5**, 1085.
237. Filella, I., Peñuelas J., Llusià, J. (2006) Dynamics of the enhanced emissions of monoterpenes and methyl salicylate, and decreased uptake of formaldehyde, by *Quercus ilex* leaves after application of jasmonic acid. *New Phytol.* **169**, 135.
238. Peñuelas, J., Filella, I., Llusià, J. (2007) Methyl salicylate fumigation increases monoterpene emission rates. *Biologia Planarium* **51**, 372.
239. Peñuelas, J., Filella, I., Seco, R., Llusià, J. (2009) Increase in isoprene and monoterpene emissions after re-watering of droughted *Quercus ilex* seedlings. *Biologia Plantarum.* **53**, 351.
240. Fillela, I., Peñuelas, J., Seco, R. (2009) Short-chained oxygenated VOC emissions in Pinus halepensis in response to changes in water availability. *Acta Physiol. Plant.* **31**, 311.
241. Karl, T., Guenther, A., Turnipseed, A. *et al.* (2008) Chemical sensing of plant stress at the ecosystem scale. *Biogeosciences* **5**, 1287.
242. Šimpraga, M., Verbeeck, H., Demarcke, M. *et al.* (2011) Clear link between drought stress, photosynthesis and biogenic volatile organic compounds in *Fagus sylvatica* L. *Atmos. Environ.* **45**, 5254.
243. Croft, K. P. C., Jüttner, F., Slusarenko, A. (1993) Volatile products of the lipoxygenase pathway evolved from *Phaseolus vulgaris* (1.) leaves Inoculated with *Pseudomonas syringae* pv *phaseolicola. Plant Physiol.* **101**, 13.
244. D'Auria, J. C., Pichersky, E., Schaub, A. *et al.* (2007) Characterization of a BAHD acyltransferase responsible for producing the green leaf volatile (*Z*)-3-hexen-1-yl acetate in *Arabidopsis thaliana. Plant J.* **49**, 194.
245. Brown, P., Watts, P., Märk, T. D., Mayhew, C. A. (2010) Proton transfer reaction mass spectrometry investigations on the effects of reduced electric field and reagent ion internal energy on product ion branching ratios for a series of saturated alcohols. *Int. J. Mass Spectrom.* **294**, 103.
246. Karl, T., Fall, R., Jordan, A., Lindinger, W. (2001) On-line analysis of reactive VOCs from urban lawn mowing. *Environ. Sci. Technol.* **35**, 2926.
247. Karl, T., Guenther, A., Jordan, A. *et al.* (2001) Eddy covariance measurement of biogenic oxygenated VOC emissions from hay harvesting. *Atmos. Environ.* **35**, 491.

248. de Gouw, J. A., Howard, C., Custer, T. G., Fall, R. (1999) Emissions of volatile organic compounds from cut grass and clover are enhanced during the drying process. *Geophys. Res. Lett.* **26**, 811.
249. Karl, T., Harren, F., Warneke, C. *et al.* (2005) Senescing grass crops as regional sources of reactive volatile organic compounds. *J. Geophys. Res.* **110**, D15302.
250. Eller, A. S. D., Sekimoto, K., Gilman, J. B. *et al.* (2011) Volatile organic compound emissions from switchgrass cultivars used as biofuel crops. *Atmos. Environ.* **45**, 3333.
251. Crespo, E., Graus, M., Gilman, J. B. *et al.* (2013) Volatile organic compound emissions from elephant grass and bamboo cultivars used as potential bioethanol crop. *Atmos. Environ.* **65**, 61.
252. von Dahl, C. C., Hävecker, M., Schlögl, R., Baldwin, I. T. (2006) Caterpillar-elicited methanol emission: a new signal in plant–herbivore interactions? *Plant J.* **46**, 948.
253. Bartolome, J., Peñuelas, J., Filella, I. *et al.* (2007) Mass scans from a proton transfer mass spectrometry analysis of air over Mediterranean shrubland browsed by horses. *J. Environ. Biol.* **28**, 697.
254. Schausberger, P., Peneder, S., Jürschik, S., Hoffmann, D. (2012) Mycorrhiza changes plant volatiles to attract spider mite enemies. *Functional Ecology.* **26**, 441.
255. Oikawa, P. Y., Giebel, B. M., da Silveira Lobo O'Reilly Sternberg, L. *et al.* (2011) Leaf and root pectin methylesterase activity and $^{13}C/^{12}C$ stable isotopic ratio measurements of methanol emissions give insight into methanol production in Lycopersicon esculentum. *New Phytol.* **191**, 1031.
256. Brilli, F., Cicciolo, P., Frattoni, M. *et al.* (2009) Constitutive and herbivore-induced monoterpenes emitted by *Populus* × *euroamericana* leaves are key volatiles that orient *Chrysomela populi* beetles. *Plant Cell Environ.* **32**, 542.
257. Schaub, A., Blande, J. D., Graus, M. *et al.* (2010) Real-time monitoring of herbivore induced volatile emissions in the field. *Physiologia Plantarum* **138**, 123.
258. Danner, H., Samudrala, D., Cristescu, S. M., van Dam, N. M. (2012) Tracing hidden herbivores: time-resolved non-invasive analysis of belowground volatiles by Proton-Transfer-Reaction Mass Spectrometry (PTR-MS). *J. Chem. Ecol.* **38**, 785.
259. van Dam, N. M., Samudrala, D., Harren, F. J. M., Cristescu, S. M. (2012) Real-time analysis of sulfur containing volatiles in *Brassica* plants infested with root feeding *Delia radicum* larvae using Proton Transfer Reaction-Mass Spectrometry. *AOB Plants*. doi:10.1093/aobpla/pls021.
260. Crespo, E. (2012) *Proton transfer reaction-mass spectrometry, applications in life sciences* PhD dissertation. Radboud University, Nijmegen.
261. Crespo, E., Hordijk, C. A., deGraaf, R. M. *et al.* (2012) On-line detection of root-induced volatiles in *Brassica nigra* plants infested with *Delia radicum* L. root fly larvae. *Phytochemistry.* **84**, 68.
262. Tani, A., Kato, S., Kajii, Y., Wilkinson, M. *et al.* (2007) A proton transfer reaction mass spectrometry based system for determining plant uptake of volatile organic compounds. *Atmos. Environ.* **41**, 1736.
263. Seco, R., Peñuelas, J., Filella, I. (2008) Formaldehyde emission and uptake by Mediterranean trees *Quercus ilex* and*Pinus halepensis*. *Atmos. Environ.* **42**, 7907.
264. Tani, A., Hewitt, C. N. (2009) Uptake of aldehydes and ketones at typical indoor concentrations by houseplants. *Environ. Sci. Technol.* **43**, 8338.

6

PTR-MS in the Food Sciences

6.1 Background

Volatile compounds play a fundamental role in food science and technology. In part, this is because they are released when foods and drinks are consumed, and are hence involved in aroma and flavour which are intimately linked to our enjoyment, palatability and perception of food. Aroma volatiles that reach the olfactory epithelium,[1] via orthonasal and retronasal pathways, play not only a major role in our enjoyment of food (and drink), but also allow us to assess food quality or to detect subtle changes in food products as a result of modifications in ingredients and production. Volatile organic compounds (VOCs) are also emitted directly from food which can, for example, give us information about ripening processes (e.g. cheese or fruits) and the effects of storage.

Given the importance of VOCs that are emitted during the consumption of food, a crucial area of food science research is to determine what and when volatile compounds are delivered to our olfactory senses during eating and drinking. Knowledge of this leads to a better understanding of the aroma impact of food products and their acceptability to the human palate, which in turn may lead to improved food products for the consumer. A particular difficulty of this type of research is that no two individuals are the same owing to differences in chewing behaviour and perception. The consumption process and variation in individuals (e.g. resulting from differences in saliva content and sensory perception) therefore requires investigation whilst monitoring the VOCs in real time.

In addition to food research dealing with the perception of flavour, other activities using proton transfer reaction mass spectrometry (PTR-MS) have involved headspace analysis. This has been used, for example, to monitor how volatile emissions change as a function of time (e.g. shelf life, ageing, post-harvest storage, ripening and fermentation), for various types of classification (e.g. differentiating foods/drinks by their variety and/or by geographical origin), for food/drink quality and control (e.g. freshness and adulteration), determining

[1] This is the layer of sensory cells inside the nasal cavity which are involved in our sense of smell.

Proton Transfer Reaction Mass Spectrometry: Principles and Applications, First Edition. Andrew M. Ellis and Christopher A. Mayhew.

the effects of additives (e.g. switching sweeteners for sugar) and for monitoring food/drink processing (e.g. for rapid information on the automatic control of food production).

Independent of whether food research is associated with volatile compounds released during consumption or related to headspace measurements, real-time and high-time resolution analytical techniques are valuable for processing information quickly and for exploring events that happen relatively rapidly, such as flavour release during food and drink consumption. As a consequence of this, a number of real-time mass spectrometric techniques have been used by food scientists for VOC analysis, including mass spectrometry-based e-noses (MS-e-noses),[2] atmospheric pressure chemical ionization mass spectrometry (APCI-MS), selected ion flow tube-MS (SIFT-MS) and of course PTR-MS. All of these techniques have their advantages and disadvantages and Biasioli *et al.* have provided a review and comparison of these techniques [1]. In this chapter, the focus of discussion will be predominantly on the use of PTR-MS.

In a review of the application of PTR-MS to food science and technology, Biasioli *et al.* [2] have stated that the technology 'is an accurate, highly sensitive, direct-injection technique that allows for the rapid characterization of food products and for the monitoring of processes in food science and technology and agro-industry, without any pre-treatment'. This statement appropriately summarizes the benefits of PTR-MS for investigating volatile emissions from food and drink. Although static headspace measurements have their role to play in the food sciences, the primary strength of PTR-MS is its ability to monitor rapid changes in trace VOC concentrations (with an ultimate time resolution of $\sim$100 ms) within the complex chemical matrix of food, that is, it has the ability to investigate the dynamic release of aromas and therefore rapid changes in composition.

PTR-MS measurements can rarely be used to unambiguously establish the identity of a volatile compound when analysing emissions from a complex chemical environment, although high-resolution PTR-MS and switchable reagent ion sources are beginning to make progress in this regard. Thus a signal at a nominal m/z could have contributions from various VOCs in the form of protonated parent molecules, fragment ions, product ions from isotopologues or cluster ions. Isobaric compounds cannot be distinguished from mass measurements of the protonated parent molecule when a low mass resolution instrument is used, and structural isomers can of course never be mass spectrometrically distinguished by their protonated m/z value independent of the resolution of the mass spectrometer. However, differentiation of isobaric and isomeric compounds may be possible on the basis of different fragmentation behaviour. For example, PTR-MS studies by Buhr *et al.* of various food flavour compounds at a fixed E/N have indicated the potential of using fragmentation ion patterns to discriminate isobaric and even isomeric compounds [3]. A more detailed

[2] An 'electronic nose' consists of a sampling system, an array of chemical gas sensors which are used for the detection of selected VOCs, and a processing system. The sensors are often solid-state devices whose physical properties, for example, electrical conduction, are altered by adsorption of certain types of molecules. An MS-e-nose also consists of sampling, detection and analysis, but instead of using a sensor array a mass spectrometer is employed to record mass spectral profiles which are used in the analysis [1]. Any mass spectrometer and ionization method can be used, including electron impact ionization (EI). According to the description given above, both PTR-MS and SIFT-MS could be considered to be types of MS-e-noses. However, MS-e-noses do not attempt to provide chemical identification of the VOCs, but instead simply provide a spectral fingerprint. Although both PTR-MS and SIFT-MS can be and are used for spectral fingerprinting, they are also used to provide some chemical identification and hence are not generally considered to be MS-e-noses.

PTR-MS fragmentation study by Aprea *et al.* involving 21 selected esters, dissolved in water and alcoholic solutions, has investigated the fragmentation behaviour as a function of E/N (from ca. 95 Td up to about 140 Td), that is, as a function of mean collision energy [4]. The aim here was to find an alternative to the mass resolution problem in order to provide a basis for the identification and quantification of esters in the headspace of alcoholic beverages. Esters are common flavour compounds in food and drinks, and in particular in wine. Aprea *et al.* found that the fragmentation patterns of methyl 2-methyl-butanoate, ethyl butanoate and propyl propanoate (all of which have the molecular formula $C_6H_{12}O_2$) are significantly different from each other, and hence these isomers could in theory be distinguished using PTR-MS. However, although there is some evidence that fragmentation patterns may aid in VOC assignments in favourable cases, it seems unlikely that this could be used as a practical means to distinguish between these compounds when present in a complex VOC mixture, as would usually be the case for a food or drink. Other analytical techniques or procedures are therefore often required if a high level of confidence is to be placed on the identification of a given m/z to a given VOC or number of VOCs. An example of this is to use reagent ions other than H_3O^+, which may lead to different reaction pathways for different isomers.

6.2 Combined GC–MS and PTR-MS Studies for Food Analysis

GC–MS is a two-dimensional technique, combining temporal separation by passage over a chromatographic column with mass spectrometry. As such GC–MS is a tempting alternative to PTR-MS, but this comes with the loss of temporal information owing to the need for pre-concentration and subsequent GC separation. Furthermore, different columns and conditions are required in order to detect a wide-range of VOCs at the sensitivity required. There is also the question about the capabilities of GC–MS for measuring certain VOCs because of their loss in the GC-column or during a preparation step (which is not an issue for PTR-MS because no sample preparation is required) [5]. All of these issues and other aspects of GC techniques were raised earlier in Section 1.3.1.

Nevertheless, GC–MS still has a major role to play in VOC analysis because they can be used to classify chemical compounds within a complex aroma via separation in the GC and then by ionization in the mass spectrometer. Having achieved this identification, the advantages of the fast analytical capabilities of PTR-MS technology can then be mobilized to monitor temporal changes of the VOCs.

A comparison with a reference technique often provides the information required to identify VOCs that make a dominant contribution to specific ion peaks observed in PTR-MS [6]. For example, GC-based analyses have been used to identify PTR-MS mass peaks corresponding to benzene formation [7] or VOCs being released from a number of food products including cheese [8], coffee [9], white truffles [10], raspberries [11], fruit juices [12] and cheese crackers [13]. These studies highlight that GC–MS analysis can be beneficial in the exploratory phase of any study, but thereafter PTR-MS can be used with confidence for rapid real-time measurements.

Generally, GC–MS sampling is done separately rather than in parallel with PTR-MS measurements, which may result in important VOC information being lost, especially if there are time-dependent changes in VOC concentrations. At the Nestlé Research Centre in Switzerland, Lindinger *et al.* have developed what is arguably the most reliable analytical system for the unequivocal identification of VOCs in complex chemical mixtures by combining PTR-MS and GC–MS in a single instrument [14]. This instrument retains the major advantage of PTR-MS, namely its ability for undertaking dynamic VOC release measurements, whilst permitting unambiguous identification of many VOCs. In this apparatus, the headspace gas is split into two distinct flows for fast processing by PTR-MS and later offline investigation by GC–MS and GC–PTR-MS detection. The performance of this dual-technique methodology was illustrated by analysing VOCs from the headspace of freshly made espresso coffee, which is a very VOC-rich environment. Naturally, the types of VOCs emitted and their abundances will be dependent on the sort of coffee used and its temperature. However, that is not of relevance here since the initial study by Lindinger *et al.* was a proof-of-principle investigation to demonstrate the power of their approach. The researchers observed many different types of ions arising from the large variety of VOCs being emitted from the coffee, but to demonstrate the performance of the instrument only nine representative ion signals in the PTR-MS mass spectrum were fully analysed; these were at *m*/*z* 45, 68, 73, 75, 80, 87, 95, 101 and 111. From the GC measurements, the majority of these ions consisted of contributions from several VOCs. This is illustrated in Figure 6.1, which summarizes the results obtained. Only for the ions at *m*/*z* values of 68, 75, 80 and 95 was the contribution relatively pure (>86% contribution from a single VOC) and therefore they could be assigned to arise predominantly from pyrrole (96%), methyl acetate (86%), pyridine (99%) and 2-methylpyrazine (92%), respectively. Although it should be appreciated that these GC–MS measurements are only valid for the time window in which the measurements were taken, this study by Lindinger *et al.* nevertheless demonstrated how powerful a combined PTR-MS/GC–MS analytical instrument is for providing details on the contributions of various VOCs to a given *m*/*z* observed in a PTR-MS mass spectrum. This is especially valuable for coffee investigations given that coffee aroma is made up of many hundreds of individual VOCs.

Using the GC-PTR-MS apparatus described above, Lindinger and co-workers also investigated the formation of acrylamide (2-propenamide, C_3H_5NO), which is believed to result from a reaction between asparagines and reducing sugars. Acrylamide has been classified as a potential carcinogen and it is formed when carbohydrate-rich foods are subjected to high-heat treatment, which is a major concern given the amount of fried foods that are eaten, particularly in the western world. Acrylamide is therefore an important compound to unambiguously identify during food processing. The experiments to investigate the formation of acrylamide were done in real time with potatoes heated to between 120°C and 170°C using a Maillard reaction[3] model system consisting of asparagines and either fructose or glucose [15]. Lindinger *et al.* showed that *m*/*z* 72 in their PTR-MS spectra arose exclusively from protonated acrylamide. The research team also demonstrated that fructose is more effective at the production of acrylamide than was glucose. In a follow-up study by researchers at the Nestlé laboratories the formation of acrylamide in crystalline model systems based

[3] A Maillard reaction is a chemical reaction that occurs between sugars and compounds possessing a free amino group in food, particularly amino acids. This reaction is initiated by a high temperature and causes changes in the food, such as colour (e.g. the browning of a crust of bread) and flavour. A Maillard reaction model system is an artificial mixture of sugars and proteins selected to investigate Maillard reaction processes without the complexity of using real food.

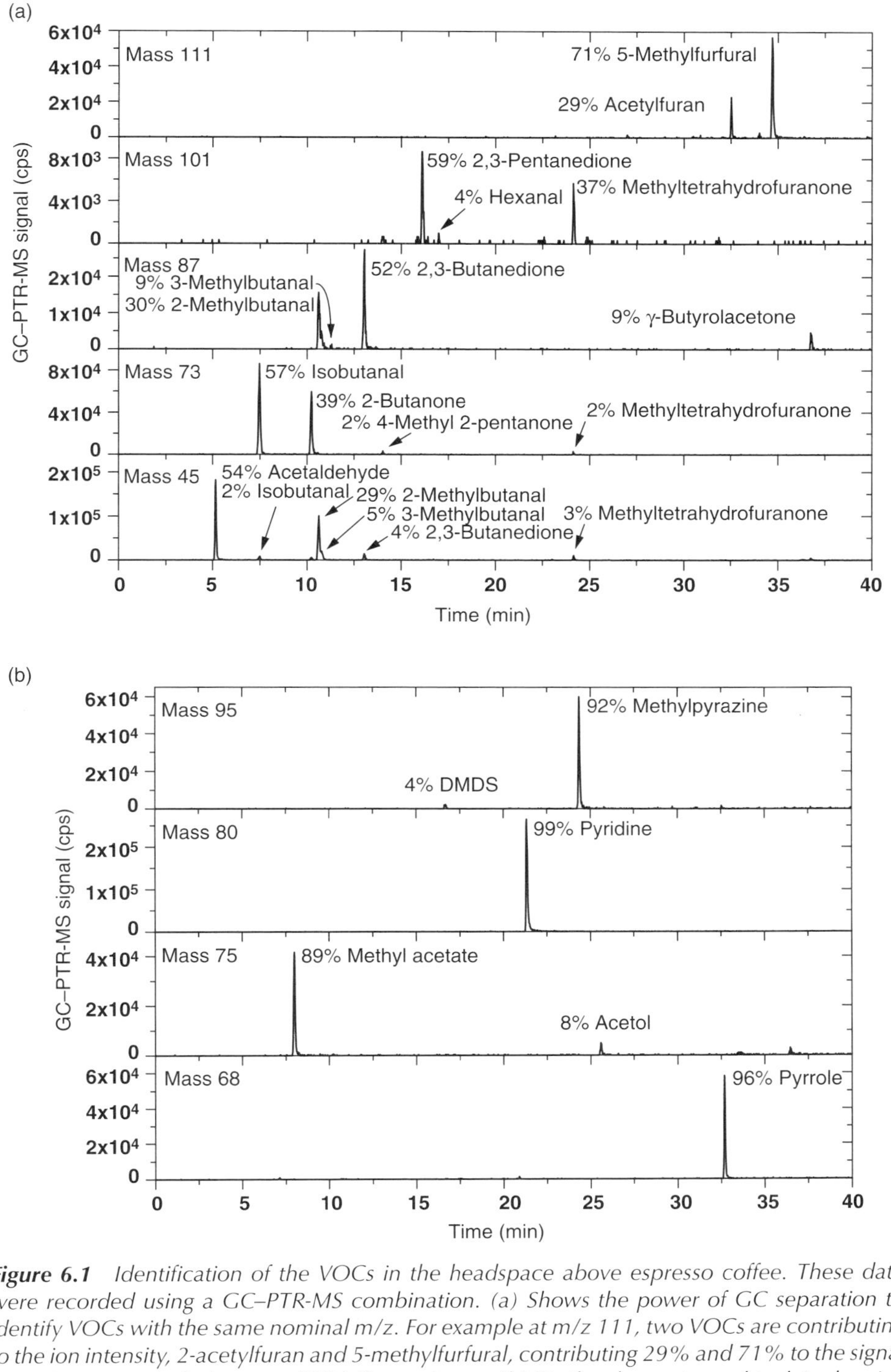

Figure 6.1 *Identification of the VOCs in the headspace above espresso coffee. These data were recorded using a GC–PTR-MS combination. (a) Shows the power of GC separation to identify VOCs with the same nominal m/z. For example at m/z 111, two VOCs are contributing to the ion intensity, 2-acetylfuran and 5-methylfurfural, contributing 29% and 71% to the signal intensity, respectively, in a pure PTR-MS experiment. (b) Similar chromatographic data showing features at m/z values where only a single VOC dominates. Reprinted from Reference 14 with permission.*

on asparagines and reducing sugars was investigated under low-moisture reaction conditions [16]. This work demonstrated that the choice of ingredients, their physical state (i.e. whether a solid or a liquid), the reaction temperature and the reaction time all had an effect on the formation of acrylamide during the cooking process. In particular, and in agreement with their first study, the researchers discovered that more acrylamide was produced when fructose was used compared to galactose and glucose.

The PTR-MS/GC–MS combination was also employed to investigate the thermal treatment of simplified chemical systems that might model the production of furan and methyl furan in real foods [17]. This is of interest because furan and methyl furan are also potential carcinogens, and levels of these compounds are found to increase in foods that are thermally processed. Märk *et al.* found substantial quantities of these chemicals in the headspace of ascorbic acid and linolenic acid under dry roasting conditions, with ascorbic acid resulting in the greatest production of furan ($\sim$10 mmol furan per mol of ascorbic acid) [17]. A large number of VOCs were observed resulting from the thermal treatment of ascorbic and linolenic acids. However, the dominant ions were observed at m/z 47, 69 and 97 which, with the aid the GC–MS measurements, were identified to be protonated formic acid, furan and furfural, respectively.

Although the PTR-MS/GC–MS system developed by Lindinger and co-workers shows considerable potential for food analysis studies, it has only been used so far on a limited basis elsewhere. This may be because the complex setup is too expensive for many groups to reproduce and/or because no commercial apparatus is currently available. To partially overcome the first problem, Pozo-Bayón *et al.* developed a simplified and hence cheaper version of the instrument by interfacing a PTR-MS instrument with a GC–MS system through a sampling port [6]. For the GC–MS measurements, which were used for compound assignment, the VOCs were purged and trapped. This instrument was applied to the investigation of the headspace above cheese crackers in order to monitor how the volatile profile changes during early shelf life storage (defined as the first 4 months), and hence how flavour may change as a function of time [6, 13]. By applying multivariate statistical analysis to the mass spectral data obtained from the PTR-MS measurements, it was found that the samples could be classified according to their age (shelf life) [13].

Even with this simplified instrument, PTR-MS/GC–MS systems are still not in regular use, possibly because the real advantage of PTR-MS in the Food Sciences is not necessarily in the identification of the VOCs but in its real-time rapid monitoring and fingerprinting capabilities. The latter's capability can be used for fast classification of foods (e.g. geographical origin) or for determining food quality and this is the subject of the next section.

6.3 Mass Spectral Fingerprinting

Although the identification of VOCs may be necessary for certain applications, compound identification is not always required and a more holistic approach has its advantages. Simple VOC patterns, which we will refer to rather loosely as *fingerprinting*, together with appropriate statistical analysis, may provide more than sufficient information for certain analytical food applications. Comparison with other techniques, such as sensory analysis by a panel of human tasters, may also be valuable and provide confirmatory information when using PTR-MS for food analysis (the food's quality, origin, etc.). Examples where this approach has been employed include the classification of food samples based on geographical location (i.e. the origin of the product, see Section 6.5.1), production processes, food type identification (e.g. classifications of different types of fat) or being able to distinguish the variety of the same food type [2].

The types of statistical analytical methods required in this application are often multivariate methods. These statistical procedures are called multivariate when the property being measured, for example, the location of the food, is being related to several variables (such as the signal levels in different *m/z* channels) in the analysis. Multivariate statistical methods can be broadly divided into two types: (1) unsupervised, which means that no *a priori* knowledge of the samples to be classified is required and (2) supervised, which requires *a priori* knowledge about the samples [18]. A good example of an unsupervised method is principal component analysis (PCA) [19–27], which looks for patterns in a block of data that depend on different variables. PCA provides a useful tool to explore and visualize information, and in particular to identify patterns in complex data, and it is therefore widely used. Applications of PCA in food science will be presented later in this chapter.

In food analysis, partial least squares discrimination is a frequently encountered supervised multivariate statistical analysis method [22, 28–35]. In addition to this procedure, another method that is now being used to discover patterns or subtle correlations in large data sets is that of data mining, which has considerable potential for PTR-TOF-MS investigations because of the large amounts of spectral data this instrumental technique generates [24, 36–38]. Basically, data mining combines computer science with statistical analysis in an attempt to find patterns in large data sets, such as those obtained from TOF mass spectra. However, whichever multivariate method is employed, care must be taken to understand the meaning and limitations of any analysis procedure, and in particular a proper validation is a critical step when using supervised methods.

6.4 Flavour Release and Perception

The first published application of PTR-MS in food science was in 1995 by the Austrian group led by Werner Lindinger (the founder of PTR-MS). For this pioneering work, a prototype PTR-MS instrument was used to analyse methanol and ethanol in the breath of human volunteers after they had consumed either alcoholic drinks or fruits [39]. It was found that the methanol concentration increased dramatically from natural levels of ~0.4 ppmv up to about 2 ppmv a few hours after eating almost a half a kilogram of fruit. This was found to be comparable to concentrations reached after drinking 100 mL of brandy containing 24% ethanol and 0.19% methanol by volume. A subsequent study by the same research group reported investigations on the endogenous production of methanol after the consumption of fruit, which they attributed to bacterial degradation of pectin in the colon [40]. Both of these early studies pose interesting health issues because, as Lindinger *et al.* pointed out in their work [40], methanol is believed to be a precursor for liver cirrhosis.

The next study involving food and PTR-MS, which was also by Lindinger and his colleagues, was a report on the time dependence of the concentrations of various compounds in human breath over a period of ~32 h after the consumption of 38 g of raw garlic [41]. The compounds monitored were acetone (protonated signal at *m/z* 59, $C_3H_6OH^+$), dimethyl sulfide (*m/z* 63, $^{12}C_2H_7{}^{32}S^+$), allyl methyl sulfide (*m/z* 89, $^{12}C_4H_9{}^{32}S^+$), diallyl sulfide (*m/z* 115, $^{12}C_6H_{11}{}^{32}S^+$), allyl methyl disulfide (*m/z* 121, $^{12}C_4H_9{}^{32}S_2{}^+$), diallyl disulfide (*m/z* 147, $^{12}C_6H_{11}{}^{32}S_2{}^+$) and diallyl trisulfide (*m/z* 179, $^{12}C_6H_{11}{}^{32}S_3{}^+$). The assignment of a VOC to a given *m/z* was principally based on isotopic abundances. However, without some separation measurement these identifications can only be tentative, and this should be kept in mind in the following review of their results. For this experiment breath sampling was not done online, but instead breath samples were collected into Tedlar bags every 20 min (for more detail on Tedlar bags see Section 7.2.8.1). These sample collections started 1 h before consumption of the garlic and continued for a time span of 8 h, which was then followed by a 14-h break during the night, and then continued for an additional 9 h the following day. Based on the time dependence of the concentrations, it was found that the VOCs could be classified into two distinct groups. The first group, which included allyl methyl disulfide, diallyl sulfide, diallyl disulfide and diallyl trisulfide, are sulfide compounds that were also observed in the headspace air above crushed garlic. The concentrations of these compounds were found to rise and reach a maximum shortly after ingestion of the garlic and then decline to the baseline within 2–3 h. The second group consisted of allyl methyl sulfide, dimethyl sulfide and acetone. The fact that allyl methyl sulfide and dimethyl sulfide are not observed above crushed garlic must mean that they are products of garlic metabolism in the human body. The concentrations in the breath of those compounds in the second group were found to reach much higher values than those in the first group but, unlike the first group, they were found to increase gradually after the garlic was eaten. The allyl methyl sulfide concentration was found to reach a maximum (~900 ppbv) after about 4–5 h and then declined slowly, so much so that the baseline was still not reached after 32 h of measurements. In comparison, the concentration of dimethyl sulfide kept increasing from a baseline of ~20 ppbv and reached a plateau level of about 80 ppbv after 32 h of observations.

A further interesting observation from this early food study was an observed increase in acetone levels following ingestion of the garlic. Although acetone is found in the headspace of crushed garlic, in concentrations of ~2 ppmv, in breath samples the concentration was found to rise above the baseline level to about 5 ppmv some 24 h after garlic consumption. This was attributed to enhanced metabolism of fatty components in the bloodstream [41]. Allicin, which is the main compound in the extract from crushed garlic, was also observed in the headspace air above garlic but interestingly it was not observed in breath samples, presumably because it is rapidly metabolized in the body.

Although not done in real time, these early investigations illustrated that PTR-MS measurements could play a major role in analysing the consumption and metabolism of food. These studies laid the foundation for the development of strategies and methodologies for online analysis and optimization of dynamic food processing which opened the door for wider applications of PTR-MS in the food industry. These early studies also indicated the potential for using PTR-MS for simple headspace analysis and one of the most successful applications of this has been in the investigation of aromas coming from coffee.

6.4.1 Drinks

6.4.1.1 Coffee

The roasting of coffee beans is the most crucial process in achieving a good-quality coffee. Being able to consistently provide the same flavour profile roast after roast in order to achieve a given quality specification is a major processing goal for coffee manufacturers.

The complexity of the aromas released by coffee beans during roasting and also from a cup of coffee was made apparent in the first detailed investigations of the headspace of roasted coffee beans and coffee by PTR-MS. This early work involved a collaboration between the research group of Lindinger in Innsbruck, Austria and the Nestlé Research Centre in Lausanne, Switzerland [42–44], a collaboration already discussed briefly in Section 6.2.

The first study by the research team was very straightforward: PTR-MS was simply employed to monitor the VOC emissions released from ground roasted coffee beans after water at a relatively low temperature of 18°C was added to produce the familiar aroma of coffee we are all familiar with [42]. The following volatile compounds were monitored: acetaldehyde, acetone/propanal, methyl ether ketone/methylpropanal, ethyl formate/methyl acetate, acetic acid, methanol, 3-methylbutanal/furan, methyl furan, 5-methylfural, 2,3-butanedione/2-methylbutanal and furfural.

This initial work was really a proof-of-principle study in which the intention was to show that a multitude of VOCs could be monitored in coffee aroma, rather than to draw any profound conclusions. This coffee aroma study was published in a review paper in 1998 [42]. Although in itself the work reported is of little scientific significance, it instigated a major research programme and is one which is still continuing strongly today. In part this continuing interest in coffee aromas is associated with the variety and complexity of the VOCs released, but predominantly it is because of commercial interest: after crude oil, coffee is said to be the most valuable commodity exported by developing countries [45]. A better understanding of the roasting processes and any resulting improvements in taste and smell means not only a better product for the consumer, but in addition potentially increased consumption and hence higher profits for a company. Thus a mechanism to control the quality or to enhance coffee aroma and flavour is therefore of major commercial benefit.

The next coffee research programmes employing PTR-MS focused on flavour formation and release and therefore investigated the complete process of coffee making, dealing with emissions from the green bean and its subsequent roasting through to the aromas released by cups of coffee [43]. Emissions from the green bean were found to be dominated by alcohols, but aldehydes, hydrocarbons and organic acids were also observed to be released in high abundances. The roasting measurements consisted of two separate studies. In one study, a batch of coffee beans was roasted. Following an initial drying phase, with the emissions being dominated by water and methanol, the concentrations of acetic acid, acetaldehyde, pyridine and methylbutanal rapidly increased and reached a maximum at medium coffee roasting. In the second study, only six beans were roasted. This yielded sporadic bursts of furan, butanal and 2,3-pentanedione and more smoothly varying changes in concentrations of other VOCs, such as pyridine and pyrazine.

The research team also investigated the headspace of coffee extracts prepared from beans roasted to different levels. Their real-time investigation of coffee roasting gave the first insight into the complex chemical processes taking place. In a subsequent study that investigated the headspace above a filter brew of Colombian coffee (Arabica, medium roast) in more detail [46], the complexity of the VOC emissions was further demonstrated. This is illustrated in Figure 6.2 and shows the results of headspace analysis of a coffee brew. Most noticeable is the complexity of the mass spectrum, suggesting that many different VOCs were released. In addition, this work demonstrated the potential for using PTR-MS as a food process monitoring device in environments where high humidity and temperatures are present. Given that numerous spectral peaks were observed and the fact that PTR-MS only

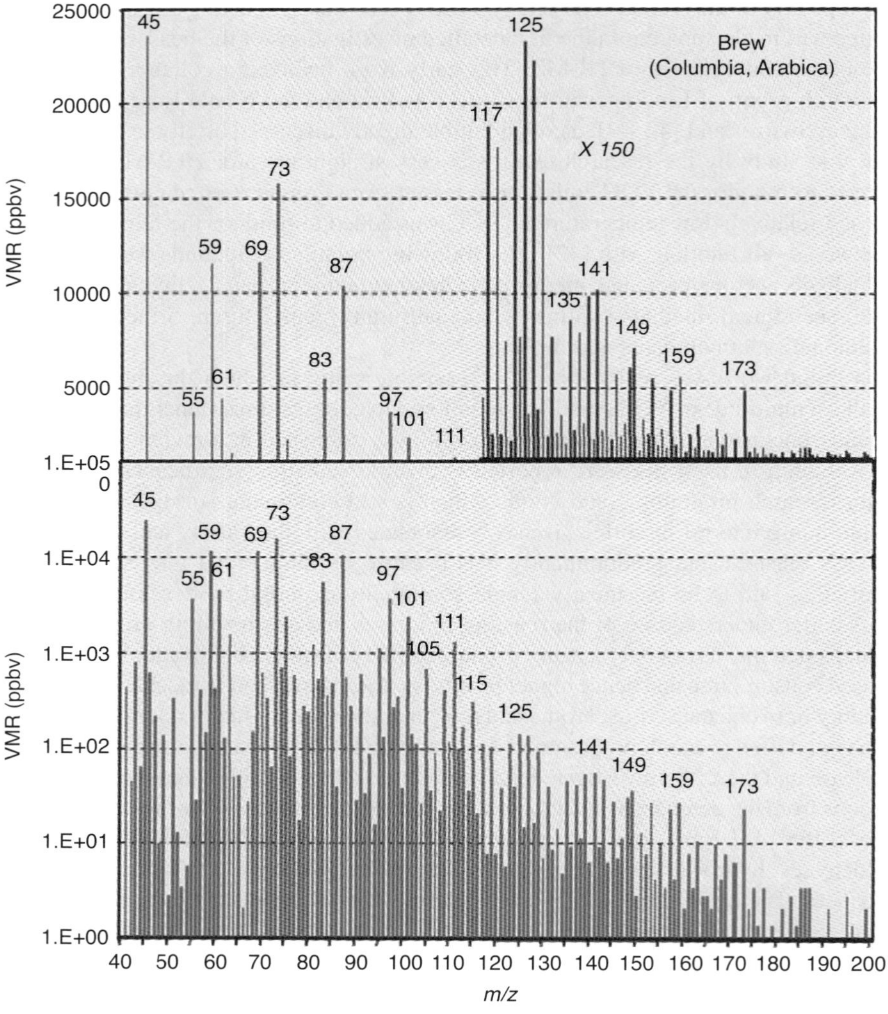

Figure 6.2 *PTR-MS mass spectra resulting from the headspace sampling above coffee maintained at 40°C. The bottom spectrum is the same as the top spectrum, except the vertical scale is logarithmic for the former. VMR = volume mixing ratio. Reproduced with permission from [46]. Copyright © 2003, Elsevier.*

characterizes compounds by *m*/*z*, there is a problem in assigning a particular *m*/*z* to a given VOC. To help overcome this, the research team used additional information to aid in the identification of the VOCs. This included: (1) switching reagent ion from H_3O^+ to $NH_4{}^+$; (2) use of collision-induced dissociation to change the product-ion patterns (fragmentation studies) and (3) using time-dependent measurement of air–liquid partitioning, which is linked to the determination of Henry's law coefficient (see also Section 9.1). By analysing all of this information, Lindinger *et al.* were able to identify over 60 different VOCs in the headspace above a coffee brew, ranging from *m/z* 42 (protonated acetonitrile) through to *m/z* 191, which the researchers assigned to (*E*)β-damascenone ($C_{13}H_{18}O$). Given the complexity of the mass spectrum, the assignments of the various VOCs represented a major achievement but the procedures adopted were time-consuming and are not easily applied to other systems, and they are therefore not practical for routine rapid food/drink monitoring and analysis.

Later coffee work, involving the Nestlé Product Technology Centre at Orbe in Switzerland and the Nestlé Research Centre, used a combination of PTR-MS, GC/PTR-MS and GC–MS to investigate coffee samples subjected to different roasting degrees and ground to different particle sizes under dry and wet 'stripping' conditions. This refers to the 'stripping' of VOCs from dry, roasted and ground coffee with 740 sccm of dry nitrogen at 90°C ('dry stripping') or from pre-wetted coffee to a 50% w/w water content with 740 sccm of water-saturated nitrogen at 90°C ('wet stripping'). The findings from these experiments were then combined with sophisticated diffusion models to learn more about the mechanism of flavour release [9]. Also at Nestlé, Lindinger *et al.* developed a robust and reproducible model to predict the flavour attributes of espresso coffee that linked VOC measurements by PTR-MS with sensory profiling data obtained from a trained panel of tasters [47]. Sensory analysis by a trained panel is currently the most direct and favoured procedure for characterizing the perceived quality of food and drink. However, it is by no means straightforward and it is also expensive and time-consuming. The replacement of sensory panels with an automated, reliable and highly reproducible analytical technique, such as PTR-MS, should therefore be considered to be a key objective in the food sciences. Before that is possible, it is essential to investigate the relationship between the types and concentrations of VOCs being emitted (PTR-MS and GC–MS studies) and sensory perception (panel analysis).

Given the economic importance of coffee, Yeretzian and colleagues in Switzerland have been developing an automated process control tool for coffee roasting. From their comprehensive and thorough studies, it was concluded that a consistent and accurate targeted roasting can be achieved by monitoring the VOCs in the off-gas of a drum roaster using PTR-TOF-MS [48–50]. Figure 6.3 provides a schematic representation of the experimental analytical approach being developed by Yeretzian and co-workers [48,49]. Using this technique for the online monitoring of coffee roasting together with PCA of the PTR-TOF-MS data, the work by Wieland *et al.* has provided another step forward towards a real-time process roasting control for a consistent roast profile [50]. The use of a high-resolution PTR-TOF-MS also resulted in a higher confidence in the identification of the VOCs being released by the coffee beans without the need for separate GC–MS measurements.

6.4.1.2 *Tea*

Given the extensive research carried out on the emission of VOCs from coffee, one might expect a similar interest in tea. In many countries, tea is a popular hot beverage and in the United Kingdom it is in fact more popular than coffee. According to the UK Tea Council, approximately twice as much tea is drunk daily in the United Kingdom than coffee. Despite the popularity of tea (at least by the British!), there has only been one reported investigation of tea using PTR-MS at the time of writing, and this only explored the influence of teabag shape on the brewing of green tea [51]. By monitoring the aroma intensity above green tea, it was shown (perhaps unsurprisingly) that, over a 4-min brewing period, a pyramid teabag

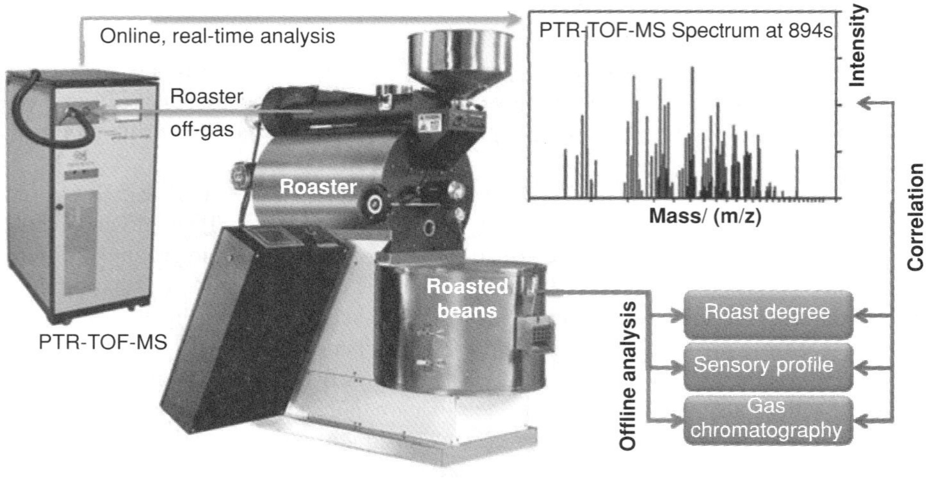

Figure 6.3 *Schematic representation of the experimental analytical approach developed by Yeretzian* et al. *for real-time process control of coffee roasting [48–50]. Reproduced with permission from [49]. Copyright © Swiss Chemical Society, 2012.*

delivered approximately twice the quantity of aroma compounds compared to a standard double-chamber teabag. Given the paucity of investigations, applications of PTR-MS to explore tea quality, tea mixtures (blends) and processing procedures would be a beneficial and fruitful area of research.

6.4.1.3 Carbonated Drinks

In addition to tea and coffee, carbonated soft drinks are extremely popular with the public, with global annual sales of ~$150 billion. Just as for tea, it is therefore somewhat surprising that there has only been one reported PTR-MS study at the time of writing, and that investigated the effects of carbonation on aroma release from liquid systems using an artificial throat [52]. An artificial throat, which is designed to mimic the act of swallowing followed by exhalation, was employed because the highest aroma release for liquids occurs in the first expiration after swallowing. (A mechanical device such as an artificial throat also has advantages over that of *in vivo* human measurements because of its overall lower costs, it is less time-consuming and it offers greater reproducibility.) Six aroma compounds with differing physicochemical properties were used in this study: benzaldehyde, ethyl butyrate, ethyl propionate, isoamyl acetate, limonene and 2-nonanone with protonated *m/z* values at 103, 107, 117, 131, 137 and 143, respectively. These compounds were added either individually or as mixtures into distilled or carbonated water. The researchers employed two-way analysis of variance to interpret their results in order to determine the effects of carbonation on the release of aroma compounds under static (equilibrium) and dynamic (*in vivo* release after swallowing) conditions. The PTR-MS instrument had a quadrupole mass spectrometer and was operated in selected ion mode, that is, full mass scans were not used. Instead the mass spectrometer was set to transmit ions at *m/z* 103, 107, 117, 131, 137 and 143 for quasi-continuous monitoring. Generally, carbonation was found to increase the release of aroma compounds in both static and in dynamic testing conditions, with the most volatile and hydrophobic compounds being the most strongly affected. Release of the flavour compounds was also found to increase if they were added as a mixture, rather than as individual compounds. The conclusion was that CO_2 is crucial in creating enhanced release of flavourings from carbonated liquid systems.

6.4.1.4 Fruit Juices

Discriminatory analysis is a technique used to distinguish between and assess the quality of food products, often of the same type which, for example, may have gone through different processes or have different origins. This analysis can be done by a panel of trained people who, using their senses for aroma and taste, can differentiate between different food types. However, as mentioned in an earlier section a sensory panel is an expensive and time-consuming approach for food and drink analysis, as well as being unavoidably subjective. An alternative approach is to use an instrument to detect volatiles followed by an objective statistical analysis.

A study of fruit juices has employed both a sensory panel and a PTR-MS analysis of emitted VOCs to investigate changes in red orange juice after the application of different types of stabilizing treatments (flash-pasteurized, pasteurized and high pressure) [20]. PTR-MS was used for measurements at m/z values of 33, 43, 45, 47, 59, 61, 73, 95, 97, 111, 117, 119, 137 and 155 for both treated and untreated red orange juice. The data obtained were then subjected to a PCA to extract the necessary information. This enabled the effects of different treatments to be separated on the basis of the emitted VOC profile. On the other hand, the trained judges were unable to discriminate between pressure-stabilized juice and untreated juice, thus demonstrating the ability of PTR-MS to provide a rapid and objective separation of orange juices subjected to different pre-treatments, with what appears to be a greater accuracy than was obtained from a sensory panel.

In a related study, Gasperi *et al.* used PTR-MS, GC–MS and a sensory analysis panel consisting of 22 trained judges to investigate the impact of CO_2/N_2O supercritical pasteurization on the quality of fresh apple juice [12]. This type of pasteurization is an alternative to the more standard thermal or high-pressure techniques. Compared to thermal treatment, supercritical pasteurization does not have a significant impact on the nutritional and physicochemical properties of food. It is also easier to control compared to the high-pressure treatments. Although it was found that the CO_2/N_2O treatments resulted in a dramatic reduction of total VOCs in the headspace (approximately a factor of 2 over a 40-min treatment time as determined by PTR-MS), with the highest change being observed for some esters (acetates) and aldehydes, which could have an effect on the flavour, the discriminative sensory analysis results indicated only a moderate effect of the supercritical pasteurization on the quality of the apple juice.

6.4.1.5 Wine

Fast and non-invasive analysis of VOCs has the potential to determine the quality and origin of wine. This is particularly important given the huge commercial market in wine and the potential for adulteration and production of fake products. Boscaini *et al.* were the first to use PTR-MS for the characterization of wine using direct headspace analysis and mass spectral fingerprints [53]. Owing to the high ethanol content of wine it was not possible to use the PTR-MS instrument in a conventional way because H_3O^+ became depleted by reaction with the large quantity of ethanol present in the headspace vapour above the wine. This produced not only protonated ethanol monomers but also protonated ethanol clusters, $C_2H_5OH_2{}^+(C_2H_5OH)_n$ ($n = 1$ or 2), in the drift tube. To overcome this problem the solution adopted was to exploit this outcome and use the protonated ethanol and its dimer as the proton transfer source for VOC analysis. However, in order to obtain a stable reagent ion signal and distribution, it was found that a stream of ethanol-saturated nitrogen had to be continuously added to the wine headspace and that was then used as the carrier gas. This ensured that the abundances of protonated ethanol and ethanol cluster ions were completely unaffected by any changes in ethanol content in the wine samples, and hence stable reagent ion signals were obtained. Four wine varieties were investigated and specific correlations were found between the wine variety and the m/z of the product ions, as demonstrated by the successful discrimination between the four wine varieties using PCA.

Despite this success, there are significant disadvantages with the technique used by Boscaini *et al.* [53]. To begin with, the proton affinity of ethanol (776 kJ mol^{-1}) is higher than water (691 kJ mol^{-1}), which means that fewer VOCs can accept a proton from protonated ethanol than from hydronium because their proton affinities may lie above that of water but below that of ethanol. Ligand switching reactions (see Section 2.2.4.2) with the $C_2H_5OH_2{}^+(C_2H_5OH)_n$ ions are another problem because they result in a more complicated mass spectral footprint (e.g. signals from mixed VOC/ethanol cluster ions) than would occur using H_3O^+ as the proton source.

To overcome these issues, an alternative method for PTR-MS headspace analyses of wine samples, which avoids the high ethanol concentration and therefore uses H_3O^+ as the reagent ion, has been devised. This approach heavily dilutes the headspace above the wine by a factor of $\sim$40 using pure nitrogen so that the ethanol volume mixing ratio was significantly reduced [23]. Although concerns were raised in the initial wine study that dilution would cause many trace VOCs to fall below the detection limit of the instrument, this was not a problem because the researchers used a more sensitive PTR-MS instrument. This higher sensitivity instrument was tested on eight different wine samples produced from two different grape varieties, Pinot Noir and Cabernet Sauvignon. The resulting mass spectra were simpler than those obtained using protonated ethanol and its clusters as reagent ions. Mirroring the earlier study, PCA of selected m/z ion signals was found to be sufficient to distinguish between the wine varieties. It was even shown that wine could be discriminated according to the different harvesting times and the location where the grapes were grown. It was also discovered that the wine samples derived from grapes at higher altitude sites and later harvesting dates displayed higher signal intensities for a number of VOCs when compared to the grapes harvested earlier and in lowland samples. Importantly, the PTR-MS results obtained were found to correspond with the ratings obtained from a tasting panel. These findings illustrated the subtle information that may be extracted from PTR-MS headspace analyses of wines which may ultimately lead to a rapid and standard procedure for wine analysis and quality control.

6.4.1.6 Vodka

Unlike the wine studies mentioned above, which investigated wine quality, Déléris *et al.* investigated the impact of swallowing on the dynamics of aroma release and hence on the perception of flavour during the consumption of flavoured vodka [54]. PTR-MS was used for nose-space analysis in order to evaluate aroma release into the nasal cavity.[4]

In this study, air passing out of the nose was sampled using a nosepiece that consisted of two inlets, one for each nostril. The PTR-MS measurements were compared with an evaluation of aroma perception by a panel using two protocols, namely spitting out or swallowing the vodka. When the vodka was spat out, the panel perceived a 'fruity' note, whereas the panel found that swallowing resulted in more complex perceptions with a 'fruity' note being first perceived, followed by a 'warm' note to be replaced by another 'fruity' note and then by a 'green' note. The 'fruity' note was lower when the vodka was swallowed, which was found to be correlated with a decrease in the rate of aroma reaching the olfactory receptors in the nose. The PTR-MS data could also be used to partly account for the difference found in perception. For example, the 'warm' perception just after swallowing correlated with the signal at *m/z* 47 (protonated ethanol) being enhanced, meaning that the amount of ethanol in the nasal cavity was greater when the vodka was swallowed than when it was spat out.

6.4.1.7 Infant Formula

It has been suggested that food preferences may well have been influenced by early flavour experiences [55]. If a baby is not breast-fed, one of the earliest foods he or she experiences is an infant formula (a manufactured alternative to breast milk). However, infant formulas differ considerably in flavour and therefore they may well have lasting consequences on what types of flavours are preferred by the child as he or she grows. There have been very few studies that report the characterization of flavour in infant formulas. Using a combination of PTR-MS and GC–MS, van Ruth *et al.* have investigated the VOCs in the headspace above 13 infant formulas, with the PTR-MS providing the 'fingerprinting' data and GC–MS the identification of the VOCs [56]. By subjecting the PTR-MS mass spectral data to a multivariate analysis of variance (MANOVA), which is a statistical procedure used to compare multivariate means of several groups, alongside PCA, it was possible to use the PTR-MS results to discriminate between different brands, type and physical form of the infant formulas.

6.4.2 Food

6.4.2.1 Cheese

The first reported use of PTR-MS for food, rather than for drink, perception was by Gasperi *et al.* [19]. This was also the first food study that compared classical sensory analysis with VOC headspace analysis by PTR-MS. The foods chosen for investigation were seven varieties of Italian mozzarella cheese, for which measurements were made while they were held at a temperature of 36°C. By using a multivariate statistical data analysis approach,

---

[4] The idea behind PTR-MS in-nose (nose-space) studies is that they allow the researcher to investigate the link between aroma release and concentrations during food/drink consumption and the resulting perception by the olfactory receptors. This gives a more realistic analysis of human perception than can be achieved by a simple headspace analysis because the influence of the human mouth on flavour release is avoided. *In vivo* nose-space analysis also provides a method to investigate the influence of food composition, food texture and mastication on flavour release. Owing to its high sensitivity, dynamic range and rapid response time, PTR-MS analysis is well suited for nose-space measurements.

it was shown that the two methods (PTR-MS and flavour profile analysis using a panel consisting of eight judges) achieved comparable sample discrimination. By finding the similarities between VOCs recorded by PTR-MS and the odour/aroma profile of a trained sensory panel, this study was the first to demonstrate that PTR-MS VOC fingerprinting could be directly linked to sensory characterization.

Boscaini *et al.* reported another study dealing with cheese, but in this case focusing on three Italian grana (hard) cheeses: Grana Padano, Parmigiano Reggiano and Grana Trentino [8]. In this study, the research team combined gas chromatography–olfactometry (GC–O) measurements with PTR-MS and PCA. Once again, PTR-MS was used simply to sample the headspace above each of the cheeses. Fifty distinct peaks (protonated parent and fragment ions) were identified in the PTR-MS mass spectral profile from volatiles emitted by the cheeses, covering an *m/z* range of 27–173. A tentative assignment to specific VOCs based on literature data and *m/z* values was then attempted. Seven of the *m/z* values were associated with VOC concentrations higher than 1 ppmv, with the highest at *m/z* 45 (assigned to be acetaldehyde and fragments of heavier aldehydes), followed by *m/z* 59 (acetone, propanal), *m/z* 47 (ethanol), *m/z* 61 (acetic acid and a fragment of ethyl acetate), *m/z* 43 (various fragments) and *m/z* 89 (butyric acid, acetoin, 2-methylpropionic acid). Other VOC concentrations fell in the range of 1–100 ppbv and one of the most intense was observed at *m/z* 145, which the researchers assigned to be protonated ethyl hexanoate, a compound that produces a distinctive fruity note in food. Although assignment of the VOCs being emitted by the cheeses is important, it was not essential for use in separating the cheese varieties. Significant differences were found in the volatile spectral profiles (intensities and *m/z*) which were fed into a PCA treatment to clearly distinguish between the three cheeses. The importance of this work is related to that of quality control, since it makes it possible to establish that the cheese is actually from a particular region. Protected destination of origin has a high commercial value and therefore it is important to be able to determine real from fake food products quickly and reliably.

A more systematic study involving a parmesan-like hard cheese, focusing on Grana Padano, was reported by Biasioli *et al.* [28]. In comparison to the earlier study described above [8], a more sophisticated statistical analysis of the data (partial least square regression) was used to calibrate PTR-MS results with sensory data. The headspace above 20 samples of 'Trentingrana', the variety of Grand Padano produced in Trentino (northern Italy), was compared with a sensory assessment by a panel of eight selected judges trained to recognize, describe and quantify basic tastes and odours of cheese. Several *m/z* features in the PTR-MS mass spectrum could be attributed to specific flavour attributes. Taken together with an earlier study carried out on strawberry cultivar discrimination [29], which showed that PTR-MS mass spectral fingerprints could be successfully coupled with data compression and class modelling methods to provide a fast and sensitive tool to identify the cultivar of strawberry fruits (basically discrimination based on genetic features), Biasioli *et al.* concluded that PTR-MS fingerprinting has the potential of bridging the gap between sensory assessment of quality and genetic control of agro-industrial products.

Following on from the original cheese study, there have been many subsequent studies utilizing PTR-MS for investigations into the flavour perception of foods. Examples include a series of PTR-MS investigations by van Ruth *et al.*, who studied vegetables, including red peppers [57], to explore the influence of saliva on temporal VOC release, and red kidney and soya beans [58, 59], with specific VOCs being identified by GC–MS. All three reported studies by van Ruth and colleagues evaluated the VOCs released *in vitro* (in a model mouth system) and *in vivo* (in-nose). Comparison of the results obtained from the in-nose measurements with those from a model mouth system, which provides a more well-defined, simplified and controllable environment than found in a real mouth, yielded good agreement.

6.4.2.2 *Bread*

Heenan *et al.* have used PTR-MS to compare sensory characteristics (defined by odour) and VOC vapour composition in the characterization of fresh bread flavour, with the aim of improving the understanding of how bread freshness is perceived [30]. Twenty bread varieties, consisting of 10 speciality breads and 10 commercial breads (all equally fresh in terms of elapsed time from baking), were targeted in this work. Partial least squares regression modelling was used to determine the relationships between the sensory characteristics and the VOC measurements. It was shown that VOC information can be applied to model and characterize the sensory properties of different bread types and that it can be used to predict consumer perception of the freshness of different bread varieties. Thus, VOC analysis provides promising opportunities for the rapid profiling and screening of bread quality and freshness.

6.4.2.3 *Onions*

The fast time resolution of PTR-MS has also been put to good effect by Løkke *et al.* to continuously investigate the dynamics of VOCs emitted from freshly cut onions (*Allium cepa* L.) for a period of 120 min [60]. The researchers compared their PTR-MS results with those obtained using GC. The data from PTR-MS, which were used to estimate the concentrations of the VOC compounds in the headspace above freshly cut onions, were subjected to PCA so that the main variations in the VOC emissions as a function of time could be captured. The headspace VOC composition was found to change rapidly as a result of reactive sulfurous compounds emitted from the cut onion, which apparently resulted from cell damage caused when cutting the onions. To illustrate this, Figure 6.4 presents approximate concentrations of propanal, hydrogen sulfide, methanethiol, propanethiol, dipropyl disulfide, dimethyl thiosulfinate and dipropyl thiosulfinate as a function of time starting from when the onions were first cut. For comparison, Figure 6.5 shows headspace mass spectra recorded at 1, 10 and 60 min after cutting. This investigation demonstrated the use of PTR-MS as a powerful analytical tool to investigate the dynamics of VOC emissions from freshly cut fruits and vegetables.

6.4.2.4 *Wheys*

Whey is used in a number of food products, either in production (e.g. ricotta cheese) or as an additive. In comparison to the nose-space investigations mentioned above, work by Gallardo-Escamilla *et al.* on liquid and fermented wheys dealt only with the headspace volatile composition [61, 62]. Using PCA, the VOC data obtained by PTR-MS could be used to differentiate the samples, that is, the mass spectra of the VOCs emitted resulted in sufficient information to provide a reliable fingerprint of each sample. Furthermore, by applying a simple statistical approach, sensory characteristics such as ‘rancid’, ‘caramelized

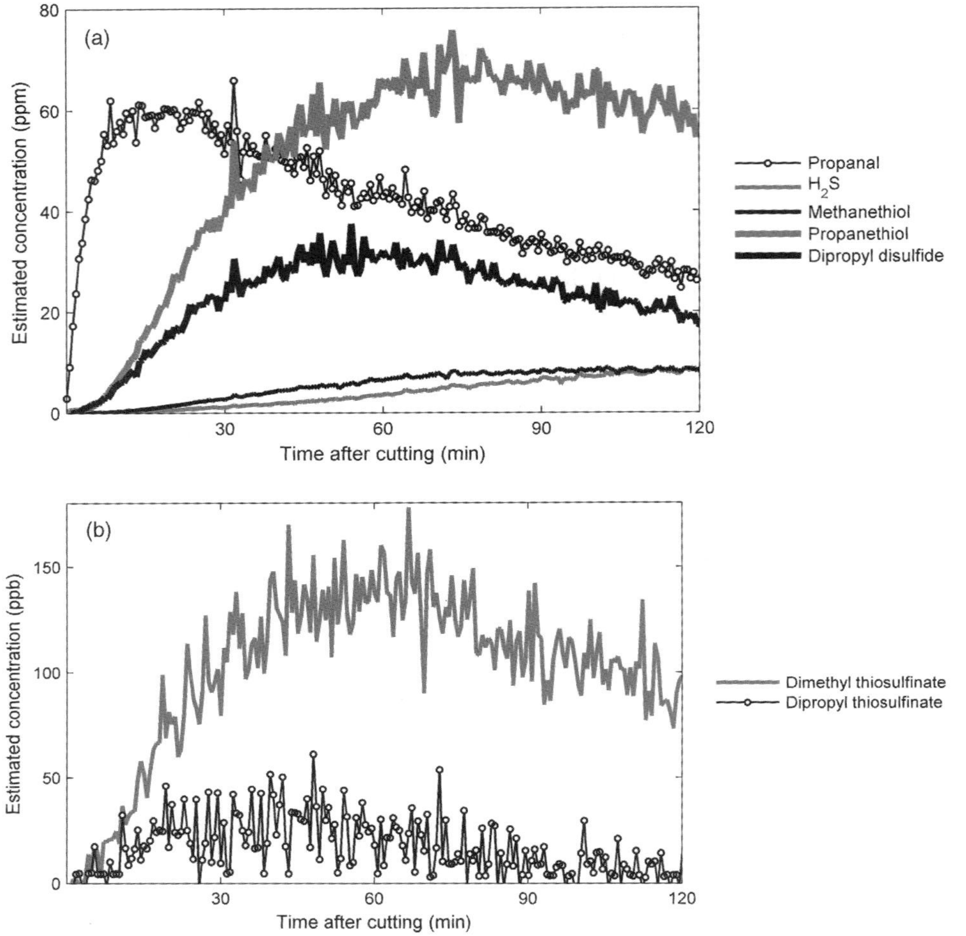

Figure 6.4 *Estimated concentrations of (a) various VOCs in ppm and (b) thiosulfinates in ppb released after the cutting of a raw onion, as followed by PTR-MS (0–120 min). Reproduced from http://www.mdpi.com/1424-8220/12/12/16060.*

milk' and 'sweet' could be correlated with the presence and/or the absence of specific VOCs. The same researchers also assessed the impact of hydrocolloids on the partition of flavour compounds in the headspace of fermented whey [21].

6.4.2.5 Fruit

During fermentation the aroma compounds released from fruit change as a function of time. Boamfa *et al.* investigated this for the fermentation of apples [63]. The aldehydes, alcohols, acids, acetates and C_6 compounds released by four apple cultivars (Elstar, Jonaglod, Granny

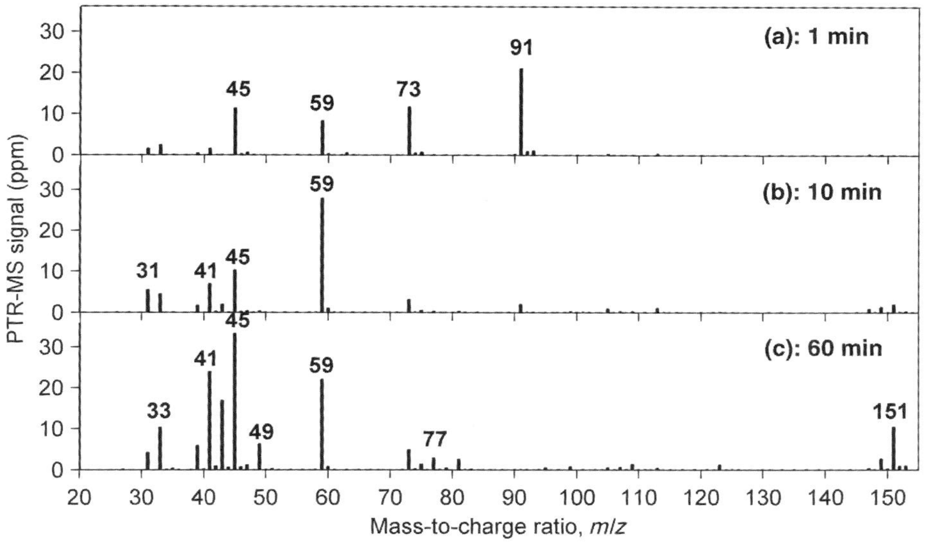

Figure 6.5 PTR-MS mass spectra of VOCs emitted by freshly cut onion after a delay of (a) 1 min, (b) 10 min and (c) 60 min. These data represent an average of three consecutive scans and are corrected for dilution factors. Signals from protonated water clusters, O_2^+ and their isotopes have been removed for clarity. Reproduced from [60].

Smith and Pink Lady) were monitored under short anaerobic (24 h) and post-anaerobic conditions. In the same paper, another fermentation study was reported whereby PTR-MS and CO laser-based photoacoustic measurements of alcoholic fermentation products (acetaldehyde and ethanol) from young rice plants were compared. Although good agreement for the acetaldehyde concentrations was obtained, poor agreement was obtained for ethanol concentrations. This was attributed to ethanol adhering to the nylon tube used to transport chemicals to the detector. Although this can be a problem, it is important to realize that considerable care must be taken when deriving ethanol concentrations from PTR-MS data because the protonated parent signal intensity (m/z 47) is strongly dependent on the E/N value used owing to fragmentation (see also Section 2.2.4.1) [64].

Finally, we comment that PTR-MS has been used to draw a link between the volatile compound release and the genetic factors through the emission of various esters from cultivated strawberries [65]. Esters, which are produced via esterification of alcohols in the presence of acyl-CoA by the enzyme alcohol acyl-transferase (AAT), are among the most important constituents of a strawberry aroma and are largely responsible for the sweet smell associated with strawberries. By detecting the esters in the headspace above strawberries by PTR-MS, a correlation was found between the expression of an AAT gene, total AAT activity, and the presence of esters. The PTR-MS data presented in this work represented a subset of a major data set collected over a 3-year period (2002–2004) to characterize the fruit quality of strawberry genotypes in northern Italy.

6.4.3 Flavour Release: Food Texture, Composition and Physiological Effects

Other studies dealing with flavour release and perception have dealt with the effects of food texture and composition on flavour release, and many of these studies have been involved with dairy products. So far we have discussed cheeses, but work has also been reported on whey, yogurts and custards [61,62, 66–73]. In the first of these studies Roberts *et al.* spiked water, skimmed milk, semi-skimmed milk and full fat milk with five aroma compounds, all of which are soluble in these liquids, in order to compare nose-space (PTR-MS) and sensory panel (perceived) intensity ratings for the evaluation of flavour absorption by fat [66]. This was the first study to use PTR-MS to investigate the influence of food composition (here fat content) on flavour release. The goal of the study was to improve our understanding of the relationship between physical stimuli and flavour perception. The five aroma compounds and the three concentrations (increasing by a factor of 4 each time) used for each liquid were β-damascenone (3.125, 12.5 and 50 mg L^{-1}), hexanal (1, 4 and 16 mg L^{-1}), ethyl butyrate (1, 4 and 16 mg L^{-1}), benzaldehyde (1, 4 and 16 mg L^{-1}) and 2,3-butanedione (10, 40 and 160 mg L^{-1}). The PTR-MS (nose-space) measurements showed that only a small quantity of the aroma compounds added to the drinks were released during consumption. For water, it was found that the compound volatility in water was the driving factor in compound release, that is, the higher the volatility, the greater the release. It was also found that the release of the flavour compounds depended to some extent on the amount of fat content and the ability of compounds to dissolve in fat (lipophilicity). β-Damascenone, which has the highest lipophilicity of all the compounds used in the study, showed marked decreases in emission concentration as the fat content increased for all three concentrations used. For the high and medium aroma concentrations (12.5 and 50 mg L^{-1}), the percentage loss in nose-space concentrations for full fat milk relative to water was measured to be similar at ∼70%. For hexanal, statistically significant changes were observed at the high (16 mg L^{-1}) and medium (4 mg L^{-1}) concentrations for full fat milk, showing an approximate decrease of 38% and 21% in nose-space concentration, respectively, relative to water. Similarly, decreases of 49% and 30% in the concentrations of ethyl butyrate were observed for full fat milk at the highest concentration (16 mg L^{-1}) and at the next highest concentration (4 mg L^{-1}) of ethyl butyrate, respectively. The release of benzaldehyde was found to decrease with increasing fat content for the two highest concentrations, with the decrease in emission being ∼29% for the highest concentration and 39% for the next highest concentration. At the lowest concentration of benzaldehyde (1 mg L^{-1}), only semi-skimmed milk and full fat milk showed a statistically significant change in compound release relative to water, with the decrease being 26% and 39%, respectively. For, 2,3-butanedione, which has a low lipophilicity, no significant dependence on fat content on the release of aroma compounds was observed.

It was found that increasing concentrations of the odour compounds in the nose-space correlated with increasing sensory perception ratings. In comparison to the PTR-MS nose-space measurements, the headspace measurements showed a greater decrease in concentration with increasing fat content. This work therefore suggested that compound absorption by fat based on headspace measurements would be overestimated. In conclusion, this study showed that the release of lipophilic compounds is greatly suppressed by increasing fat levels in milk [66].

In-nose PTR-MS studies to investigate the influence of food texture on flavour release were conducted by van Ruth *et al.* [67]. The effect of flavour emissions on the viscosity (thickness) of custard desserts (basically on the concentration or type of carboxymethyl celluose) was investigated. The custards were flavoured with 15 volatile compounds to provide an artificial strawberry flavour. It was observed that the firmer custards were found to generate higher in-nose flavour concentrations than the softer varieties. In comparison, headspace analysis of the custards using PTR-MS showed opposite behaviour, whereby a firmer texture resulted in lower headspace concentrations. This difference was explained by the fact that firmer custards are likely to be chewed more, and the more extensive the chewing, the greater the frequency of the opening between the mouth and the pharynx (velum), which leads to the observed higher in-nose flavour concentrations. van Ruth *et al.* also evaluated the influence of water on the volatile flavour retention at four intranasal locations in nine subjects during the consumption of custard desserts. Observed differences between the locations indicated various degrees of retention by the watery mucus in the nasal tract [72].

PTR-MS nose-space evaluation of custards was further studied by Aprea *et al.* [71], who also specifically investigated the effects of texture and oral processing on the release of strawberry flavour in custards. The in-nose concentrations of strawberry flavour compounds during the consumption of the custard were significantly affected by the oral processing protocol involved (free chewing or imposed). This demonstrated how *in vivo* measurements can provide information about individual consumers.

In addition to the influence of food composition and texture on flavour release, PTR-MS has also been used to investigate the physiological effects that can determine the release of flavour volatiles from food and their influence on olfactory perception during the consumption of food. These include saliva flow, chewing, swallowing, tongue movements, mouth volume, nasal mucosal tissues and breathing pattern. Buettner and Beauchamp have provided a review of 'peripheral' factors that can influence flavour perception during food consumption [74].

The first reported PTR-MS study investigating physiological effects on flavour release was by van Ruth and Buhr [57]. In that study, a model mouth was used to evaluate the effects of (artificial) saliva on the volatile flavour release from rehydrated red bell peppers as a function of time. It was found that saliva significantly reduced the headspace concentration of VOCs tentatively assigned to 2-methylpropanal (protonated signal at m/z 73), 2-methylbutanal and 3-methylbutanal (m/z 87) and hexanal (m/z 83). Model mouth (and *in vivo* (in-nose)) PTR-MS measurements have also been used by the same research group to investigate the influence of gel strength and structure on aroma release from pectin gels [75,76]. This work showed that both the strength and structure of the pectin-containing systems play a significant role in the quantity of aroma compounds that are released and that mastication and saliva have large influences on how much of the aroma compound is released from the mouth to the nose [75].

In a follow-up study, Boland *et al.* investigated the physicochemical release of 11 flavour compounds from gelatin, starch and pectin gels [76]. Flavour release was found to be significantly affected by the rigidity of the gel, with the most rigid (gelatin) showing the lowest flavour release. However, the study showed that other factors were also involved in flavour release and these factors were linked to matrix-volatile interactions. For example, there was a greater release of hydrophilic compounds from pectin gels than from starch gels, whilst the opposite was found to be true for hydrophobic compounds. This work was supplemented by in-nose PTR-MS and sensory analyses to investigate the influence of the texture of gelatin and pectin gels on strawberry flavour release and perception [77]. It was found that both the type of hydrocolloid and the rigidity of the sample greatly affected flavour release and perception. Gelatin gels were shown to have higher static headspace concentrations of strawberry flavour, but decreased flavour release compared to the pectin gels.

Through the simultaneous application of sensory and PTR-MS methods, Déléris *et al.* investigated the dynamics of aroma release during the consumption of candies (sweets) in terms of a candy's texture and gelatin content and the eating technique used (i.e. whether the sweet was allowed to melt in the mouth or whether chewing was involved) [78]. PTR-MS was used to monitor ions at *m/z* 83 (Z-3-hexen-1-ol), *m/z* 87 (diacetyl), *m/z* 93 (c-decalactone), *m/z* 145 (ethyl hexanoate) and *m/z* 171 (c-decalactone). The effects of changing the gelatin content contained in the sweets (0%, 2%, 5% and 15% of gelatin) on the ion signals at *m/z* 83, 87 and 145, which are associated with VOCs that impart green, butter and strawberry notes in food, respectively, were investigated. It was found that the highest *in vivo* release for all three of these aroma compounds occurred for candies containing 2% gelatin, although in comparison it was found that gelatin does not have any significant effect on the headspace/product partition and diffusion properties of the aroma compounds. The results relating to the eating technique showed that chewing resulted in initial release rates that were higher and reached maximal intensities quicker than when allowing the candies to simply melt in the mouth. Chewing also resulted in a greater mixing ratio of aroma compounds, although signal persistence was less.

A study by Mei *et al.* used nose-space measurements to investigate the influence of yogurt composition, flavoured by ethyl butanoate, ethyl 3-methylbutanoate, (*Z*)-hex-3-enol, 2-methylbutanoic acid, 5-hexylhydro-2(3*H*)-furanone and 3-methyl-3-phenylglycidic acid ethyl ester, on the aroma release in the mouth during eating [68]. These particular flavour compounds were used because an appropriate mixture of these compounds provides an artificial strawberry flavour. The aroma release of the ethyl butanoate, (*Z*)-hex-3-enol and ethyl 3-methylbutanoate was found to be suppressed by sweeteners, with high-fructose corn syrup having the greatest effect, while the addition of thickening agents was found to have no significant effect on the aroma release profiles.

The retention and release of aromas from food is dependent on the physicochemical properties of the aroma compounds and their interactions with food components and these interactions play an important role in flavour perception. To study this, Tietz *et al.* used PTR-MS headspace analysis to investigate the release of single aroma compounds, namely hexanal ($C_6H_{12}O$) and menthone ($C_{10}H_{18}O$) (or a blend of these two compounds), added in low concentrations (typically 6×10^{-3} mg/100g) to an aqueous starch solution (starch concentration of 4 g/100g) [79]. Low aroma concentrations were used in this study in order to simulate that found in food. Both hexanal and menthone are known to form complexes with starch. For the system doped with hexanal, product ions at *m/z* values of 101 (4%), 83 (47%) and 55 (100%) were observed. The ion at *m/z* 55 was considered to have contributions from an ion resulting from dissociative proton transfer to hexanal, as well as the protonated water trimer (which is produced in large quantities in the drift tube owing to the high humidity of the headspace being sampled). Owing to this complication the next most intense ion signal at *m/z* 83 (loss of water from the protonated parent molecule) was therefore selected as the marker for hexanal. Menthone resulted in product ions being produced at *m/z* values of 155 (7%), 137 (15%), 95 (18%) and 81 (60%), with the product ion branching percentages being provided in parentheses. Although the protonated ion at *m/z* 155 is observed, the researchers selected *m/z* 95 as representative for menthone because this peak could be exclusively assigned to menthone. In comparison to aroma-water systems, the PTR-MS work found that the rates of aroma release from starch-aroma systems were reduced owing to strong starch-aroma interactions. Tietz *et al.* reported another headspace investigation using PTR-MS to study the release of menthone from a menthone-starch system [80]. In this study the impact of α-amylase addition on starch hydrolysis was investigated. Alpha-amylase is an enzyme found in saliva and in the gut that breaks down starch. Although an enhancement in menthone release was expected on the addition of α-amylase, none was actually observed. In fact after 60 minutes of starch hydrolysis only a moderate increase in the headspace intensity of menthone was noted.

Mestres *et al.* used nose-space extraction to monitor the influence of whey protein gel texture on volatile release patterns during mastication [69,70]. The retronasal aroma release was followed and compared to sensory perception of overall aroma intensity, with the objective to find a connective link between physicochemical release patterns and perceived sensory impressions. Gel texture was found to be correlated with the *in vivo* physicochemical release patterns and to the corresponding retronasal aroma perception.

Owing to their low caloric values, artificial sweeteners as sugar replacements in food and drink represent a big business. However, the addition of sweeteners influences the release of volatile flavour compounds so that the taste of the food may be changed, which may not be acceptable to the consumer. To understand this process the physicochemical interactions between the volatile flavour compounds and the non-volatile constituents of a food matrix need to be investigated. Towards this goal, Siefarth *et al.* investigated the dynamic headspace release of four volatile flavour compounds (cis-3-hexen-1-ol, benzaldehyde, ethyl butanoate and butyl isovalerate) from pure water and various low-viscosity aqueous solutions of sucrose, maltitol, erythritol, polydextrose and oligofructose [81]. Flavour release from the aqueous solutions was found to increase with increasing hydrophobicity and/or volatility of the flavour compound. Also, the addition of sucrose, polyols and bulking agents affected flavour release from aqueous solutions relative to pure water. In comparison, no correlation was observed between the flavour release and the viscosity of the solutions. In a follow-up study, the flavour released from specific high-viscosity polyol and bulking agent solutions was characterized [82]. In this case, a correlation between viscosity of the solutions and flavour release was discovered, which suggested complex matrix-volatile interactions in the high-viscosity solutions.

The studies by van Ruth and co-workers [75,76], discussed earlier, highlight the importance of mastication and how the release of VOCs during eating differs from that associated with the headspace above uneaten foodstuff. However, the first study to explore this was a collaboration involving Lindinger's group in Innsbruck and Nestlé, which compared the VOC composition in the nose during mastication of ripe and unripe bananas to that produced from the simple headspace above the fruit [83]. Ripe bananas were found to produce a much broader and more intense array of volatile esters than unripe bananas, as well as producing higher yields of compounds such as hexanal. Significant differences were found between the headspace and the nose-space measurements for both ripe and unripe bananas, indicating that mastication has an important impact on the aroma profile.

The importance of the mastication rate on dynamic flavour release was further analysed by a combined model mouth/PTR-MS system for sunflower oil [84]. This study showed that an increase in the rate of mastication brought about an increase in the proportion of the more hydrophilic compounds (such as 2-butanone) released into the air relative to more hydrophobic compounds (such as 2-heptanone). In another flavour-related study involving oils, Landy *et al.* compared the relative effects of water-in-oil emulsions versus microemulsions on the release of VOCs [85]. This work indicated that water-in-oil emulsion and microemulsions have the potential of being aroma delivery systems which improve retronasal or orthonasal perception when diluted in food products. *In vivo* VOC release patterns in palm oil have been investigated by Lasekan and Otto [86]. The quantities of the compounds exhaled in the breath changed significantly with time after the consumption of palm oil. The most prominent compound they claimed to have detected was ethanol, followed by acetic acid, in addition to acetoin, methyl acetate and 2-phenyl ethanol.

Mastication is, of course, the key process leading to the release of flavours from chewing gum. For this product, a consumer ideally requires something that will sustain its taste over a long period. PTR-MS has been used for *in vivo* investigations of aroma release (mint flavour) from three different types of encapsulated flavour systems. The aim here was to establish the release rates of menthol and menthone from the chewing gum and the maximum abundance these compounds reach in the gas phase, as well as the time taken to reach that maximum [87]. It was found that the release profile of the mint aroma was greatly affected by the type of encapsulation system, and that it is possible to accelerate or delay flavour release and perception. Furthermore, the researchers noted that the menthol and menthone release curves showed a good correlation with mint sensory perception as a function of time. Information of this sort, derived from PTR-MS, can aid in the development of improved food products by specifically changing flavour profiles to increase the enjoyment for the consumer.

Wagyu beef (from Japanese black cattle) is known worldwide for its excellent taste, which results from its marbling characteristics (i.e. the meat contains intramuscular fat) that enhance the flavour, tenderness and juiciness of the meat. To investigate the released aroma volatiles of cooked Wagyu beef during mastication, breath taken from both oral and nasal cavities of a single panellist was investigated using PTR-MS measurements and analysed using PCA [88]. Twelve volatile compounds were monitored: 1-butanol (*m/z* 57), acetic acid (*m/z* 61), 1-pentanol (*m/z* 71), 1-hexanol (*m/z* 85), diacetyl (*m/z* 87), ethyl acetate (*m/z* 89), hexanal (*m/z* 101), ethylbenzene (*m/z* 107), 7-octen-4-ol (*m/z* 111, which is a fragment ion corresponding to an elimination of H_2O from the protonated parent molecule) and three lactones (*m/z* 157, 171 and 185). The study showed that ethyl acetate had the highest concentrations in the oral cavity followed by acetic acid, butanol, pentanol and diacetyl. In the nasal cavity ethyl acetate, butanol and hexanol were also detected. Compounds such as ethyl acetate, butanol, acetic acid and diacetyl were found to be released in high levels during mastication, which the researchers attribute to partitioning into the saliva. Acetic acid and diacetyl, which are water soluble compounds, were not found in any substantial quantities in the nasal cavity.

High-resolution PTR-TOF-MS has been used by Heenan and colleagues to monitor flavour release from strawberry flavoured cereal bars with varying sugar composition *in vitro* (semi-static headspace measurements) and *in vivo* (dynamic nose-space measurements) [89]. The artificial strawberry flavour used consisted of 17 aroma compounds. For the headspace measurement, the concentration of the flavour VOCs released from the cereal bar matrix was found to be significantly affected by the sugar composition of the cereal bars. In particular, substitution of glucose syrup solids with the sweetener polydextrose increased the release of acetaldehyde, alkyl esters (ethyl butyrate and ethyl methyl butyrate), menthol and menthone. In the case of the *in vivo* investigations, it was observed that the maximum signal intensity and total areas under each mass spectral peak (which is proportional to the specific VOC concentration) were affected by the glucose syrup to polydextrose ratio. The conclusion the researchers drew from their investigations was that flavour release is influenced by the occurrence of '*salting out*', whereby the volume of free water is reduced as the ratio of polydextrose to glucose syrup is increased. *In vivo* release, however, is also influenced by oral processing, both directly because of the physicochemical processes taking place in the mouth and, indirectly, because sugar substitution affects texture and consequently bolus formation and swallowing. Ting *et al.* present a similar investigation on a real food product (six commercial apple cultivars: Fuji, Granny Smith, Golden Delicious, Jonagold, Morgen Dallago and Red Delicious) and correlated this with genetic traits and texture analysis [90].

6.5 Food Classification, Food Quality and Food Control

Food quality is related to the complex flavour profile and its evolution as we eat or drink. Consequently, in one sense we have already covered food quality above. In this section, however, we will focus on the quality and control of food *prior* to consumption.

Anonymous fingerprinting applications of PTR-MS, that is, where the analysis is not concerned with identifying specific VOCs, but instead merely uses the mass spectral peaks as some form of information, have been successfully applied in several studies of food quality and control. A few examples have already been met earlier in this chapter in a different context, but in this section the emphasis is on food classification. A particularly interesting and perhaps unexpected application of this relates to determining the geographical origin of food and drink products.

6.5.1 Geographical Location

Often consumers use geographical origin as a basis for food and drink quality, and therefore being able to guarantee the authenticity of the origin of food and drink products and their traceability is an important quality control issue. In combination with appropriate statistical techniques, PTR-MS measurements have been shown to be capable of providing geographical identification for several food products, including fresh white truffles [10], butter [31], olive oil [32], sea roe [91], dry-cured ham [92,93] and cumin cheeses [94].

6.5.1.1 White Truffles

For the white truffle study the researchers investigated the VOC emissions from 18 white truffles (*Tuber magnatum Pico*) selected from six regions in Italy (three from each region). The VOCs used in this headspace analysis were methanol (protonated peak at *m/z* 33), acetaldehyde (*m/z* 45), ethanol (*m/z* 47), methanethiol (*m/z* 49), dimethyl sulfide (*m/z* 63), dimethyl disulfide (*m/z* 95 + 79*), dimethyl sulfoxide (*m/z* 79), dimethyl sulfone (*m/z* 95), bis(methylthio)methane (*m/z* 61*), 2(3*H*)-furanone + dihydro-3,5-dimethyl (*m/z* 115), 2-acetyl-5-methylfuran + (methylthio)dimethyl sulfoxide (*m/z* 125), 3-ethyl-4-methyl-3-penten-2-one + 3,4-dimethyl-3-hexen-2-one (*m/z* 127), p-cymene (*m/z* 93* + 135), benzothiazole (*m/z* 136), limonene (*m/z* 81* + 137), 2,4-octanedione (*m/z* 143) and tris(methylthio)methane (*m/z* 107*), where the *m/z* refers to the signal recorded for the given VOC, which is either the protonated parent or a fragment ion (as indicated by an

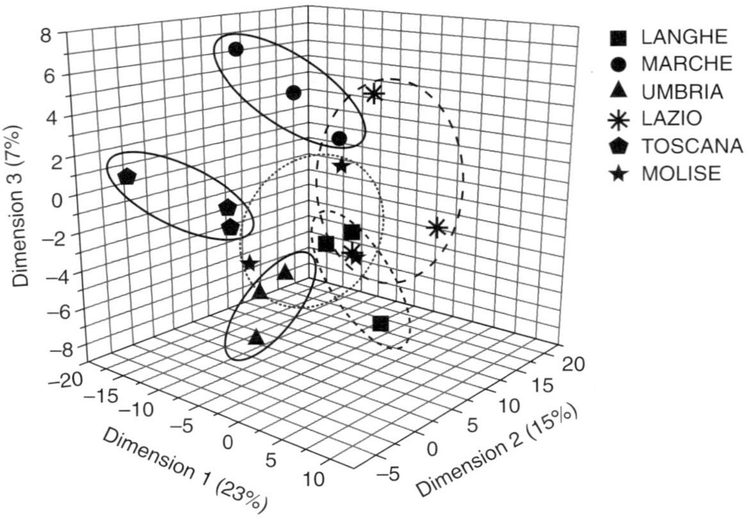

Figure 6.6 *First three principal component analysis scores (correlation) for the analysis of truffles taken from six regions in Italy (Langhe, Marche, Umbria, Lazio, Toscana and Molise). The results for truffles taken from Marche, Umbria and Toscana are particularly well separated. Reproduced with permission from [10]. Copyright © 2007 John Wiley & Sons, Ltd.*

asterisk). Solid Phase Microextraction (SPME)[5]–GC–MS was used to support the assignments of the PTR-MS mass spectral features. Dimethyl sulfide dominated the emissions with concentrations ranging from ~41 ppmv (Toscana) up to 71 ppmv (Lazio), followed by bis(methylthio)methane with a concentration range of ~1.1 ppmv (Toscana) to 4.1 ppmv (Langhe). Methanethiol was found to have a concentration range of about 60 ppbv (Lazio) to 310 ppbv (Marche). Methanol was also found to have a large range of concentrations, spanning from ~12 ppbv (Langhe) up to 234 ppbv (Lazio). Other observed VOCs showed a smaller range in concentrations between the regions and were present at generally <100 ppbv. A PCA treatment of the mass spectra revealed that the truffles are partly clustered according to the location they were harvested from. This is illustrated in Figure 6.6, which shows that truffles from the regions of Marche, Umbria and Toscana are well separated by PCA from the other locations, while the truffles from Lazio and Molise partially overlap with those from the Langhe area.

6.5.1.2 Butter

A study of VOC emissions from butter evaluated whether PTR-MS headspace measurements combined with a partial least square-discriminant analysis (a multivariate statistical technique) could be used to predict the origin of various butters that were produced in

[5] SPME is a fast solvent-free sample extraction and preparation technique, whereby the steps of extraction, pre-concentration and transfer all occur in one device. In brief, the technique extracts substances from an analyte, which can be a gas or liquid, into an absorbing layer. The absorbed compounds can then be transferred into an inlet system of a chromatograph by rapid thermal desorption. For more details, see 'Solid Phase Microextraction: Theory and Practice' by the inventor of the technique, J. Pawliszyn (Wiley-VCH, Inc., 1997).

Europe [31]. Eighty-three commercial butters from eighteen European countries and from three European regions (i.e. east, north/west and south Europe) were used in the study. The researchers showed that on average 76% of the butters could be correctly classified according to the region. For the determination of the country of origin the average success rate was higher, being 88%.

6.5.1.3 *Olive Oil*

Araghipour *et al.* used various partial least square-discriminant analysis models to determine the potential of PTR-MS measurements to classify olive oil according to geographical origin (country, region and district) [32]. One hundred and ninety-two olive oil samples from five European countries (Cyprus, France, Greece, Italy and Spain) were investigated. The greatest success corresponded to classifying olive oil according to the country of origin (86% success rate). The researchers also managed to separate Italian samples of olive oil according to the regions with some success (74% success rate). On the district scale, the analysis was not so successful with only a 52% success rate being obtained.

6.5.1.4 *Roe*

Roe from sea urchin (*Evechinus chloroticus*) was collected from two different and extreme locations in New Zealand [91]. One region was in the north (Matheson Bay) and the other was in the south (Doubtful Sound). By analysing 35 distinct m/z ion signals from PTR-MS, which showed significant differences between locations, the researchers found that it was possible to discriminate between roe from the two areas. This is illustrated in Figure 6.7, which presents a PCA biplot of the mean scores and loadings. Furthermore, by measuring the signal from five spectral mass peaks the gender of southern sea urchins could be determined.

6.5.1.5 *Dry-Cured Ham*

A study involving dry-cured ham investigated the differences in the headspace VOC composition between hams of four protected designations of origin (PDO) from two countries, Spain (*Dehesa de Extremadura*) and Italy (*Prosciutto di San Daniele*, *Prosciutto di Parma* and *Prosciutto Toscano*) [92]. Using PCA of the mass spectral peaks the researchers were able to discriminate the hams according to their geographical region and production process (and therefore determine the final quality of the dry-cured hams). Furthermore, by using data mining methods it was possible to create classification models with a high success rate in cross validation (e.g. leave out a sample and check if the classification model obtained from the remaining samples can be used to successfully classify the sample left out). This investigation exploited the high mass resolution of PTR-TOF-MS (a PTR-TOF-MS 8000 from IONICON Analytik GmbH was employed) together with literature data available on dry-cured ham headspace volatile composition, to provide a tentative identification of the VOCs emitted. It was found that 2-ketones, aldehydes and alcohols were more abundant in the Spanish cured ham, while the Italian cured ham (and especially Parma and San Daniele) had the highest concentrations of terpenes.

In a follow-up study by the same research group, this methodology was used to investigate the effects of different pig rearing systems on the quality of the final product of dry-cured Iberian ham. In this work, a switchable reagent ion source was used to generate H_3O^+,

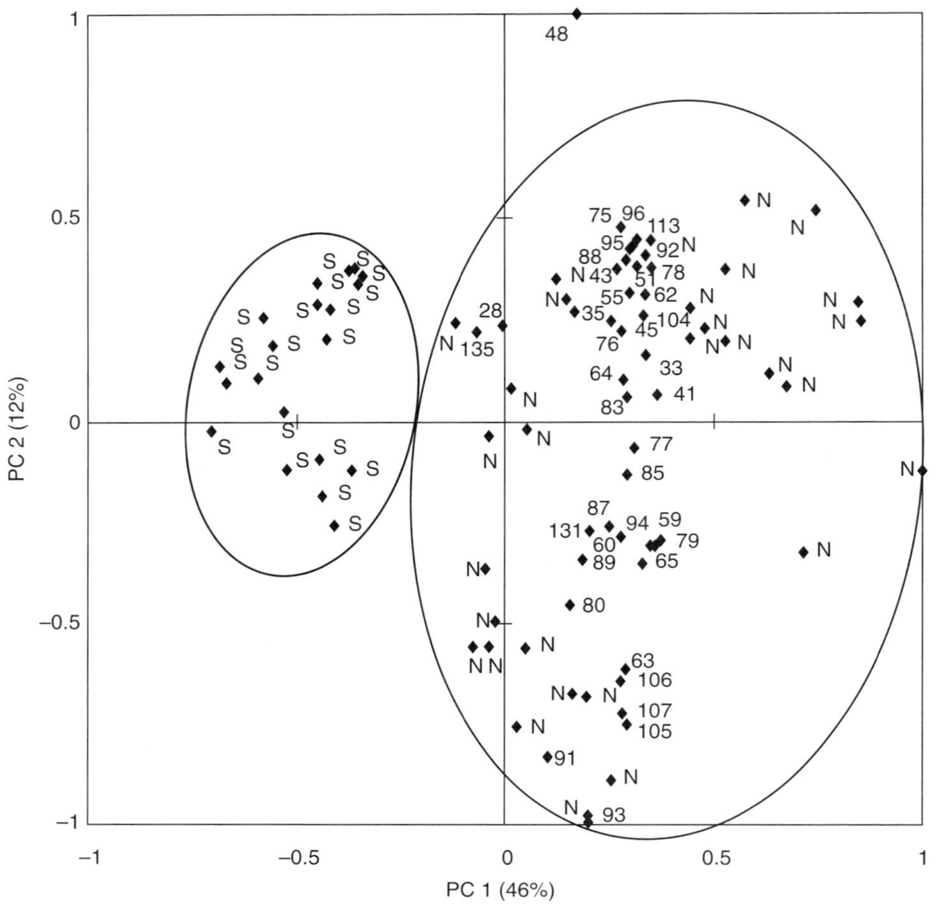

Figure 6.7 *Principal component analysis biplot of the mean scores and loadings (numbers represent m/z values) for PTR-MS data extracted from sea urchin roe taken from northern (N) and southern (S) locations in New Zealand. Reproduced from [91] with permission from Elsevier.*

O_2^+ or NO^+ as the reagent ions in order to improve the analytical performance and, in particular, to separate contributions of aldehydes from ketones. It was found that for pigs fattened outdoors on acorn and pasture, which gives the *Montanera* hams, higher concentrations of aldehydes and ketones and lower concentrations of sulfur-containing compounds were observed compared with the volatiles being emitted by *Campo* hams, which came from pigs fattened outdoors on a high-oleic (i.e. high in monounsaturated fats) concentrated feed [93].

6.5.1.6 Cumin Cheese

Another investigation of PDO-registered food involved a study of cumin cheese from the Netherlands [94]. The headspace VOCs from 30 cumin samples of artisan cheese from Leiden were compared with those from 29 cumin samples of various commercial Dutch brands without PDO protection. The study used both high-sensitivity PTR-MS for the fingerprinting of VOCs and high-resolution PTR-TOF-MS for the assignment of these VOCs and for distinguishing isobaric compounds. Sixty of the ions derived from the headspace of the cheese, ranging in m/z from 27 up to 96 and detected by the high-sensitivity (low mass resolution) PTR-MS instrument, were used in the fingerprint analysis. Of these 60 ions, 27 were significantly higher and 9 significantly lower in concentration for the artisan cheese when compared to those found from the commercial cheeses. These VOC fingerprints were subjected to a partial least squares-discriminant analysis which resulted in 96% of the PDO-protected cheeses and 100% of the commercial cumin cheese samples being correctly classified, that is, excellent separation between the two classes of cheese was obtained by the techniques used. A consumer test provided sensory evaluation of the cheeses and rated the overall flavour of the PDO cumin cheese higher than that of the more commercial varieties. This study yet again demonstrates the potential of using PTR-MS fingerprinting for the authentication of the identity, quality and integrity of food products.

6.5.2 Food Classification and Quality

PTR-MS fingerprinting of VOC emissions in combination with statistical analysis has also been used for the classification of food products, which is important in quality control. Studies have included animal fats (milk, cow fat, pig fat and poultry fat) and vegetable oils (coconut, palm and palm kernel) [95] (with an overall 89% of the samples being correctly classified), butter and butter oils [33] (with 84% of the samples being successfully classified into butter and butter oil classes), strawberries [96] (successful classification of strawberries according to the type of cultivar (variety)), animal fats, fish oil and recycled cooking oils [97] (with 92% of the samples being correctly identified from the PTR-MS mass spectra) and olive oils [22,98].

In Section 6.5.1, it was mentioned that PTR-MS fingerprinting has been used in the classification of olive oils [22]. In that study, the volatiles emitted from extra-virgin oil and defective (rancid) samples, as described by a sensory panel, were compared during thermo-oxidation. Oxidation over time and/or through thermal treatment results in an increase in VOC emissions, such as hexanal, which adversely affects odour and taste and therefore the quality of a food product. Using mass spectral fingerprints coupled with multivariate analysis techniques, it was demonstrated that it is possible to discriminate between normal and rancid oils, in agreement with results from the panel of trained judges. This study demonstrated the possibility of using PTR-MS and multivariate analysis as an automated means for distinguishing between healthy and defective oils. Another, and more recent, study of olive oils has investigated the classification of monovarietal (single variety) extra-virgin olive oils [98]. The paper describes the mass spectral fingerprinting of 30 samples of extra-virgin olive oils from 5 different varieties of olive fruit (*Arbequina*, *Cornicabra*, *Frantoio*, *Hojiblanca* and *Picual*) and the subsequent partial least square-discriminant analysis. This analysis showed that the VOCs emitted from these oils can be used in the assessment of product authenticity, that is, all five varieties could be distinguished from each other.

As a last example of the use of fingerprinting of VOCs for cultivar characterization, PTR-MS has also been used to explore the influence of genetics on food. Zini and co-workers used the VOCs emitted from ripe apples to draw a link to their genetic constitution [99].

Analysis using a technique known as quantitative trait loci was able to link specific peaks in the mass spectrum to particular regions of the genome.

6.5.3 Food Freshness and Ripening

6.5.3.1 Meat Degradation

An interesting application of PTR-MS, which was recognized early on by the inventors of the technique, is its use for monitoring the deterioration of food. The objective of the initial studies was simply to demonstrate the advantage of the rapid, real-time capabilities of PTR-MS for measurements of food freshness or spoilage when compared to bacteriological tests, which can take days to complete. In the earliest studies Lindinger and co-workers investigated meat degradation associated with microbial growth by using simple headspace analysis [42,100]. Various compounds were monitored, including methanethiol, dimethylsulfide, acetaldehyde, dimethylamine and methanol. Although the organosulfur compound dimethyl sulfide (C_2H_6S) showed significant increases in concentration as the meat decayed, methanethiol (CH_3SH) showed the most dramatic change. Up to 27 hours after the meat had been purchased and maintained at room temperature, the methanethiol was found to be at a background level of approximately a few pptv. However, after 27 hours the concentration of this compound showed a sharp rise, eventually reaching levels above 2 ppmv after 60 hours. The principal conclusion from these early studies was that headspace analysis has the potential to provide an indication of the degree of freshness or degradation of meat.

These very preliminary studies were followed up with more detailed work by Mayr *et al.*, who investigated the evolution of microbial spoilage for air- and vacuum-packaged meat (beef, pork and poultry) over a number of days [101, 102]. PTR-MS measurements showed that the time dependence for changes in the levels of a number of VOCs mirrored the bacterial growth curve. However these VOCs, which have the potential for acting as decay markers, were found to differ depending on the packaging conditions. For example, large differences in VOC emissions were observed depending on whether the meat was air- or vacuum-packed, which seemed to result from different types of bacteria living under aerobic and anaerobic conditions with different metabolic activities. Under aerobic conditions *Pseudomonas* flourishes, whereas under anaerobic conditions it is believed that the microfloras are composed mainly of *Lactobacilli*. A mass spectral peak at m/z 63, assigned by the researchers to be protonated dimethyl sulfide, was found to be the most clear-cut marker of decay for air-packaged beef and pork. The concentrations of C_4-esters were found to increase strongly under aerobic conditions, but stayed nearly constant (beef) or showed little enhancement (poultry) under anaerobic conditions. Lactic acid bacteria (so-called because they transform sugars into lactic acid) were found to produce acetaldehyde and ethanol, with ethanol being found to be the typical spoiling compound marker for vacuum-packed meat.

Shortly after taking these degradation measurements, a study was undertaken to determine if the rate of microbial contamination of meats could be reduced by exposing the meat to ozone [103]. Using the ion intensity at m/z 63 as a diagnostic marker of bacterial activity, it was observed that meat treated with ozone showed strongly reduced volatile emissions compared to untreated meat. This could be taken to suggest that the microbial activity had been suppressed by ozone treatment. However, these findings were not conclusive because independent microbial counting measurements showed no reduction in bacterial numbers. To explain these conflicting results, it was proposed that the ozone treatment inhibited physiological activity of the bacteria, but that the ozone concentration and length of treatment used were insufficient to cause bacterial death. In fact it was found that even high ozone doses did not result in sufficient reduction in microbial counts. If this is correct, then it raises doubts about the use of ozone as a meat treatment process, or alternatively it may indicate that measurement of VOC emissions is of limited utility in assessing bacterial levels in meat products. However, the latter proposal is doubtful given that there have been a number of PTR-MS [104–110] and other soft chemical ionization [111–118] studies that provide strong evidence to suggest that VOCs can be used as markers of bacterial growth. These bacterial studies, in agreement with the work by Mayr *et al.*, show that the VOCs emitted from different bacteria cultures are sufficiently different (m/z and intensities) such that discrimination of different bacteria should be possible [104–107]. More studies investigating the effects of ozone on the microbial spoilage of meat are therefore clearly needed before anything conclusive can be stated.

6.5.3.2 Fruit and Vegetables: Ripening, Storage and Monitoring

Food spoilage is also related to the shelf life of food. In particular, many fruits and vegetables have a short shelf life. PTR-MS has the potential for assessing the quality of fruit and vegetables whilst in storage, during shipment and on the shelf. Recently, Raseetha *et al.* have attempted to use PTR-MS to assess the quality of broccoli (*Brassica oleracea* L. *italica*), the rationale for this study being that broccoli is well known to have a short shelf life [119]. The relationship between the volatiles emitted and changes in the activity of enzymes that influence the quality of broccoli during post-harvest storage was investigated. By incubating crushed broccoli florets at different time intervals, the effects of wounding on VOC emission and enzyme activity were simulated. Using a partial least square regression model, a good correlation between the m/z spectral fingerprints and peroxidase enzymatic activity in broccoli florets was found. This correlation was maintained over several days of storage and incubation time after crushing.

Another study that dealt with post-harvest storage was the application of high-resolution PTR-TOF-MS to investigate apple ripening [26]. It is well known that as apples ripen, a variety of chemical and structural changes occur, which result in an increased production of aroma volatiles. Three apple cultivars (Golden Delicious, Braeburn and Gold Rush) were used for the study and the VOCs emitted from each were monitored during 25 days of post-harvest shelf life ripening. In addition to monitoring the product ions produced by reaction with H_3O^+, the impurity ion $O_2{}^+$ from the ion source was used to monitor ethylene via a non-dissociative charge transfer reaction resulting in $C_2H_4{}^+$ formation observed at m/z 28.031, which was well resolved from another ion at m/z 28.019 (protonated hydrogen cyanide). The discrimination between these two ions is shown in Figure 6.8, which provides another illustration of the benefits of using a high-resolution mass spectrometer. The detection of ethylene and accurate measurements of its concentration are important because of the role it plays in the formation of esters and alcohols. All apples showed a burst of ethylene post harvesting, with Golden Delicious apples being found to have a higher ethylene emission rate than those determined for Gold Rush and Braeburn. The researchers found that VOC production as a function of time was strongly cultivar dependent. Multivariate

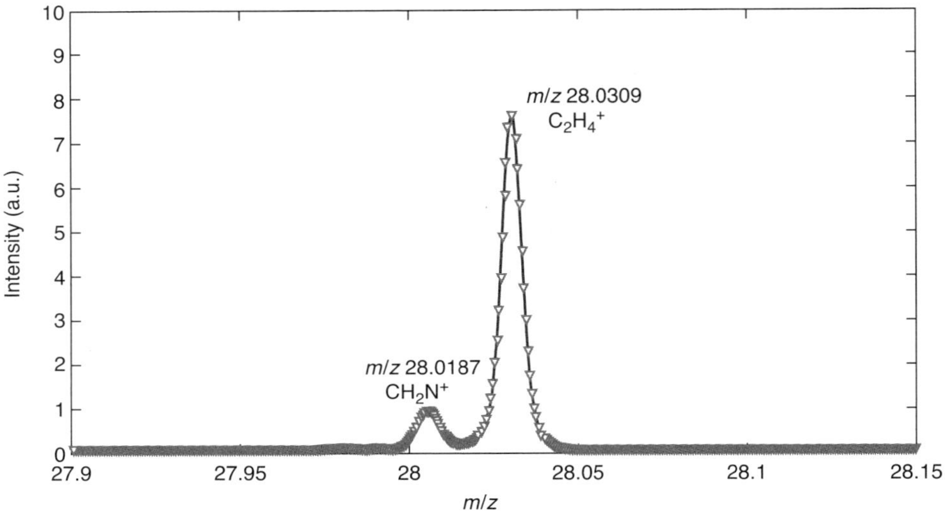

Figure 6.8 Part of the mass spectrum recorded for the headspace above an apple. This illustrates the high-resolution capabilities of PTR-TOF-MS by showing the separation of two compounds with the same nominal m/z value of 28 for the protonated compound. Reproduced with permission from [26]. Copyright © 2010 John Wiley & Sons, Ltd.

analysis of the VOC emission levels deduced from the PTR-TOF-MS measurements was used to visualize a mass spectral data set comprising several hundred peaks. From this, the researchers showed that the most abundant VOCs emitted as the apples were maturing could be separated into three major groups: (a) high *m/z* esters, terpenoids and methanol, (b) low *m/z* esters together with ethylene and 2-propanone/propanal and (c) carbonyl compounds, esters of intermediate molecular weight and *m/z* 149.095 (tentatively assigned to be protonated estragole).

PTR-MS has also been successfully used to monitor the changes in volatile profiles for 14 tomato varieties during ripening and storage [120]. The effect of cold storage was found to be clearly reflected in the volatile profile, and by using multivariate statistical analysis (PCA and cluster analysis) of the mass spectral peaks an unambiguous separation between the tomato varieties was possible. Furthermore, it was demonstrated that spectral fingerprinting and subsequent analysis permitted separation between the different tomato types (round truss, cocktail and cherry tomatoes). The researchers proposed that PTR-MS can rapidly and easily monitor and quantify VOCs related to ripening and/or senescence (biological ageing) such as methanol, ethanol, acetaldehyde, hexanal and hexenals. This has the potential for using measurements of volatile emissions for improving the breeding of new fruit and vegetable cultivars.

The influence of olive fruit ripening and oil extraction under different nitrogen regimes on the volatile profiles of extra-virgin olive oils have been described by Vezzaro *et al.* [121]. By applying PCA to the full mass spectral signature, they showed that the stage of ripening has a stronger effect on the volatile profiles than the use of nitrogen during oil extraction.

Boschetti *et al.* monitored the post-harvest ageing of strawberries, blackberries, raspberries, blueberries and white and red currants by the detection of VOCs in the headspace of the fruits [122]. The evolution of the VOCs during ageing was found to vary according to the type of berry in terms of different trends in emissions and relative concentrations. However, the quality of the fruit could be assessed by correlating relative concentrations of methanol and ethanol and other VOCs, such as esters and aldehydes. It was found that the specificity of each fruit could be exploited to assess the ripening stage or ageing by monitoring only a few selected volatile marker compounds.

6.5.3.3 Ripening of Cheese

The ripening of cheese has also been investigated by PTR-MS. A study by Aprea *et al.* [27] monitored the effects of ageing of Trentingrana cheese (which belongs to the cheese family Grana Padano) on the VOCs being emitted by means of online direct headspace analysis. This study was a follow-up of an earlier study by the same group which investigated the correlation between PTR-MS spectral fingerprinting and flavour profiling by a sensory panel [28] (see Section 6.4.2.1). Ripening was found to correlate strongly with the production of esters, which are known common constituents of cheese flavour, at m/z 89 (ethyl acetate/butanoic acid), 103 (methyl butanoate/isovaleric acid), 117 (hexanoic acid/C_6-esters) and 145 (ethyl hexanoate). These ion signals increased with ripening for the majority of the cheese samples. Using a combination of the mass spectral data with partial least squares analysis a systematic means of distinguishing ripe from young cheeses was found. As is always the case with PTR-MS investigations, it should be appreciated that without separate GC–MS measurements or the use of high mass resolution spectrometers for the separation of isobaric compounds (VOCs with the same nominal m/z but different molecular formula) the assignment of VOCs to a given m/z can only be tentative.

Another PTR-MS study on Trentingrana cheese investigated the influence of milk storage conditions on VOC profiles [24], and thereby in turn how these affect the quality of the cheese. Unsupervised multivariate analysis provided evidence to suggest that the cheese samples can be separated and classified on the basis of spectral fingerprinting, and that supervised data mining methods provided good classification. By using high-resolution PTR-TOF-MS together with isotopic abundances, the investigators were able to identify many of the VOCs being emitted by the cheese with extremely high levels of confidence.

6.5.4 Process Monitoring and Biochemical Processing

By using PTR-MS as an analytical fingerprinting tool, it should be possible for the quality of a food/drink product to be monitored, controlled and guaranteed during processing. As with coffee roasting, discussed in Section 6.4.1, controlling food processing requires good time resolution in the measurements.

PTR-TOF-MS has been proposed for use as a non-invasive quality control technology for monitoring the lactic acid fermentation of milk [123]. This work has demonstrated the outstanding ability of the instrument to monitor and identify VOCs which are either formed or depleted during the fermentation process. These compounds (in order of abundance) include acetaldehyde, acetone, diacetyl and acetoin. The time evolution of these four compounds as a function of incubation time is shown in Figure 6.9. Notice from these data the accuracy to which the m/z of a spectral peak can be determined when using

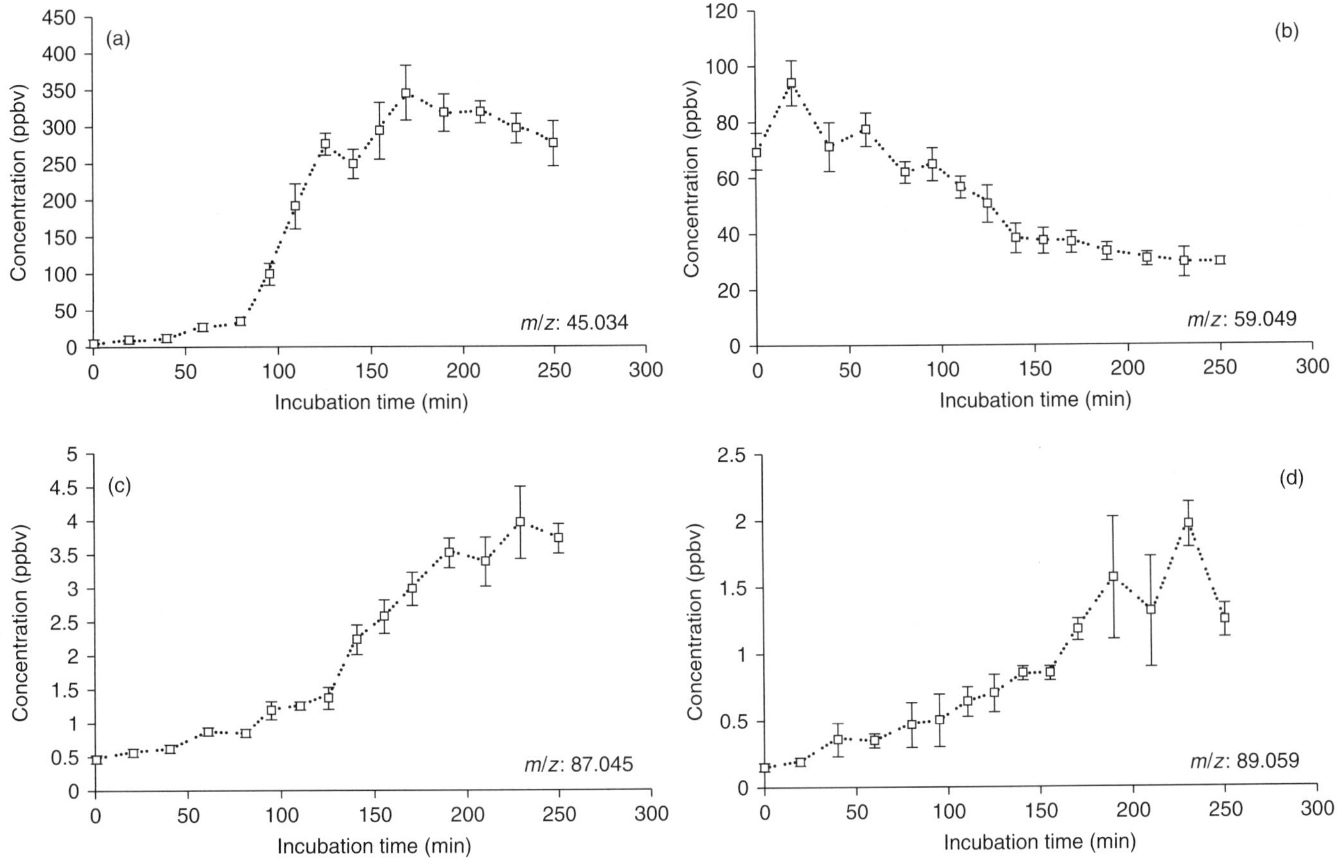

Figure 6.9 *Time evolution of the concentrations of (a) acetaldehyde, (b) acetone (c) diacetyl and (d) acetoin during lactic acid fermentation. Reproduced with permission from [123]. Copyright © 2010 John Wiley & Sons, Ltd.*

a high-resolution PTR-TOF-MS instrument, which leads to a high level of confidence in the assignment of an observed spectral mass peak to a given volatile compound. For example, protonated CO_2 (detected at *m/z* 44.997 (exact mass 44.998 Da)[6]), which is often but only weakly observed in PTR-MS instruments,[7] is well resolved from that of protonated acetaldehyde (detected at *m/z* 45.034 (exact mass 45.034)), and protonated diacetyl (detected at *m/z* 87.049 (exact mass 87.044)) is well separated from the isobaric-protonated compounds 2-pentanone and pentanal (detected at *m/z* 87.079 (exact mass 87.080)). That diacetyl can be distinguished from other possible compounds is important because it is a key compound responsible for a buttery flavour. In addition to the four major VOCs investigated, the study also provided information about the evolution of some minor VOCs including acetic acid, 2,3-pentanedione and ethanol. Dimethyl sulfide (detected at *m/z* 63.028 (exact mass 63.026)) and furfural (detected at *m/z* 97.028 (exact mass 97.028)) were also identified when the fermented milk became tainted as a result of severe heat treatment. This study highlights the potential of using VOC markers, such as dimethyl sulfide, as indicators of possible malpractices in the manufacturing steps.

In another study involving the same research group, PTR-TOF-MS was used to analyse the influence of milk-based supplements on the texture and the headspace concentration of endogenous VOCs in yogurt [25]. The effects of milk fat, solid non-fats and modified starch concentrations were investigated. The addition of modified tapioca starch was found to have no effect on the concentrations of the monitored VOCs, other than ethanol. In comparison, milk fats and solid non-fats significantly reduced the concentrations of several flavour VOCs in the headspace, including acetaldehyde, ketones (2-propanone, 2-butanone, 2-heptanone) and diketones (2,3-butanedione, 2,3-pentanedione). Solid non-fats were found to increase the headspace concentrations of flavour compounds (ethanol and acetoin). In addition to this work, the effects of milk treatment (both high pressure and transglutaminase treatments) on yoghurt have been investigated by Tsevdou *et al.* [124].

In a collaborative and comprehensive study involving one of the leading companies in the area of phytopharmaceuticals (Bionorica), Jaksch *et al.* developed an automated and objective method for the determination of the quality of various herb extracts based on PTR-MS headspace measurements [125]. The headspace of 83 batches of herb extract, of which 8 were known to be of 'bad' quality, was measured. The method employed is capable of separating out the consumer's acceptability of what is considered to be a 'good' or 'bad' extract. Following a field study at Bionorica, it was suggested that this method could be used for controlling and optimizing the production process – just as is now starting to be realized with coffee roasting [49].

A recently highlighted application of PTR-MS is its use in bioprocess monitoring [78, 106, 126, 127], which is important in the pharmaceutical industry. Although monitoring

---

[6] By 'exact mass', the authors are referring to accurate values for the *m/z* of the protonated molecules, which here are given to three decimal places.

[7] That protonated CO_2 is observed in PTR-MS is a surprise given its low proton affinity (541 kJ mol^{-1}), making proton transfer from H_3O^+ endothermic by 150 kJ mol^{-1}. Consequently, this ion cannot be formed in the drift tube because it would react with water and be lost from the system. It can only be formed in the region between the exit of the drift tube and the region where the ions are pulsed into the TOF tube. The energy needed to drive this reaction comes from the kinetic energy gained by the H_3O^+ ions in the lower pressure environment of the ion transfer optics region from the various voltage drops between lens elements.

of production processes in terms of quality and specification of the final product can be done offline, this is far from ideal because fast, real-time *in situ* measurements are required for advanced process control. Improved process control would result in more reliable and hence more cost-effective production. A potential non-invasive method to use in the bioprocess industry is to monitor the VOCs that are emitted as a by-product of microbial growth, which could be used as markers of the biological activity [106]. However, Clementschitsch and Bayer suggest that the applicability of sensor systems in bioprocess monitoring is limited with respect to online and *in situ* measurements because of a number of factors that are involved in such processing [126]. For example, there is a need for aseptic (sterile) conditions, which require that any sensor used for *in situ* measurements must not contaminate the bioprocess. Furthermore, a high number of analytes are usually involved with an extended dynamic range and therefore a sensor must be able to detect the majority of these for maximum information. Finally, any detector may be required to operate for long periods of time (potentially weeks) without any opportunity for recalibration. Given all of these requirements, PTR-MS is particularly well suited for monitoring bioprocessing via headspace analysis. This potential application has been reported in papers by Luchner *et al.* [127] and Singer *et al.* [128]. These studies report the development of a suitable sampling system that couples PTR-MS to a bioreactor for the detection of VOCs in the gas being released to improve the control of fermentation. Luchner *et al.* illustrated the value and reproducibility of this approach using three identical cultivations with a recombinant *Escherichia coli* strain. These studies have successfully demonstrated the use of PTR-MS for measuring VOC emissions from a fermentation process and highlight its potential use in advanced process control.

6.6 Outlook for Food Science and Technology Applications

PTR-MS is a powerful non-invasive tool for food research, food technology and for the food industry. The bulk of activities to-date has involved flavour studies and food quality. The high-time resolution capabilities have been used with great effect to monitor the rapid changes in VOC flavour profiles during the consumption of foods. There is still much more to do in terms of the analysis of food matrices, together with the further development of in nose-space analysis for individual flavour perception studies.

Despite the potential disadvantage of not being able to provide unique compound identification, PTR-MS can be used directly for monitoring food in quality control and provides excellent input data for chemometric analysis and data mining [25]. Owing to its real-time and rapid analytical capabilities, PTR-MS also has the ability to be employed in flavour analysis at the individual level during eating and drinking which, as Yeretzian *et al.* state [129], has the potential of directing research to ultimately match what a consumer wants with what he or she actually needs, and thereby in the future deliver nutritional value and health through food products that are preferred by the consumer. This is a major challenge for the food industry, but if it is achieved it would represent a major step forward since it would increase the value to consumers by personalizing flavour, diet and health. However, for this to be realized, a deep understanding of flavour perception and preference at the individual level is required. This is simply because the same food may register different aromas to different people, due to the specificities of their in-mouth environment and other factors discussed in this chapter. Moreover, a person may eat food in a different manner, leading to variations in the aroma profile reaching the nose.

It has been shown in this chapter how mass spectral fingerprinting in combination with appropriate statistical analysis can be used to provide classification of foods and drinks based on geographical location, variety, processing and time of harvesting, from which details on quality and authenticity can be determined. With the advent of high mass resolution PTR-MS instruments, the identification of VOCs has greatly improved and higher confidence in compound assignments can be made. This in turn will improve our understanding of the biochemical processes occurring within foods and drinks which is crucial for product development and quality control in the food industry.

Finally, we conclude by mentioning that PTR-MS studies are now beginning to provide a means to link food science and medical science through investigations of diet, health and the incidence of diet-related diseases. As an example of this, Aprea *et al.* have used high-resolution PTR-TOF-MS to analyse the breath in rats with non-alcoholic steatohepatitis (NASH) induced by a high-fat diet [130]. Such studies are difficult owing to factors such as high inter-individual variability in VOC emissions and stress-induced VOCs resulting from the breath sampling procedure. Nevertheless, by using controls (i.e. rats without NASH fed on a standard diet) the researchers claim that they could identify markers in their mass spectra corresponding to methanol, sulfur compounds and ammonia, which are directly related to diet and specific pathologic conditions. Providing that these results are confirmed then the prospects for non-invasive breath analysis as a medical diagnostic tool becomes closer to reality. The use of PTR-MS for medical science, and in particular for breath analysis, is discussed in more detail in the next chapter.

References

1. Biasioli, F., Yeretzian, C., Märk, T. D. *et al.* (2011) Direct-injection mass spectrometry adds the time dimension to (B)VOC analysis. *Trends Anal. Chem.* **30**, 1003.
2. Biasioli, F., Yeretzian, C., Gasperi, F., Märk, T. D. (2011) PTR-MS monitoring of VOCs and BVOCs in food science and technology. *Trends Anal. Chem.* **30**, 968.
3. Buhr, K., van Ruth, S., Delahunty, C. (2002) Analysis of volatile flavour compounds by proton transfer reaction-mass spectrometry: fragmentation patterns and discrimination between isobaric and isomeric compounds. *Int. J. Mass Spectrom.* **221**, 1.
4. Aprea, E., Biasioli, F., Märk, T. D., Gasperi, F. (2007) PTR-MS study of esters in water and water/ethanol solutions: fragmentation patterns and partition coefficients. *Int. J. Mass Spectrom.* **262**, 114.
5. Warneke, C., de Gouw, J. A., Kuster, W. C. *et al.* (2003) Validation of atmospheric VOC measurements by proton-transfer-reaction mass spectrometry using a gas-chromatographic pre-separation method. *Environ. Sci. Technol.* **37**, 2494.
6. Pozo-Bayón, M.-Á., Schirlé-Keller, J.-P., Reineccius, G. A. (2008) Determining specific food volatiles contributing to PTR-MS ion profiles using GC-EI-MS. *J. Agric. Food Chem.* **56**, 5278.
7. Aprea, E., Biasioli, F., Carlin, S. *et al.* (2008) Monitoring benzene formation from benzoate in model systems by proton transfer reaction-mass spectrometry. *Int. J. Mass Spectrom.* **275**, 117.

8. Boscaini, E., van Ruth, S., Biasioli, F. *et al.* (2003) Gas chromatography-olfactometry (GC–O) and proton transfer reaction-mass spectrometry (PTR-MS) analysis of the flavor profile of grana padano, parmigiano reggiano, and grana trentino cheeses. *J. Agric. Food Chem.* **51**, 1782.
9. Mateus, M. L., Lindinger, C., Gumy, J. C., Liardon, R. (2007) Release kinetics of volatile organic compounds from roasted and ground coffee: online measurements by PTR-MS and mathematical modeling. *J. Agric. Food Chem.* **55**, 10117.
10. Aprea, E., Biasioli, F., Carlin, S. *et al.* (2007) Rapid white truffle headspace analysis by proton transfer reaction mass spectrometry and comparison with solid-phase microextraction coupled with gas chromatography/mass spectrometry. *Rapid Commun. Mass Spectrom.* **21**, 2564.
11. Aprea, E., Biasioli, F., Carlin, S. *et al.* (2009) Investigation of volatile compounds in two raspberry cultivars by two headspace techniques: solid-phase microextraction/gas chromatography-mass spectrometry (SPME/GC–MS) and proton-transfer reaction-mass spectrometry (PTR-MS). *J. Agric. Food Chem.* **57**, 4011.
12. Gasperi, F., Aprea, E., Biasioli, F. *et al.* (2009) Effects of supercritical CO_2 and N_2O pasteurization on the quality of fresh apple juice. *Food Chem.* **115**, 129.
13. Pozo-Bayón, M.-Á., Martín-Álvarez, P. J., Reineccius, G. A. (2009) Monitoring changes in the volatile profile of cheese crackers during storage using GC–MS and PTR–MS. *Flavour Fragr. J.* **24**, 133.
14. Lindinger, C., Pollien, P., Ali, S. *et al.* (2005) Unambiguous identification of volatile organic compounds by proton-transfer reaction mass spectrometry coupled with GC/MS. *Anal. Chem.* **77**, 4117.
15. Pollien, P., Lindinger, C., Yeretzian, C., Blank, I. (2003) Proton transfer reaction mass spectrometry, a tool for on-line monitoring of acrylamide formation in the headspace of maillard reaction systems and processed food. *Anal. Chem.* **75**, 5488.
16. Robert, F., Vuataz, G., Pollien, P. *et al.* (2004) Acrylamide formation from asparagine under low-moisture maillard reaction conditions. 1. Physical and chemical aspects in crystalline model systems. *J. Agric. Food Chem.* **52**, 6837.
17. Märk, J., Pollien, P., Lindinger, C. *et al.* (2006) Quantitation of furan and methylfuran formed in different precursor systems by proton transfer reaction mass spectrometry. *J. Agric. Food Chem.* **54**, 2786.
18. Lendl, B., Karlberg, B. (2005) Advancing from unsupervised, single variable-based to supervised, multivariate-based methods: a challenge for qualitative analysis. *Trends Anal. Chem.* **24**, 488.
19. Gasperi, F., Gallerani, G., Boschetti, A. *et al.* (2001) The mozzarella cheese flavour profile: a comparison between judge panel analysis and proton transfer reaction mass spectrometry. *J. Sci. Food Agric.* **81**, 357.
20. Biasioli, F., Gasperi, F., Aprea, E. *et al.* (2003) Fingerprinting mass spectrometry by PTR-MS: heat treatment vs. pressure treatment of red orange juice – a case study. *Int. J. Mass Spectrom.* **223–224**, 343.
21. Gallardo-Escamilla, F. J., Kelly, A. L., Delahunty, C. M. (2007) Mouthfeel and flavour of fermented whey with added hydrocolloids. *Int. Dairy J.* **17**, 308.
22. Aprea, E., Biasioli, F., Sani, G. *et al.* (2006) Proton transfer reaction-mass spectrometry (PTR-MS) headspace analysis for rapid detection of oxidative alteration of olive oil. *J. Agric. Food Chem.* **54**, 7635.

23. Spitaler, R., Araghipour, N., Mikoviny, T. *et al.* (2007) PTR-MS in enology: advances in analytics and data analysis. *Int. J. Mass Spectrom.* **266**, 1.
24. Fabris, A., Biasioli, F., Granitto, P. M. *et al.* (2010) PTR-TOF-MS and data-mining methods for rapid characterisation of agro-industrial samples: influence of milk storage conditions on the volatile compounds profile of Trentingrana cheese. *J. Mass. Spectrom.* **45**, 1065.
25. Soukoulis, C., Biasioli, F., Aprea, E. *et al.* (2012) PTR-TOF-MS analysis for influence of milk base supplementation on texture and headspace concentration of endogenous volatile compounds in yogurt. *Food Bioprocess Technol.* **5**, 2085.
26. Soukoulis, C., Cappellin, L., Aprea, E. *et al.* (2013) PTR-ToF-MS, a novel, rapid, high sensitivity and non-invasive tool to monitor volatile compound release during fruit post-harvest storage: the case study of apple ripening. *Food Bioprocess Technol.* **6**, 2831.
27. Aprea, E., Biasioli, F., Gasperi, F. *et al.* (2007) Assessment of Trentingrana cheese ageing by proton transfer reaction-mass spectrometry and chemometrics. *Int. Dairy J.* **17**, 226.
28. Biasioli, F., Gasperi, F., Aprea, E. *et al.* (2006) Correlation of PTR-MS spectral fingerprints with sensory characterisation of flavour and odour profile of "Trentingrana" cheese. *Food Qual. Pref.* **17**, 63.
29. Biasioli, F., Gasperi, F., Aprea, E. *et al.* (2003) Coupling proton transfer reaction-mass spectrometry with linear discriminant analysis: a case study. *J. Agric. Food Chem.* **51**, 7227.
30. Heenan, S. P., Dufour, J.-P., Hamid, N. *et al.* (2009) Characterisation of fresh bread flavour: relationships between sensory characteristics and volatile composition. *Food Chem.* **116**, 249.
31. Maçatelli, M., Akkermans, W., Koot, A. *et al.* (2009) Verification of the geographical origin of European butters using PTR-MS. *J. Food Compos. Anal.* **22**, 169.
32. Araghipour, N., Colineau, J., Koot, A. *et al.* (2008) Geographical origin classification of olive oils by PTR-MS. *Food Chem.* **108**, 374.
33. van Ruth, S. M., Koot, A., Akkermans, W. *et al.* (2008) Butter and butter oil classification by PTR-MS. *Eur. Food Res. Technol.* **227**, 307.
34. Spitaler, R., Araghipour, N., Mikoviny, T. *et al.* (2007) PTR-MS in enology: advances in analytics and data analysis. *Int. J. Mass Spectrom.* **266**, 1.
35. Cappellin, L., Aprea, E., Granitto, P. *et al.* (2012) Linking GC-MS and PTR-TOF-MS fingerprints of food samples. *Chemom. Intell. Lab. Syst.* **118**, 301.
36. Cappellin, L., Biasioli, F., Granitto, P. M. *et al.* (2011) On data analysis in PTR-TOF-MS: from raw spectra to data mining. *Sens. Actuat. B.* **155**, 183.
37. Cappellin, L., Soukoulis, C., Aprea, E. *et al.* (2012) PTR-ToF-MS and data mining methods: a new tool for fruit metabolomics. *Metabolomics* **8**, 761.
38. Granitto, P. M., Biasioli, F., Furlanello, C., Gasperi, F. (2008) Efficient feature selection for PTR-MS fingerprinting of agroindustrial products. In: *Artificial Neural Networks – ICANN 2008*, vol. 5164, *Lecture Notes in Computer Science Series*, Springer-Verlag, pp. 42–51.
39. Taucher, J., Lagg, A., Hansel, A. *et al.* (1995) Methanol in human breath. *Alcohol. Clin. Exp. Res.* **19**, 1147.

40. Lindinger, W., Taucher, J., Jordan, A. *et al.* (1997) Endogenous production of methanol after the consumption of fruit. *Alcohol. Clin. Exp. Res.* **21**, 939.
41. Taucher, J., Hansel, A., Jordan, A., Lindinger, W. (1996) Analysis of compounds in human breath after ingestion of garlic using proton-transfer-reaction mass spectrometry. *J. Agric. Food Chem.* **44**, 3778.
42. Lindinger, W., Hansel, A., Jordan, A. (1998) On-line monitoring of volatile organic compounds at pptv levels by means of proton-transfer-reaction mass spectrometry (PTR-MS) medical applications, food control and environmental research. *Int. J. Mass Spectrom. Ion Processes* **173**, 191.
43. Yeretzian, C., Jordan, A., Badoud, R., Lindinger, W. (2002) From the green bean to the cup of coffee: investigating coffee roasting by on-line monitoring of volatiles. *Eur. Food Res. Technol.* **214**, 92.
44. Lindinger, C., Labbe, D., Pollien, P. *et al.* (2008) When machine tastes coffee: instrumental approach to predict the sensory profile of espresso coffee. *Anal. Chem.* **80**, 1574.
45. Pendergrast, M. (2009) Coffee second only to oil. *Tea Coffee Trade J.* **181**, 38.
46. Yeretzian, C., Jordan, A., Lindinger, W. (2003) Analysing the headspace of coffee by proton-transfer-reaction mass-spectrometry. *Int. J. Mass Spectrom.* **223–224**, 115.
47. Lindinger, C., Yeretzian, C., Blank, I. (2009) When machine tastes coffee: successful prediction of coffee sensory profiles by instrumental methods based on on-line PTR-MS. *Chimia* **63**, 292.
48. Yeretzian, C., Wieland, F., Gloess, A. N. *et al.* (2012) On-line process control of the roast degree of coffee. *New Food* **15**, 22.
49. Wieland, F., Gloess, A. N., Keller, M. *et al.* (2012) On-line process control of the roast degree of coffee. *Chimia* **66**, 443.
50. Wieland, F., Gloess, A. N., Keller, M. *et al.* (2012) Online monitoring of coffee roasting by proton transfer reaction time-of-flight mass spectrometry (PTR-ToF-MS): towards a real-time process control for a consistent roast profile. *Anal. Bioanal. Chem.* **402**, 2531.
51. Jublot, L. (2011) Pushing boundaries of flavour analysis with GC × GC–TOFMS and PTR-MS. *New Food* **14**, 19.
52. Pozo-Bayón, M.-Á., Santos, M., Martín-Álvarez, P. J., Reineccius, G. (2009) Influence of carbonation on aroma release from liquid systems using an artificial throat and a proton transfer reaction–mass spectrometric technique (PTR–MS). *Flavor Fragr. J.* **24**, 226.
53. Boscaini, E., Mikoviny, T., Wisthaler, A. *et al.* (2004) Characterization of wine with PTR-MS. *Int. J. Mass Spectrom.* **239**, 215.
54. Déléris, I., Saint-Eve, A., Guo, Y. *et al.* (2011) Impact of swallowing on the dynamics of aroma release and perception during the consumption of alcoholic beverages. *Chem. Senses* **36**(8), 701.
55. Menella, J. A., Jagnow, C. P., Beauchamp, G. K. (2001) Prenatal and postnatal flavor learning by human infants. *Pediatrics* **107**, e88.
56. van Ruth, S. M., Floris, V., Fayoux, S. (2006) Characterisation of the volatile profiles of infant formulas by proton transfer reaction-mass spectrometry and gas chromatography–mass spectrometry. *Food Chem.* **98**, 343.

57. van Ruth, S. M., Boscaini, E., Mayr, D. *et al.* (2003) Evaluation of three gas chromatography and two direct mass spectrometry techniques for aroma analysis of dried red bell peppers. *Int. J. Mass Spectrom.* **223**, 55.
58. van Ruth, S. M., Dings, L., Buhr, K., Posthumus, M. A. (2004) In vitro and in vivo volatile flavour analysis of red kidney beans by proton transfer reaction–mass spectrometry. *Food Res. Int.* **37**, 785.
59. van Ruth, S. M., Dings, L., Aprea, E., Odake, S. (2005) Comparison of volatile flavour profiles of kidney beans and soybeans by GC–MS and PTR-MS. *Food Sci. Technol. Res.* **11**, 63.
60. Løkke, M. M., Edelenbos, M., Larsen, E., Feilberg, A. (2012) Investigation by real time proton transfer reaction-mass spectrometry (PTR-MS) of volatiles emitted from freshly cut onions (*Allium cepa* L.). *Sensors* **12**, 16060.
61. Gallardo-Escamilla, F. J., Kelly, A. L., Delahunty, C. M. (2005) Sensory characteristics and related volatile flavor compound profiles of different types of whey. *J. Dairy Sci.* **88**, 2689.
62. Gallardo-Escamilla, F. J., Kelly, A. L., Delahunty, C. M. (2005) Influence of starter culture on flavor and headspace volatile profiles of fermented whey and whey produced from fermented milk. *J. Dairy Sci.* **88**, 3745.
63. Boamfa, E. I., Steeghs, M. M. L., Cristescu, S. M., Harren, F. J. M. (2004) Trace gas detection from fermentation processes in apples; an intercomparison study between proton-transfer-reaction mass spectrometry and laser photoacoustics. *Int. J. Mass Spectrom.* **239**, 193.
64. Brown, P., Watts, P., Märk, T. D., Mayhew, C. A. (2010) Proton transfer reaction mass spectrometry investigations on the effects of reduced electric field and reagent ion internal energy on product ion branching ratios for a series of saturated alcohols. *Int. J. Mass Spectrom.* **294**, 103.
65. Carbone, F., Mourgues, F., Biasioli, F. *et al.* (2006) Development of molecular and biochemical tools to investigate fruit quality traits in strawberry elite genotypes. *Mol. Breeding* **18**, 127.
66. Roberts, D. D., Pollien, P., Antille, N. *et al.* (2003) Comparison of nosespace, headspace, and sensory intensity ratings for the evaluation of flavor absorption by fat. *J. Agric. Food Chem.* **51**, 3636.
67. van Ruth, S. M., de Witte, L., Uriarte, A. R. (2004) Volatile flavor analysis and sensory evaluation of custard desserts varying in type and concentration of carboxymethyl cellulose. *J. Agric. Food Chem.* **52**, 8105.
68. Mei, J. B., Reineccius, G. A., Knighton, W. B., Grimsrud, E. P. (2004) Influence of strawberry yogurt composition on aroma release. *J. Agric. Food Chem.* **52**, 6267.
69. Mestres, M., Moran, N., Jordan, A., Buettner, A. (2005) Aroma release and retronasal perception during and after consumption of flavored whey protein gels with different textures. 1. In vivo release analysis. *J. Agric. Food Chem.* **53**, 403.
70. Mestres, M., Kieffer, R., Buettner, A. (2006) Release and perception of ethyl butanoate during and after consumption of whey protein gels: relation between textural and physiological parameters. *J. Agric. Food Chem.* **54**, 1814.
71. Aprea, E., Biasioli, F., Gasperi, F. *et al.* (2006) *In vivo* monitoring of strawberry flavour release from model custards: effect of texture and oral processing. *Flavour Fragrance J.* **21**, 53.

72. van Ruth, S. M., Frasnelli, J., Carbonell, L. (2008) Volatile flavour retention in food technology and during consumption: juice and custard examples. *Food Chem.* **106**, 1385.
73. Frasnelli, J., van Ruth, S., Kriukova, I., Hummel, T. (2005) Intranasal concentrations of orally administered flavors. *Chem. Senses* **30**, 575.
74. Buettner, A., Beauchamp, J. (2010) Chemical input – sensory output: diverse modes of physiology–flavour interaction. *Food Qual. Prefer.* **21**, 915.
75. Hansson, A., Giannouli, P., van Ruth, S. J. (2003) The influence of gel strength on aroma release from pectin gels in a model mouth and in vivo, monitored with proton-transfer-reaction mass spectrometry. *Agric. Food Chem.* **51**, 4732.
76. Boland, A. B., Buhr, K., Giannouli, P., van Ruth, S. M. (2004) Influence of gelatin, starch, pectin and artificial saliva on the release of 11 flavour compounds from model gel systems. *Food Chem.* **86**, 401.
77. Boland, A. B., Delahunty, C. M., van Ruth, S. M. (2006) Influence of the texture of gelatin gels and pectin gels on strawberry flavour release and perception. *Food Chem.* **96**, 452.
78. Déléris, I., Saint-Eve, A., Dakowski, F. *et al.* (2011) The dynamics of aroma release during consumption of candies of different structures, and relationship with temporal perception. *Food Chem.* **127**, 1615.
79. Tietz, M., Buettner, A., Conde-Petit, B. (2008) Interaction between starch and aroma compounds as measured by proton transfer reaction mass spectrometry (PTR-MS). *Food Chem.* **108**, 1192.
80. Tietz, M., Buettner, A., Conde-Petit, B. (2008) Changes in structure and aroma release from starch–aroma systems upon α-amylase addition. *Eur. Food Res. Technol.* **227**, 1439.
81. Siefarth, C., Tyapkova, O., Beauchamp, J. *et al.* (2011) Influence of polyols and bulking agents on flavour release from low-viscosity solutions. *Food Chem.* **129**, 1462.
82. Siefarth, C., Tyapkova, O., Beauchamp, J. *et al.* (2011) Mixture design approach as a tool to study in vitro flavor release and viscosity interactions in sugar-free polyol and bulking agent solutions. *Food Res. Int.* **44**, 3202.
83. Mayr, D., Märk, T., Lindinger, W. *et al.* (2003) Breath-by-breath analysis of banana aroma by proton transfer reaction mass spectrometry. *Int. J. Mass Spectrom.* **223–224**, 743.
84. van Ruth, S. M., Buhr, K. (2004) Influence of mastication rate on dynamic flavour release analysed by combined model mouth/proton transfer reaction–mass spectrometry. *Int. J. Mass Spectrom.* **239**, 187.
85. Landy, P., Pollien, P., Rytz, A. *et al.* (2007) Model studies on the release of aroma compounds from structured and nonstructured oil systems using proton-transfer reaction mass spectrometry. *J. Agric. Food Chem.* **55**, 1915.
86. Lasekan, O., Otto, S. (2009) In vivo analysis of palm wine (*Elaeis guineensis*) volatile organic compounds (VOCs) by proton transfer reaction mass spectrometry. *Int. J. Mass Spectrom.* **282**, 45.
87. Sostmann, K., Potineni, R., Blancher, G., *et al.* (2010) Expressions of multidisciplinary flavour research, in *Proceedings of the 12th Weurman Symposium*, Interlaken,

Switzerland, 2008, (Eds. I. Blank, M Wüst, C. Yeretzian) Zürcher Hochschule für Angewandte Wissenschaften, Winterthur, pp. 154–156.

88. Odake, S., Shimamura, T., Akuzawa, R. *et al.* (2006) Volatile compounds of Wagyu (Japanese black cattle) beef analysed by PTR-MS. *Dev. Food Sci.* **43**, 29.
89. Heenan, S., Soukoulis, C., Silcock, P. *et al.* (2012) PTR-TOF-MS monitoring of in vitro and in vivo flavour release in cereal bars with varying sugar composition. *Food Chem.* **131**, 477.
90. Ting, V. J. L., Soukoulis, C., Silcock, P. *et al.* (2012) In vitro and in vivo flavor release from intact and fresh-cut apple in relation with genetic, textural, and physicochemical parameters. *J. Food Sci.* **77**, 1226.
91. Phillips, K., Niimi, J., Hamid, N. *et al.* (2010) Sensory and volatile analysis of sea urchin roe from different geographical regions in New Zealand. *LWT – Food Sci. Technol.* **43**, 202.
92. del Pulgar, J. S., Soukoulis, C., Biasioli, F. *et al.* (2011) Rapid characterization of dry cured ham produced following different PDOs by proton transfer reaction time of flight mass spectrometry (PTR-ToF-MS). *Talanta* **85**, 386.
93. del Pulgar, J., Soukoulis, C., Carrapiso, A. I. *et al.* (2013) Effect of the pig rearing system on the final volatile profile of Iberian dry-cured ham as detected by PTR-ToF-MS. *Meat Sci.* **93**, 420.
94. Galle, S. A., Koot, A., Soukoulis, C. *et al.* (2011) Typicality and geographical origin markers of protected origin cheese from the Netherlands revealed by PTR-MS. *J. Agric. Food Chem.* **59**, 2554.
95. van Ruth, S. M., Villegas, B., Akkermans, W. *et al.* (2010) Prediction of the identity of fats and oils by their fatty acid, triacylglycerol and volatile compositions using PLS-DA. *Food Chem.* **118**, 948.
96. Granitto, P. M., Biasioli, F., Aprea, E. *et al.* (2007) Rapid and non-destructive identification of strawberry cultivars by direct PTR-MS headspace analysis and data mining techniques. *Sens. Actuat. B.* **121**, 379.
97. van Ruth, S. M., Rozijn, M., Koot, A. *et al.* (2010) Authentication of feeding fats: classification of animal fats, fish oils and recycled cooking oils. *Animal Feed Sci. Technol.* **155**, 65.
98. Ruiz-Samblás, C., Tres, A., Koot, A. *et al.* (2012) Proton transfer reaction-mass spectrometry volatile organic compound fingerprinting for monovarietal extra virgin olive oil identification. *Food Chem.* **134**, 589.
99. Zini, E., Biasioli, F., Gasperi, F. *et al.* (2005) QTL mapping of volatile compounds in ripe apples detected by proton transfer reaction-mass spectrometry. *Euphytica* **145**, 269.
100. Lindinger, W., Hansel, A., Jordan, A. (1998) Proton-transfer-reaction mass spectrometry (PTR-MS): on-line monitoring of volatile organic compounds at pptv levels. *Chem. Soc. Rev.* **27**, 347.
101. Mayr, D., Margesin, R., Klingsbichel, E. *et al.* (2003) Rapid detection of meat spoilage by measuring volatile organic compounds by using proton transfer reaction mass spectrometry. *Appl. Environ. Microbiol.* **69**, 4697.
102. Mayr, D., Margesin, R., Schinner, F., Märk, T. D. (2003) Detection of the spoiling of meat using PTR–MS. *Int. J. Mass Spectrom.* **223–224**, 229.

103. Jaksch, D., Margesin, R., Mikoviny, T. *et al.* (2004) The effect of ozone treatment on the microbial contamination of pork meat measured by detecting the emissions using PTR-MS and by enumeration of microorganisms. *Int. J. Mass Spectrom.* **239**, 209.
104. Critchley, A. D. J., Elliott, T. S., Harrison, G. *et al.* (2004) The proton transfer reaction mass spectrometer and its use in medical science: applications to drug assays and the monitoring of bacteria. *Int. J. Mass Spectrom.* **239**, 235.
105. Mayrhofer, S., Mikoviny, T., Waldhuber, S. *et al.* (2006) Microbial community related to volatile organic compound (VOC) emission in household biowaste. *Environ. Microbiol.* **8**, 1960.
106. Bunge, M., Araghipour, N., Mikoviny, T. *et al.* (2008) On-line monitoring of microbial volatile metabolites by proton transfer reaction-mass spectrometry. *Appl. Environ. Microbiol.* **74**, 2179.
107. O'Hara, M., Mayhew, C. A. (2009) A preliminary comparison of volatile organic compounds in the headspace of cultures of *Staphylococcus aureus* grown in nutrient, dextrose and brain heart bovine broths measured using a proton transfer reaction mass spectrometer. *J. Breath Res.* **3**, 027001.
108. Kai, M., Crespo, E., Cristescu, S. M. *et al.* (2010) *Serratia odorifera*: analysis of volatile emission and biological impact of volatile compounds on *Arabidopsis thaliana*. *Appl. Microbiol. Biotechnol.* **88**, 965.
109. Crespo, E., Cristescu, S. M., deRonde, H. *et al.* (2011) Proton transfer reaction mass spectrometry detects rapid changes in volatile metabolite emission by *Mycobacterium smegmatis* after the addition of specific antimicrobial agents. *J. Microbiol. Methods* **86**, 8.
110. Crespo, E., de Ronde, H., Kuijper, S. *et al.* (2012) Potential biomarkers for identification of mycobacterial cultures by proton transfer reaction–mass spectrometry analysis. *Rapid Commun. Mass Spectrom.* **26**, 679.
111. Custer, T. G., Wagner, W. P., Kato, S. *et al.* (2003) Potential of on-line CIMS for bioprocessing monitoring. *Biotechnol. Prog.* **19**, 1355.
112. Wang, T., Smith, D., Španěl, P. (2004) Selected ion flow tube, SIFT, studies of the reactions of H_3O^+, NO^+ and $O_2{}^+$ with compounds released by *Pseudomonas* and related bacteria. *Int. J. Mass Spectrom.* **233**, 245.
113. Scotter, J. M., Langford, V. S., Wilson, P. F. *et al.* (2005) Real-time detection of common microbial volatile organic compounds from medically important fungi by selected ion flow tube-mass spectrometry (SIFT-MS). *J. Microbiol. Methods* **63**, 127.
114. Carroll, W., Lenney, W., Wang, T. *et al.* (2005) Detection of volatile compounds emitted by pseudomonas aeruginosa using selected ion flow tube mass spectrometry. *Pediatr. Pulmonol.* **39**, 452.
115. Allardyce, R. A., Langford, V. S., Hill, A., Murdoch, D. R. (2006) Detection of volatile metabolites produced by bacterial growth in blood culture media by selected ion flow tube mass spectrometry (SIFT-MS). *J. Microbiol. Methods.* **65**, 361.
116. Scotter, J. M., Allardyce, R. A., Langford, V. S. *et al.* (2006) The rapid evaluation of bacterial growth in blood cultures by selected ion flow tube–mass spectrometry (SIFT-MS) and comparison with the BacT/ALERT automated blood culture system. *J. Microbiol. Methods.* **65**, 628.

117. Dolch, M. E., Hornuss, C., Klocke, C. *et al.* (2012) Volatile organic compound analysis by ion molecule reaction mass spectrometry for Gram-positive bacteria differentiation. *Eur. J. Clin. Microbiol. Infect. Dis.* **31**(11), 3007.
118. Dolch, M. E., Hornuss, C., Klocke, C. *et al.* (2012) Volatile compound profiling for the identification of gram-negative bacteria by ion molecule reaction-mass spectrometry. *J. App. Microbiol.* **113**, 1097.
119. Raseetha, S., Heenan, S. P., Oey, I. *et al.* (2011) A new strategy to assess the quality of broccoli (*Brassica oleracea* L. *italica*) based on enzymatic changes and volatile mass ion profile using proton transfer reaction mass spectrometry (PTR-MS). *Innovat. Food Sci. Emerg. Tech.* **12**, 197.
120. Farneti, B., Cristescu, S. M., Costa, G. *et al.* (2012) Rapid tomato volatile profiling by using proton-transfer reaction mass spectrometry (PTR-MS). *J. Food Sci.* **55**, C551.
121. Vezzaro, A., Boschetti, A., Dell'Anna, R. *et al.* (2011) Influence of olive (*cv* Grignano) fruit ripening and oil extraction under different nitrogen regimes on volatile organic compound emissions studied by PTR-MS technique. *Anal. Bioanal. Chem.* **399**, 2571.
122. Boschetti, A., Biasioli, F., van Opbergen, M. *et al.* (1999) PTR-MS real time monitoring of the emission of volatile organic compounds during postharvest aging of berryfruit. *Postharvest Biol. Technol.* **17**, 143.
123. Soukoulis, C., Aprea, E., Biasioli, F. *et al.* (2010) Proton transfer reaction time-of-flight mass spectrometry monitoring of the evolution of volatile compounds during lactic acid fermentation of milk. *Rapid Commun. Mass Spectrom.* **24**, 2127.
124. Tsevdou, M., Soukoulis, C., Cappellin, L. *et al.* (2013) Monitoring the effect of high pressure and transglutaminase treatment of milk on the evolution of flavour compounds during lactic acid fermentation using PTR-ToF-MS. *Food Chem.* **138**, 2159.
125. Jaksch, D., Hartungen, E., Mikoviny, T. *et al.* (2004) Quality control of herb extract using PTR-MS. *Int. J. Mass Spectrom.* **239**, 203.
126. Clementschitsch, F., Bayer, K. (2006) Improvement of bioprocess monitoring: development of novel concepts. *Microb. Cell Fact.* **5**, 1.
127. Luchner, M., Gutmann, R., Bayer, K. *et al.* (2012) Implementation of proton transfer reaction-mass spectrometry (PTR-MS) for advanced bioprocess monitoring. *Biotechnol. Bioengin.* **109**, 3059.
128. Singer, W., Herbig, J., Gutmann, R. *et al.* (2011) Applications of PTR-MS in medicine and biotechnology. *Am. Lab.* 43, 34–37.
129. Yeretzian, C., Pollien, P., Lindinger, C., Ali, S. (2004) Individualization of flavor preferences: toward a consumer-centric and individualized aroma science. *Comp. Rev. Food Technol.* **3**, 152.
130. Aprea, E., Morisco, F., Biasioli, F. *et al.* (2012) Analysis of breath by proton transfer reaction time of flight mass spectrometry in rats with steatohepatitis induced by high-fat diet. *J. Mass Spectrom.* **47**, 1098.

7

PTR-MS in the Medical Sciences

7.1 Background

Invasive medical procedures, which can cause significant stress to a patient, are commonly used in the diagnosis and monitoring of disease, as well as in the monitoring of drug consumption and therapies. Usually this involves the collection of biological material which is then investigated in clinical laboratories. These investigations are often time-consuming and expensive. One of the most common biological specimens used for medical diagnosis is blood. However, blood is a complex biological matrix and the preparation of a collected blood specimen and its analysis is often complicated and requires great care.

A major challenge in medicine is to develop reliable and simple non-invasive procedures that enable better and faster diagnosis and monitoring of diseases. Of all the possible applications of PTR-MS, that associated with the health sciences perhaps holds the greatest potential for having a significant impact on our society in terms of the quality of life. Arguably, this application captures the public imagination far more than any other because of its possible use for the early diagnosis and monitoring of disease, without the need for invasive procedures (as occurs for biopsies and blood sampling). Yet to date it is the one major application of PTR-MS that so far has had the least success. This is not due to any inherent fault with the PTR-MS technology or for the want of trying, but rather from the lack of progress in developing reliable and reproducible sampling techniques and the inherent VOC profile variations associated with any biological sample (both *inter-individual* and *intra-individual*). Despite these obstacles, the rewards of such research are potentially high. It is therefore not surprising that a number of groups around the world are exploring the applications of PTR-MS in the medical sciences.

Non-invasive PTR-MS sampling for medical applications has been used for headspace analysis above urine [1] and fluids obtained from the gut during colonoscopy [2], the *in vitro* discrimination of cancerous and non-cancerous cell lines [3] and for studies of emissions from the skin [4–6]. PTR-MS has also been used for investigations involving occupational health exposure in various medical environments [7–10]. Other medical-related work has

Proton Transfer Reaction Mass Spectrometry: Principles and Applications, First Edition. Andrew M. Ellis and Christopher A. Mayhew.

been associated with investigating VOCs emitted from different kinds of bacterial cultures [11–19], with the ultimate objective of being able to provide a sensitive analytical technique to identify *in vivo* infection quickly, and thereby enable the correct antibiotic treatment to be administered more rapidly. However, by far the largest number of investigations have been associated with breath analysis [20–58] and the investigations and development of breath sampling techniques for PTR-MS [59–65]. It is therefore with breath analysis, which has the potential of detecting VOCs associated with diseases and from environmental exposure to toxic chemicals, that we shall start this account of the applications of PTR-MS in the fields of medical science and biotechnology.

7.2 Breath Analysis

Molecular biomarkers of disease and drugs present in the bloodstream can pass through the barrier between the blood capillaries and alveoli in the lungs and then be carried out in the breath. The exhaled volatiles on breath provide a non-invasive window into the metabolic processes occurring within the body and to any drugs present in the bloodstream [66]. Changes in metabolic processes as a result of abnormal physiologies and/or biochemistries, such as those which occur in disease, drug/medication intake or environmental exposure to toxic chemicals, may alter the trace molecular profile of the breath. In the advanced stages of some diseases, volatiles are at such high concentrations that the nose can actually detect them: examples include the fruity acetone-like odour associated with uncontrolled diabetes mellitus, the musty fish reek of hepatic failure and the urine-like smell of kidney failure.

A major challenge is to provide sensitive and reliable analytical techniques to detect VOCs in the breath at an early stage of the disease, since this is usually when it is most treatable. Another potential application of PTR-MS and breath analysis is to quickly assess a drug overdose. Successful utilization of this technique could help to determine the drug taken and its concentration in the blood, details of which are needed rapidly if treatment is to be effective and hence life-saving.

Hundreds of VOCs have been identified in human breath at concentrations ranging from parts-per-trillion-by-volume (pptv) to parts-per-million-by-volume (ppmv). VOCs therefore have the potential to be clinically useful for the early detection, monitoring and diagnosis of diseases, as well as for investigating physiological disorders and for therapeutic monitoring. Similarly, many drugs and/or their metabolites will be present on breath at minute concentrations that can potentially be detected using PTR-MS techniques. Breath analysis also has the potential for being used to assess environmental exposure to toxic compounds.

From the above discussion, it is clear that breath contains useful trace gas markers, in the form of VOCs, which could be used as fingerprints to monitor different processes in the human body. To help unlock the messages that volatiles on exhaled breath can provide requires the application of highly reliable, sensitive and selective trace gas detection techniques. A specific breath marker related to a specific disease is the ideal. However, this is unlikely to be the case for the majority of diseases or conditions, where it is more probable that a range of VOCs with varying intensities will have to be used. It still needs to be ascertained whether there are unique patterns of VOCs, made up of specific and/or non-specific biomarkers, for a particular disease that can be discerned from the complex chemical environment of breath.

Another difficulty with breath analysis is that the way a patient breathes can alter the abundance of some VOCs, so standardization of breath sampling techniques and data analysis procedures need to be agreed upon and adopted for breath analysis to progress. However, it should be noted that it is not only exhalation kinetics that can have an influence on VOC concentrations in the breath; other factors such as heart rate and diet can also be important and so breath analysis is by no means straightforward. Furthermore, robust algorithms for identifying and modelling VOC patterns need to be developed which take into account and correct for exogenous VOCs inhaled from the ambient air along with VOCs in breath coming directly from the gut and the oral cavity.

Basically, breath sampling can be divided into those measurements which compare cohorts of healthy volunteers with patients suffering from a given disease and those whereby a patient acts as their own control (essentially personalized medicine), with the latter overcoming issues of inter-individual variability caused by lifestyle decisions (e.g. nutrition, physical activity). Independent of the approach, detailed multivariate statistical analysis will undoubtedly be required to extract information from what is a complex, multi-parameter dataset. Multivariate statistical techniques, such as principal component analysis and partial least squares discriminant analysis, have already been successfully used in applying PTR-MS to the food sciences, as described in the previous chapter, and these can equally well be employed to help in breath analysis (see reference [55]). In this way, by adopting what are essentially pattern-recognition techniques, rather than trying to identify every single VOC, breath 'fingerprinting' could provide a suitable method for discriminating between healthy and diseased patients. However, a word of warning is required here, because so-called 'voodoo correlations' can occur when using VOC patterns owing to the high chance of coincidental correlations occurring when handling a large number of measured variables [67].

Given all of the problems associated with analytical methods for breath sampling and analysis, it is not surprising that to date the majority of research associated with breath analysis and PTR-MS has been limited to 'proof-of-principle' trials which may not have been using ideal breath sampling procedures. Often these studies have also involved an insufficient number of patients and provided little detailed analysis. In-depth clinical trials to substantiate the effectiveness of VOCs as selective biological indicators of disease are required, but these are expensive and time-consuming.

Although breath analysis faces many challenges, there is evidence in the literature to suggest that breath analysis has considerable potential for medical applications [68]. Notable is the pioneering work of Phillips, which has provided evidence for the presence of identifiable VOCs in the breath related to lung and breast cancers [69–73]. Using mass spectrometric techniques to analyse these VOCs, Phillips and colleagues have claimed that it is possible to discriminate between patients with and without lung cancer [69]. However, the analytical techniques used by Phillips rely on offline gas chromatography–mass spectrometry (GC–MS). This requires the trapping and concentration of large volumes of breath, which is not only time-consuming to collect but also difficult for fragile patients to provide. Furthermore, transportation to and analysis in specialized laboratories may be required, introducing further delay and increased scope for artefacts and errors. When concentrating breath samples, issues of VOC surface adsorption, contamination and permeability arise and these are effects that can affect different VOCs in different ways. Another problem associated with GC–MS is its inherent reliance on chromatographic columns, which makes it difficult for one type of column to be able to handle a large range of VOCs (e.g. hydrocarbons, oxygenated VOCs and organosulfur compounds). In these respects, PTR-MS is a more promising analytical technique for breath analysis. However, as discussed in other sections of this book, it is really only a tool that is good for monitoring. In isolation, it cannot be used for the unambiguous identification of chemical compounds. Thus GC–MS and PTR-MS complement each other and there is therefore a role for both techniques in the fascinating world of breath analysis.

There are various soft chemical ionization mass spectrometric analytical techniques being used for breath analysis, and after PTR-MS the most popular is SIFT-MS (and most notably the work by Smith and Španěl [74] and their co-workers). In this book, we will only concentrate on those studies that have been undertaken with PTR-MS. However, for the interested reader who wishes to know more about the results from a wide range of analytical instruments, we comment that the International Association for Breath Research (IABR), which was founded in 2005, is establishing a database of volatile substances found in breath (both for humans and animals), as well as from skin, urine, faeces and flatulence. Once available, this database will be accessible through the website http://iabr.voc-research.at. Furthermore, disease markers in breath and their potential diagnostic properties are also comprehensively discussed in a book edited by Marczin and Yacoup [75].

By now it will come as no surprise to those readers who have followed some of the historical narrative in this book that the first investigations involving PTR-MS and breath analysis were by Werner Lindinger and his colleagues [20–27, 76–78]. In their first published study dealing with VOCs in breath, the researchers did not in fact use a PTR-MS instrument, but instead used a forerunner which was effectively a hybrid of SIFT-MS and PTR-MS. This selected ion flow drift tube (SIFDT) system operates differently from PTR-MS in that ions are initially generated by electron impact in an ionization source and then the reagent ion of interest (which in the case of H_3O^+ results from the reaction of H_2O^+ with H_2O in the ionization source) is selected using a quadrupole mass filter prior to injection into the reaction region (see Section 1.4.1) [20]. In addition, the drift tube region of the instrument used by Lindinger and co-workers operated under very different operating conditions than those now used routinely in conventional PTR-MS instruments; for example, helium was the main buffer gas and the reduced electric field was set at $\sim$50 Td. The instrument also had far less sensitivity than is currently available with modern PTR-MS instruments. Furthermore, it should be noted that in this early work breath sampling was not done in real time. Instead, breath samples were first collected into special gas-carrier bags (Plastigas) maintained at a temperature between 340 and 350 K in order to avoid condensation. Despite all of these differences, this work was, nevertheless, the first to investigate the reactions of H_3O^+ ions with breath and the technology used ultimately led to the development of PTR-MS. These early studies therefore warrant some discussion.

In the first study, Lagg *et al.* demonstrated the applicability of using H_3O^+ as a reagent ion to monitor methanol, ethanol and acetone in human breath [20]. Methanol and acetone were monitored as a function of time after volunteers had consumed 0.4 L of wine. Using the fact that the rate of decline in ethanol concentration in blood as a function of time in a healthy person is approximately constant and is independent of absolute blood alcohol concentration, good agreement between the measured and calculated alcohol breath concentrations for the duration of the measurements was obtained. In addition to their ethanol study, Lagg *et al.* reported acetone levels of 53 healthy volunteers against that of a person suffering from diabetes. Lagg *et al.* noted that acetone levels for the healthy subjects were usually ranged between 1 and 3 ppmv, whereas for the person suffering from diabetes the concentration of acetone was measured to be 20 ppmv. Although care must be taken before drawing any conclusions from such a limited study (one patient!), this preliminary work provided some indication of the potential of PTR-MS for breath analysis (for a more detailed discussion on breath acetone, see Section 7.2.3). The SIFDT apparatus was further used by Lindinger and co-workers in PTR-MS mode in another study to investigate methanol in human breath for a number of volunteers consuming alcoholic drinks and various amounts of fruit [79, 80].

In 1998, Lindinger *et al.* published a review on PTR-MS which included a summary of their medical applications up to that point in time and which also demonstrated advances in improved VOC detection sensitivity (by then to a few pptv) compared to that reached with their prototype instrument [77]. In that review, the online monitoring of the fast enzymatic conversion of isopropanol into acetone was demonstrated.

Factors such as contamination, dead space (both instrumental and physiological) and loss of analyte lead to significant variations in the determination of VOC concentrations, and hence considerably complicate breath analysis. This raises concerns about the reproducibility and accuracy of published results, especially those that have been obtained using non-validated breath sampling procedures. Furthermore, in none of the studies to date, either using GC–MS or PTR-MS, has breath analysis been tested for its ability to discriminate between patients suffering from the same type but different forms of disease. For example, while some studies apparently give good discrimination between lung cancer patients and normals (control groups), this may simply be an indicator of any type of cancer. In addition, there is often little agreement between studies as to which VOCs constitute an appropriate discriminating set. For example, using solid phase micro extraction (SPME) and GC–MS techniques Deng *et al.* found hexanal and heptanal in the blood and breath of lung cancer patients [81]. However, in a lung cancer study using breath analysis by PTR-MS carried out by Wehinger *et al.* [37], no comment was made about these particular VOCs. This illustrates the challenge in drawing any definitive conclusions from studies with regards to breath VOCs and disease.

Having provided some brief historical background and discussions of potential pitfalls relating to the use of PTR-MS in the analysis of human breath, it is now appropriate that we turn our attention to consider some specific PTR-MS breath analysis examples and we shall start with volatiles observed in the breath of smokers. Although smoking in itself is of course not a disease, smoking can lead to serious illness such as cancer, heart disease, stroke and various lung diseases, with the result that the average lifespan of smokers is shorter than non-smokers. It is one of the most preventable causes of morbidity (ill health) and mortality (death) in the world. Furthermore, smoking introduces VOCs into the breath of healthy smokers which are not in the breath of healthy non-smokers. These VOCs could be mistakenly assigned to diseases in unhealthy smokers and therefore they need to be eliminated in any studies on the detection of diseases.

7.2.1 Smoking and Breath Volatiles

By 1995, Lindinger's group had developed a prototype PTR-MS instrument which had far greater sensitivity than their SIFDT apparatus (see previous Section). One of the first applications of this new instrument was to investigate the breath of smokers and non-smokers [21, 24, 76]. The research team noted that in comparison to non-smokers, acetonitrile was elevated in the samples of breath taken from smokers and they therefore suggested that this compound can be used as a breath marker for smokers [21]. In addition to finding acetonitrile on the breath of smokers in higher concentrations than found for non-smokers, the same was also found to be true for benzene. However, unlike acetonitrile, which was found to be at an elevated level in the breath of smokers for a considerable time (up to at least a week after smoking a cigarette [21, 25]), benzene concentrations increased dramatically during smoking but returned to levels associated with non-smokers within approximately 30 min. Analysis showed that the concentration of breath acetonitrile was directly proportional to the number of cigarettes smoked per day and that acetonitrile is mainly lost through exhalation rather than, for example, being lost through urination. Lindinger and co-workers also investigated the breath of non-smokers who stayed in a smoke-laden environment for approximately 8 hours [24]. The measurements indicated that by breathing the air in this environment, a quantity of acetonitrile is delivered to non-smokers which is equivalent to that received by smoking one to two cigarettes. However, this does not acknowledge that tar is the major problem associated with smokers, so this equivalence has to be used with some caution.

Subsequent work by Lirk *et al.* aimed to more accurately quantify recent smoking behaviour by measuring exhaled concentration levels of acetonitrile in 268 volunteers (48 smokers and 220 non-smokers) attending a local health fair [31]. Again smokers were shown to have elevated levels of acetonitrile in their breath and it was therefore confirmed that analysis of acetonitrile on breath can be used as a procedure to determine if someone had recently been smoking.

Another detailed study relating to smoking was undertaken by Kusch *et al.* [82]. The breath of 370 volunteers (of which 81 were smokers, 210 were non-smokers and 79 were ex-smokers) was collected in Tedlar bags and analysed within 12 hours. The idea was to interpret breath data from patients who not only smoke but also suffer with respiratory disease and/or lung cancer (see Section 7.2.4), so that VOCs on breath that are a result of smoking are not mistakenly assigned to be markers for lung disease. The researchers claim that seven ions were found to be useful for differentiating between smokers and non-smokers. These ions, observed at *m/z* 28, 42, 69, 79, 93, 97 and 123, were tentatively assigned to derive from protonated hydrogen cyanide, acetonitrile, isoprene, benzene/dimethysulfoxide, toluene, dimethylfuran/dimethylpyrazole and *N,N*-dimethyl-pyridineamine/methoxymethyl-benzene, respectively, with acetonitrile being confirmed as the clearest marker for smoking. Given that the seven ions mentioned above have been shown to occur as a result of smoking, it may well be impossible to tell if any are also associated with patients who are suffering from lung cancer or chronic obstructive pulmonary disease (COPD), because such patients are often smokers. Therefore, the authors of this work concluded that considerable care must be taken in using signal intensities at the above *m/z* values for the analysis of breath from patients who smoke. Two other ions observed at *m/z* 31 and 33, which are usually associated with formaldehyde and methanol, respectively, showed no significant differences in their intensities between smokers and non-smokers.

7.2.2 Isoprene in Breath

Isoprene (C_5H_8) is an abundant endogenous hydrocarbon contained in exhaled breath (typically at a level of hundreds of ppbv) and is considered to be linked to a number of metabolic processes in the human body, and in particular as a biomarker for cholesterol biosynthesis. The routine detection of isoprene in human breath may therefore have the potential for monitoring the effects of lipid-lowering therapy.

Early on in the development of PTR-MS, the technique was used to investigate isoprene in exhaled breath [23]. In the first study of breath isoprene, Teflon® bags were used to collect the central portion of a single breath exhalation. The difference in the concentration of isoprene in the breath of adults and children was explored, as were diurnal variations. The PTR-MS measurements on young children (aged 4–6 years old) found a much lower average isoprene concentration in their breath (~100 ppbv) than for adults (~2.4 times higher). The range of concentrations was also significantly different. For the adult population, the isoprene concentration ranged from ~50 ppbv up to more than 1000 ppbv. In comparison, all the isoprene concentrations recorded in the breath of children were less than 200 ppbv. This study also suggested a diurnal variation in isoprene levels associated with sleep or wakefulness. However, following a reinvestigation of isoprene levels in breath using online PTR-MS methods, it was suggested that the observed changes in isoprene concentrations could be a result of changes in heart rate [26]. For example, it was observed that at the start of exercise isoprene levels in breath rapidly increased (within seconds), although this does not necessarily establish cause and effect. The reason given for this increase was that the rise in the isoprene concentration resulted simply from enhanced blood flow caused by the increase in heart rate due to exercise. This leads to more isoprene exchange in the lungs and hence its quantity is increased in the breath. However, rather than reaching a maximum abundance and then remaining at that level, the isoprene concentrations in the breath were found to sharply decrease after several minutes and went down after ~20 min to levels lower than found at the start of exercise. A conclusion from this work is that breath isoprene is only useful as a non-invasive marker of cholesterologenesis providing breath samples are taken at constant heart rate and under what the researchers called 'standard conditions' (which means that volunteers should not undertake any energetic activity for at least one hour before breath is sampled). A series of sleep monitoring studies by Amann *et al.* provided confirmatory data that isoprene levels in breath are related to heart rate [30], as demonstrated in Figure 7.1.

Karl *et al.* studied the relationship between the concentration of isoprene found in breath and the level of cholesterol in blood and developed a reasonably successful *in vivo* gas-exchange model which had some success in simulating isoprene concentrations observed in breath [26]. Karl *et al.* also investigated lipid-lowering therapy by monitoring a person with hyperlipidemia who was on lipid-lowering medication [26]. The concentration of breath isoprene, serum cholesterol and low-density lipoprotein, declined collectively by ~35%. Thus, this study appears to have shown that cholesterol-lowering drugs result in a reduction of isoprene levels in the body. This and a follow-up study presented evidence which suggested that isoprene levels on breath showed a positive correlation with changes in cholesterol levels in blood [27]. This suggests that isoprene analysis in exhaled breath could therefore be used as a marker for cholesterologenesis and hence as an aid in controlling lipid-lowering therapy. Certainly, a non-invasive procedure to monitor cholesterol levels is highly attractive. However, in contrast to these studies, Kushch *et al.* found no significant correlation between isoprene concentrations on the breath and cholesterol levels in the blood. The suggestion here was that isoprene is in fact correlated with the rate of synthesis of cholesterol [40], so there is still some doubt about the link between isoprene levels in breath and blood cholesterol at the time of writing.

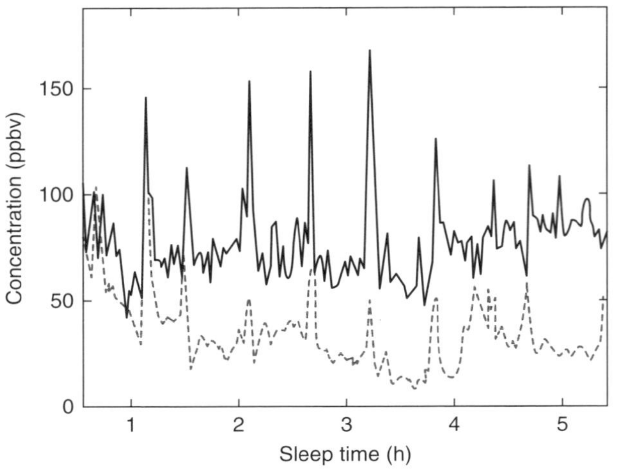

Figure 7.1 *Concentration of a compound measured at m/z 69 (assigned to protonated isoprene) as a function of time during one night for a healthy volunteer (solid line). The dashed line represents the average heart rate on an arbitrary scale. Reprinted with permission from Reference [30]. Copyright 2004, with permission from Elsevier.*

A comprehensive study involving 21 healthy volunteers by Schubert *et al.* assessed correlations between exhaled concentrations of isoprene (in addition to ammonia and acetone) with metabolism and physiological parameters [58]. Isoprene concentrations showed a close correlation with cardiac output, in agreement with previous studies, and minute ventilation (the total volume of gas in litres exhaled from the lungs per minute).

In a study by Moser *et al.*, the breath profiles of a substantial group of adult volunteers (344 in total) were investigated and no significant correlations were found between isoprene concentrations (or any other molecular mass) and age, blood pressure and body mass index [33]. That no evidence for age dependence within an adult group was observed is in agreement with the earlier work by Karl *et al.* [26]. However, a later study by Lechner *et al.*, involving 126 volunteers (consisting of 66 females and 60 males), demonstrated that this may not be the case [35]. It was found that volunteers in the 19–29 years age bracket had significantly lower concentrations of isoprene in their breath than found in older adults. This study also investigated any gender dependences and found that isoprene levels were higher in the exhaled air from male subjects than from females. Thus, based on this study, there appear to be gender- and age-specific differences in the breath levels of isoprene. However, this is in contrast to the results from the work by Kushch *et al.*, which involved 205 healthy adult volunteers [40]. The isoprene concentrations were found to range from 6 to 275 ppbv, with a geometric mean of 99 ppbv. This study found no statistically significant difference in the mean isoprene concentrations between males (mean 105 ppbv) and females (99 ppbv). They did, however, find a dependence on age in the male population, but in this study ageing led to a decrease in measured isoprene concentration.

An investigation of offline isoprene levels in the breath of patients suffering from end-stage renal disease following haemodialysis (HD) found that statistically significant increases in isoprene concentrations occurred after HD compared to those levels found before HD [29]. This, as with all the other previous studies mentioned so far, assumed that the ion signal detected at *m/z* 69 results only from isoprene, and that no other VOC is contributing to this, which may not be the case in practice. The mean ratio of the post- and pre-dialysis isoprene level was determined to be 1.84 ± 1.41. The cause of this increase is uncertain. Increases in isoprene levels have been associated with oxidative stress. However, this is conjecture and the work by Lirk *et al.* found no evidence for this. As indicated by the error associated with the ratio, the researchers observed large inter-individual variability in the isoprene levels. No correlation with the circadian rhythm (biological processes that follow a 24-hour cycle) was found that could account for this and it was therefore proposed that large inter-individual variability results from differences in respiratory frequency and/or tidal volume.

Later investigations of breath isoprene (and acetone) and development of detailed mathematical models dealing with the dynamics of isoprene (and acetone) in breath have been presented by King and colleagues [48, 56, 83–86]. Arendacká *et al.* have also used offline PTR-MS to investigate the observed variability issues in determining isoprene concentrations in breath [39]. Herbig *et al.* found high variability in breath-by-breath isoprene concentrations which they attributed to exhalation depth and influences of hyperventilation (which results in low values of isoprene) [61].

From the discussions above, it is clear that there is much uncertainty in the relationship between the concentrations of isoprene found in exhaled breath and factors such as age and cholesterol. Further investigations are required in order to resolve these issues.

7.2.3 Acetone in Breath

Another common VOC found in exhaled breath is acetone (2-propanone). This compound is present in human breath at even greater concentrations than found for isoprene, with levels ranging from a few hundreds of ppbv up to several ppmv. The presence of acetone on breath produces a fruity odour which can be a sign of ketoacidosis[1] associated with diabetes. It is tempting to state that the analysis of acetone on the breath could be used as a diagnostic tool for diabetes. However, this is not possible owing to the large variability in acetone concentrations that can occur for any individual throughout the course of the day. In addition, there are differences between individuals owing to possible factors other than diabetes such as different metabolisms, age, gender, stress, health and diseases.

Blood glucose levels are currently used to monitor diabetes. However, blood glucose levels do not make it possible to distinguish between hyperglycemia (high glucose levels in the blood) caused by overeating or that resulting from insufficient insulin. Nevertheless, if a base (normal) acetone concentration level is determined for an individual it should be possible to use changes in acetone levels combined with information on blood glucose to monitor a prescribed diet regime and hence manage diabetes more effectively [87]. For example, when a diabetic is classified as being hyperglycaemic but 'normal' breath acetone levels are monitored, then it can be assumed with some confidence that the hyperglycemia must be a result of overeating and hence an improved diet is required, rather than treating a hyperglycemic patient with more insulin. This resulted in Crofford *et al.* stating over a quarter of a century ago that '*you don't treat overeating with insulin but with better dietary*

[1] Ketoacidosis is where the blood has become acidic owing to an excess of ketones in the body.

management' [87]. On the other hand, if a diabetic presents with hyperglycemia and, in addition, has elevated levels of acetone relative to the 'normal', then that reveals that the patient has had insufficient insulin and the dose needs to be adjusted accordingly. Given this information, it is perhaps surprising that breath acetone levels are not also routinely measured together with the standard blood glucose level to provide a much greater control on the management of diabetes, especially when the number of overweight diabetics is rising in many countries.

From the above discussion, there is a clear need for a simple test to monitor acetone levels in exhaled breath. However, there is a potential problem for determining this owing to the high solubility of acetone in water. Given the moisture present in lungs, the trachea and the mouth, the recorded concentration in exhaled breath will be less than in the alveoli owing to acetone dissolving into mucous membrane linings as it passes through the respiratory tract. This has been confirmed by measurements undertaken by King *et al.*, who have shown that the use of a buffered end-tidal (BET) sampler (see Section 7.2.8.2 and Reference 61) with PTR-MS underestimates the concentration of highly soluble blood-borne VOCs [86]. The researchers claim that, in view of their finding, acetone concentrations in the breath will underestimate the underlying alveolar concentration by a factor of up to 1.5.

An issue to address, which was referred to above, is to determine what factors can affect acetone levels in exhaled breath. In an attempt to provide an answer to this, a comprehensive study of acetone levels in breath was undertaken by Schwarz *et al.* [4]. This investigation sampled and analysed the breath of 243 adult volunteers, 215 (83 males and 132 females) of whom had fasted overnight before sampling and 28 (12 males and 16 females) who had no dietary control. The age distributions of the volunteers who had fasted were 23–84 (mean 52.4 years) and 19–89 (mean 51.4 years) for the males and females, respectively. For the subjects with no dietary control the age distributions were 27–70 (mean 51.4 years) for the males and 23–69 (mean 47.3 years) for the females. Very few adults were in the age group 18–25. None of the adult volunteers were suffering from diabetes. In addition to adults, Schwarz *et al.* also included 44 healthy children between the ages of 5 and 11 in their study (24 boys and 20 girls). The acetone concentration (determined by measuring the ion intensity at m/z 59) for the fasted volunteers had a range of 177–2441 ppbv, with a median of 628 ppbv, whereas for the group with no dietary control it had a smaller span of 281–1246 ppbv, with a median of 544 ppbv. Nevertheless, there is no significant difference in the acetone levels between the two groups. Also, no substantial gender differences in acetone levels were observed, although ageing was found to be associated with a slight increase of acetone in the fasted females. Body mass index showed no correlation with acetone levels. In comparison to the adult measurements, the children showed much lower concentrations, with a median of 263 ppbv.

7.2.4 Lung Studies: Cancer and Emphysema

The early detection of lung diseases, and especially cancer, is crucial for improved survival rates. Breath analysis could potentially provide early diagnosis in a simple, routine and non-invasive way. However, a word of warning is required when considering using PTR-MS in clinical environments for investigating potential markers for any disease. Using GC–MS measurements, Petri *et al.* found that, in comparison to healthy controls, which were matched for age and smoking history, the breath of patients suffering with lung cancer showed elevated concentrations of *o*-toluidine (C_7H_9N) [88]. This appears to have been corroborated in a PTR-MS study by Rieder *et al.*, who investigated the breath of patients with different types of tumours. In their study, an ion signal at *m/z* 108, which was tentatively assigned to be protonated *o*-toluidine, was found to be elevated in comparison to age-matched healthy controls [27]. It was therefore suggested that *m/z* 108 could be used as a marker for cancers, such as lung cancers. However, this was later shown to be incorrect by Amann *et al.*, who found no such correlation and showed no difference between healthy volunteers and patients suffering with carcinoma [30]. Apparently various exogenous VOCs present in the clinical environment were contributing to the ion signal at *m/z* 108 in the study by Rieder *et al.* Thus without compound separation, the PTR-MS signal at *m/z* 108 cannot be used as a marker for lung cancer. This highlights the fact that considerable care must be taken in the assignment of VOCs in breath samples and that environmental factors need to be taken into account, especially in clinical environments where plenty of unusual VOCs may be present. In breath analysis using PTR-MS, it is all too easy to assume that a particular *m/z* is associated with a VOC linked to a given disease.

A study by Wehinger *et al.* reported a PTR-MS analysis of breath from 17 patients suffering from lung cancer and compared these findings with those from the breath of 170 healthy controls [37]. The key results obtained were that ions detected at *m/z* 31 (tentatively assigned to protonated formaldehyde) and *m/z* 43 (considered to be a fragment of protonated isopropanol, although many other compounds can produce $C_3H_7^+$) were at significantly higher levels in the breath of patients suffering from primary lung cancer when compared to those found in the breath of the healthy controls. It was concluded that these two ions could be used to provide a discriminant protocol to determine if a person is suffering from primary lung cancer. However, the single breath sampling technique used was not ideal because measured VOC concentrations are dependent on the length of exhalation, and particularly for those VOCs with low solubility in water [60].

Another PTR-MS study dealing with lung cancer, which also included GC–MS measurements, demonstrated that breath analysis has considerable potential as a diagnostic tool [45]. In that study, involving 220 lung cancer patients and 441 healthy volunteers, isoprene, acetone and methanol exhibited decreased concentrations in the breath of lung cancer patients when compared with healthy controls. Another study which analysed breath samples in Tedlar bags collected from 79 lung cancer patients (23 females and 56 males) showed that the decline of isoprene concentrations in the breath of lung cancer patients was inversely correlated with concentrations for an immune activation marker, neopterin [89]. This suggests that immune activation may play a role in the decrease in isoprene and is probably related to lipid metabolic changes. The breath isoprene level was also found to correlate significantly with total cholesterol and low-density lipoprotein cholesterol levels in the blood but not with high-density lipoprotein.

Cristescu *et al.* have used PTR-MS measurements of breath to investigate VOCs in the breath of current and former heavy smokers participating in a lung cancer screening trial to determine if VOC profiles could be used to detect emphysema [90]. This was an extremely detailed investigation that combined computed tomography (CT) scans, pulmonary function tests and breath samples from 204 volunteers. Using in-depth statistical analysis, they discovered that the VOC profiles obtained from the breath samples provided a potential diagnostic tool for emphysema, airflow limitation and/or gas-exchange impairment. An ion signal observed at *m/z* 87, which the researchers tentatively assigned to be protonated methylbutanal, is a proposed marker for the condition. Ions detected at *m/z* 52 (possibly resulting from protonated chloramine) and *m/z* 135 (considered to be from protonated alkylbenzene) seem to be linked to obstructive disease and ions detected at *m/z* 122 to impaired diffusion capacity. A potential application of this work is in the development of new pulmonary function tests.

7.2.5 Other PTR-MS Breath Studies

7.2.5.1 Crohn's Disease and Ulcerative Colitis

The breath from people suffering from inflammatory bowel diseases (Crohn's disease and ulcerative colitis) and irritable bowel syndrome has been studied using PTR-MS with the view to diagnosing gastrointestinal diseases [2]. In the same investigation, PTR-MS was also used to detect the VOC content of the headspace air above fluid taken from the gut of these patients during colonoscopy. Significant differences in the mass spectra were found for both the fluid headspace and in breath when compared with that obtained from healthy volunteers.

7.2.5.2 Carbohydrate Malabsorption

Carbohydrate malabsorption is a condition in which a person does not have the ability to absorb or digest certain carbohydrates owing to the lack of one or possibly more intestinal enzymes. This is a debilitating disease because it leads to diarrhoea, abdominal distension and flatulence. The disease can be detected by monitoring H_2 in the breath resulting from the bacterial metabolism of carbohydrates. Although a breath test based on H_2 measurement is available for diagnosis, Amann *et al.* investigated the breath of patients suffering from this disease with the aim to establish patterns of VOCs which are characteristic for malabsorption syndromes [30]. Seventeen outpatients (in the age range of 20–50 years) were examined. None of the patients showed signs of inflammatory bowel disease, chronic disease or infectious diseases and none of these patients were under any form of medication. The patients suffering from the disease were diagnosed using the standard H_2 breath test. For these subjects, ions at *m/z* 63 showed elevated intensities in comparison to those found in the breath of healthy individuals. The researchers tentatively assigned *m/z* 63 to either protonated ethanethiol or dimethylsulfide. It was proposed that a signal at *m/z* 63 in a mass spectrum could provide an alternative marker to H_2 for carbohydrate malabsorption, but further work is required to establish this definitively.

7.2.5.3 High Mass-Resolution PTR-TOF-MS Breath Studies

Herbig *et al.* were the first researchers to use PTR-MS based on a time-of-flight mass spectrometer (TOF-MS) for online breath analysis [47]. This instrument was used in conjunction with the BET breath sampling technique (see also Section 7.2.8.2), which was developed by the same research team [61]. With their highly sensitive PTR-TOF-MS instrument, it proved possible to measure all detectable VOCs up to an *m/z* value of 320 from a single exhalation and with a limit-of-detection in the sub-ppbv range. To illustrate the time response of this PTR-TOF-MS instrument, Figure 7.2(a) shows several inhalation/exhalation cycles in direct online sampling while Figure 7.2(b) shows one exhalation in a BET sampling experiment for $H_3O^+(H_2O)$ (measured using its isotopologue at *m/z* 39.033, to avoid detector

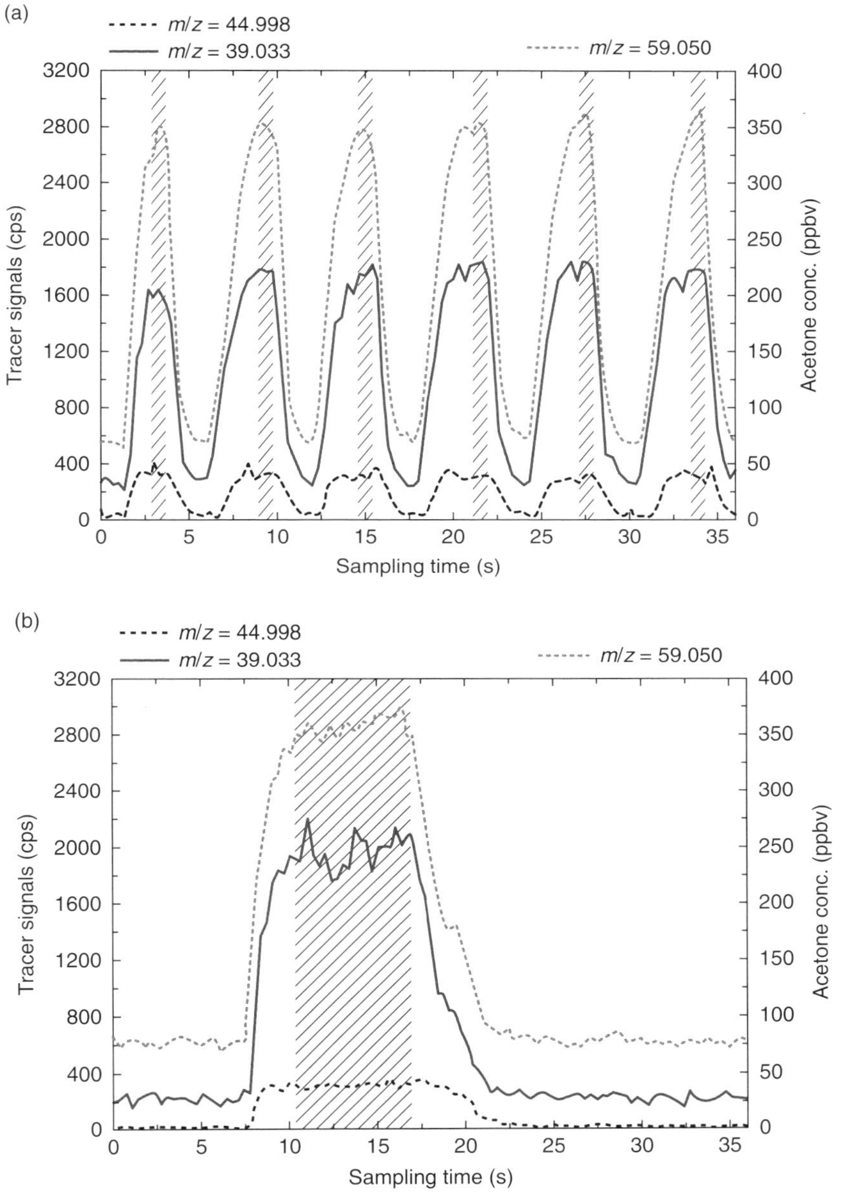

Figure 7.2 *Signal levels for three ions measured using PTR-TOF-MS for (a) a series of inhalations/exhalations using direct online sampling, with the shaded areas marking the sampling time for the end-tidal fraction, and (b) a BET end-tidal measurement from one direct online sampling exhalation. The three species highlighted are* $H_3O^+(H_2O)$ *(measured via its isotopologue at m/z 39.033 owing to detection saturation of the ion peak at m/z 37.029),* CO_2H^+ *(m/z 44.998) and protonated acetone (m/z 59.050). Reprinted with permission from reference [47].*

saturation effects), CO_2H^+ (*m/z* 44.998)[2] and protonated acetone (*m/z* 59.050). A clear advantage of using a TOF-MS in comparison to a quadrupole MS instrument is its high mass resolution, which can enable the separation of nominally isobaric compounds and thereby lead to greater confidence in the identification of a specific VOC. Importantly, an improved identification of the VOCs in breath provides a new and more in-depth understanding of the metabolic processes occurring within the human body as a result of disease. This opens up a completely new dimension in the application of PTR-MS to breath analysis. Another distinct advantage of using a TOF-MS is that the complete mass spectrum is recorded and hence is accessible, whereas for a quadrupole MS a scan is required or the relevant *m/z* values need to be pre-selected (which means important VOC information could be lost). The use of a TOF-MS is therefore particularly beneficial in breath sampling applications when scans are not feasible or when it is not known what VOCs should be monitored.

7.2.5.4 *Kidney Function and PTR-MS*

Using the high-resolution PTR-TOF-MS instrument mentioned in the previous section and in Section 3.5.4.6., Kohl *et al.* have investigated the VOC constituents in the breath of uremic patients with kidney malfunction [91]. Breath samples (642 in total from 96 volunteers) were taken and analysed just before and after kidney transplantation. From a correlation analysis, one particular VOC was found to be elevated in kidney patients, which correlated with blood serum creatinine (the accepted marker for assessing kidney function). The peak of the ion signal was measured to be at *m/z* 115.1113. Assuming that this corresponds to a protonated parent ion and not a fragment ion then it must have derived from a neutral species with a mass of 114.1035 Da. The rather precise molecular mass extracted is considered to be consistent with the molecular formula $C_7H_{14}O$. Further investigations with a PTR-MS instrument equipped with a triple quadrupole tandem mass spectrometer, in order to study fragmentation of the associated protonated parent ion, suggested that $C_7H_{14}O$ is probably a C_7 ketone or a branched C_7 aldehyde. However, more detailed fragmentation studies are required to confirm this assignment. The overall conclusion was that breath analysis has the potential as a non-invasive real-time monitoring tool for determining kidney function in a clinical setting.

7.2.5.5 *Liver Disease*

Liver disease poses a major public health issue, encompassing not only physical and psychological morbidity and mortality but also incurring significant societal costs. In the United Kingdom alone, deaths from chronic liver diseases have increased eightfold in men aged 35–44 and sevenfold in women over the past three decades.

One of the main difficulties with liver disease is that the liver can still function even when a quite large proportion of it is already damaged. Often patients do not present symptoms until the disease is quite advanced. Even then, diagnosis is difficult and symptoms are often general and can be mistaken for other causes. Many liver diseases are clinically silent and are only disclosed at a stage where hopes of recovery are slim. Invasive medical procedures are commonly used in the diagnosis and monitoring of liver disease and its complications,

---

[2] See Section 6.5.4 for an explanation as to why traces of CO_2H^+ can be observed in PTR-MS mass spectra.

such as cirrhosis and cancer. Such invasive procedures result in stress for the patient, pose a risk, have significant sample and interpretive errors, and are expensive. There is thus a need for a non-invasive test that can be used at an early stage or for high-risk groups even before symptoms occur. The liver is the central organ involved in metabolism and will be the source of many VOCs that enter the bloodstream. Subtle changes in the metabolic processes as a result of disease will be passed into blood and partition from the bloodstream via the alveolar pulmonary membrane into the lungs, and then be carried out in the breath. Thus the application of PTR-MS to study the breath of patients suffering from liver disease is an ideal area for research.

Recently, Morisco *et al.* have investigated the potential of using a high-resolution PTR-TOF-MS for the real-time analysis of breath VOC patterns in the diagnosis of liver cirrhosis [92]. This was only a proof-of-principle study involving just 26 volunteers, 12 of whom were suffering from liver cirrhosis of varying severities and different causes. The remaining 14 volunteers were the healthy controls. Although limited in scope, 12 VOCs were identified which were found to be significantly different in the breath of the cirrhotic patients: four ketones (2-butanone, 2- or 3-pentenone, a C8-ketone and a C9-ketone), a monoterpene (tentatively identified to be limonene), a terpene-related compound (tentatively assigned to be p-cymene), four sulfur or nitrogen containing compounds, and two alcohols (heptadienol, methanol). Thus there is good evidence to suggest that breath profiling of VOCs can be used to discriminate cirrhotic from healthy patients. However, if it is to be of practical use this VOC profiling must be able to distinguish between healthy patients and those suffering from the early stages of cirrhosis even before symptoms occur. Early detection permits earlier intervention in disease treatment, thereby offering the potential of patients receiving prompt treatment and improving the chances of recovery. This would also reduce overall costs of treatment and better utilization of scarce hospital resources. In addition, patients will benefit from swifter, more accurate and less intimidating procedures.

7.2.6 Drug Monitoring and Pharmacokinetics Using Breath Analysis and PTR-MS

The potential of breath analysis as an online analytical method for pharmacokinetic and environmental exposure studies has been reviewed by Beauchamp [93]. It was the early work by Lindinger's research team that demonstrated the online detection capabilities of PTR-MS for monitoring breath VOCs associated with compounds that had been ingested or administered, and hence the potential of investigating the uptake and distribution of exogenous compounds introduced into the body. For example, Lindinger *et al.* investigated the enzymatic conversion of propan-2-ol into acetone after oral ingestion of 0.5 g of propan-2-ol (diluted in water) or when taken intravenously [78]. It was suggested that these kinds of processes could be used to provide non-invasive tests of organ functions in the body.

Such work was taken further by Harrison *et al.* [28], who for the first time illustrated the use of real-time breath monitoring for drugs directly injected into the bloodstream and also showed its application during surgical procedures. The commonly used anaesthetic, propofol (2,6-diisopropyl phenol), was investigated on the breath of patients undergoing surgery. The problem with any intravenous anaesthetic is that it is not possible to rapidly determine the blood plasma concentration. The question proposed by the researchers was whether a breath test could indirectly achieve this. Through a simple

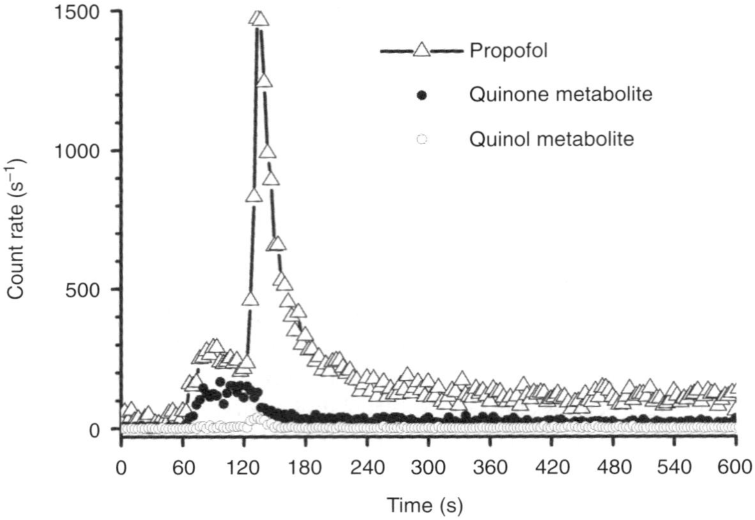

Figure 7.3 *Real-time measurements of protonated propofol and two protonated metabolites for the first 10 min of expired breath sampling from a patient undergoing surgery. Reproduced from [28] with permission from Oxford University Press.*

proof-of-principle study, the aim was to demonstrate that propofol injected into the bloodstream could be detected on the exhaled breath. Figure 7.3 illustrates that this is indeed the case. This diagram shows the protonated propofol signal intensity for a particular patient from the start of surgery (time $t = 0$) and also two of its metabolites for a period of up to 10 min. The results suggested that propofol monitoring might be feasible in an analogous manner to that for volatile anaesthetics. Given the importance of such research for monitoring plasma concentrations of intravenous anaesthetics, the preliminary investigations by Harrison *et al.* have been continued and substantially extended by other research groups [94–99], although only one of these, by Takita *et al.* [95], used PTR-MS. The work by Takita *et al.* went much further than the proof-of-principle study reported by Harrison *et al.* in that they investigated the relation between exhaled and plasma propofol concentrations. The good correlation found between the two suggested that PTR-MS can be used for real-time propofol monitoring in a clinical setting.

The usefulness of PTR-MS to monitor administered drugs has also been demonstrated by Erhart *et al.* [46] and by Beauchamp *et al.* [49]. Erhart *et al.* analysed breath samples from 27 patients suffering from epilepsy who were being treated with valproic acid (VPA: 2-propylpentanoic acid, but also known as dipropylacetic acid). VPA is a commonly used antiepileptic drug. As a result of this therapy, all patients were found to have elevated concentrations of 3-heptanone ($C_7H_{14}O$) in their breath, which was assumed to have resulted from the metabolism of valproic acid. The researchers found little correlation of 3-heptanone with VPA concentrations in serum or in the daily dose of the drug, and this may therefore limit the applicability of breath analysis as a sensitive non-invasive monitor of VPA therapy. Enhanced levels of 3-heptanone have also been observed from an offline analysis of breath sampled from patients suffering from propionic acidaemia (a rare metabolic disorder) [100].

Beauchamp *et al.* investigated the concentrations of eucalyptol (1,8-cineole) in the breath of 11 healthy adult volunteers at 1-min intervals following ingestion of capsules containing eucalyptol [49]. These capsules dissolved in the gut from where the transfer of eucalyptol into the blood occurred. BET sampling was used in this study (see Section 7.2.8.2). Although eucalyptol was observed to be present in the breath, the actual time at which eucalyptol was detected varied considerably, ranging from 1 hour and 6 min to 4 hours and 48 min, as did maximum breath concentrations, ranging from 106 ppbv to 1598 ppbv. These variations were found to occur not only for inter-subject studies but also for intra-subject investigations, that is, the same volunteer would show differences in onset times and concentrations when the experiment was repeated. No correlation with general physiology or gender was found. Eucalyptol was still found to be present in breath ~25 hours after the capsule was swallowed. That drugs can remain in the human body and hence be detected in breath after long periods of time is not a surprising observation, but it is important to take this into consideration when exploiting VOC profiles to distinguish between healthy and disease states, that is, a difference in VOC patterns could simply be a result of medication. The study by Beauchamp *et al.* has highlighted the potential of using PTR-MS breath analysis for pharmacokinetic studies. However, further studies associated with drug consumption and breath analysis are required to fully establish the potential of PTR-MS in this field of research.

Halbritter *et al.*, in what was very much a pilot but seminal study, have investigated the potential of using non-invasive PTR-MS breath gas analysis for the screening of gestational diabetes mellitus (GDM) in 52 pregnant women [55]. This work compared the breath results with standard oral glucose tolerance tests (OGTT). An OGTT determines the concentration of blood glucose measured immediately before and following ingestion of a glucose solution. This process is invasive and time-consuming, whereas a breath analysis procedure would be non-invasive and potentially rapid. The researchers collected end-tidal breath from the women during OGTT using a BET sampler (see section 7.2.8.2). The standard OGTT diagnosis classified women into various groups according to whether they had GDM (8 women), impaired glucose tolerance (12 women) or normal glucose tolerance (32 women). The success of this breath analysis technique relies on parameterising the time dependence of selected (responsive) VOCs during at least the first hour of the OGTT, with breath samples being taken every 6–10 minutes using the BET sampler. Owing to the small number of patients, the application of unsupervised statistical methods, such as PCA, to the breath results could not distinguish the subgroups found using the OGTT diagnosis test. However, by using the supervised multivariate analysis of variance (MANOVA) method the breath results did correlate with the OGTT, that is, the women could be discriminated according to whether they had GDM, impaired glucose tolerance or normal glucose tolerance. Consequently, breath analysis shows some potential as a replacement technique for GDM screening. This could lead to a significant improvement in the diagnosis of gestational diabetes as the OGTT is quite onerous. For example, OGTT is extremely time-consuming, taking several hours and requires several blood samples taken over a period of time. Moreover, OGTT is not suitable for women suffering from severe pregnancy sickness. This is particularly true for women who are suffering from Hyperemesis Gravidarum, because they are unable to tolerate the glucose drink and may be too ill to even attend a clinic for the test.

Oral glucose tests have also been investigated by Krug *et al.*, this time for 15 healthy male volunteers [101]. In fact, this was just one part of a more extensive and comprehensive study to analyse the dynamics of the human metabolome in response to environmental stimuli, including 36 hours of fasting, oral lipid tests, liquid test meals, physical exercise and cold stress in addition to the oral glucose tests. Not only was breath analysis using PTR-MS investigated but also blood, urine and breath condensates were analysed using both mass spectrometric and nuclear magnetic resonance-based techniques. This comprehensive study has provided a unique and large data set which should be of considerable use for other metabolomic studies.

Since all PTR-MS breath studies dealing with drugs have so far been restricted to a small number of patients, more extensive investigations are required before any definitive conclusions can be drawn. Nevertheless, the preliminary work illustrates the potential of using PTR-MS to monitor drugs taken into the body (either orally or directly injected into the blood) and its application in pharmacokinetics. These studies also raise the question as to whether breath analysis could be routinely used to check if a patient is taking his/her medication. This may seem a trivial application. However, not only are considerable resources wasted as a result of patients not taking their medications but also patients are risking their health. A quick and simple non-invasive procedure, such as breath analysis, is needed to monitor pharmaceutical products, their uptake and distribution within a patient and potentially any side-effects.

7.2.7 Breath VOC Levels Measured Using PTR-MS versus Blood Concentrations

It is generally assumed that the breath concentrations of endogenous VOCs correlate with those in the blood and hence that end-expired breath is equilibrated with arterial blood. This means that VOCs in breath should be a reflection of their concentrations in blood. However, few studies have investigated this assumption. O'Hara *et al.* compared breath concentrations of acetone and isoprene in the breath and in blood from samples taken from healthy volunteers (six males and four females) [41, 44]. These VOCs were selected because they are always present in breath in trace quantities and because they have very different solubilities in blood (and as mentioned earlier solubility is known to alter the dynamics of gas exchange). The breath samples were collected in bags using a rebreathing technique that had previously been developed by O'Hara *et al.* [60] and the samples were then analysed within minutes using a PTR-Quad-MS. Both peripheral venous and arterial blood samples were collected and independently analysed by a commercial company by means of solid-phase micro-extraction and GC–MS. The sample-mean radial arterial blood/breath ratios (ranges) for acetone and isoprene were found to be 580 (280–1060) and 0.47 (0.22–0.77), respectively. Although limited in the number of volunteers, the study resulted in some interesting observations. The most important was that it indicated that there is indeed a direct correlation between VOC concentrations in alveolar air and in radial arterial blood, at least for acetone and isoprene. For acetone, levels in breath also correlated with peripheral venous levels. However, levels of isoprene in breath showed no correlation with that in peripheral venous blood. Although a peripheral venous blood sample is easier to collect than an arterial blood sample, this lack of correlation with isoprene should be noted by researchers wishing to compare blood and breath levels of VOCs.

Another interesting observation was that the breath concentrations taken over a period of time showed far less inter-subject variability than the concentrations determined from the blood measurements, which is speculated to be due to a large volume of blood having flowed through the lungs over the observation period. This inter-subject variability is illustrated

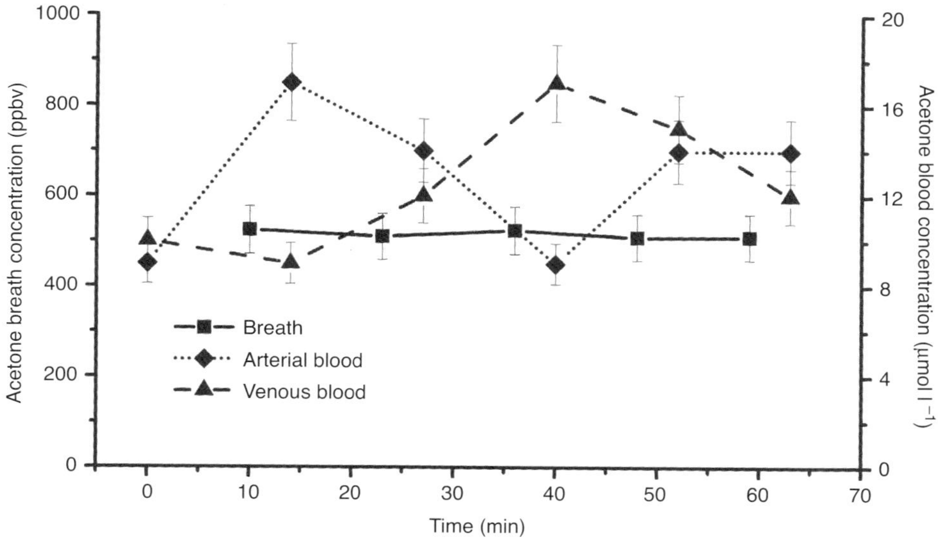

Figure 7.4 *Concentrations of acetone in the breath and blood for an individual volunteer plotted as a function of time. The lines drawn connecting the data points are visual guides only. Error bars represent a conservative experimental error of 10%. Note the much larger variability in the blood concentration measurements, which is not associated with any experimental uncertainties, compared to the breath acetone concentrations. Reproduced from [41] with permission from Elsevier.*

in Figure 7.4 for acetone for a particular volunteer. These studies suggest that breath VOC measurements may provide a more consistent measure than analysis of a single aliquot of blood for investigating underlying physiological function or pathology within individuals. However, it remains to be seen whether the conclusions made from this study are applicable to other endogenous VOCs beyond acetone and isoprene.

7.2.8 Breath Sampling and PTR-MS

Given that breath analysis has considerable potential to be used as a non-invasive diagnostic technique, it is surprising that no systematic procedure and protocols have evolved for breath sampling by PTR-MS. Without this, poorly designed and random breath sampling methods will limit the reproducibility and applicability of any analytical technique, and therefore ultimately limit confidence in the use of breath analysis. Thus key to the adoption of PTR-MS, or any other analytical technique for that matter, for breath analysis is a reproducible, reliable, easy-to-use, easy-to-handle, cheap and rapid sampling process, either for offline or online measurements, followed by standardized analysis procedures. The breath analysis community need to agree on one or more standardized sampling and analysis procedures: one procedure may not suffice owing to differences in the physical properties of various VOCs, for example, solubility.

At the time of writing, no standard procedure has been adopted. In the next few sections, we discuss possible breath collection procedures, both for online and offline sampling.

7.2.8.1 Offline Breath Sampling

As mentioned numerous times in this book, the major advantages of PTR-MS technology compared to GC–MS include its real-time and online measuring capabilities. However, real-time online measurements are not always practical and therefore sampling procedures have been adopted so that analysis can be done offline. Much of the earliest work on breath analysis using PTR-MS was done offline using Teflon bags (see, for example, Reference 22). However, Teflon bags are expensive, fragile and permeable in both directions and are therefore not ideal for clinical use.

The sampling of breath into bags has been investigated by Harren's group in the Netherlands [59, 65], but in their case Tedlar was the chosen material. The first study by this group investigated the suitability of black-layered Tedlar bags for breath collection of mixed expiratory air and subsequent analysis with PTR-MS [65]. Tedlar bags were found to emit VOCs, with ions being detected at *m/z* 88 and 95. The species responsible for *m/z* 88 was assigned to protonated *N,N*-dimethylacetamide (C_4H_9NO), which is a compound used in the production of Tedlar film, while *m/z* 95 was attributed to protonated phenol. These compound assignments were confirmed by an independent GC–MS analysis. An issue associated with bags is chemical degradation, surface adsorption and VOC loss resulting from diffusion through the bags. The half-lives of various VOCs in Tedlar bags, namely methanol, acetaldehyde, acetone, isoprene, benzene, toluene and styrene, were therefore also investigated by this group, with the half-lives being found to be between 5 and 13 days. Although these half-lives are relatively long, they still highlight that it is best to analyse the breath samples as quickly as possible after collection for the most reliable, contaminant-free measurements. In a second paper by the same group, Harren and co-workers developed an offline breath sampling and analysis method that is suitable for large screening studies [59]. For this work, a breath collection procedure was adopted which was based on a standard recommended by the American Thoracic Society for the offline sampling of exhaled NO. It was found that consecutive fillings of five black-layered Tedlar bags by the same person gave reproducible results to within a 12% relative standard deviation for methanol, acetaldehyde, acetone and water. However, isoprene was found to be more variable with a 30% relative standard deviation.

Beauchamp *et al.* also investigated the use of Tedlar bags for breath sample collection, whereby a sample bag was filled with a mixture of VOCs (containing alcohol, nitrile, aldehyde, ketone, terpene and aromatic compounds) in known concentrations and air was then sampled regularly from that bag over a 70-hour period [62]. As expected, the concentrations of these compounds declined over time owing to the various loss processes mentioned above. The researchers concluded that accurate results can only be obtained providing measurements are made within 10 hours of sample collection. The same contaminant ions at *m/z* 88 and 95 seen in the work by Harren and co-workers [65] were also observed by Beauchamp *et al.*

Thekedar *et al.* have investigated breath sampling using Teflon bags [63, 64]. In their first study, the variability of exhaled breath sampling was explored [63]. In a subsequent study, the investigation of breath samples in bags was taken a step further by monitoring the effects of different breath sampling-specific parameters on the breath VOC profiles [64]. The concentration of the exhaled VOCs was found to depend on factors such as the rate of exhalation, whether multiple or single exhalations were used, different volumes of exhalation, holding breath and differences in surrounding air conditions.

An investigation by O'Hara *et al.* found that end-expired levels of many VOCs (i.e. VOCs in breath samples that are representative of alveolar air) from a single expirogram are highly variable [60]. To try and rectify this, a multiple cycle isothermal rebreathing protocol was developed. The term isothermal refers to the fact that the rebreathed air is maintained at temperatures near to that of the body. Rebreathing procedures increase the alveolar VOC concentrations in the breath sample and address the variability observed in online sampling of low solubility gases, because it ensures that gas exchange equilibrium is established in the entire respiratory tract. To characterize this technique, three important breath VOCs with different solubilities were investigated: isoprene, acetone and methanol. It was found that the concentration of these volatile compounds increased with the number of rebreaths until a plateau value was reached after approximately 20 rebreaths. A key conclusion from this work was that uncontrolled single exhalations are not suitable for the reliable measurement of isoprene in breath. Although this rebreathing procedure showed good repeatability, the problem is that whilst it works well with healthy volunteers, it has not been demonstrated in a clinical setting using sick patients, who may not be able to provide a multiple breath sample or be able to follow simple instructions. Furthermore, the effects of changes in O_2 and CO_2 levels, which will change the pH of the aqueous breathing system, are unknown, as too are their effects on the structure of the alveoli. Thus this rebreathing procedure is likely to find only limited use as a clinical tool.

7.2.8.2 *Online Breath Sampling*

A much easier method for PTR-MS breath sampling, and which is online, has been developed by Herbig et *al.* [61]. This is the so-called buffered end tidal (BET) sampling method already mentioned in section 7.2.6. This sampling system is designed so that it buffers only the end-tidal fraction of the breath (which is more commonly referred to as the alveolar breath that comes from deep within the lungs), and thus the time for sampling of the end-tidal phase is greatly extended. To use this system, the volunteer exhales through a buffer tube for just one complete exhalation. The advantage of this technique is that it is relatively easy to use, since only single breaths are required and it aids in optimizing the collection of alveolar breath. It is probably the most reliable and easy-to-use breath sampling technique currently available for online measurements. A schematic drawing of the BET device is provided in Figure 7.5(a) and a typical exhalation profile obtained using such a system is presented in Figure 7.5(b). A photograph of the BET apparatus is shown in Figure 7.5(c).

7.2.9 PTR-MS and Breath Analysis: Requirements and Future Directions

To date breath analysis has seen only limited application in a clinical setting, for example, monitoring of $^{13}CO_2/^{12}CO_2$ ratios in breath following ingestion of ^{13}C-labelled substances – a procedure used to diagnose infection with *Helicobacter pylori*. Without detailed and careful fundamental research and in-depth clinical studies, breath analysis cannot become a routine clinical technique for non-invasive detection, prognosis and monitoring.

The accuracy and hence viability of breath analysis as a technique for the definitive screening of diseases at the molecular level still needs to be confirmed. There is thus a real need to search for volatile markers on breath and to verify if specific patterns of biomarkers

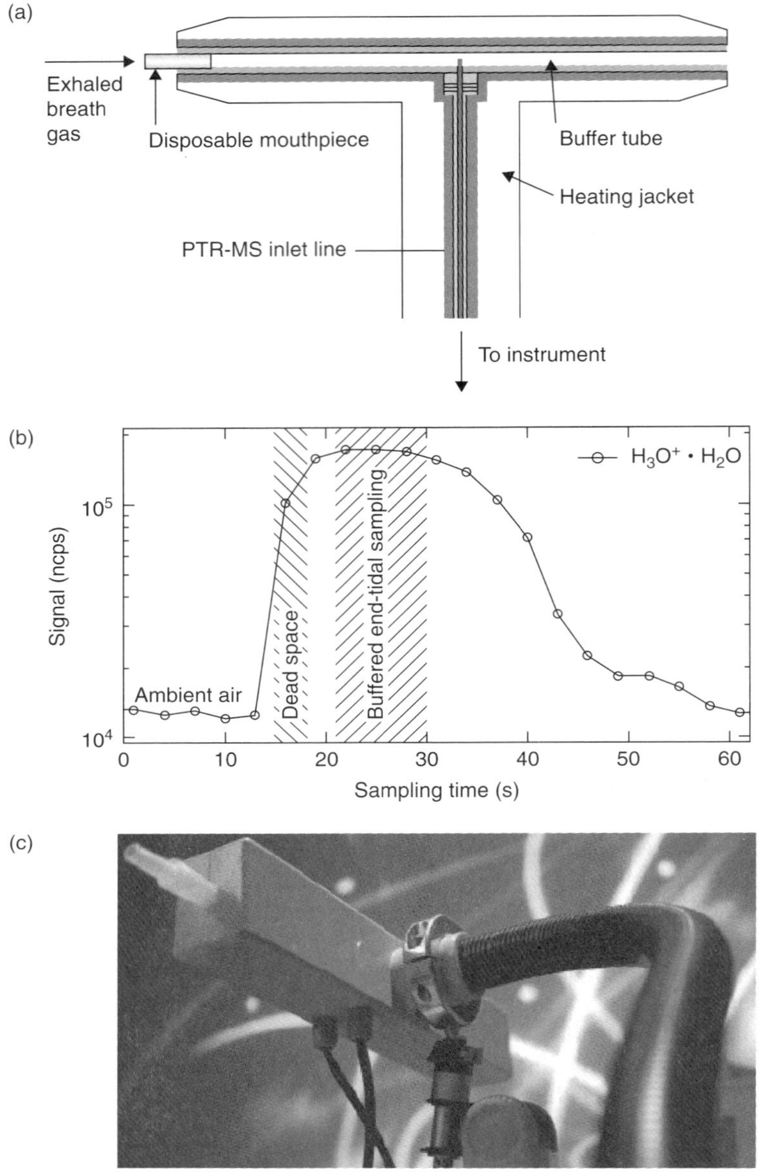

Figure 7.5 (a) A schematic representation of a buffered end-tidal (BET) online breath sampling device. The end-tidal fraction is buffered in the heated buffer tube. The analytical instrument draws the sample gas through its inlet line from the middle of the buffer tube. (b) Typical exhalation profile for BET sampling measured continuously with PTR-MS for the protonated water dimer. The sharp increase marks the onset of the exhalation. The dead space (first shaded area) is sampled shortly after the onset and is omitted in the data analysis. The following plateau resembles the buffered sampling of the end-tidal fraction. The second shaded area marks the interval available for BET data taking, which is stable for about 10 s and then slowly decays to the room-air concentration. Reproduced from [61]. © IOP Publishing. Reproduced by permission of IOP Publishing. All rights reserved. (c) A photograph of the BET system showing the heated inlet line coupled directly to the PTR-MS, which was kindly provided by Alex Braun. Reproduced with permission from Alex Braun.

can be used to (a) identify various diseases with sufficient sensitivity and specificity so that they are clinically significant, that is, to determine if there are differential molecular profiles between different diseases, (b) assess patients throughout therapy and (c) identify markers of drug/radiotherapy toxicity. This has the potential of ultimately leading to a new generation of advanced chemical detectors deployed to diagnose or screen for disease at the point of care without the need for specialist laboratories. However, before breath analysis can be introduced into clinical practice a number of crucial technical and scientific issues first need to be addressed and these, many of which have already been voiced in the literature (see, for examples, References 34, 67, 101–105), are summarized below:

(i) The development and application of appropriate breath sampling methods applied under carefully controlled conditions to provide reliable and reproducible results. (The actual sampling method will almost certainly be to some extent VOC specific and ideally the sampling method should be representative of end-tidal breath.)
(ii) The development and standardization of breath data analysis protocols. In particular to develop and test suitable statistical analysis and modelling methodologies, including multivariate statistical methods. Examples are principal component analysis or discriminant function analysis, and possibly pattern recognition algorithms to analyse the mass spectra obtained in order to provide discriminatory procedures that can be adopted for patients suffering with a given disease.
(iii) The determination of the normal VOC ranges for healthy controls (normal ranges etc), with dependences (if any) on age, gender, ethnicity, body mass index, diet, oral hygiene, environment, health, medication, smoking and drinking habits. Such studies are essential to determine ordinary biological variation. This will generate limits on normal concentration ranges for various biomarkers.
(iv) The evaluation of patterns of VOCs in the breath of patients suffering with a given disease and statistically compare them to healthy controls in order to determine if breath analysis can provide discrimination between healthy and ill patients.
(v) The undertaking of large-scale and in-depth clinical validation programmes involving patients suffering from a given disease in order to determine the accuracy and viability of breath analysis for screening, diagnosing and monitoring.
(vi) The correct identification and determination of the concentrations of VOCs in the background air.
(vii) The investigation into whether people/patients can act as their own control, that is, breath samples taken in routine regular medical examinations or to check for the effects of treatment.
(viii) The correct identification of VOCs using GC–MS or identification of the breath VOCs with a high level of confidence by using high-mass resolution PTR-TOF-MS instruments. This will provide an improved understanding of the biochemical pathways of VOC production in the human body.
(ix) The compliance of all devices (e.g. breath sampling and detection systems) with medical device standards.
(x) The development of low-cost VOC detectors, such as ion mobility spectrometers (see Section 1.3.2), to make VOC detection more widely available.

To date the majority of research associated with breath analysis has been limited to small feasibility trials without any detailed follow-up [104]. It will be necessary to revisit, revise, reproduce and validate these preliminary measurements. In-depth clinical studies and research are required to identify conclusively the key chemical compounds in the breath of patients either suffering from various diseases or using certain drugs and to determine if they can be accurately and reliably used as a clinical differential diagnostic tool.

By introducing new instrumentation and by developing new measuring and analysis protocols, there are real possibilities that many forms of disease could be screened for or detected in their earliest, and hence most treatable, stages. This would result in considerable benefits for society and have a direct impact on the quality of life. Breath analysis also has the potential of offering significant cost and operational benefits to healthcare providers and improved care for patients. Patients will also benefit from swifter, more accurate and less intimidating procedures.

7.3 Online PTR-MS Measurements of Volatile Emissions from Microbial Cultures

It is well known that microbes, such as bacteria and fungi, emit characteristic odours that the human nose can smell, for example, an 'earthy' smell is associated with fungal-contaminated environments, a 'fishy' smell is associated with bacterial vaginosis, or a 'grape-like' smell is associated with *Pseudomonas aeruginosa*. These smells are a result of metabolic processes leading to an emission of characteristic VOCs from the microbes. Consequently, the analytical detection of these VOCs is a potential means for rapid microbial identification and is an interesting and fruitful area of research.

7.3.1 Bacteria

As well as being by-products of metabolic processes, VOCs emitted are also released from bacteria as a means of communication between bacteria [106]. However, the role played by bacterial VOC emissions for communication or in response to the environment is not well understood. Thus, just as has been found for complex organisms such as plants, it would appear that even simple organisms, such as bacteria, have also evolved to 'sense' airborne VOCs and to take appropriate actions in response to the stimulus they evoke. Understanding such processes may also assist in the development of drugs to fight bacterial infection.

Given that different microbial cell cultures emit characteristic aromas, sampling of the headspace above microbes should lead to different PTR-MS mass spectral information according to the type of cell cultures, and hence provide a basis for assignment and discrimination. There have been a number of PTR-MS studies interrogating the headspace above bacterial cultures [11–18, 107, 108]. To date the studies have involved *in vitro* measurements. However, arguably the greatest potential of PTR-MS in this field is to exploit its *in situ* real-time capabilities to undertake *in vivo* breath analysis in order to rapidly identify bacterial infection. Real-time non-invasive diagnosis of bacterial infection in the body would enable the correct antibiotic treatment to be administered more rapidly. This is crucial for the effective and early management of bloodstream infections because it has been well documented that appropriate early treatment results in lower mortality rates.

The first studies of microbial VOC emissions measured using PTR-MS were undertaken by Werner Lindinger's group in Austria [11, 107, 108]. However, these were associated with meat spoilage rather than investigating VOC emissions from specific bacteria, and these studies have already been discussed in detail in Chapter 6. In a proof-of-principle study, Critchley *et al.* were the first to report the use of PTR-MS for *in vitro* headspace analysis of volatile emissions from specific microbial cultures [12]. In their study, air was sampled above pure blood agar cultures of *P. aeruginosa* and *Streptococcus milleri*. Encouragingly, the two spectral profiles obtained were dramatically different and this highlighted that it should be possible to discriminate between different types of bacterial infection using mass spectral 'fingerprint' analysis. This early work is illustrated in Figure 7.6. To obtain the PTR-MS mass spectra presented in this figure, background mass spectra were first acquired of the headspace air over uninoculated blood agar in the same container. The background spectra were then averaged and subtracted from the averaged spectra recorded from each cell culture. As can be seen from Figure 7.6, the difference between the two mass spectra is dramatic. Whilst a number of mass peaks are common, it should be noted that the intensity scales are very different. In addition, there are a number of different mass peaks in the two spectra. Although only a pilot study, the work by Critchley *et al.* demonstrated the potential of PTR-MS for this type of online analysis and its use for discriminating different bacterial species. Subsequent research by other groups, described below, has consistently shown that distinctive pattern of VOCs are emitted by different bacterial species.

Profiling of VOC emissions from different strains of *Escherichia coli*, *Klebsiella*, *Citrobacter*, *P. aeruginosa*, *Staphylococcus aureus* and *H. pylori* was performed in an *in vitro* study by Lechner *et al.* [19]. In agreement with the earlier microbial profiling based on VOC emissions by Critchley *et al.*, the work by Lechner and co-workers showed substantial differences in the mass spectral patterns between the various microbial cultures. Furthermore, certain m/z ions were associated with specific bacteria, an observation that provided further support for the idea that the analysis of VOC emissions from bacteria using PTR-MS could be used to differentiate bacterial strains *in vitro*.

Bunge *et al.* investigated the headspace above *E. coli*, *Shigella flexneri*, *Salmonella enterica* and *Candida tropicalis* bacterial cultures [14]. It was found that *C. tropicalis* could be uniquely identified by several characteristic marker ions, while *E. coli* and *S. enterica* could be discriminated from each other and from *S. milleri* by specific marker ions. The temporal profiles of the VOC concentrations were also investigated for the first time in this work. It was concluded that temporal effects during cell growth need to be taken into account for the characterization or differentiation of cell types based on VOC emissions.

O'Hara *et al.* also investigated the temporal profiles of bacterial emissions [15]. However, in their study they used just one type of microbe, namely *S. aureus*, and investigated the dependence of the temporal profile of VOCs on the type of growth medium. Three types of broth feed for bacteria were used in this study; nutrient, dextrose and brain heart bovine. The important conclusion from this work was that the VOCs emitted by *S. aureus* in each medium were observed to be the same. However, the time-dependent mass spectra showed marked differences and intensities between the different growth media. This is in contrast to the work reported in the doctoral work by Steeghs, who undertook *in vitro* measurements of *Mycobacterium tuberculosis*, *Mycobacterium avium* and *Mycobacterium kansassii* using PTR-MS [109]. The volatiles emitted from these bacteria were found to be strongly dependent on the medium in which they were cultured. The difference in observations between the work of O'Hara *et al.* and Steeghs requires further investigation. From their investigation O'Hara *et al.* concluded that, given the high

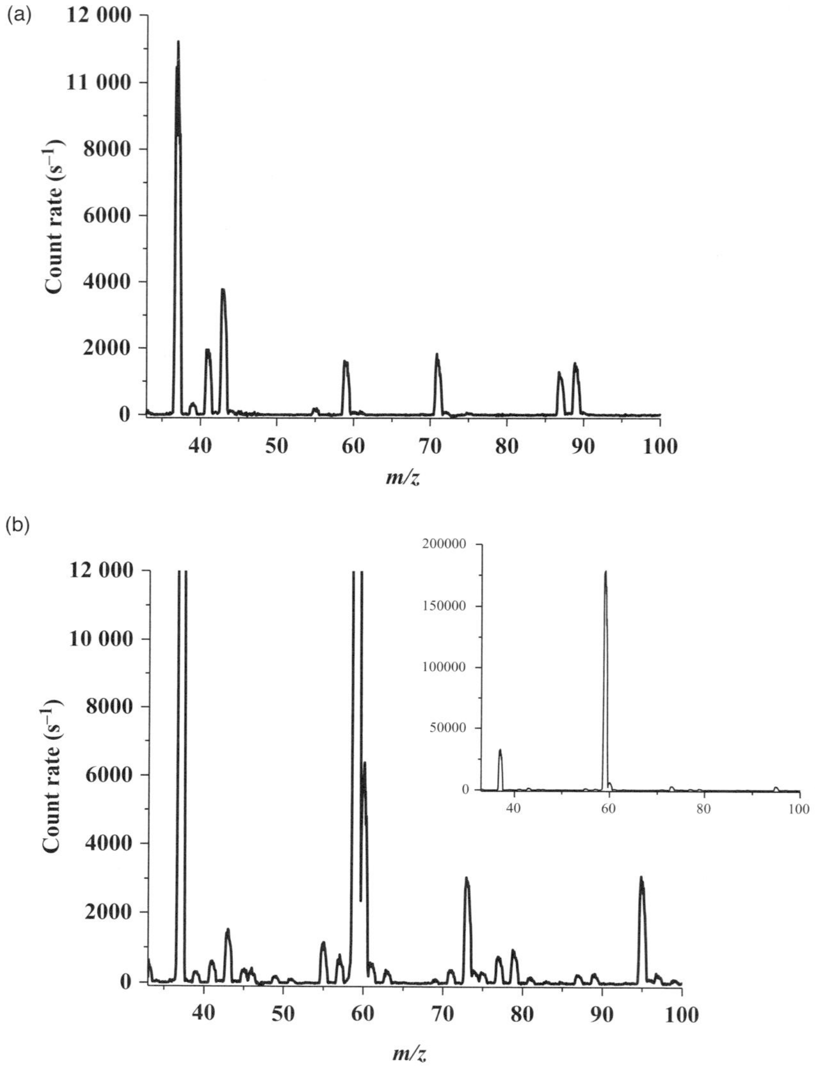

Figure 7.6 *Mass spectra resulting from the reaction of H_3O^+ with the headspace emitted from a cell culture containing (a)* Streptococcus milleri *and (b)* P. aeruginosa. *Reproduced from [12] with permission from Elsevier.*

sensitivity of the PTR-MS instrument and the high concentrations of the VOCs emitted from the *S. aureus* culture, even after short incubation times it should be possible to diagnose infection from *in vivo* measurements. This suggests that the use of VOC emissions as a signature of specific infections is plausible and it might ultimately be achievable through breath analysis, although much more research will be needed if this is to become a viable proposition.

The limitation of all the above studies is that the identification of VOCs leading to the observed ions is often speculative. By combining GC–MS and PTR-MS, Kai *et al.* provided a comprehensive profile of VOCs emitted from the rhizobacterium (root associated bacterium) *Serratia odorifera* [16]. Sodorifen ($C_{16}H_{26}$, a bicyclic polymethylated diene) was found to be the dominant compound emitted, as determined from the GC–MS measurements. This compound was not detected in the PTR-MS measurements because the mass range was limited to 20–160 Da. Volatiles observed by PTR-MS included dimethyl disulfide, dimethyl trisulfide, methanethiol, terpenoids, 2-phenylethanol and various other aromatic compounds. Specific compound assignments were partly based on isotopic pattern analysis of the PTR-MS mass spectral profiles, but predominantly through the use of GC–MS. Over a period of 96 hours and with measurements taken at 45-min intervals, the emission profiles of methanol (at *m/z* 33), ethanol (at *m/z* 47), methanethiol (at *m/z* 49), benzene (at *m/z* 79), dimethyl disulfide (at *m/z* 95), dimethyl trisulfide (at *m/z* 127) and monoterpenes (at *m/z* 137) were all monitored. A sharp rise in emissions occurred during logarithmic bacterial growth, that is, the phase during which the bacterial population undergoes exponential growth.

Crespo *et al.* performed a very interesting set of experiments whereby they studied, in real-time, the VOCs in the headspace of actively growing cultures of paired ciprofloxacin sensitive and resistant bacterial strains of *Mycobacterium smegmatis* [17]. These measurements were made before and after the addition of the antibiotics ciprofloxacin and gentamicin, which are two antimicrobial agents that have very different mechanisms of action. In addition to using a standard quadrupole PTR-MS instrument, an ion trap mass spectrometer was also used so that MS/MS techniques could be employed to aid in the identification of specific compounds. For the first 24 hours of growth, little change in the VOCs emitted from the culture and the growth medium was observed. Thereafter, the levels of acetaldehyde and ethanol dropped below that released from the medium alone, and once depleted acetone was found to increase dramatically, which the researchers attributed to a change in metabolism. These findings are illustrated in Figure 7.7. The VOCs emitted in response to each antibiotic were also measured. In particular, it was found that an ion signal at *m/z* 73 (associated with unidentified C_4 compounds) doubled in intensity within 10 hours after the addition of ciprofloxacin and an ion signal at *m/z* 45 (protonated acetaldehyde) tripled in intensity within 4 hours after the addition of gentamicin. Figure 7.8 illustrates the effect of adding ciprofloxacin 29 hours after inoculation to the susceptible strain (A), the resistance strain (B) and to the medium on specific ions (*m/z* 57, 75, 83 and 101). A clear effect was observed for strain A within hours, whereas strain B did not show any response (the minor response, which occurred at the time point when acetone levels increased dramatically, was attributed to a change in metabolism or mild stress). Importantly this work has demonstrated that headspace analysis can potentially provide a rapid drug susceptibility test.

Crespo *et al.* have also carried out an online investigation of the VOC emissions from two slow-growing non-tuberculosis mycobacterial species, *M. avium* and *Mycobacterium kansasii*, in addition to a non-pathogenic fast growing species, namely *M. smegmatis* [18]. As found with the other studies investigating bacterial emissions, the differences in the PTR-MS mass spectral profiles obtained were sufficient to allow discrimination between the microbial cultures. In particular, three ions at *m/z* 63, 77 and 95, which the researchers attributed to protonated dimethyl sulfide, propanethiol and dimethyl disulfide, respectively, were only observed to be emitted from *M. avium*. GC measurements were

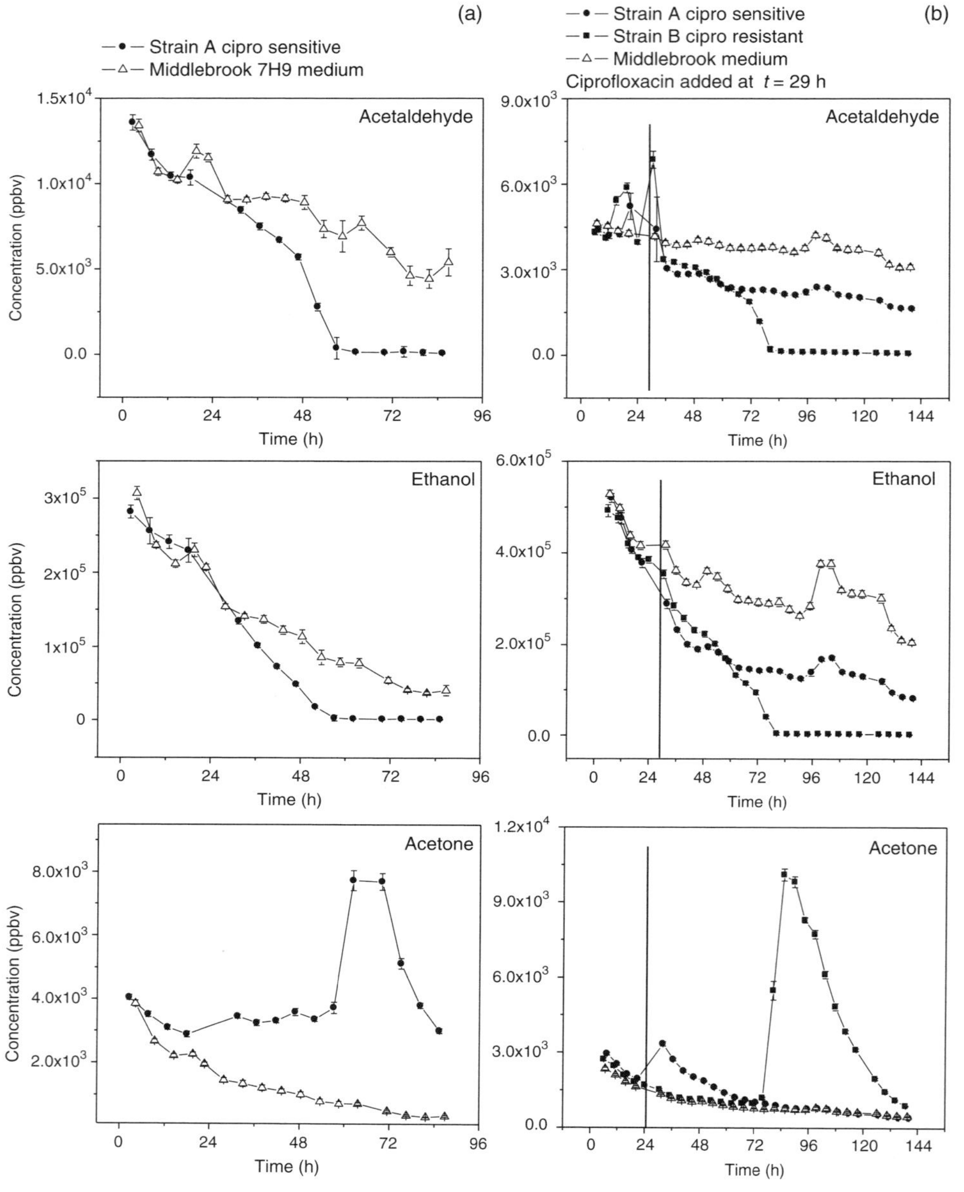

Figure 7.7 *Panel (a): time-resolved emission profiles of acetaldehyde, ethanol and acetone from* M. Smegmatis *cultures during growth (filled circles) and medium alone (empty triangles). Panel (b): time-resolved profiles of acetaldehyde, ethanol and acetone before and after the addition of ciprofloxacin at* t = *29 hours. The resistant strain B shows the same behaviour with and without ciprofloxacin, while the sensitive strain A shows a reduction in the rate of decrease in emissions when compared with the resistant strain B. Reproduced from [17] with permission from Elsevier.*

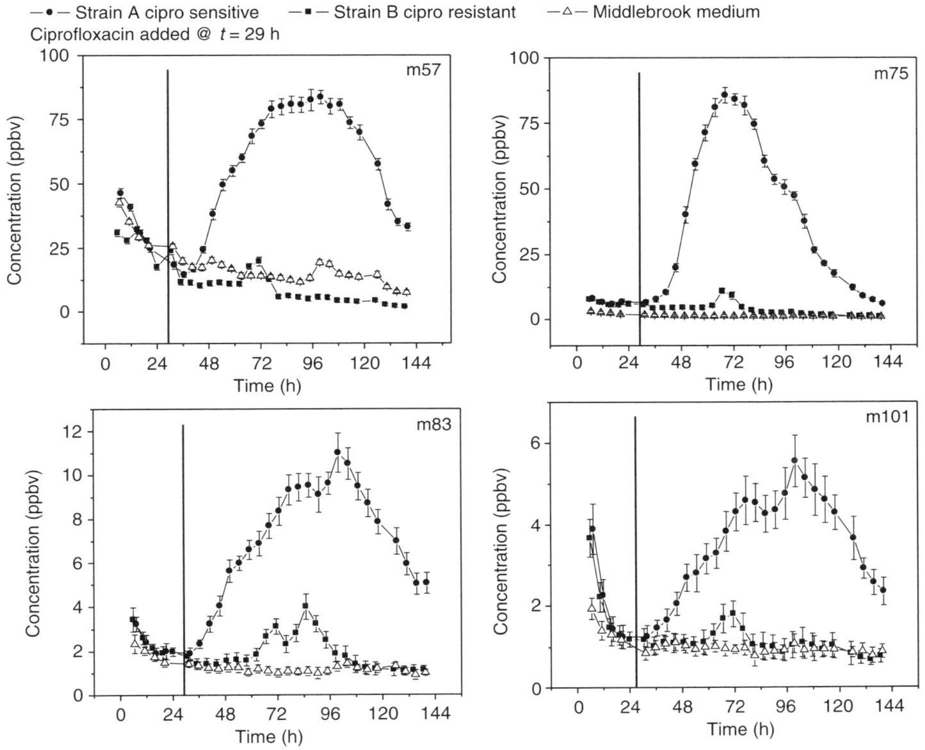

Figure 7.8 *Effect of ciprofloxacin on* M. smegmatis *headspace showing time profiles of VOC emissions for the most significant masses at m/z 57, 75, 83 and 101 after the addition of ciprofloxacin (7 μg mL^{-1}) at t = 29 h (black bar) to strain A (ciprofloxacin sensitive, filled squares) and strain B (ciprofloxacin resistant, filled circles and Middlebrook 7H9 medium alone, empty triangles). For the cultures 5 mL of bacteria with a density of 1.7 × 10^8 bacteria/mL were inoculated in a 75 mL of medium at the start (t = 0 h). Reproduced from [17] with permission from Elsevier.*

later used to confirm the assignments of *m/z* 63 and 95 but for *m/z* 77 no propanethiol was detected, presumably because of the lower sensitivity of the GC system compared with PTR-MS. In addition to mass spectral profiling, a number of ions of different *m/z* values with high intensities were selected to be monitored during bacterial growth for up to 7 days. From such studies, temporal dependences of the VOC emissions could be determined. For example, it was found that whilst methanol emissions in *M. avium* increased with increasing mycobacteria growth, that from *M. smegmatis* and *M. kansasii* did not. Thus methanol production appears to be a good marker for *M. avium* growth.

The unpleasant odours and potentially pathogenic microorganisms that result from domestic waste collection and dumps are related in some respects to health issues and health risks associated with microbial cultures. Mayrhofer *et al.* investigated whether the VOCs emitted as a result of decaying biowaste are related to the presence of specific microorganisms [13]. The succession of microbial communities during 16 days of storage in organic waste collection bins was followed by using denaturing gradient gel electrophoresis (DGGE) in parallel with a classical cultivation and isolation approach. Over this 16-day period, approximately 60 different bacterial species and 20 different fungal species were isolated. Then, by using co-inertia analyses[3] of the VOC concentrations recorded using PTR-MS with the DGGE band intensities, both positive and negative correlations of band intensities with the measured VOCs were observed. The positive correlations indicated the enhancement of microbial growth, whereas the negative correlations corresponded to microbial inhibition by the respective VOC. By studying the VOC emissions from a pure culture, *Lactococcus lactis*, a positive correlation was found between microbial growth and VOC emissions associated with ions at *m/z* 89 (tentatively assigned to protonated butyric acid), *m/z* 73 (possibly protonated butanone), *m/z* 69 (considered to result from protonated isoprene) and *m/z* 63 (tentatively assigned to be protonated dimethysulfide). (Without GC–MS measurements or the availability of high-resolution PTR-TOF-MS spectra, the actual identification of the VOCs leading to these ions is of course uncertain.)

7.3.2 VOC Emissions from Fungi

The first use of PTR-MS to investigate VOCs emitted from fungi was by Ezra *et al.* [110]. This study focused on VOC production from an endophytic fungus, *Muscodor albus*, which is used to both inhibit and kill other microorganisms. Two Petri dishes were employed, with both containing potato dextrose agar. One of these was inoculated with *M. albus* and the other was kept as a control. Both of these dishes were sealed and incubated at 23°C. After 6 days, the headspace of the two Petri dishes was sampled for initial identification of VOCs. The research team then investigated the emission of selected VOCs as a function of time, temperature and light exposure, and also from an *in situ* experiment with *M. albus* in soil. It was found that the most abundant VOC observed under the conditions used was ethanol, as registered by signals recorded at *m/z* 47 (the protonated parent, MH^+), *m/z* 65 ($MH^+(H_2O)$) and *m/z* 83 ($MH^+(H_2O)_2$). In a PTR-MS study of ethanol by Brown *et al.*, no significant clustering of protonated ethanol with water was observed at any *E/N* value between 90 and 140 Td [111]. Unfortunately, Ezra *et al.* make no reference to the operating conditions used for their PTR-MS measurements. However, based on a comparison with the results of Brown *et al.*, the observation of water cluster ions indicates that the air being sampled was highly humid. Another ion signal was also observed at *m/z* 93, which was tentatively assigned to the protonated ethanol dimer rather than protonated toluene owing to the very high concentrations of ethanol being emitted by the fungus.

Using PTR-MS, Strobel *et al.* investigated emissions from a filamentous fungus called *Gliocladium roseum* [112]. *G. roseum* was found to emit a volatile short chain hydrocarbon mixture not dissimilar to that found in diesel. As a result of this the researchers coined the term 'myco-diesel' for the volatiles being emitted to highlight the potential of generating biodiesel from fungi. Related to this, Mallette *et al.* have used PTR-MS (in conjunction with SPME GC–MS) to investigate the volatile profile of *Ascocoryne sarcoides*, which is a species of fungus in the Heliotiaceae family, over a growth cycle [113]. PTR-MS measurements permitted temporal resolution investigations whereas SPME GC–MS allowed

[3] This is an analytical procedure for matching two sets of data that have been previously analysed by a standard method, for example, PCA. It is used to investigate species-environment relationships.

compound identification. The PTR-MS results showed that the VOC emissions were dominated by compounds that gave rise to high signal intensities at *m/z* 45 and 47, which were identified by GC–MS to result from protonated acetaldehyde and ethanol, respectively. 'Fuel' VOCs were also observed, including nonanal, 1-octen-3-ol, 1-butanol, 3-methyl- and benzaldehyde.

Ethanol fermentation within lichens under oxygen-deprived conditions has been shown by PTR-MS to emit acetaldehyde and ethanol. The researchers concluded that ethanol fermentation represents both a source of atmospheric VOC emissions and a so far unconsidered carbon loss route for lichens [114].

7.3.3 Concluding Remarks on Microbial Emissions

In vitro headspace analysis by PTR-MS has shown that there are sufficient differences in the mass spectral profiles from different microbial cultures to permit rapid discrimination. Table 7.1 provides a summary of the various product ions (collated by *m/z*) observed, together with tentative assignments of the VOCs that give rise to these ions and the bacterial/microbial cultures that resulted in the volatile emissions. Whether the actual VOCs emitted are dependent on the growth material the microbial cultures are feeding from is uncertain and further investigations are needed to determine this.

Although the results from the various PTR-MS studies are extremely encouraging for the use of headspace analysis above microbial cultures, the main question to address is whether current mass spectrometric analytical techniques have sufficient sensitivity to identify an infection in an asymptomatic person by the *in vivo* analysis of the VOCs on their breath in real-time, so that pre-emptive medicine can be administered before an infection becomes serious. Although the answer to this can only be obtained from further research, the results obtained from the *in vitro* studies are encouraging.

Another important issue to address is the definitive identification of the VOCs emitted from microbial cultures, for which there has only been a limited amount of work to date. Such assignments would shed light on the metabolic processes occurring in the cultures, which are of considerable interest to microbiologists and may aid in the development of new antibiotics. However, we comment that these are not necessarily required to ascertain the type of bacterial infection, that is, a 'fingerprinting' approach may be more than sufficient for use in discriminatory procedures.

7.4 Other Medical Applications

The use of PTR-MS in medical science is not just restricted to breath analysis and online *in vitro* measurements of microbial cultures. PTR-MS has considerable potential for application in other medical areas, such as the analysis of the headspace above liquid and solid biological samples (e.g. urine, faeces or sputum samples) or detecting emissions from skin (a potential application being the early detection of skin cancers), but surprisingly there have been only a limited number of such studies at the time of writing. Furthermore, with the development of direct aqueous injection inlet systems (see Section 9.2), PTR-MS is not just limited to headspace analysis but now also has the potential to analyse liquid samples directly.

Table 7.1 *Observed product ion m/z (following proton transfer) and tentative VOC assignment associated with the headspace above bacterial cultures*

m/z	Assignment	Microbial culture
33	Methanol[b,c,e]	Meat spoilage (aerobic bacteria, *Pseudomonas* spp., *Enterobacteriaceae*, lactic acid bacteria, *Enterococcus* spp.)[b]; *E. coli*,[c] *S. flexneri*,[c] *S. enterica*,[c] *C. tropicalis*,[c] *S. odorifera*,[e] *M. kansasii*,[f] *M. avium*,[f] *M. smegmatis*[f]
35	Hydrogen sulfide[e]	*S. odorifera*[e]
39	Furan and/or a fragment ion of isoprene[d]	*S. milleri*,[a] *P. aeruginos*,[a] *E. coli*,[c] *C. tropicalis*,[c] *S. aureus*[d]
41	Propyne, propadiene and/or cyclopropene[d]	*S. milleri*,[a] *P. aeruginosa*,[a] *E. coli*,[c] *S. flexneric*,[c] *C. tropicalis*,[c] *S. aureus*,[d] *M. kansasii*,[f] *M. avium*[f]
43	Fragment ion of acetic acid, propylene, propan-1-ol or propan-2-ol[d]	*S. milleri*,[a] *P. aeruginosa*,[a] *E. coli*,[c] *S. flexneri*,[c] *S. enterica*,[c] *C. tropicalis*,[c] *S. aureus*,[c] *S. odorifera*,[e] *M. kansasii*,[f] *M. avium*[f]
45	Acetaldehyde[c–f]	*P. aeruginosa*,[a] *E. coli*,[c] *S. flexneri*,[c] *S. enterica*,[c] *C. tropicalis*,[c] *S. aureus*,[d] *S. odorifera*,[e] *M. kansasii*,[f] *M. avium*,[f] *M. smegmatis*[f]
47	Ethanol[b–f] and/or formic acid[d]	Meat spoilage (aerobic bacteria, *Pseudomonas* spp., *Enterobacteriaceae*, lactic acid bacteria, *Enterococcus* spp.),[b] *E. coli*,[c] *S. flexneri*,[c] *S. enterica*,[c] *C. tropicalis*,[c] *S. aureus*,[d] *S. odorifera*[e]
49	Methanethiol[b–e]	*P. aeruginosa*,[a] meat spoilage (aerobic bacteria, *Pseudomonas* spp., *Enterobacteriaceae*, lactic acid bacteria, *Enterococcus* spp.),[b] *E. coli*,[c] *S. flexneri*,[c] *S. enterica*,[c] *C. tropicalis*,[c] *S. aureus*,[d] *S. odorifera*[e]
51		*P. aeruginosa*,[a] *M. kansasii*,[f] *M. avium*[f]
53		*E. coli*,[c] *C. tropicalis*[c]
55	C_4 compound (e.g. cyclobutene, 1,2-butadiene, 1,3-butadiene, 1-methyl-cycloproprene, or 2-butyne)[d]	*P. aeruginosa*,[a] *S. milleri*,[a] *S. aureus*[d]
57	Fragment ion of either 1-butanol,[b,c] 2-methylpropan-2-ol,[b] or 3-methyl-1-propanol[c]	*P. aeruginosa*,[a] meat spoilage (aerobic bacteria,*Pseudomonas* spp., *Enterobacteriaceae*, lactic acid bacteria, *Enterococcus* spp.),[b] *E. coli*,[c] *S. flexneri*,[c] *S. enterica*,[c] *C. tropicalis*[c]
59	Acetone[c,d,f]	*S. milleri*,[a] *P. aeruginosa*,[a] *E. coli*,[c] *S. flexneri*,[c] *C. tropicalis*[c], *S. aureus*[d], *M. kansasii*[f], *M. avium*[f], *M. smegmatis*[f]

Table 7.1 *(Continued)*

m/z	Assignment	Microbial culture
61	Acetic acid,[c,d,e(?)] 1-propanol (?)[d]	*P. aeruginosa*,[a] *E. coli*,[c] *S. flexneri*,[c] *S. enterica*,[c] *C. tropicalis*,[c] *S. aureus*,[d] *S. odorifera*[e]
63	Dimethyl sulfide[b,d–f]	*P. aeruginosa*,[a] meat spoilage (aerobic bacteria, *Pseudomonas* spp., *Enterobacteriaceae*, lactic acid bacteria, *Enterococcus* spp.),[b] *S. aureus*,[d] *S. odorifera*,[e] *M. avium*[f]
64		*S. odorifera*[e]
69	Isoprene[d]	*P. aeruginosa*,[a] *S. enterica*,[c] *C. tropicalis*,[c] *S. aureus*[d]
71		*S. milleri*,[a] *P. aeruginosa*,[a] *E. coli*,[c] *S. flexneri*,[c] *S. enterica*,[c] *C. tropicalis*,[c] *S. odorifera*[e]
73	2-butanone,[c,d] butanal,[d] methyl ethyl ketone,[d] or a fragment of 3-hydroxy-2-butanone[c]	*P. aeruginosa*,[a] *S. enterica*,[c] *C. tropicalis*,[c] *S. aureus*[d]
75	Methyl acetate[b]	*P. aeruginosa*,[a] meat spoilage (aerobic bacteria,*Pseudomonas* spp., *Enterobacteriaceae*, lactic acid bacteria, *Enterococcus* spp.),[b] *E. coli*,[c] *S. flexneri*,[c] *C. tropicalis*[c]
77	Given the *E/N* used by Critchley *et al.* this could be a benzene fragment; propanethiol or methyl ethyl disulfide[f]	*P. aeruginosa*,[a] *M. avium*[f]
79	Benzene (?)[e]	*P. aeruginosa*,[a] *S. odorifera*[e]
87		*S. milleri*,[a] *P. aeruginosa*,[a] *E. coli*,[c] *S. flexneri*,[c] *S. enterica*,[c] *C. tropicalis*,[c] *S. odorifera*[e]
89	2-methyl-1-butanol[c] or 3-methyl-1-butanol[c]	*S. milleri*,[a] *P. aeruginosa*,[a] *E. coli*,[c] *S. flexneri*,[c] *S. enterica*,[c] *C. tropicalis*[c]
91	Diethyl sulfide, thioacetic acid methyl ester, 2,3-butanediol[b]	Meat spoilage (aerobic bacteria, *Pseudomonas* spp., *Enterobacteriaceae*, lactic acid bacteria, *Enterococcus* spp.),[b] *E. coli*,[c] *C. tropicalis*[c]
93	Toluene[b]	Meat spoilage (aerobic bacteria, *Pseudomonas* spp., *Enterobacteriaceae*, lactic acid bacteria, *Enterococcus* spp.),[b] *E. coli*,[c] *C. tropicalis*,[c] *S. odorifera*[e]
95	Dimethyl disulfide[b,e,f]	*P. aeruginosa*,[a] meat spoilage (aerobic bacteria, *Pseudomonas* spp., *Enterobacteriaceae*, lactic acid bacteria, *Enterococcus* spp.),[b] *C. tropicalis*,[c] *S. odorifera*,[e] *M. avium*[f]
97		*P. aeruginosa*,[a] *S. odorifera*[e]

(continued)

Table 7.1 (Continued)

m/z	Assignment	Microbial culture
103		*E. coli*,[c] *S. enterica*,[c] *C. tropicalis*[c]
105		*C. tropicalis*[c]
109		*S. enterica*[c]
113	1-Octanol (fragment), octene[b]	
117		*E. coli*,[c] *S. flexneri*,[c] *S. enterica*,[c] *C. tropicalis*[c]
118	Indole,[c] heteroaromatic compounds[e]	*E. coli*,[c] *S. flexneri*,[c] *S. odorifera*[e]
119		*E. coli*,[c] *S. flexneri*,[c] *C. tropicalis*[c]
123	2-Phenylethanol[e]	*S. odorifera*[e]
127	2,3-Dimethyl trisulfide,[b,e] nonene[b]	Meat spoilage (aerobic bacteria, *Pseudomonas* spp., *Enterobacteriaceae*, lactic acid bacteria, *Enterococcus* spp.),[b] *S. enterica*,[c] *C. tropicalis*,[c] *S. odorifera*[e]
129	3-Ethyl-4-methyl hexane, ethyl tiglate, nonane, octanone[b]	Meat spoilage (aerobic bacteria, *Pseudomonas* spp., *Enterobacteriaceae*, lactic acid bacteria, *Enterococcus* spp.)[b]
131	Isoamyl acetate, 1-octanol[b]	Meat spoilage (aerobic bacteria, *Pseudomonas* spp., *Enterobacteriaceae*, lactic acid bacteria, *Enterococcus* spp.)[b]
132		*E. coli*,[b]
135	Aromatic compound[e]	*C. tropicalis*,[c] *S. odorifera*[e]
137		*S. odorifera*[e]
143	Dimethyl octane[b]	Meat spoilage (aerobic bacteria, *Pseudomonas* spp., *Enterobacteriaceae*, lactic acid bacteria, *Enterococcus* spp.)[b]
145		*E. coli*,[c] *S. enterica*[c]
149		*S. odorifera*[e]

[a]Critchley *et al.*; *m/z* values only reported [12].

[b]Mayr *et al.* [107]. These researchers tentatively identified the VOCs based on the *m/z* value of the product ions, fragmentation patterns, isotopic ratios, and published VOC compositions of spoiling meat.

[c]Bunge *et al.* [14]. The tentative assignments of VOCs are based on information available in the literature and on isotopic abundances.

[d]O'Hara *et al.* [15]. The tentative assignment of VOCs is based on information available in the literature and on isotopic abundances.

[e]Kai *et al.* [16]. To aid in assignment Kai *et al.* used a combination of GC–MS and PTR-MS. Those compounds showing (?) indicate a tentative assignment.

[f]Crespo *et al.* [18]. *Mycobacterium avium* and *Mycobacterium kansasii*, and a non-pathogenic fast growing species, *Mycobacterium smegmatis*, in Middlebrook M7H9 culturing media were followed online. To aid in the identification volatiles were collected in Tenax tubes and analysed with a gas chromatograph equipped with a flame photometric detector. Identification of VOCs was also based on isotopic ratios and CID results.

7.4.1 Urine Headspace Analysis

Non-invasive PTR-MS sampling for medical applications has been used for the analysis of urine from non-smokers and habitual smokers [1]. Earlier breath analysis studies had already demonstrated that the acetonitrile level is elevated in the breath of smokers. The aim of this later study was to ascertain whether acetonitrile could also be detected in the headspace of urine and, if so, whether elevated levels are found for smokers. The study included 101 volunteers (57 men and 44 women) and the conclusion was that elevated acetonitrile levels are found in the urine of smokers. The mean (standard deviation) acetonitrile concentration in the headspace above urine of 46 non-smokers was 3.7 (1.8) ppbv, whereas for heavy smokers (defined as someone who smokes more than 30 cigarettes per day) it was found to be 28.0 (5.4) ppbv. An interesting point raised by the authors of this work was whether the increased levels of acetonitrile in the urine of smokers is a contributing factor leading to the known increase in bladder cancer for smokers compared to non-smokers.

7.4.2 Skin Emissions

Steeghs *et al.* performed *in vivo* online headspace measurements to study ultraviolet (UV) light-induced lipid peroxidation in human skin [4]. Emissions from the skin of 16 healthy, non-smoking male volunteers were monitored before, during and after irradiation. Five VOCs associated with skin damage by UV radiation were discovered. Using Tenax-based GC pre-separation combined with PTR-MS, two ions detected at m/z 45 and m/z 59 were identified to be protonated acetaldehyde and propanal, respectively. Other signals measured at m/z 73 and m/z 87 could not be associated with specific compounds with any reliability. The fifth VOC, ethene, was detected using laser-based photoacoustic trace gas detection. The researchers concluded that acetaldehyde and propanal can be used as biomarkers for lipid peroxidation.

In another study using real-time *in vivo* trace sensing of ethylene, propanal and acetaldehyde from human skin, the effect of supplemental intake of systemic antioxidants on UV-induced lipid peroxidation was explored [5]. No significant effect of antioxidant intake was found. A later study monitored the production of volatiles from the skin as a result of reactions of ozone with skin lipids [6]. Compound assignments included acetone, 6-methyl-5-hepten-2-one, 2,6-dimethyl-2,6-undecadien-10-one (geranyl acetone), decanal and bifunctional compounds containing carbonyl, carboxyl and α-hydroxy ketone groups. The major bifunctional products identified were four dicarbonyls: 4-oxopentanal, 1,4-butanedial, 4-methyl-8-oxo-4-nonenal and 4-methyl-4-octene-1,8-dial. Some of these compounds, such as the dicarbonyls, are considered to be respiratory irritants, and therefore the observation of these compounds has health implications following exposure to ozone.

7.4.3 VOC Emissions from Human Cells

Headspace analysis with PTR-MS has also been used to good effect for the *in vitro* discrimination of cultured human cancerous and non-cancerous cell lines by Brunner *et al.* [3]. Two pairs of cancerous and non-cancerous human cell lines were used. The non-cancerous pair consisted of lung epithelium cells, A-549, and retinal pigment epithelium cells, hTERT-RPE1 (cultured in different growth media). The second pair consisted of squamous lung carcinoma cells ELPC and immortalized human bronchial epithelial cells BEAS2B (cultured in identical media). A number of VOCs were identified but the most intriguing ion signal was at *m/z* 45, which was tentatively assigned protonated acetaldehyde. The results seemed to show significant consumption of acetaldehyde by the cancerous cells but not by the non-cancerous cells. A multivariate statistical analysis was applied to 42 selected VOC markers, from which Brunner *et al.* demonstrated that they could separate cancerous and non-cancerous cell lines from each other.

7.4.4 VOCs in Clinical Environments

Finally, we briefly reflect on the work by Rieder and colleagues, who have used PTR-MS for investigations involving occupational health exposure to volatile anaesthetics in medical settings [7–10]. In their first study, the air quality at a surgical and urological post-anaesthetic care unit was investigated, where patients exhale volatile anaesthetics for some time after the operation [7]. The mean levels of the volatile anaesthetics sevoflurane and isoflurane in the urological post-anaesthesia care unit were found to be 15.9 ppbv and 9.5 ppbv, respectively. These concentrations are significantly less than the maximum limits recommended by international guidelines, and therefore it was concluded that because of the high ventilation capacity the 24-hour occupational burden by anaesthetic gases at this care unit is relatively low. In a second study, sevoflurane concentrations were monitored in an operating theatre [8]. Intra-operative sevoflurane concentrations were found to be five times lower than the occupational safety limit requirements and 1000 times lower only 5 min after the removal of the airway device with the patient awake. In a third study the exposure of operating staff to sevoflurane was investigated by screening the breath of 40 staff members before an operation, and at 0-, 1-, 2- and 3-hour intervals after duty and before commencing duty on the following day [9]. The conclusion from that study was that significant concentrations of sevoflurane may be continuously present in persons routinely exposed to sevoflurane, the health effects of which are unknown. From these studies, the researchers propose that PTR-MS could be used as an intelligent sensor to control ventilation systems in order to reduce the exposure of personnel to volatile anaesthetics. Lirk *et al.* have provided a review of the occupational exposure to volatile agents in medical environments [10].

References

1. Pinggera, G. M., Lirk, P., Bodogri, F. *et al.* (2005) Urinary acetonitrile concentrations correlate with recent smoking behaviour. *BJU Int.* **95**, 306.
2. Lechner, M., Colvin, H. P., Ginzel, C. *et al.* (2005) Headspace screening of fluid obtained from the gut during colonoscopy and breath analysis by proton transfer reaction-mass spectrometry: a novel approach in the diagnosis of gastro-intestinal diseases. *Int. J. Mass Spectrom.* **243**, 151.
3. Brunner, C., Szymczak, W., Höllriegl, V. *et al.* (2010) Discrimination of cancerous and non-cancerous cell lines by headspace-analysis with PTR-MS. *Anal. Bioanal. Chem.* **397**, 2315.

4. Steeghs, M. M. L., Moeskops, B. W. M., van Swam, K. *et al.* (2006) On-line monitoring of UV-induced lipid peroxidation products from human skin in vivo using proton-transfer reaction mass spectrometry. *Int. J. Mass Spectrom.* **253**, 58.
5. Moeskops, B. W. M., Steeghs, M. M. L., van Swam, K. *et al.* (2006) Real-time trace gas sensing of ethylene, propanal and acetaldehyde from human skin in vivo. *Physiol. Meas.* **27**, 1187.
6. Wisthaler, A., Weschler, C. J. (2010) Reactions of ozone with human skin lipids: sources of carbonyls, dicarbonyls, and hydroxycarbonyls in indoor air. *Proc. Nat. Acad. Sci.* **107**, 6568.
7. Rieder, J., Prazeller, P., Boehler, M. *et al.* (2001) Online monitoring of air quality at the postanesthetic care unit by proton-transfer-reaction mass spectrometry. *Anesth. Analg.* **92**, 389.
8. Rieder, J., Keller, C., Brimacombe, J. *et al.* (2002) Monitoring pollution by proton-transfer-reaction mass spectrometry during paediatric anaesthesia with positive pressure ventilation via the laryngeal mask airway or uncuffed tracheal tube. *Anaesthesia* **57**, 663.
9. Summer, G., Lirk, P., Hoerauf, K. *et al.* (2003) Sevoflurane in exhaled air of operating room personnel. *Anesth. Analg.* **97**, 1070.
10. Lirk, P., Bodrogi, F., Rieder, J. (2004) Medical applications of proton transfer reaction-mass spectrometry: ambient air monitoring and breath analysis. *Int. J. Mass Spectrom.* **239**, 221.
11. Jaksch, D., Margesin, R., Mikoviny, T. *et al.* (2004) The effect of ozone treatment on the microbial contamination of pork meat measured by detecting the emissions using PTR-MS and by enumeration of microorganisms. *Int. J. Mass Spectrom.* **239**, 209.
12. Critchley, A. D. J., Elliott, T. S., Harrison, G. *et al.* (2004) The proton transfer reaction mass spectrometer and its use in medical science: applications to drug assays and the monitoring of bacteria. *Int. J. Mass Spectrom.* **239**, 235.
13. Mayrhofer, S., Mikoviny, T., Waldhuber, S. *et al.* (2006) Microbial community related to volatile organic compound (VOC) emission in household biowaste. *Environ. Microbiol.* **8**, 1960.
14. Bunge, M., Araghipour, N., Mikoviny, T. (2008) On-line monitoring of microbial volatile metabolites by proton transfer reaction-mass spectrometry. *Appl. Environ. Microbiol.* **74**, 2179.
15. O'Hara, M., Mayhew, C. A. (2009) A preliminary comparison of volatile organic compounds in the headspace of cultures of *Staphylococcus aureus* grown in nutrient, dextrose and brain heart bovine broths measured using a proton transfer reaction mass spectrometer. *J. Breath Res.* **3**, 027001.
16. Kai, M., Crespo, E., Cristescu, S. M. *et al.* (2010) *Serratia odorifera*: analysis of volatile emission and biological impact of volatile compounds on *Arabidopsis thaliana*. *Appl. Microbiol. Biotechnol.* **88**, 965.
17. Crespo, E., Cristescu, S. M., de Ronde, H. *et al.* (2011) Proton transfer reaction mass spectrometry detects rapid changes in volatile metabolite emission by *Mycobacterium smegmatis* after the addition of specific antimicrobial agents. *J. Microbiol. Methods* **86**, 8.

18. Crespo, E., de Ronde, H., Kuijper, S. *et al.* (2012) Potential biomarkers for identification of mycobacterial cultures by proton transfer reaction-mass spectrometry analysis. *Rapid Commun. Mass Spectrom.* **26**, 679.
19. Lechner, M., Fille, M., Hausdorfer, J. *et al.* (2005) Diagnosis of bacteria in vitro by mass spectrometric fingerprinting: a pilot study. *Current Microbiol.* **51**, 267.
20. Lagg, A., Taucher, J., Hansel, A., Lindinger, W. (1994) Applications of proton transfer reactions to gas analysis. *Int. J. Mass Spectrom. Ion Processes* **134**, 55.
21. Jordan, A., Hansel, A., Holzinger, R., Lindinger, W. (1995) Acetonitrile and benzene in the breath of smokers and non-smokers investigated by proton transfer reaction mass spectrometry (PTR-MS). *Int. J. Mass Spectrom. Ion Processes* **148**, L1.
22. Warneke, C., Kuczynski, J., Hansel, A. *et al.* (1996) Proton transfer reaction mass spectrometry (PTR-MS): propanol in human breath. *Int. J. Mass Spectrom. Ion Processes* **154**, 61.
23. Taucher, J., Hansel, A., Jordan, A. *et al.* (1997) Detection of isoprene in expired air from human subjects using proton-transfer-reaction mass spectrometry. *Rapid. Comm. Mass Spectrom.* **11**, 1230.
24. Prazeller, P., Karl, T., Jordan, A. *et al.* (1998) Quantification of passive smoking using proton-transfer-reaction mass spectrometry. *Int. J. Mass Spectrom.* **178**, L1.
25. Karl, T., Jordan, A., Hansel, A. *et al.* (1998) Benzene and acetontrile in smokers and nonsmokers. *Ber. Nat.-Med. Verein Innsbruck.* **85**, 7.
26. Karl, T., Prazeller, P., Mayr, D. *et al.* (2001) Human breath isoprene and its relation to blood cholesterol levels: new measurements and modelling. *J. Appl. Physiol.* **91**, 762.
27. Rieder, J., Lirk, P., Ebenbichler, C. *et al.* (2001) Analysis of volatile organic compounds: possible applications in metabolic disorders and cancer screening. *Wien. Klin. Wochenschr.* **113**, 181.
28. Harrison, G. R., Critchley, A. D. J., Mayhew, C. A. *et al.* (2003) Real-time breath monitoring of propofol and its volatile metabolites during surgery using a novel mass spectrometric technique: a feasibility study. *Brit. J. Anaesth.* **91**, 797.
29. Lirk, P., Bodrogi, F., Raifer, H. *et al.* (2003) Elective haemodialysis increases exhaled isoprene. *Nephrol. Dial. Transplant* **18**, 937.
30. Amann, A., Poupart, G., Telser, S. *et al.* (2004) Applications of breath gas analysis in medicine. *Int. J. Mass Spectrom.* **239**, 227.
31. Lirk, P., Bodrogi, F., Deibl, M. *et al.* (2004) Quantification of recent smoking behaviour using proton transfer reaction-mass spectrometry (PTR-MS). *Wien Klin. Wochenschr.* **116(1)**, 21.
32. Amann, A., Schmid, A., Scholl-Bürgi, S. *et al.* (2005) Breath analysis for medical diagnosis and therapeutic monitoring. *Spectroscopy Europe* **17**, 18.
33. Moser, B., Bodrogi, F., Eibl, G. *et al.* (2005) Mass spectrometric profile of exhaled breath – field study by PTR-MS. *Respir. Physiol. Neurobiol.* **145**, 295.
34. Miekisch, W., Schubert, J. K. (2006) From highly sophisticated analytical techniques to life-saving diagnostics: technical developments in breath analysis. *Trends Anal. Chem.* **25**, 665.
35. Lechner, M., Moser, B., Niederseer, D. *et al.* (2006) Gender and age specific differences in exhaled isoprene levels. *Respir. Physiol. Neurobiol.* **154**, 478.

36. Amann, A., Španěl, P., Smith, D. (2007) Breath analysis: the approach towards clinical applications. *Mini-Rev. Med. Chem.* **7**, 115.
37. Wehinger, A., Schmid, A., Mechtcheriakov, S. *et al.* (2007) Lung cancer detection by proton transfer reaction mass spectrometric analysis of human breath gas. *Int. J. Mass Spectrom.* **265**, 49.
38. Keck, L., Hoeschen, C., Oeh, U. (2008) Effects of carbon dioxide in breath gas on proton transfer reaction-mass spectrometry (PTR-MS) measurements. *Int. J. Mass Spectrom.* **270**, 156.
39. Arendacká, B., Schwarz, K., Štolc Jr, S. *et al.* (2008) Variability issues in determining the concentration of isoprene in human breath by PTR-MS. *J. Breath Res.* **2**, 037007.
40. Kushch, I., Arendacká, B., Štolc, S. *et al.* (2008) Breath isoprene – aspects of normal physiology related to age, gender and cholesterol profile as determined in a proton transfer reaction mass spectrometry study. *Clin. Chem. Lab. Med.* **46**, 1011.
41. O'Hara, M. E., Clutton-Brock, T. H., Green, S. *et al.* (2009) Mass spectrometric investigations to obtain the first direct comparisons of endogenous breath and blood volatile organic compound concentrations in healthy volunteers. *Int. J. Mass Spectrom.* **281**, 92.
42. Schwarz, K., Filipiak, W., Amann, A. (2009) Determining concentration patterns of volatile compounds in exhaled breath by PTR-MS. *J. Breath Res.* **3**, 027002.
43. Schwarz, K., Pizzini, A., Arendacká, B. *et al.* (2009) Breath acetone – aspects of normal physiology related to age and gender as determined in a PTR-MS study. *J. Breath Res.* **3**, 027003.
44. O'Hara, M. E., Clutton-Brock, T. H., Green, S., Mayhew, C. A. (2009) Endogenous volatile organic compounds in breath and blood of healthy volunteers: examining breath analysis as a surrogate for blood measurements. *J. Breath Res.* **3**, 027005.
45. Bajtarevic, A., Ager, C., Pienz, M. *et al.* (2009) Non-invasive detection of lung cancer by analysis of exhaled breath. *BMC Cancer* **9**, 348.
46. Erhart, S., Amann, A., Haberlandt, E. *et al.* (2009) 3-Heptanone as a potential new marker for valproic acid therapy. *J. Breath Res.* **3**, 016004.
47. Herbig, J., Müller, M., Schallhart, S. *et al.* (2009) On-line breath analysis with PTR-TOF. *J. Breath Res.* **3**, 027004.
48. King, J., Kupferthaler, A., Unterkofler, K. *et al.* (2009) Isoprene and acetone concentration profiles during exercise on an ergometer. *J. Breath Res.* **3**, 027006.
49. Beauchamp, J., Kirsch, F., Buettner, A. (2010) Real-time breath gas analysis for pharmacokinetics: monitoring exhaled breath by on-line proton-transfer-reaction mass spectrometry after ingestion of eucalyptol-containing capsules. *J. Breath Res.* **4**, 026006.
50. Fedrigo, M., Hoeschen, C., Oeh, U. (2010) Multidimensional statistical analysis of PTR-MS breath samples: a test study on irradiation detection. *Int. J. Mass Spectrom.* **295**, 13.
51. Monks, P. S., Willis, K. A. (2010) Breath Analysis. *Educ. Chem.* **47**, 110.
52. Kamysek, S., Fuchs, P., Schwoebel, H. *et al.* (2011) Drug detection in breath: effects of pulmonary blood flow and cardiac output on propofol exhalation. *Anal. Bioanal. Chem.* **401**, 2093.

53. Schwöbel, H., Schubert, R., Sklorz, M. *et al.* (2011) Phase-resolved real-time breath analysis during exercise by means of smart processing of PTR-MS data. *Anal. Bioanal. Chem.* **401**, 2079.
54. Guamán, A. V., Carreras, A., Calvo, D. *et al.* (2012) Rapid detection of sepsis in rats through volatile organic compounds in breath. *J. Chromatogr. B. Anal. Technol Biomed. Life Sci.* **881**, 76.
55. Halbritter, S., Fedrigo, M., Höllriegl, V. *et al.* (2012) Human breath gas analysis in the screening of gestational diabetes mellitus. *Diabetes Technol. Ther.* **14**, 1.
56. King, J., Kupferthaler, A., Frauscher, B. *et al.* (2012) Measurement of endogenous acetone and isoprene in exhaled breath during sleep. *Physiol. Meas.* **33**, 413.
57. Kastenmüller, S.G., Stückler, F., Rist, M. J. *et al.* (2012) The dynamic range of the human metabolome revealed by challenges. *FASEB J.* **26**, 2607.
58. Schubert, R., Schwoebel, H., Mau-Moeller, A. *et al.* (2012) Metabolic monitoring and assessment of anaerobic threshold by means of breath biomarkers. *Metabolomics* **8**, 1069.
59. Steeghs, M. M. L., Cristescu, S. M., Munnik, P. *et al.* (2007) An off-line breath sampling and analysis method suitable for large screening studies. *Physiol. Meas.* **28**, 503.
60. O'Hara, M. E., O'Hehir, S., Green, S., Mayhew, C. A. (2008) Development of a protocol to measure volatile organic compounds in human breath: a comparison of rebreathing and on-line single exhalations using proton transfer reaction mass spectrometry. *Physiol. Meas.* **29**, 309.
61. Herbig, J., Titzmann, T., Beauchamp, J. *et al.* (2008) Buffered end-tidal (BET) sampling – a novel method for real-time breath-gas analysis. *J. Breath Res.* **2**, 037008.
62. Beauchamp, J., Herbig, J., Gutmann, R., Hansel, A. (2008) On the use of Tedlar bags for breath-gas sampling and analysis. *J. Breath Res.* **2**, 046001.
63. Thekedar, B., Szymczak, W., Höllriegl, V. *et al.* (2009) Investigations on the variability of breath gas sampling using PTR-MS. *J. Breath Res.* **3**, 027007.
64. Thekedar, B., Oeh, U., Szymczak, W. *et al.* (2011) Influences of mixed expiratory sampling parameters on exhaled volatile organic compound concentrations. *J. Breath Res.* **5**, 016001.
65. Steeghs, M. M. L., Cristescu, S. M., Harren, F. J. M. (2007) The suitability of Tedlar bags for breath sampling in medical diagnostic research. *Physiol. Meas.* **28**, 73.
66. Phillips, M. (1992) Breath tests in medicine. *Sci. Am.* **267**, 74.
67. Miekisch, W., Herbig, J., Schubert, J. K. (2012) Data interpretation in breath biomarker research: pitfalls and directions. *J. Breath Res.* **6**, 036007.
68. Amann, A., Smith, D. (Eds) (2005) *Breath Analysis for Clinical Diagnosis and Therapeutic Monitoring*, World Scientific.
69. Phillips, M., Gleeson, K., Hughes, J. M. B. *et al.* (1999) Volatile organic compounds in breath as markers of lung cancer: a cross-sectional study. *Lancet* **353**, 1930.
70. Phillips, M., Cataneo, R. N., Cummin, A. R. C. *et al.* (2003) Detection of lung cancer with volatile markers in the breath. *Chest* **123**(6), 2115.
71. Phillips, M., Cataneo, R. N., Ditkoff, B. A. *et al.* (2006) Prediction of breast cancer using volatile biomarkers in the breath. *Breast Cancer Res. Treat.* **99**(1), 19.

72. Phillips, M., Altorki, N., Austin, J. H. M. *et al.* (2008) Detection of lung cancer using weighted digital analysis of breath biomarkers. *Clin. Chim. Acta.* **393**(2), 76.
73. Phillips, M., Cataneo, R. N., Saunders, C. *et al.* (2010) Volatile biomarkers in the breath of women with breast cancer. *J. Breath Res.* **4**, 026003.
74. Smith, D., Španěl, P. (1996) The novel selected-ion flow tube approach to trace gas analysis of air and breath. *Rapid Commun. Mass Spectrom.* **10**, 1183.
75. Marczin, N., Yacoub, M. H. (Eds) (2002) *Disease Markers in Exhaled Breath: Basic Mechanisms and Clinical Applications*, 346 of NATO Science Series, IOS Press.
76. Hansel, A., Jordan, A., Holzinger, R. *et al.* (1995) Proton transfer reaction mass spectrometry: on-line trace gas analysis at the ppb level. *Int. J. Mass Spectrom. Ion Processes* **149–150**, 609.
77. Lindinger, W., Hansel, A., Jordan, A. (1998) On-line monitoring of volatile organic compounds at pptv levels by means of proton-transfer-reaction mass spectrometry (PTR-MS): medical applications, food control and environmental research. *Int. J. Mass Spectrom. Ion Processes* **173**, 191.
78. Lindinger, W., Hansel, A., Jordan, A. (1998) Proton-transfer-reaction mass spectrometry (PTR.MS): on-line monitoring of volatile organic compounds at pptv levels. *Chem. Soc. Rev.* **27**, 347.
79. Taucher, J., Lagg, A., Hansel, A. *et al.* (1995) Methanol in human breath. *Alcohol. Clin. Exp. Res.* **19**, 1147.
80. Lindinger, W., Taucher, J., Jordan, A. *et al.* (1997) Endogenous production of methanol after the consumption of fruit. *Alcohol. Clin. Exp. Res.* **21**, 939.
81. Deng, C., Zhang, X., Li, N. (2004) Investigation of volatile biomarkers in lung cancer blood using solid-phase microextraction and capillary gas chromatography–mass spectrometry. *J. Chromatogr. B. Analyt Technol Biomed Life Sci.* **808**, 269.
82. Kushch, I., Schwarz, K., Schwentner, L. *et al.* (2008) Compounds enhanced in a mass spectrometric profile of smokers' exhaled breath versus non-smokers as determined in a pilot study using PTR-MS. *J. Breath Res.* **2**, 026002.
83. King, J., Koc, H., Unterkofler, K. *et al.* (2010) Physiological modeling of isoprene dynamics in exhaled breath. *J. Theor. Biol.* **267**, 626.
84. Koc, H., King, J., Teschl, G. *et al.* (2011) The role of mathematical modeling in VOC analysis using isoprene as a prototypical example. *J. Breath Res.* **5**, 037102.
85. King, J., Unterkofler, K., Teschl, G. *et al.* (2011) A mathematical model for breath gas analysis of volatile organic compounds with special emphasis on acetone. *J. Math. Biol.* **63**, 959.
86. King, J., Unterkofler, K., Teschl, G. *et al.* (2012) A modeling-based evaluation of isothermal rebreathing for breath gas analyses of highly soluble volatile organic compounds. *J. Breath Res.* **6**, 016005.
87. Crofford, O. B., Mallard, R. E., Winton, R. E. *et al.* (1977) Acetone in breath and blood. *Trans. Am. Clin. Climatol. Assoc.* **88**, 128.
88. Preti, G., Labows, J. N., Kostelc, J. G. *et al.* (1988) Analysis of lung air from patients with bronchogenic carcinoma and controls using gas chromatography-mass spectrometry. *J. Chromatogr.* **432**, 1.
89. Fuchs, D., Jamnig, H., Heininger, P. *et al.* (2012) Decline of exhaled isoprene in lung cancer patients correlates with immune activation. *J. Breath Res.* **6**, 027101.

90. Cristescu, S. M., Gietema, H. A., Blanchet, L. *et al.* (2011) Screening for emphysema via exhaled volatile organic compounds. *J. Breath Res.* **5**, 046009.
91. Kohl, I., Beauchamp, J., Cakar-Beck, F. *et al.* (2013) First observation of a potential non-invasive breath gas biomarker for kidney function. *J. Breath Res.* **7**, 017110.
92. Morisco, F., Aprea, E., Lembo, V. *et al.* (2013) Rapid "breath-print" of liver cirrhosis by proton transfer reaction time-of-flight mass spectrometry. A pilot study. *PLOS ONE.* **8**, e59658.
93. Beauchamp, J. (2011) Inhaled today, not gone tomorrow: pharmacokinetics and environmental exposure of volatiles in exhaled breath. *J. Breath Res.* **5**, 037103.
94. Grossherr, M., Hengstenberg, A., Meier, T. *et al.* (2007) Discontinuous monitoring of propofol concentrations in expired alveolar gas and in arterial and venous plasma during artificial ventilation. *Anesthesiology* **104**, 786.
95. Takita, A., Masui, K., Kazama, T. (2007) On-line monitoring of end-tidal propofol concentration in anesthetized patients. *Anesthesiology* **106**, 659.
96. Hornuss, C., Praun, S., Villinger, J. *et al.* (2007) Real-time monitoring of propofol in expired air in humans undergoing total intravenous anesthesia. *Anesthesiology* **106**, 665.
97. Grossherr, M., Hengstenberg, A., Meier, T. *et al.* (2009) Propofol concentration in exhaled air and arterial plasma in mechanically ventilated patients undergoing cardiac surgery. *Brit. J. Anaesth.* **102**, 608.
98. Perl, T., Carstens, E., Hirn, A., *et al.* (2009) Determination of serum propofol concentrations by breath analysis using ion mobility spectrometry. *Brit. J. Anaesth.* **103**, 822.
99. Gong, Y., Li, E., Xu, G., *et al.* (2009) Investigation of propofol concentrations in human breath by solid-phase microextraction gas chromatography-mass spectrometry. *J. Int. Med. Res.* **37**, 1465.
100. Janovsky, U., Scholl-Bürgi, S., Karall, D. *et al.* (2005) Breath gas analysis in patients suffering from propionic acidaemia [Chapter 28]. In: *Breath Analysis for Clinical Diagnosis and Therapeutic Monitoring* (Eds A. Amann, D. Smith), World Scientific, Singapore.
101. Krug, S., Kastenmüller, G., Stückler, F. *et al.* (2012) The dynamic range of the human metabolome revealed by challenges. *FASEB J.* **26**, 2607.
102. Singer, W., Herbig, J., Gutmann, R., *et al.* (2011) Applications of PTR-MS in medicine and biotechnology. *Am. Lab.* **43**, 34–37.
103. Miekisch, W., Schubert, J. K., Noeldge-Schomburg, G. F. E. (2004) Diagnostic potential of breath analysis–focus on volatile organic compounds. *Clin. Chim. Acta.* **347**, 25.
104. Risby, T. H., Solga, S. F. (2006) Current status of breath research. *Appl. Phys. B.* **85**, 421.
105. Risby, T. H. (2008) Critical issues for breath analysis. *J. Breath Res.* **2**, 030302.
106. Nijland, R., Burgess, J. G. (2010) Bacteria olfaction. *Biotechnol. J.* **5**, 974.
107. Mayr, D., Margesin, R., Klingsbichel, E. *et al.* (2003) Rapid detection of meat spoilage by measuring volatile organic compounds by using proton transfer reaction mass spectrometry. *Appl. Environ. Microbiol.* **69**, 4697.
108. Mayr, D., Margesin, R., Schinner, F., Märk, T. D. (2003) Detection of the spoiling of meat using PTR-MS. *Int. J. Mass Spectrom.* **223–224**, 229.

109. Steeghs, M. (2007) Development of proton-transfer reaction mass spectrometry techniques. Detection of trace gases in biology and medicine. PhD thesis. Radboud University Nijmegen, Netherlands.
110. Ezra, D., Jasper, J., Rogers, T. *et al.* (2004) Proton transfer reaction-mass spectrometry as a technique to measure volatile emissions of *Muscodor albus*. *Plant Sci.* **166**, 1471.
111. Brown, P., Watts, P., Märk, T. D., Mayhew, C. A. (2010) Proton transfer reaction mass spectrometry investigations on the effects of reduced electric field and reagent ion internal energy on product ion branching ratios for a series of saturated alcohols. *Int. J. Mass Spectrom.* **294**, 103.
112. Strobel, G. A., Knighton, B., Kluck, K. *et al.* (2008) The production of myco-diesel hydrocarbons and their derivatives by the endophytic fungus *Gliocladium roseum* (NRRL 50072). *Microbiol.* **154**, 3319.
113. Mallette, N. D., Knighton, W. B., Strobel, G. A. *et al.* (2012) Resolution of volatile fuel compound profiles from *Ascocoryne sarcoides*: a comparison by proton transfer reaction-mass spectrometry and solid phase microextraction gas chromatography mass spectrometry. *AMB Express* **2**, 23.
114. Wilske, B., Holzinger, R., Kesselmeie, J. (2001) Evidence for ethanolic fermentation in lichens during periods of high thallus water content. *Symbiosis* **31**, 95.

8

Applications of PTR-MS to Homeland Security: The Detection of Threat Agents

8.1 Background

Homeland Security is a term originating in the United States and refers to a coordinated effort to protect a nation from terrorism in all its guises. One of the requirements for effective prevention of terrorism is the ability to detect dangerous agents such as explosives, chemical warfare agents (CWAs) and illicit drugs, reliably and rapidly and often in trace concentrations. The determined terrorist will make stringent efforts to conceal threat agents in sealed containers, meaning that the merest trace may need to be detected. Trace detection may, for example, be used to indicate either that a threat agent is present or that a person has been handling threat agent material. This detection of trace amounts is a particularly challenging task because many of the threat agents, and especially those making up solid explosives and drugs, have very low vapour pressures at room temperatures.

PTR-MS has significant strengths as a tool for Homeland Security applications. Its high sensitivity is crucial but the fact that it (usually) targets the parent molecule to produce a protonated parent ion is also important, because this leads to a higher level of confidence in compound assignments and hence a reduction in false positives when applying PTR-MS to real-world situations. In addition, PTR-MS can detect a large range of threat compounds.

In the following sections, we describe potential applications of PTR-MS in the detection of explosives, CWAs and drugs. We also comment on PTR-MS operational techniques that can be applied to improve selectivity without the need for separation processes when there is some doubt about the identity of a specific compound from the m/z value of an ion only. We will begin this account by considering explosives.

Proton Transfer Reaction Mass Spectrometry: Principles and Applications, First Edition. Andrew M. Ellis and Christopher A. Mayhew.

8.2 Explosives

For convenience, the description of the analytical detection of explosives will be divided into two sections – *Forensics* and *Environmental*. Here forensics is associated with the identification of explosives either in bulk, as a trace (e.g. on a suspect's hands, clothing, laptop or other items) or in the analysis of post-explosion residues. What we term as environmental issues relate to contamination, for example, analysis of explosives and their potential degradation products in soil and water. The environmental impact of many solid explosives is significant considering their toxicity and the large quantities of obsolete explosives that have been either buried or dumped at sea. There is therefore a need for on-site detection and identification of residues of explosives in areas suspected of such contamination. This will be discussed in more detail in Section 8.2.2. First we will consider forensic applications.

8.2.1 Forensic Issues

Analytically, PTR-MS is considered to be most useful when the protonated parent molecule is produced and detected with a significant intensity in the mass spectrum. If fragmentation occurs, then this will tend to reduce the associated detection selectivity and sensitivity of the explosive, but that of course depends on the amount of fragmentation and the types of fragment ions being produced. Low *m/z* fragment ions will be less likely to lead to a high level of confidence in identification than would high *m/z* fragment ions, since low *m/z* fragment ions are more likely to have overlapping contributions from various compounds. An example of an explosive where substantial fragmentation occurs following proton transfer from H_3O^+ is triacetone triperoxide (TATP), $C_9H_{18}O_6$, which was the first explosive to be investigated using PTR-MS technology [1]. Unlike solid explosives, TATP has a relatively high vapour pressure of $\sim 5 \times 10^{-2}$ mbar at room temperature (the highest of the most commonly used explosives [2]), and therefore it is rather easy to transfer sufficient vapour into the drift tube of a PTR-MS instrument for analysis without the need for thermal desorption techniques. However, there is simply too much fragmentation of the protonated parent molecular ion (which if observed would be recorded at *m/z* 223, the same nominal *m/z* as the protonated explosive molecule RDX (see below)) when using H_3O^+ as the reagent ion for it to be of much value for detecting TATP. When operating a PTR-MS drift-tube at an *E/N* of 141 Td, Shen *et al.* observed no protonated parent molecule [1]. Instead fragment ions were observed at *m/z* 43 (CH_3CO^+), *m/z* 74 ($C_3H_6O_2{}^+$), *m/z* 75 ($C_3H_7O_2{}^+$) and *m/z* 91 ($C_3H_7O_3{}^+$), none of which are particularly unique markers for TATP. In order to see a weak $TATPH^+$ signal, the drift tube had to be operated at 50 Td, but one of the downsides in this is that the reagent ions will be predominantly protonated water clusters.

In comparison to TATP, the vapour pressures of solid explosives such as 1,3,5-trinitrobenzene (TNB, $C_6H_3N_3O_6$), 2,4,6-trinitrotoluene (TNT, $C_7H_5N_3O_6$), pentaerythritol tetranitrate (PETN, $C_5H_8N_4O_{12}$), cyclotrimethylenetrinitramine (RDX, $C_3H_6N_6O_6$), cyclotetramethylenetetranitramine (HMX, $C_4H_8N_8O_8$) and dinitrobenzenes (DNB, $C_6H_4N_2O_4$) are much lower [2], making it far more difficult to introduce sufficient concentrations of these explosives into the PTR-MS reaction drift tube for detection. Table 8.1 lists these explosives together with their structures, vapour pressures, their protonated mass and

Table 8.1 *Comparison of the properties of some common solid explosives*

Explosive	Structure	Vapour pressure at 25°C (mbar)[a]	*m/z*[b]	PA[c] ($kJ\ mol^{-1}$)
1,2-Dinitrobenzene	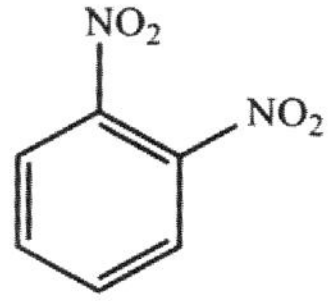			762
1,3-Dinitrobenzene	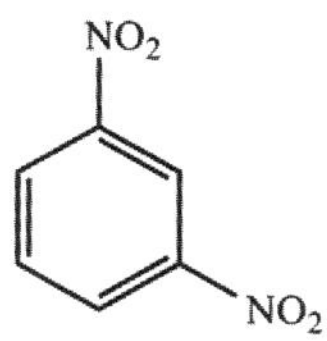	2×10^{-4}	169.02	761
1,4-Dinitrobenzene $C_6H_4N_2O_4$	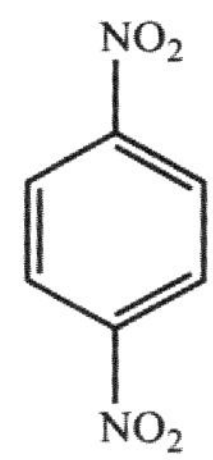			757
1,3,5-Trinitrobenzene $C_6H_3N_3O_6$	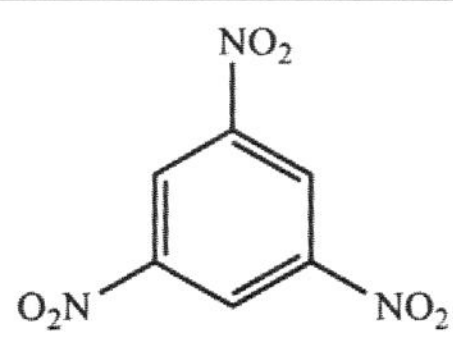	3.2×10^{-6}	214.01	722
Cyclotrimethyl-enetrinitramine (RDX) $C_3H_6N_6O_6$	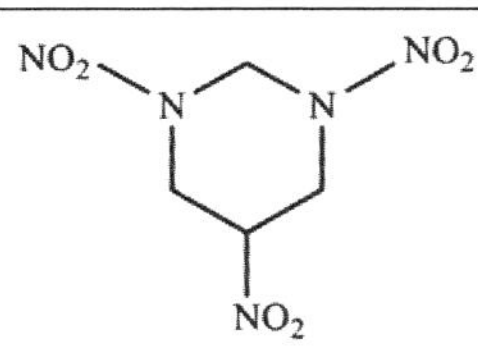	1.4×10^{-9}	223.04	771
Trinitrotoluene (TNT) $C_6H_2(NO_2)_3CH_3$	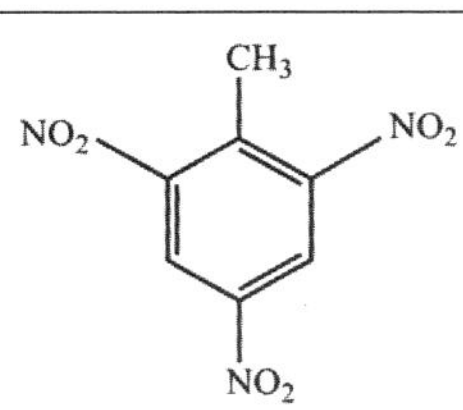	3.0×10^{-6}	228.03	744

(continued)

Table 8.1 *(Continued)*

Explosive	Structure	Vapour pressure at 25°C (mbar)[a]	m/z[b]	PA[c] (kJ mol^{-1})
Cyclotetramethyl-enetetranitramine (HMX) $C_4H_8N_8O_8$	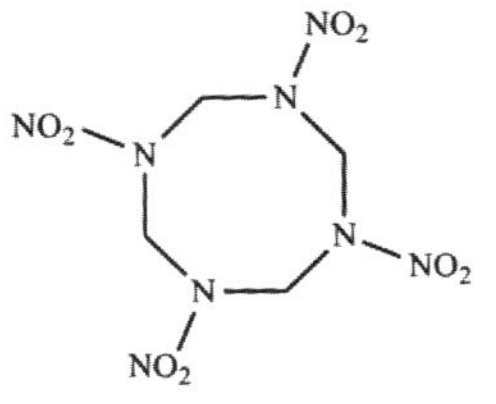	1.6×10^{-13}	297.05	765
Pentaerythritol tetranitrate (PETN) $C_5H_8N_4O_{12}$	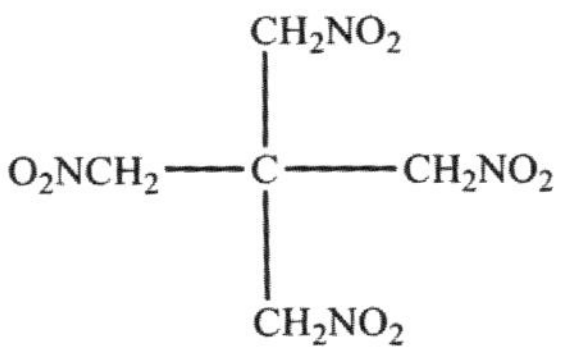	3.8×10^{-10}	317.02	739

[a]Values for vapour pressures have been taken from Reference 2 with the exceptions of TNB and DNB. For TNB, a range of values are available in the literature; the value shown has been taken from http://www.epa.gov/iris/supdocs/tnbsup.pdf. Data on DNB has been taken from http://toxnet.nlm.nih.gov/cgi-bin/sis/search/a?dbs+hsdb:@term+@DOCNO+4017 and is specific to 1,3-DNB.
[b]The m/z value refers to the protonated parent molecule.
[c]Proton affinities (PA) derived from DFT calculations (298 K) using Gaussian 09 with the GaussView 5 interface (unpublished data from the Molecular Physics Group, University of Birmingham, UK).

calculated proton affinities (determined using density functional theory calculations [3]). Confidence in the calculated proton affinities derives from the fact that the calculation of the proton affinity for H_2O using exactly the same computational procedure yields a value of 684 kJ mol^{-1}, which is close to the accepted literature value of 691 $\pm$ 3 kJ mol^{-1}. It should be further noted that it is not necessarily the absolute values which are important, but the differences in the values, and these will be generally determined with a better accuracy. Consequentially, it can be stated with some confidence that proton transfer is an exothermic channel for all of the above-mentioned solid explosives when using H_3O^+ as the source of chemical ionization. It is crucial to note from Table 8.1 that not only are the calculated proton affinities of the compounds listed all higher than that of water, but also that they are similar (all being in the range of 722–771 kJ mol^{-1}). Although the exothermicity of the proton transfer reaction will not be high (with ΔH ranging from –80 to –31 kJ mol^{-1}), this does not necessarily rule out dissociative proton transfer owing to possible reaction pathways leading to molecular elimination. The thermodynamics for the reactions of H_3O^+ with any of the explosives in Table 8.1 suggests that proton transfer will be highly efficient and most likely occurs at the collision rate [4]. Given the above, the only obstacles for the detection of these solid explosives would appear to be whether sufficient quantities of the explosive can be sampled into the drift tube of a PTR-MS instrument and if significant fragmentation occurs upon protonation, as is found for TATP, which could limit the selectivity and hence the sensitivity of detection.

A proof-of-principle study, which sampled the headspace at room temperature above small samples of the explosives RDX, TNT, PETN and HMX, demonstrated that even with explosive samples at room temperature sufficient quantities of explosive molecules could reach the reaction region of a PTR-MS to provide a detectable signal of the protonated parent molecule provided a heated inlet line system is used [5]. Although the protonated parent molecules for RDX, TNT, PETN and HMX were observed in the pilot study, in the case of RDX it has been shown that the non-dissociative proton transfer channel has a low branching ratio [6]. The dominant reaction channel is in fact loss of HONO from protonated RDX resulting in an ion fragment being observed at *m/z* 176.

$$H_3O^+ + RDX \rightarrow [RDX - HONO]H^+ + HONO + H_2O \quad (8.1)$$

This is not the only dissociative proton transfer channel observed resulting from the reaction of H_3O^+ with RDX – other (weaker) channels have been observed at *m/z* 75, *m/z* 102 and *m/z* 130 which are assigned to be $CH_3N_2O_2{}^+$, $[RDX\text{-}HONO\text{-}CH_2N_2O_2]H^+$ and $[RDX\text{-}HONO\text{-}NO_2]H^+$, respectively. This means that a series of *m/z* values could be used to provide what would effectively result in an unambiguous fingerprint of RDX.

8.2.1.1 The Unambiguous Detection of TNT

We have already mentioned numerous times in this book that the unambiguous identification of a compound from PTR-MS measurements is generally not possible because often only the nominal *m/z* is measured, and that may have contributions from a number of VOCs. However, in the case of TNT there is an unusual response for the protonated TNT signal intensity on *E/N* and this can be used to considerably enhance the selectivity. Figure 8.1

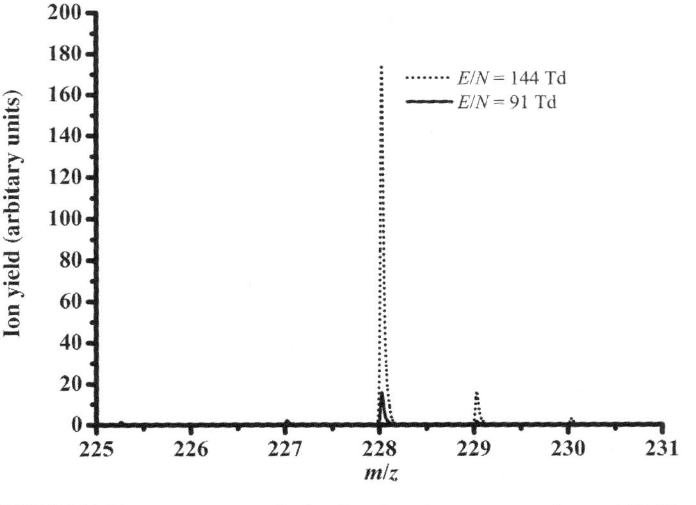

Figure 8.1 *PTR-TOF-MS spectra recorded using headspace sampling of TNT at room temperature. Ion signal intensities at two* E/N *values are shown. Note how the signal intensity increases dramatically when increasing* E/N *from 91 to 144 Td. Reproduced from [5] with permission from Elsevier.*

illustrates the mass spectrum recorded during a proof-of-principle study for TNT over the range of *m/z* 226–230 at two *E/N* values for the drift tube, 91 Td and 144 Td [5]. An interesting observation illustrated in this figure is that the signal intensity for $TNTH^+$ increases by approximately an order of magnitude in changing the reduced electric field from 91 Td to 144 Td. Such behaviour is unusual, because up to the time the pilot study was reported all other PTR-MS investigations had found that the protonated parent molecule intensity always decreased with increasing *E/N* owing to factors such as decreased reaction time and the production of fragment ions (if any) from more aggressive collision-induced dissociation. As an example of this, Figure 8.2(a) shows this 'normal' dependence on *E/N* for protonated PETN for a range of *E/N* values. The observed reduction in the $PETNH^+$ signal intensity with increasing *E/N* is steeper than would be expected from just the decrease in reaction time, which suggests that fragmentation of the protonated parent molecule is occurring. However, no other ions were identified that could be assigned to fragments derived from protonated PETN.

Following the report of the unusual dependence on the signal intensity of $TNTH^+$ on *E/N*, Sulzer *et al.* undertook a more comprehensive study of the behaviour of the protonated parent ion signal intensity of TNT over a range of *E/N* values from 90 to 250 Td,[1] the results for which are summarized in Figure 8.2(b) [7]. This image should be compared with Figure 8.2(a) for $PETNH^+$, which shows the more standard intensity dependence of a protonated parent ion signal on *E/N*. As can be seen from Figure 8.2(b), the normalized $TNTH^+$ signal increases in intensity with increasing *E/N* until a maximum is reached at ~180 Td (corresponding to a mean centre of mass collision energy of ~0.47 eV), with a gain in ion signal of approximately an order of magnitude. Beyond 180 Td, the signal follows the more normal and expected behaviour of decreasing intensity with increasing *E/N*. Similar behaviour was also reported by Sulzer *et al.* for the protonated parent ion signal for TNB, as shown in Figure 8.2(c).

Sulzer *et al.* have provided an explanation for the unusual dependence of the detection sensitivity of TNT and TNB on the reduced electric field [7]. The proposal is that as *E/N* is reduced, water clustering to $TNTH^+$ and $TNBH^+$ becomes favourable, which leads to a reduction in the signal intensities of the non-hydrated ions. The adduct ions $TNTH^+(H_2O)$ or $TNBH^+(H_2O)$ do not survive in the drift tube owing to fast bimolecular reactions with water which leads to the formation of $H_3O^+(H_2O)$ as the final product ion.

$$MH^+(H_2O) + H_2O \rightarrow H_3O^+(H_2O) + M \tag{8.2}$$

Thus the terminal product ion, $H_3O^+(H_2O)$, does not contain the explosive molecule. Furthermore, this ion is already present in considerable abundance in the drift tube and therefore its production cannot be monitored. It is this unusual reaction (8.2) that leads to the observed reduction in the sensitivity of detection for TNT and TNB as *E/N* is reduced. In confirmation of this explanation, DFT calculations indicate that Reaction 8.2 is exergonic[2] for TNB ($\Delta G_{298} = -14$ kJ mol^{-1}). Reaction 8.2 is slightly endoergonic for TNT ($\Delta G_{298} = +5$ kJ mol^{-1}) [7] but this small endoergonicity is readily overcome because the reaction will not only be kinetically driven by the high water number density present in the drift

[1] By reference to Figure 3.13 it can be seen that 90 Td corresponds to a mean centre of mass collision energy for H_3O^+ in the drift tube of approximately 0.15 eV whereas 250 Td corresponds to approximately 0.86 eV.

[2] Exergonic reactions are ones where ΔG is negative whereas endoergonic reactions have a positive ΔG.

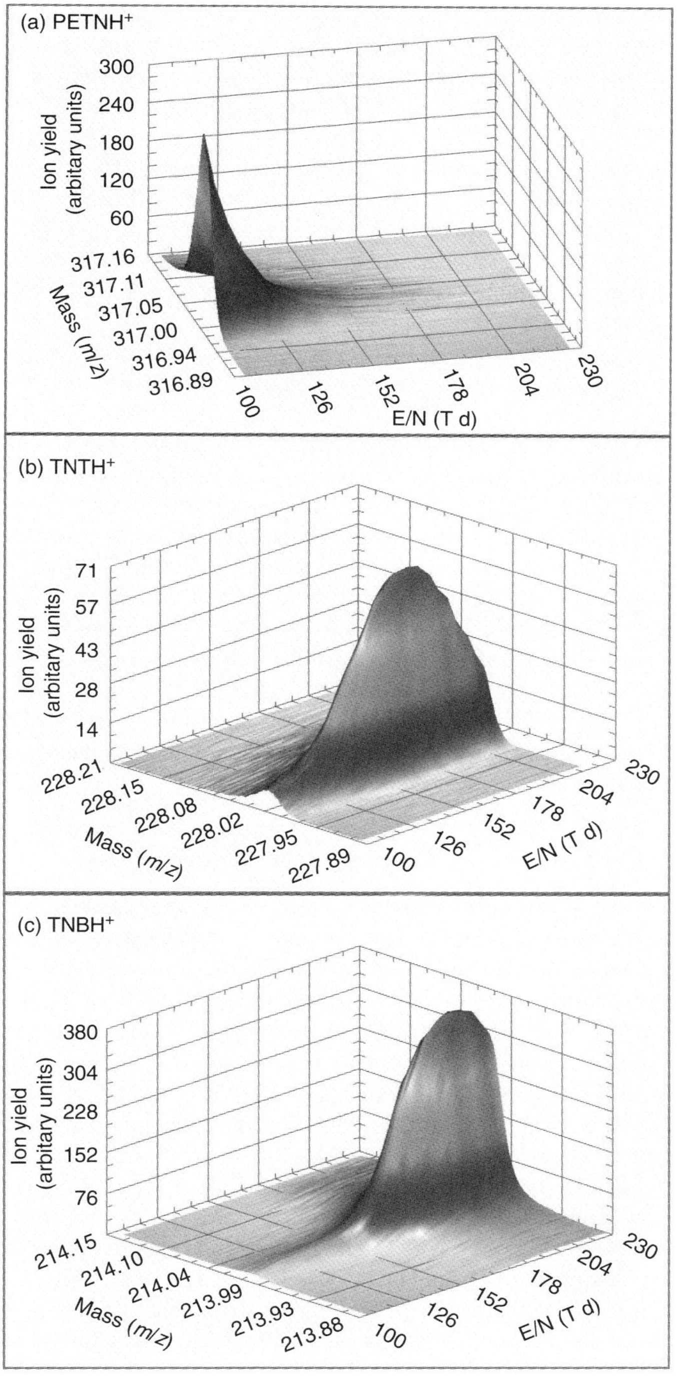

Figure 8.2 *Ion signal intensities for protonated (a) PETN, (b) TNT and (c) TNB as a function of E/N recorded using a PTR TOF-MS instrument. (a) Courtesy of Phillip Sulzer (Ionicon Analytik GmbH). Kindly provided by Dr Philipp Sulzer of Ionicon Analytik GmbH. (b, c) Reproduced with permission from [7]. Copyright 2012, American Chemical Society.*

tube but also because the effective temperature of the ions in the drift tube is far higher than 298 K (see Section 3.4.4). Nevertheless, Sulzer *et al.* conclude that the differences in the ergonicity are consistent with the shallower fall-off of the protonated parent ion signal intensity at low *E*/*N* observed for TNT than for TNB (compare Figure 8.2(b) with Figure 8.2(c)).

If the above explanation is correct then the production of $MH^+(H_2O)$, where M = TNT or TNB, and its reaction with water will increase as the humidity is increased, which in turn will result in more rapid production of $H_3O^+(H_2O)$ as a terminal product ion. Furthermore, it can be expected that the *E*/*N* value at which the MH^+ intensity reaches a maximum will shift to higher values as the humidity in the drift tube is increased. Experiments by Sulzer and colleagues have confirmed this proposal [7].

Although chemically interesting in their own right, from an analytical perspective the observations for the detection sensitivity of protonated TNT and TNB have two important practical implications when using PTR-MS as an analytical tool. The first is that the drift tube of a PTR-MS instrument must be set at an appropriate *E*/*N* value in order to maximize the sensitivity of detection. In particular, operating at standard *E*/*N* values of between 110 and 140 Td (which is the most common operational range) results in a much lower detection efficiency compared to an *E*/*N* where the maximum intensity is observed. The second practical and more significant implication is that the $TNTH^+$ and $TNBH^+$ sensitivity dependences on *E*/*N* could be used to provide a potentially useful discriminatory analytical procedure for the elimination of false positives with regard to the detection of TNT and TNB using PTR-MS technology. This discrimination can be achieved by simply providing fast switching between *E*/*N* values (e.g. from 180 to 110 Td) whilst monitoring the $TNTH^+$ (*m*/*z* 228) or $TNBH^+$ (*m*/*z* 213) signal intensities. Fast switching of *E*/*N* is easily achieved by keeping *N* constant and rapidly changing the voltage applied across the drift tube in order to change the electric field strength.

8.2.1.2 High Mass Resolution PTR-TOF-MS Measurements of TNT

As mentioned above, an extremely important aspect of any analytical technique for threat agent detection is to have a high level of confidence in detection. Time-of-flight mass spectrometers have substantially higher mass resolution than quadrupole-based devices and are capable of separating ions that nominally have the same mass (isobaric compounds). In a recent PTR-MS study of explosives, an unidentified compound was found to be present in the sampled air which produced a product ion at the same nominal mass as that for protonated TNT (*m*/*z* 228) [8]. Using a high mass resolution PTR-TOF-MS, it was found that the ion associated with the contaminant compound had an *m*/*z* of 228.22, which was clearly separated from protonated TNT at *m*/*z* 228.025. Working with such a high mass resolution clearly has great potential as a means of suppressing the risk of false positives. However, high-resolution mass spectrometric instruments are not available to all researchers and are unlikely to see widespread use in actual security environments owing to the prohibitive cost of such instruments. Consequently, to make PTR-MS cost effective other techniques need to be considered to help in the assignment of an unknown compound so that low mass resolution instruments can be used with little reduction in the confidence of an assignment. One technique has been described above for TNT and TNB. However, the ion chemistry involved that results in the unambiguous assignment for TNT and TNB is unusual. Consequently, other techniques need to be considered and one of these is to use other types of reagent ions.

8.2.1.3 Reagent Ion Switching and Explosives Detection

Studies have shown that it is possible to switch between reagent ions within seconds, which is mirrored in the immediate production of different ions owing to changes in the ion–molecule chemistry occurring in the reaction (drift) region of a PTR-MS instrument [6]. In terms of reaction with TNB, for example, the ion–molecule process changes from non-dissociative proton transfer with H_3O^+ as the reagent ion to non-dissociative charge transfer when using $O_2{}^+$ as the reagent ion,

$$O_2{}^+ + \text{TNB} \rightarrow \text{TNB}^+ + O_2 \tag{8.3}$$

that is, the product ion changes by one atomic mass unit in going from TNBH^+ (*m/z* 213) resulting from the reaction with H_3O^+ to TNB^+ (*m/z* 212) resulting from the reaction with $O_2{}^+$. The corresponding reaction with TNT is more dramatic. Although the reaction with $O_2{}^+$ does lead to a non-dissociative charge transfer channel being observed, with an ion being detected at *m/z* 227, the dominant reaction pathway is found to be dissociative charge transfer resulting in an ion at *m/z* 210 corresponding to the elimination of OH from TNT^+, that is

$$O_2{}^+ + \text{TNT} \rightarrow [\text{TNT} - \text{OH}]^+ + \text{OH} + O_2 \tag{8.4a}$$

$$\rightarrow \text{TNT}^+ + O_2 \tag{8.4b}$$

The branching percentage for Reaction 8.4(a) has been found to be greater than 90% for all *E/N* values investigated (90–250 Td) [6].

Although the recombination energy of $O_2{}^+$ (12.07 eV) is more than sufficient for charge transfer to be an energetically feasible pathway for reaction with PETN (IE = 10.4 eV) or RDX (IE = 9.8 eV),[3] interestingly no such reaction is observed. By comparison NO^+ reacts with both of these compounds, although not via a bimolecular process. Instead adduct formation occurs to give MNO^+ product ions, but this is only observed at low *E/N* (<100 Td). At higher *E/N*, no adduct ions are observed owing to collision-induced dissociation.

These investigations on the switching of the reagent ion have shown how it might be used to significantly enhance selectivity of a PTR-MS, which would be beneficial to Homeland Security applications, especially when one considers that the enhancement of selectivity can be controlled by a relatively simple automated procedure to provide rapid switching of gases/vapours entering the ion source, and hence provide rapid switching of the reagent ions. (Rapid in this context refers to a change of one reagent ion to another in tens of seconds, that is, within a timescale that would be tolerated by security personnel and the travelling public). If needed, this switching of reagent ions could be done in conjunction with a change in the *E/N* to enhance the selectivity for compounds such as TNT and TNB. Although these are relatively simple methods to improve selectivity, their importance lies

---

[3] These ionization energies (IE) in fact refer to appearance energies determined by Schramm *et al.* [9]. For PETN, this corresponds to an ion with an *m/z* value of 46 (which presumably corresponds to $NO_2{}^+$). For RDX, the observed fragment ion is at *m/z* 128 (presumably corresponding to the loss of two HONO molecules from the charged parent ion).

in the fact that it would be possible to use a low mass resolution PTR-MS instrument in a security environment to keep the cost of the analytical device at an affordable level.

8.2.1.4 PTR-MS and the Detection of Traces of Explosives

So far we have only discussed the potential of PTR-MS to detect solid explosives with no mention of the limit of detection. For 'field' applications, the usefulness of the technique will depend on its ability to detect trace residues left on hands, clothing, computers, etc. Such a detection methodology is considerably aided by the use of appropriate sampling, pre-concentration, thermal desorption units and heated inlet systems in order to maximize the concentration of explosives reaching the analytical device and hence enhance its ability to detect traces of explosives. For the well-developed IMS techniques, this is usually achieved by either swipe or portal sampling followed by the efficient thermal desorption and transfer of residual material collected on the swab or filter. No comparable technology has yet been developed for direct use with PTR-MS. However, an improvised sampling and pre-concentration system has yielded significant improvements in the signal intensities for solid explosives [5, 8]. The technique involved passing high flows of air at room temperature over trace quantities of explosives deliberately placed on a surface, with the air flow being filtered through a fine stainless steel wire mesh for several seconds to collect small quantities of the explosive via adsorption. The mesh was then resistively heated at a temperature of about 150°C, resulting in a boost in the concentration of the explosive entering the drift tube. This simple pre-concentration technique has been used to enhance the signal strengths for several explosives present in trace amounts, including HMX (the compound with the lowest vapour pressure of all explosives). The major problem with this set-up was found to be memory effects, that is, the explosive remained in the system for long periods of time, so that fast cycle times (pre-concentration, desorption, transfer, ionization, detection and removal of explosives) were not possible. Improved thermal desorption and pre-concentration systems therefore need to be developed if PTR-MS is going to be a viable technology for use within Homeland Security. A particularly crucial point with regards to any sampling system is that no cold-spots must be present in the transfer and inlet lines if sufficient concentrations of an explosive are to make it to the reaction chamber. Another important issue is the temperature used in any thermal desorption systems. This must be kept below that which causes major thermal decomposition or degradation of compounds but high enough to limit surface adsorption. In the case of TNT, PETN, RDX and HMX, a recent IMS study has indicated that the optimal temperatures to be used are 80°C, 100°C, 160°C and 200°C, respectively [10]. Above these temperatures, this study provided evidence to suggest that these explosives will decompose rapidly. However, this needs to be confirmed by temperature-dependent PTR-MS investigations.

8.2.2 Environmental Aspects and Explosives

For decades, obsolete explosives and munitions have been either buried in the ground or dumped at sea. This can be a significant environmental problem because many explosives are toxic (e.g. TNT poisoning can lead to aplastic anaemia or toxic jaundice). There is therefore a need for on-site detection and identification of traces of explosives in areas suspected of such contamination.

An obvious application is to monitor the quality of groundwater to make sure it has not been contaminated. To illustrate the potential of PTR-MS for this kind of work, Jürschik *et al.* placed ~1 mg of TNT into a glass vial containing 50 mL of water without stirring, shaking or heating the sample [8]. Over a period of several months, small quantities of the water were taken periodically and injected into a PTR-MS via a direct aqueous injection (DAI) technique [11]. (The DAI technique is discussed in more detail in Chapter 9.) The researchers demonstrated that even after one night of immersion sufficient TNT could be detected in the water, despite the fact that TNT is almost insoluble in water. Over time the concentration of TNT in water increased and was found to reach a 'saturation' level after ~2 months.

8.3 Chemical Warfare Agents and Toxic Industrial Chemicals

Unlike explosives, CWAs and weaponizable toxic industrial chemicals are not destructive to property (unless long-term contamination occurs), but are targeted to kill, injure, incapacitate or cause panic. Four main scenarios for the detection of CWAs should be considered. One is to detect and identify the agents when they are being produced in illegal chemical laboratories. A second is to detect them before deployment, for example, trace quantities on hands or objects. The third is to quickly identify the chemical agent once it is deployed, a situation requiring accurate identification which is vital not only for the protection of first responders but also to provide rapid and effective treatment of any casualties in order to minimize fatalities. The fourth application is to monitor contamination after an event. Rapid detection and high confidence in the identification of trace quantities of chemical threats using PTR-MS could form the basis for a more timely warning system capability than is currently provided by existing analytical technologies.

Towards this goal, a rugged and compact CWA and Toxic Industrial Compound PTR-MS monitor for heating, ventilation and air conditioning systems has been developed [12]. However, the first reported use of PTR-MS for CWA detection was by Cordell *et al.* in collaboration with the Defence Science and Technology Laboratory (DSTL) at Porton Down, United Kingdom [13]. This study explored two real CWAs, sarin (a nerve agent) and mustard gas (a blister agent) and a number of CWA simulants – 2-chloroethyl ethyl sulfide (2-CEES), triethyl phosphate (TEP) and dimethyl methylphosphonate (DMMP). The work with the CWA simulants showed only a modest decrease in sensitivity with increased humidity, thereby revealing that humidity would not be a hindrance to analysis. The subsequent measurement of mustard gas, bis(2-chloroethyl) sulfide, showed that the protonated parent ion is readily detected at m/z 159, 161 and 163 (three peaks owing to $^{35}Cl/^{37}Cl$ combinations in approximately the ratio of 9:6:1). On the other hand, sarin led only to fragment ions being observed, except at very low E/N values.

The early work reported by Cordell *et al.* was followed by a more extended investigation by Petersson *et al.* of CWA simulants using a high-resolution PTR-TOF-MS [14]. The simulants investigated were DMMP, diethyl methylphosphonate (DEMP), diisopropyl methylphosphonate (DIMP) and dipropylene glycol monomethyl ether (DPM). The research team focused particularly on the effects of changing E/N on the product ion branching ratios. For both DMMP (protonated mass at m/z 125.04) and DEMP (protonated mass m/z 153.07), the protonated parent molecule was by far the dominant ion observed over the whole of the *E*/*N* range covered in the study. For DIMP, the ion yield of the protonated parent molecule was found to drastically decrease as *E*/*N* was increased and only at relatively low *E*/*N* values (<110 Td) was the protonated parent ion signal observed to dominate. The major fragment ions observed for this simulant were found to be at *m*/*z* 139.05 and *m*/*z* 97.01, resulting from the loss of one or two propene molecules, respectively, from the protonated parent molecule. For DPM (protonated mass *m*/*z* 149.12), the protonated parent molecule was never the most abundant product ion at any *E*/*N* value investigated. The dominant product ion from the reaction of H_3O^+ with DPM, over an extended *E*/*N* range, was that observed at *m*/*z* 73.07, which results from the breakage of bond next to the oxygen in the propoxy part of the protonated parent molecule.

8.4 Narcotics

The first and so far only reported use of PTR-MS for the detection of illicit and prescribed drugs was made by Agarwal *et al.* [15]. This study described the capabilities of PTR-MS for detecting a number of commonly used narcotics, including *N*-methyl-3,4-methylenedioxyamphetamine ($C_{11}H_{15}NO_2$, ecstasy), morphine ($C_{17}H_{19}NO_3$), codeine ($C_{18}H_{21}NO_3$), cocaine ($C_{17}H_{21}NO_4$, and its salt) and heroin ($C_{21}H_{23}NO_5$). The procedure adopted in their study involved direct sampling of the headspace above ~50 mg of each of the drugs placed in glass vials, usually maintained at room temperature.

Ecstasy, cocaine and heroin are illegal in most countries. Although morphine and codeine are prescription rather than illicit drugs, they were included in the study by Agarwal *et al.* because they are major components of opium. Morphine is more generally used as an analgesic to relieve severe pain, while codeine is an antitussive (cough suppressant) and is partly metabolized to morphine in the body.

As found for the PTR-MS studies dealing with solid explosives, the protonated parent molecule of the aforementioned drugs is readily detected in each case, but it is not the only or even the dominant product ion observed. Figure 8.3 provides examples of mass spectra for (a) ecstasy and (b) heroin over *m*/*z* ranges where product ions were observed, namely *m*/*z* 150–230 and *m*/*z* 300–380, respectively. Table 8.2 provides a summary of the product ions for all of the drugs investigated by Agarwal *et al.* and their associated branching ratios at an *E*/*N* of 120 Td.

Agarwal *et al.* did not calculate the proton affinities of the drugs investigated owing to their complexity. However, the proton affinities of the basic functional groups (subspecies) were calculated using DFT methodology at the (B3LYP6-31 + G(d,p)) level, which is useful because a number of these subspecies are common to various drugs. For example, morphine, codeine, cocaine and heroin all contain an *N*-methyl piperidine ring. Although useful, it must be appreciated that the proton affinities are only used to provide a guide to the possible routes of a proton transfer process – the actual proton affinity of a particular site within the complex molecular structure will be different from that calculated. The results of these calculations are summarized in Table 8.3, from which we can see that proton transfer from H_3O^+ ($PA(H_2O) = 691$ kJ mol^{-1}) to any of the drugs is thermodynamically allowed. Furthermore, given that the values of the proton affinities are substantially greater than that of water, it is not surprising that dissociative proton transfer channels are observed. In order to explain the observed fragmentation pathways, the researchers proposed that protonation

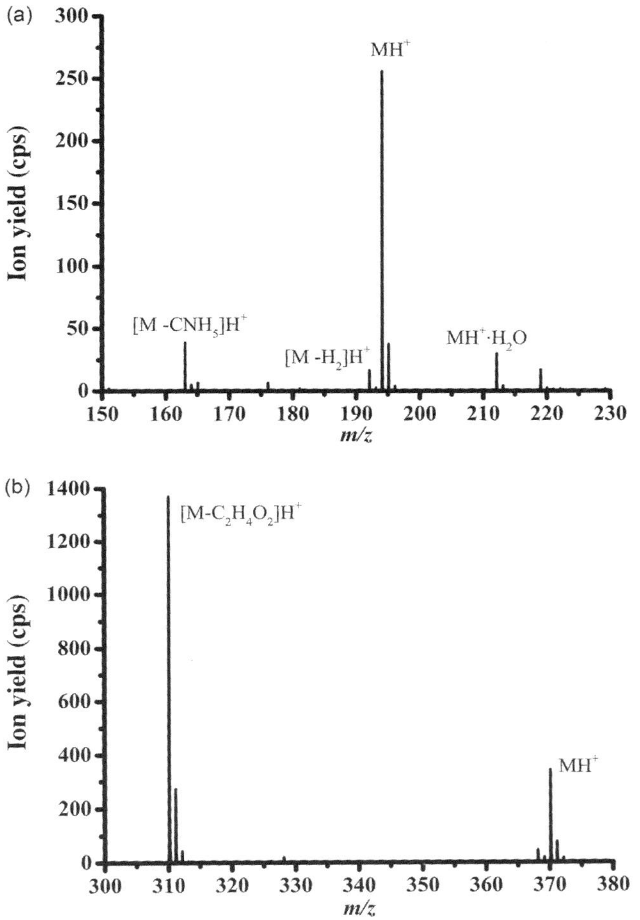

Figure 8.3 *PTR-MS mass spectra obtained from sampling the headspace above (a) ecstasy and (b) heroin over* m/z *ranges where product ions have been observed. Note that for the heroin spectrum the drug was heated to 70°C in order to obtain higher concentrations in the drift tube of the PTR-MS so that the fragmentation channels could be more easily detected. The two mass spectra were recorded at an* E/N *value of 120 Td and represent the mean of nine individual scans; each obtained using a 2-s integration time. With kind permission from Springer Science and Business Media.*

must occur on sites other than the most thermodynamically favourable. For example, in the case of heroin a substantial fraction of the protonation must occur at the acetate groups [16], which have proton affinities between 827 and 848 kJ mol^{-1}, rather than the *N*-methyl piperidine site, which has a greater proton affinity of 964 kJ mol^{-1}. Whilst protonation to another site from a more thermodynamically favoured site could occur by intramolecular proton migration, it was concluded that the relatively rigid and complex structures of the drug molecules render this extremely unlikely. Thus, it was proposed that protonation on

Table 8.2 *Type of drug, molecular formula and structure, and product ions and their branching ratios (percentages) recorded with a PTR-MS drift tube operating at a reduced electric field of 120 Td*

Drug	Product ions (%)	*m/z*
Ecstasy ($C_{11}H_{15}NO_2$) 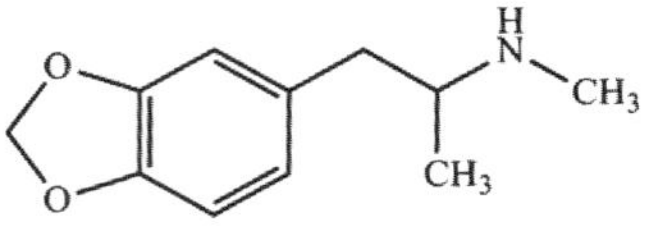	$M(H_3O^+)$ (6) MH^+ (68) $[M-H_2]H^+$ (7) $[M-CNH_5]H^+$ (19)	212 194 192 163
Morphine ($C_{17}H_{19}NO_3$) 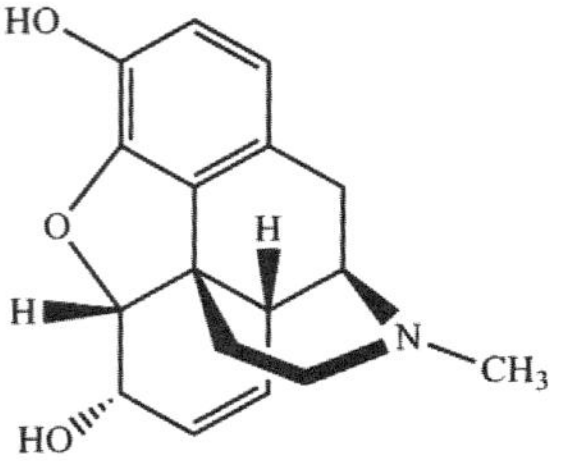	MH^+ (31) $[M-H_2]H^+$ (17) $[M-H_2O]H^+$ (52)	286 284 268
Codeine ($C_{18}H_{21}NO_3$) 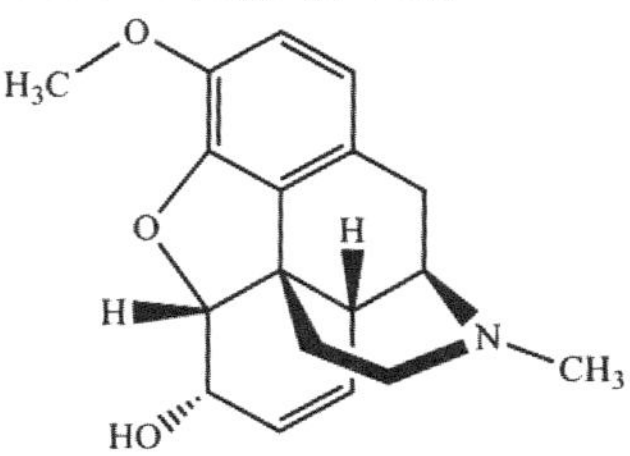	MH^+ (28) $[M-H_2]H^+$ (5) $[M-H_2O]H^+$ (64) $[M-CH_3OH]H^+$ (3)	300 298 282 268
Cocaine ($C_{17}H_{21}NO_4$) 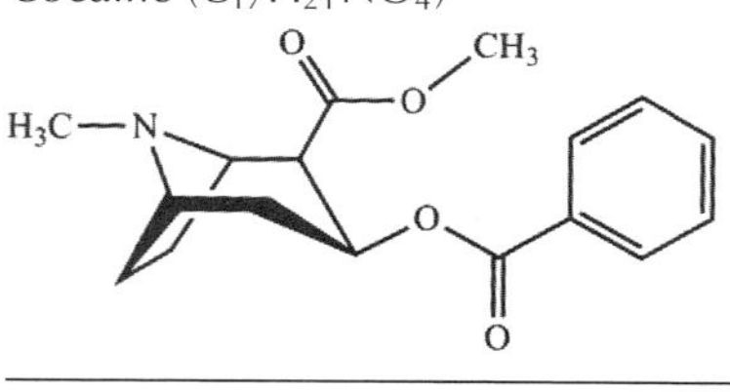	MH^+ (76) $[M-CH_3OH]H^+$ (6) $[M-C_7H_6O_2]H^+$ (18)	304 272 182
Heroin ($C_{21}H_{23}NO_5$) 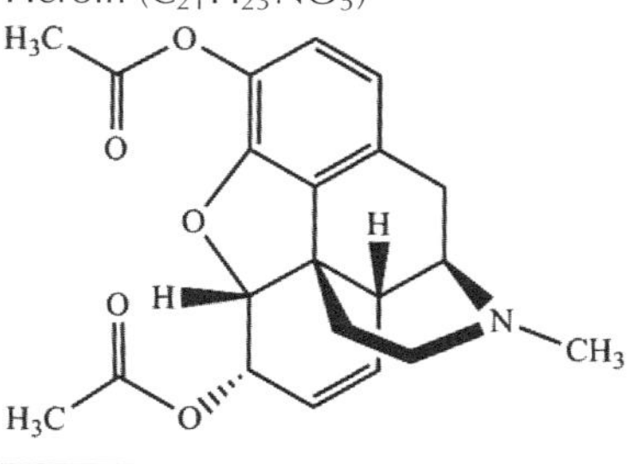	MH^+ (26) $[M-C_2H_4O_2]H^+$ (74)	370 310

Source: With kind permission from Springer Science and Business Media.
M is used in the product ion column to represent the parent drug molecule.

Table 8.3 *Calculated proton affinities (PA) of the sub-species (protonation sites) involved in drug protonation reactions*

Sub-species	PA (kJ mol^{-1})
1,3-Benzodioxole[a]	747
Isopropyl methylamine[a]	948
Phenol[b]	753
N-methyl piperidine[b,c,d,e]	964
Cyclohex-2-en-1-ol[c,e]	858
Anisole[c]	782
Methyl acetate[d]	827
Methyl benzoate[d,e]	839
Phenyl acetate[e]	848

The numerical values have been derived from DFT calculations at the B3LYP/6-31 + G(d,p) level [3].
[a] Present in ecstasy.
[b] Present in morphine.
[c] Present in codeine.
[d] Present in cocaine.
[e] Present in heroin.

the less basic sites must occur during the initial encounter between the drug molecule and the protonating species. In other words protonation is kinetically driven rather than thermodynamically driven. The site where the proton attaches therefore holds the key to interpreting the resulting mass spectra.

8.5 Date Rape Drugs

The term 'date rape drug' is a colloquial expression for any kind of substance that can be administered to an unknowing victim in order to disable a person's resistance, and usually for the purpose of sexual abuse. Sometimes these substances are also used to knock out and subsequently rob a person. The ability to detect such drugs rapidly and with a high level of confidence in identification would have great appeal for law enforcement agencies.

Jürschik *et al.* have presented proof-of-principle measurements for the detection of 'date rape drugs' spiked into various drinks (water, tea and wine) [17]. The researchers concentrated on investigating four drugs, chloral hydrate, tricholorethanol, γ-butyrolactone and butanediol. For this particular study quadrupole-based PTR-MS and high-resolution PTR-TOF-MS instruments were used together with two different sampling methods – dynamic headspace sampling or direct liquid injection [11]. Only with the dynamic headspace sampling could drug contamination of a drink be detected in a time frame of seconds. It was therefore concluded that this method is the most promising for the use of PTR-MS as a fast, sensitive and selective monitor for the detection of drinks contaminated with a date rape drug. This work has demonstrated that a PTR-MS instrument could verify extremely quickly (compared to GC/MS, which is highly selective but slow) whether or not a suspicious drink has been spiked with a drug.

8.6 Ion Mobility Mass Spectrometry and PTR-MS: A Brief Comparison for Homeland Security Applications

Ion mobility spectrometers (IMS) are the most commonly used analytical instruments for the detection of threat agents because they are robust, portable, easy to use, inexpensive and sensitive (see also Section 1.3.2). Given the current widespread use of IMS systems in areas of high security, for example, at airports, it is useful to provide a brief comparison of IMS and PTR-MS technologies, and to highlight their advantages and disadvantages. Both PTR-MS and IMS are drift tube based technologies that in positive ion mode use protonated molecular species for the reagent ion. There are, of course, differences. IMS relies on injecting pulses of ions into the drift region (which is separate from the reaction region) and then the different arrival times of the ions are used to provide selectivity. Furthermore, IMS operates with the drift tube at high pressures (typically one atmosphere) so that protonated water clusters are the dominant reagent ions. Given that the proton affinities of $(H_2O)_n$ clusters ($n > 1$) are greater than that of the H_2O monomer, this limits the range of chemicals that can be detected, which can be beneficial, for example, by eliminating certain VOCs. However, it can equally well lead to limitations. In positive ion mode IMS cannot easily detect solid explosives (TNT, PETN, HMX, etc.) and so for the detection of such compounds an IMS operates in negative ion mode. Either IMS systems may be switched in polarity or a dual-chamber IMS system can be used, with one chamber operated in negative ion mode and the other in positive ion mode, thereby increasing the applicability of IMS. PTR-MS has to date only been operated in positive ion mode[4] and usually relies on H_3O^+ as the reagent ion, although as we have seen earlier in this chapter and elsewhere in this book there are exceptions to this.

With the H_3O^+ monomer as a reagent ion, rather than protonated water clusters, a greater range of chemicals can be detected in positive ion mode (including explosives). Furthermore, the use of a mass spectrometer in PTR-MS provides higher selectivity and hence a higher confidence in compound assignments. Naturally, the selectivity of an IMS could be greatly increased by operating the drift tube with the electronic shutter (see Section 1.3.2) continuously open and coupling it directly to a mass spectrometer. IMS-MS systems have not been commercially developed for routine analytical purposes, having only seen action in research environments to investigate the complex ion–molecule chemistry occurring within high-pressure drift tube systems. However, if this became more available it does have a distinct advantage of being able to provide an additional dimension in comparison to PTR-MS, namely it can be used to provide ion mobility separation in addition to mass spectra. A disadvantage of IMS-MS, in comparison to PTR-MS, is the additional pumping required because of the need to sample ions from a high-pressure (one atmosphere) drift tube environment into the far lower pressure chamber used to house the mass spectrometer.

[4] A negative-ion proton transfer chemical-ionization mass spectrometer has been developed by Veres *et al.* [18], which uses $CH_3C(O)O^-$ as the reagent ion for the measurement of gas-phase organic acids in the atmosphere. However, this instrument uses a flow tube (operated at 32 Torr) as the reaction chamber, not a drift tube.

8.7 Future Directions

The rapid identification of trace quantities of threat agents with a high level of confidence is of great importance to the military and emergency responders, as well as for applications in checkpoint security areas such as airports, harbours and train stations. In this chapter, we have illustrated the capabilities of PTR-MS for the rapid detection of a broad range of threat agents (explosives, CWAs and illicit drugs). We have shown that the high sensitivity of this instrument permits the relatively easy detection of these compounds in trace quantities. Furthermore, the high mass resolution now achievable with some PTR-TOF-MS instruments results in a high level of confidence in the identification of a specific compound through the identification of the *m/z* of the protonated parent molecule. However, in a real-life situation, where air is frequently being sampled from a complex chemical environment, the presence of signals arising from unknown interferents may result in some uncertainty in identification and quantitative analysis. Also, high mass resolution instruments are currently too expensive for routine security use, and therefore if PTR-MS is to be adopted for Homeland Security cheaper instruments are needed. Alternatively, a security area could use IMS as the first screening procedure and then have just one high mass resolution PTR-TOF-MS available to confirm or refute alarms.

As mentioned previously in Section 8.2.1.1, Sulzer *et al.* have already demonstrated that by altering *E/N* a unique signature occurs for the detection of TNT and TNB [7]. This means that lower mass resolution quadrupole mass spectrometers, such as those used in residual gas analysers, could be used in a PTR-MS instrument for security purposes instead of TOF-MS. This would deliver both a cheaper and a more compact instrument. The adoption of cost-efficient mass spectrometers and inclusion of user-friendly feedback to security personnel would result in a rapid and field-effective PTR-MS technology that can provide high selectivity and in turn provide a much higher level of confidence for the real-time and rapid analytical detection of threat agents than is currently available. However, if PTR-MS is to be adopted as a broad-based security platform instrument, any low mass resolution instrument must be able to detect many threat compounds with a high level of confidence. So far only a small number of nitroaromatic compounds have been shown to have an unusual bias dependence on *E/N*. Other explosives, such as nitroamines (e.g. HMX) or the nitrate esters (e.g. PETN and nitroglycerine), do not show this behaviour – instead their protonated parent intensities show a similar behaviour to the majority of compounds investigated with PTR-MS to date, namely decreasing protonated parent ion intensity with increasing *E/N*. Consequently, for these threat agents it will be necessary to determine other suitable measuring protocols for rapid detection with high selectivity. This means that further investigations are required to determine the effects of changes in the product ions when reagent ions are switched. Combining rapid reagent ion switching with fast changes in operational parameters (e.g. drift tube voltage and hence *E/N*) seems to be a viable way forward in the development of PTR-MS instrumentation suitable for threat agent detection.

Although PTR-MS provides a broad-based technology for the detection of threat agents, another limitation (other than cost) for its adoption is the provision of a rapid cycle time for detection, that is, the whole process of sampling, analysis and recovery (no memory effects) ideally needs to be complete within tens of seconds. No residues must remain from compounds that readily adsorb on surfaces if PTR-MS is to be adopted for security applications under realistic conditions in the field (e.g. at an airport). Suitable sampling and inlet systems still need to be developed to provide the required fast recovery times in order to eliminate possible memory effects, and thereby for the technology to be tolerated by the travelling public. For this, it should be possible to adapt the sampling technology already in use for IMS systems.

"*Trust me Sir! It's completely non-invasive.*" (Copyright Jessica Mayhew 2011)

References

1. Shen, C., Li, J., Han, H. *et al.* (2009) Triacetone triperoxide detection using low reduced-field proton transfer reaction mass spectrometer. *Int. J. Mass Spectrom.* **285**, 100.
2. Harper, R. J., Almirall, J. R., Furton, K. G. (2005) Identification of dominant odor chemicals emanating from explosives for use in developing optimal training aid combinations and mimics for canine detection. *Talanta* **67**, 313.

3. DFT calculations were performed using the Gaussian 09 programme with the GaussView interface. (Gaussian 09, Revision A.02, Frisch, M. J., Trucks, G. W., Schlegel, H. B. *et al.* Gaussian, Inc., Wallingford, CT, 2009.)
4. Bouchoux, G., Salpin, J. Y., Leblanc, D. (1996) A relationship between the kinetics and thermochemistry of proton transfer reactions in the gas phase. *Int. J. Mass Spectrom. Ion Processes* **153**, 37.
5. Mayhew, C. A., Sulzer, P., Petersson, F. *et al.* (2010) Applications of proton transfer reaction time-of-flight mass spectrometry for the sensitive and rapid real-time detection of solid high explosives. *Int. J. Mass Spectrom.* **289**, 58.
6. Sulzer, P., Agarwal, B., Jürschik, S. *et al.* (2013) Applications of switching reagent ions in proton transfer reaction mass spectrometric instruments for the improved selectivity of explosive compounds. *Int. J. Mass Spectrom.* doi:10.1016/j.ijms.2013.05.004
7. Sulzer, P., Petersson, F., Agarwal, B. *et al.* (2012) Proton transfer reaction mass spectrometry and the unambiguous real-time detection of 2,4,6-trinitrotoluene. *Anal. Chem.* **84**, 4161.
8. Jürschik, S., Sulzer, P., Petersson, F. *et al.* (2010) Proton transfer reaction mass spectrometry for the sensitive and rapid real-time detection of solid high explosives in air and water. *Anal. Bioanal. Chem.* **398**, 2813.
9. Schramm, E., Mühlberger, F., Mitschke, S. *et al.* (2008) Determination of the ionization potentials of security-relevant substances with single photon ionization mass spectrometry using synchrotron radiation. *Applied Spectrosc.* **62**, 238.
10. Najarro, M., Dávila Morris, M. E., Staymates, M. E. *et al.* (2012) Optimized thermal desorption for improved sensitivity in trace explosives detection by ion mobility spectrometry. *Analyst* **137**, 2614.
11. Jürschik, S., Tani, A., Sulzer, P. *et al.* (2010) Direct aqueous injection analysis of trace compounds in water with proton-transfer-reaction mass spectrometry (PTR-MS). *Int. J. Mass Spectrom.* **289**, 173.
12. The SPIRIT project is a collaboration involving governmental, academic and industrial sectors. It is funded by the European Commission within the Security research objectives of the 7th Framework Programme. www.infrastructure-protection.org. (20 August 2013)
13. Cordell, R. L., Willis, K. A., Wyche, K. P. *et al.* (2007) Detection of chemical weapon agents and simulants using chemical ionization reaction time-of-flight mass spectrometry. *Anal. Chem.* **79**, 8359.
14. Petersson, F., Sulzer, P., Mayhew, C. A. *et al.* (2009) Real-time trace detection and identification of chemical warfare agent simulants using recent advances in proton transfer reaction time-of-flight mass spectrometry. *Rapid Commun. Mass Spectrom.* **23**, 3875.
15. Agarwal, B., Petersson, F., Jürschik, S. *et al.* (2011) Use of proton transfer reaction time-of-flight mass spectrometry for the analytical detection of illicit and controlled prescription drugs at room temperature via direct headspace sampling. *Anal. Bioanal. Chem.* **400**, 2631.
16. Nacson, S., Harrison, A. G., Davidson, W. R. (1986) Effect of method of ion preparation on low-energy collision-induced dissociation mass spectra. *Org. Mass Spectrom.* **21**, 317.

17. Jürschik, S., Agarwal, B., Kassebacher, T. (2012) Rapid and facile detection of four date rape drugs in different beverages utilizing proton-transfer-reaction mass spectrometry (PTR-MS). *J. Mass Spectrom.* **47**, 1092.
18. Veres, P., Roberts, J. M., Warneke, C. *et al.* (2008) Development of negative-ion-proton-transfer chemical-ionisation mass spectrometry (NI-PT-CIMS) for the measurement of gas-phase organic acids in the atmosphere. *Int. J. Mass Spectrom.* **274**, 48.

9

Liquid Analysis Using PTR-MS

9.1 Determination of Henry's Law Constants Using PTR-MS

When a substance is dissolved in solution at low concentration, a proportion of the substance can evaporate. At equilibrium, the ratio of the concentration of the substance in the liquid, $c(\text{liq})$, to its partial pressure in the gas phase, $p(\text{gas})$, is given by Henry's law, viz.

$$c(\text{liq}) = H \times p(\text{gas}) \tag{9.1}$$

In the above expression, H is known as Henry's constant and is equivalent to the partition coefficient between the liquid and gas phases.[1] Henry's law is often a reasonable approximation at relatively low concentrations, that is, for near-ideal dilute solutions, where the solvent is largely unaffected by the solute. Expressed as a mole fraction, Henry's law is likely to be valid when the mole fraction of solute is $<10^{-3}$.

The liquid/gas partitioning is important in a number of contexts, as already identified elsewhere in this book, such as the partition of VOCs between the blood and the air in the alveoli of the lungs or the release of aroma VOCs from coffee into the gas phase. Determining Henry's constants is therefore valuable in attempts to understand the aforementioned processes. A number of techniques have been developed to determine Henry's constants. Early approaches sampled the gas phase above equilibrated systems, using a technique such as GC to determine gas phase concentrations. However, since this approach requires determination of the absolute number density of the volatile, this procedure is subject to a number of sources of error, including adsorption of volatiles on the inlet of the measuring device. Furthermore, the gas-measuring device must be appropriately calibrated. Also, removal of a substantial gas aliquot can disturb the gas–liquid equilibrium, thus introducing a further source of error.

[1] An alternative definition of Henry's constant, the inverse of the definition for H above, that is, $p(\text{gas}) = H \times c(\text{liq})$, is also quite commonly encountered in physical chemistry textbooks.

Proton Transfer Reaction Mass Spectrometry: Principles and Applications, First Edition. Andrew M. Ellis and Christopher A. Mayhew.

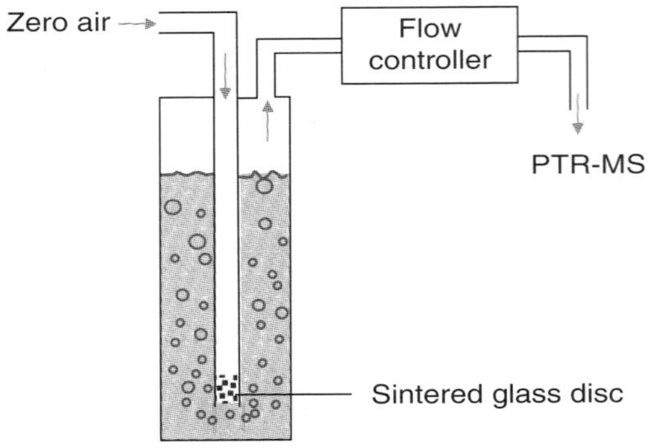

Figure 9.1 *Stripping cell configuration for determining Henry's law constants.*

It is with some of these potential flaws in mind that a dynamic method for determining Henry's law constants was developed in the late 1970s. This alternative bubbles an inert gas or clean/dry air through a liquid sample to drive dissolved species out of the liquid and into the gas phase. As the carrier gas elutes the quantity of the volatile compound gradually begins to decline and measurement of the rate of this decline can be linked to Henry's constant. A key advantage of this approach is that calibration of the GC system is no longer required. Nevertheless, this approach is still relatively tedious because the expelled gas needs periodic sampling and relatively slow analysis to determine the partial pressure of the target species.

More recently PTR-MS has been used in place of a GC system for determining Henry's constants [1]. The crucial advantage here is that PTR-MS can be used for essentially continuous monitoring, thus reducing the error margins associated with periodic sampling. Also, for systems with highly volatile dissolved VOCs,[2] the Henry's law constant can now be determined rapidly.

Figure 9.1 illustrates the apparatus employed. The solution of interest is placed in a vertically oriented cell known as a stripping cell. Clean and dry air is passed into the bottom part of the cell through a dispersing element to generate bubbles, which then travel up through the liquid at a rate determined by the flow rate of the incoming gas. The dissolved substance establishes an equilibrium concentration in the gas bubbles and when the top of the liquid layer is reached it is driven out of the system and onwards to the PTR-MS instrument for detection. By this means, the amount of dissolved substance in the liquid is gradually reduced and this is reflected in the partial pressure of the substance detected by PTR-MS.

The Henry's law constant can be related to the PTR-MS signal through the expression

$$\ln\left(\frac{\mathrm{cps}(t)}{\mathrm{cps}(0)}\right) = -\frac{F}{HVRT}t \tag{9.2}$$

[2] Highly volatile VOCs are defined as those with Henry's law constants <10 M/atm, where M = mol dm^{-3} and atm = one atmosphere of gas pressure of the specific compound.

where F is the gas flow rate (at STP), V is the liquid volume, R is the molar gas constant, T is the absolute temperature and t is the time. The quantities cps(t) and cps(0) are the PTR-MS count rates at time t and at time $t = 0$, respectively, for the protonated volatile. Since only a ratio of count rates is required, calibration is unnecessary, as mentioned earlier, and losses due to gas adsorption on various surfaces cancel.

Determination of the Henry's law constant by the above procedure is acceptable for reasonably volatile solutes and gives good agreement with carefully determined values from previous studies. However, for less volatile solutes the single stripping cell approach can take several hours to yield a measurable decline in partial pressure ($\geq$10%). Karl *et al.* have developed a solution to this in which the outlet gas from the first stripping cell is fed into the bottom of a second stripping cell [1]. With this arrangement, the gas partial pressure at the outlet of the second stripping cell *increases* with time and a much larger percentage change can be achieved in a relatively short space of time when compared to the single stripping cell experiment. A simple modification to Equation 9.2 allows Henry's law constants to be determined from the time-dependent partial pressure increase.

In their initial study, Karl *et al.* measured Henry's law constants in aqueous solution for 18 VOCs, including alcohols, aldehydes, ketones and various hydrocarbons [1]. PTR-MS has since been used to determine Henry's constants of other VOCs in aqueous solution, including carboxylic acids [2], esters [3], aromatics [4] and alkyl sulfides [5].

9.2 Analysis of Liquids

PTR-MS is designed ostensibly for the analysis of gases. However, means have been developed which allow PTR-MS to be used for the analysis of the VOC content of liquids, and in particular aqueous solutions. This section provides a brief overview of what may turn out to be an increasingly important analytical application of PTR-MS.

Simple but crude methods for determining the VOC content in an aqueous sample using PTR-MS have been around for some time. One important application is in determining the quantity of dissolved organic compounds in seawater, since the ocean surface is a significant natural source of atmospheric VOCs. An early approach used a 'purge and trap' methodology, in which a measured quantity of seawater was isolated and purged with helium to drive the organic species into a downstream cold trap [6]. Having completed the purging process, the trap was then heated to volatilize the trapped compounds and these were then driven into a PTR-MS instrument. Acetone, methanol, acetonitrile and dimethyl sulfide (DMS) were measured in seawater in this way and the estimated accuracy of this approach is $\pm$ 30%. However, there are obvious limitations to this methodology and, in particular, the purging process will be slow for compounds with high Henry constants (see previous section).

An alternative approach is the use of a membrane inlet system. Polymer membranes have been used, for several decades, for the selective introduction of molecules into the ion source region of a mass spectrometer. The membrane acts as a semi-permeable barrier and by selecting a hydrophobic material, typically a silicone polymer, an inlet can be constructed which allows organic compounds to diffuse through the membrane and into the inlet of a mass spectrometer while acting as a barrier to water. This idea has been combined with PTR-MS to detect various compounds in water [7]. A water pump is used

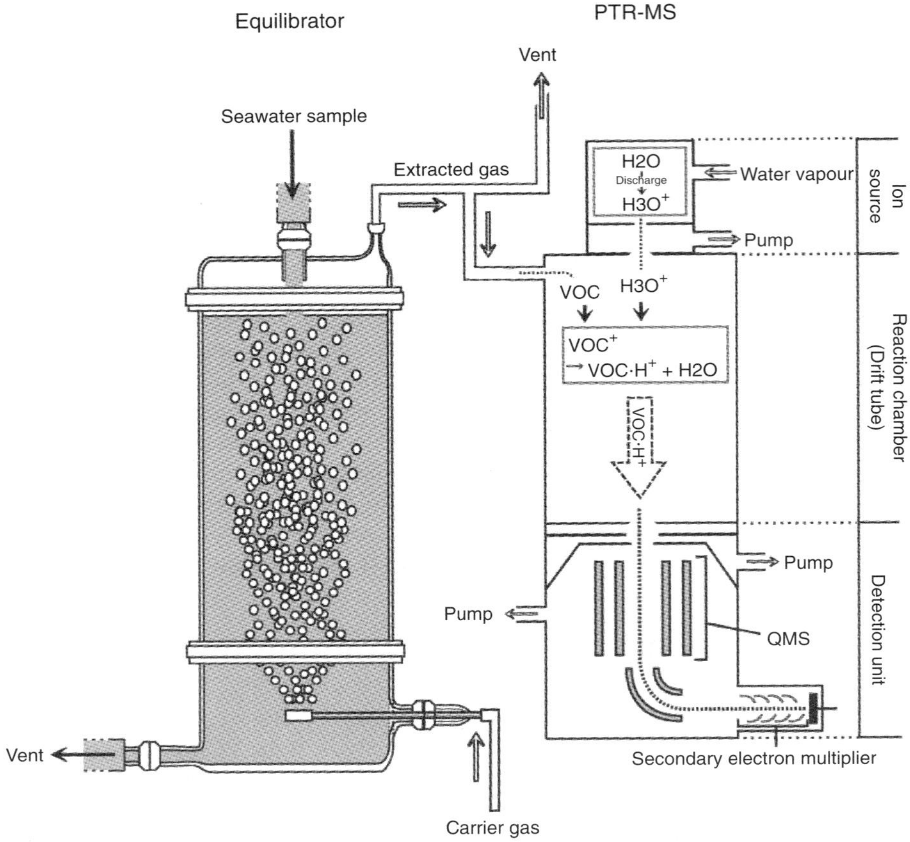

Figure 9.2 Apparatus for equilibrator inlet-PTR-MS. The main feature of the apparatus is the equilibrator cell, through which inert carrier gas is bubbled to extract dissolved VOCs into the gas phase. Adapted with permission from [8]. Copyright © 2009, American Chemical Society.

to maintain a continuous flow of the aqueous analyte over the outside of the membrane, and VOCs within the solution can then diffuse into the PTR-MS instrument. Detection sensitivities for specific VOCs as good as 100 ppt of water have been demonstrated in this approach. However, the technique has several disadvantages, including the fact that the rate of permeation can vary from compound to compound, necessitating calibration of the permeability for each specific compound.

Instead of using a membrane, Kameyama and co-workers have introduced what they term as *equilibrator inlet*-PTR-MS set-up for determining the VOC content of seawater [8, 9]. With some resemblance to the stripping cell used for determining Henry's constants (see previous section), the apparatus (see Figure 9.2) consists of an equilibrator vessel through which an inert carrier gas is bubbled to drive VOCs into the gas phase. A key difference from the Henry's constant apparatus is that the seawater is constantly replenished using a peristaltic pump. In order to obtain meaningful concentrations, it is essential that the extracted gases are in equilibrium with the dissolved gases. For those VOCs with high Henry's constants, equilibrium could be achieved relatively easily but for VOCs with

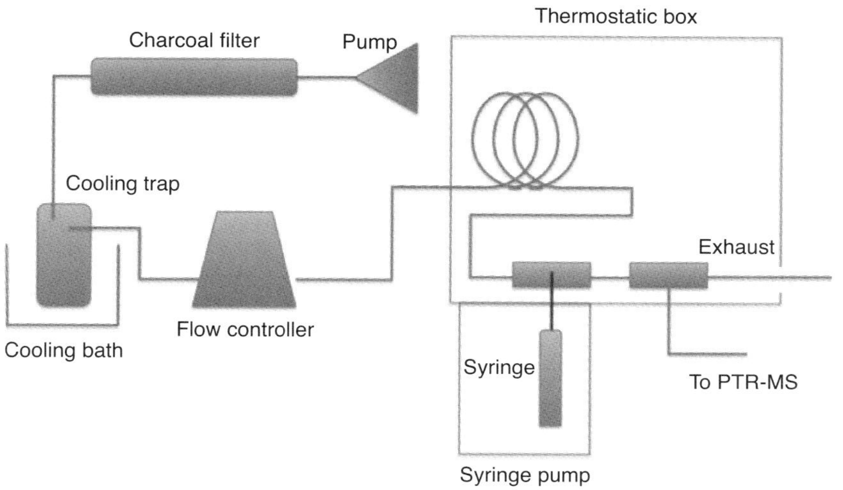

Figure 9.3 *Schematic of a direct aqueous injection apparatus for analysis of trace compounds in liquids with PTR-MS. Reproduced from [10] with permission from Elsevier.*

smaller Henry's constants, such as isoprene and propene, full equilibrium could not be reached. The reason for this is that VOCs with small Henry's constants have low solubilities and the carrier gas removes a disproportionate amount of dissolved gas from the bubbler before equilibrium is reached. If the system is not at equilibrium then a small change in the carrier gas flow rate will affect the measured VOC concentration. This provides a means of establishing if the system has reached equilibrium. Even if the system is not at equilibrium, Kameyama *et al.* showed that an *equilibration correction factor* could be employed to account for this, but this needs prior determination through measurements with different carrier gas flow rates. Response times were as short as a few minutes and the detection limits achieved by equilibrator inlet-PTR-MS were two orders of magnitude lower than the membrane inlet approach.

Yet another approach that has been developed for the analysis of aqueous solutions, and one which looks highly promising in several important respects, is a *direct aqueous injection* method [10], which is illustrated in Figure 9.3. The inlet system is rather simple, consisting of a septum and an injection syringe. Dry and warm gas which has been scrubbed of possible contaminants using a filtering system is delivered into the septum region and a small quantity of the liquid sample is continuously injected into this flowing gas, whereupon it is carried through into the inlet of a PTR-MS instrument. The temperature of the injection system and the surrounding gas lines is one essential factor in ensuring that all of the liquid is vapourized. Other critical factors include the rate of injection of the liquid and the flow rate of the carrier gas. If condensation is avoided then the volume-mixing ratio of a specific VOC measured by PTR-MS is linked to the concentration of the same compound in water, C_{VOC}, expressed as a mass fraction of the liquid, by the expression

$$\mathrm{VMR}_{\mathrm{VOC}} = \frac{C_{\mathrm{VOC}} f_{\mathrm{liq}} V_{\mathrm{m}}}{f_{\mathrm{air}} M} \tag{9.3}$$

Here f_{liq} is the liquid flow rate (injection rate), f_{air} is the flow rate of the carrier gas, M is the molecular mass of the VOC and V_m is the molar volume of the gaseous VOC.

The limit of detection for the direct aqueous injection method is comparable to that of the membrane inlet-PTR-MS but, unlike the latter, the former is not restricted to those compounds that show good permeability through the membrane. Furthermore the response time is as short as 20 s, making it possible to monitor relatively rapid changes in VOC sample concentration. An excellent illustration of the potential for direct aqueous injection is the detection of a variety of polychlorinated biphenyls and organic isocyanates in water using PTR-MS [11].

References

1. Karl, T., Yeretzian, C., Jordan, A., Lindinger, W. (2003) Dynamic measurement of partition coefficients using proton-transfer-reaction mass spectrometry (PTR-MS). *Int. J. Mass Spectrom.* **223–224**, 383.
2. von Hartungen, E., Wisthaler, A., Mikoviny, T. *et al.* (2004) Proton-transfer-reaction mass spectrometry (PTR-MS) of carboxylic acids: determination of Henry's law constants and axillary odour investigations. *Int. J. Mass Spectrom.* **239**, 243.
3. Aprea, E., Biasioli, F., Märk, T. D., Gasperi, F. (2007) PTR-MS study of esters in water and water/ethanol solutions: fragmentation patterns and partition coefficients. *Int. J. Mass Spectrom.* **262**, 114.
4. Li, J. Q., Shen, C. Y., Wang, H. M. *et al.* (2008) Dynamic measurements of Henry's law constant of aromatic compounds using proton transfer reaction mass spectrometry. *Acta Phys. Chim. Sin.* **24**, 705.
5. Schuhfried, E., Biasoli, F., Aprea, E. *et al.* (2011) PTR-MS measurements and analysis of models for the calculation of Henry's law constants of monosulfides and disulfides. *Chemosphere* **83**, 311.
6. Williams, J., Holzinger, R., Gros, V. *et al.* (2004) Measurements of organic species in air and seawater from the tropical Atlantic. *Geophys. Res. Lett.* **31**, L23S06.
7. Boscaini, E., Alexander, M. L., Prazeller, P., Märk, T. D. (2004) Membrane inlet proton transfer reaction mass spectrometry (MI-PTRMS) for direct measurements of VOCs in water. *Int. J. Mass Spectrom.* **239**, 171.
8. Kameya, S., Tanimoto, H., Inomata, S. *et al.* (2009) Equilibrator inlet-proton transfer reaction-mass spectrometry (EI–PTR-MS) for sensitive, high-resolution measurement of dimethyl sulphide dissolved in seawater. *Anal. Chem.* **81**, 9021.
9. Kameya, S., Tanimoto, H., Inomata, S. *et al.* (2010) High-resolution measurement of multiple volatile organic compounds dissolved in seawater using equilibrator inlet-proton transfer reaction-mass spectrometry (EI–PTR-MS). *Marine Chem.* **122**, 59.
10. Jürscjik, S., Tani, A., Sulzer, P. *et al.* (2010) Direct aqueous injection analysis of trace compounds in water with proton-transfer-reaction mass spectrometry (PTR-MS). *Int. J. Mass Spectrom.* **289**, 173.
11. Agarwal, B., Jürscjik, S., Sulzer, P. *et al.* (2012) Detection of isocyanates and polychlorinated biphenyls using proton transfer reaction mass spectrometry. *Rapid. Commun. Mass Spectrom.* **26**, 983.

Index

Accuracy, 122–125
Acetaldehyde, 19, 20, 29, 41, 42
  environmental sciences, 132, 134, 143–156, 159–162, 164, 169, 171–173, 177, 178, 181, 183–187, 189, 191, 195–197
  food sciences, 229, 236, 239, 244, 245, 250, 252, 255
  health sciences, 286, 293, 294, 297, 298, 301, 302
Acetic acid, 29, 41
  environmental sciences, 134, 143, 145, 154, 165, 171–173, 177, 178, 181, 189, 191, 200, 202
  food sciences, 229, 236, 243, 244, 252, 255, 286, 293
  health sciences, 298, 299
Acetoin, 236, 243, 253, 255
Acetone, 3, 16, 19, 20, 29, 37, 73, 122
  environmental sciences, 134, 142–145, 147–157, 159–162, 164–169, 171–174, 177–179, 181, 189, 190, 192, 195–197, 200, 202
  food sciences, 227, 228, 229, 236, 253, 255
  health sciences, 270, 275–277, 280, 285, 286, 298
Acetonitrile, 29, 41
  environmental sciences, 134, 142, 143, 155–161, 166–169, 172, 173
  health sciences, 271, 272, 301
Acetylene, 39
2-Acetyl-5-methylfuran, 245
Acidic sulfate aerosols, 181
Acrolein, 134
Acrylamide, 224–226
Acrylonitrile, 134, 159
Activation energy
  proton transfer reactions, 31
Adduct formation, 26, 39, 42, 139
Airborne studies, 132, 133, 142, 143, 149, 156, 167–169, 171, 172
Aircraft emissions, 164
Air quality, 157–159, 161, 164, 166, 168, 173, 197
Alcohols, 133, 141, 166, 172, 189, 193, 194
Aldehydes, 133, 139, 141, 144, 159, 173, 174, 177, 189, 193, 194, 202, 203
Alfalfa, 147, 148, 151–153, 162, 196
Aliphatic hydrocarbons, 158
Alkanes, 136, 164, 167, 174, 176
Alkylbenzene, 278
C_2-Alkylbenzenes, 162
Alkyl esters, 244
Allyl methyl disulfide, 228
Allyl methyl sulfide, 227, 228
Alveolar breath, 276, 284, 287
Amazon, 142, 143, 149, 169, 191
Ammonia, 26, 29, 43, 45
  environmental sciences, 134, 139, 165
  food sciences, 257
  health sciences, 274

Proton Transfer Reaction Mass Spectrometry: Principles and Applications, First Edition. Andrew M. Ellis and Christopher A. Mayhew.

Anaesthetics, 281–282, 302
Analogue signal processing, 103
Analogue-to-digital converter, 103
Aniline, 29, 41
Anisole, 325
Antarctica, 155
Anthropogenic emissions (VOCs), 3, 4, 131, 132, 157–166
Appearance energy, 319
Argon, 3, 4
Aromadendrene, 170
Arrhenius equation, 32
Arterial blood, 284, 285
Artificial sweeteners, 243
Artificial throat, 232
Ascocoryne sarcoides, 296
Ascorbic acid, 226
Atmospheric pressure chemical ionization (APCI), 26, 133, 156, 222

Backward trajectory analyses, 156, 168
Bacteria, 268, 290
Basicity, *see* Gas phase basicity
Beef, 244, 250
Beetle attack, 198, 199
Benzaldehyde
  environmental sciences, 161, 172, 173, 202
  food sciences, 232
  health sciences, 297
Benzene, 29, 120, 126
  environmental sciences, 135, 142, 144, 152, 155–164, 168–173, 177
  food sciences, 223
  health sciences, 272, 286, 293, 299
1,3-Benzodioxole, 325
Benzonitrile, 29, 41
1,4-Benzoquinone, 188
Benzothiazole, 245
Berries, 253
Bimolecular ion-molecule reactions
  rate law, 13
  theoretical prediction of rate coefficients, 31–38
  types, 26
Biochemical processing, 253
Biofuels, 196
Biogenic emissions (VOCs), 3, 4, 131, 132, 141–142, 143, 144, 145, 146, 147, 148, 149, 150, 151, 152, 153, 154, 159, 165, 166, 172, 176, 181, 182, 191, 192, 196, 198, 200
Biomass burning, 132, 137, 142, 155, 158, 166–169, 171, 173
Biomass plumes, 168
Biowaste, 295
Black bamboo, 197
Bladder cancer, 301
Blood analysis, 284, 285
Bradbury–Nielsen gate, 10
Brandy, 227
Brassica plants, 200
Bread, 237
Breath analysis, 5, 268–290
  acetone, 275–276
  isoprene, 272–275
  smokers, 271–272, 277
Breath sampling, 285–289
  buffered end-tidal (BET) sampling, 276, 278, 283, 287
  limitations, 287, 289
  rebreathing protocol, 287
  Tedlar bags
    food science, 228
    health sciences, 272, 277, 286
  Teflon® bags, 273, 285, 286
Broccoli, 251
Buffered end-tidal (BET) sampling, *see* Breath sampling
1,3-Butadiene, 140, 162, 188*n*-butylaldehyde, 202
Butanal, 41
  food science, 229
i-Butane, 29
Butanediol, 299, 325
2,3-Butanedione, 166, 189, 229, 255
Butanoic acid, 253
Butanol, 134, 174
1-Butanol, 244, 297
Butanone, 29, 165, 189, 254, 281, 296
Butter, 246
i-Butylaldehyde, 202

Butyric acid
  food sciences, 236
  health sciences, 296
γ –Butyrolactone, 325
i-Butyronitrile, 152

Cabbage root fly (*Delia radicum*), 200
d-Cadinene, 170
Calibration, 113–122
  dynamic methods, 117–119
  humidity effects, 120–122
  static gas, 116–117
Camphene, 159, 172
Cancers
  breast, 269
  lung, 269, 272, 276–278
Candida tropicalis, 291
Candies (sweets), 242
Carbohydrate malabsorption, 280
Carbonated drinks, 232
Carbon dioxide (CO_2), 3, 4, 15, 29, 141, 150, 174, 184, 185, 189, 192, 202
Carbon monoxide (CO), 141, 158, 163
Carboxylic acid, 41, 133
2-Carene, 172
β-Caryophyllene, 170, 188, 203
Caterpillar attack, 197
Cathode dark space, 57
Cell cultures, 290
Cell lines (cancerous versus non-cancerous), 268, 301
Centre-of-mass collision energy, 72–73
Cereal bars, 244
$CH_5{}^+$, 26
Charge transfer, 26, 45, 138–141, 146
Cheese
  cumin, 248–249
  hard cheeses, 236, 253
  headspace analysis, 235
  mozzarella, 235
  ripening, 253
Cheese crackers, 223, 226
Chemical ionization mass spectrometry (CI-MS), 25–27, 44
Chemical warfare agents, 311, 321
  bis(2-chloroethyl) sulfide (mustard gas), 321
  sarin, 321
  simulants
    2-chloroethyl ethyl sulfide, 321
    diethyl methylphosphonate, 321
    diisopropyl methylphosphonate, 321
    dimethyl methylphosphonate, 321
    dipropylene glycol monomethyl ether, 321
    triethyl phosphate, 321
Chewing gum, 244
Chloral hydrate, 327
Chloramine, 280
Chloroethane, 29
Chloromethane, 29
1-Chloro-2-methoxy-benzene, 152
Cholesterol, 275
Chronic obstructive pulmonary disease (COPD), 274
1,8-Cineole, *see* Eucalyptol
Cities and environs
  Barcelona, 159
  Boston, 163, 164
  Caracas, 158
  Houston, 158
  Innsbruck, 132, 141, 142, 157, 158, 196
  London, 160
  Los Angeles, 163, 164
  Manchester, 160
  Mexico City, 160–162
  New York, 163, 164
  Paris, 163
  Tokyo, 159
  Toronto, 162, 163
Citrobacter, 291
CLAIRE campaign, 143
Coffee, 223, 224, 228–231
  roasting, 231, 232
Co-inertia analyses, 296
Collision cross section, 31, 32, 34, 74
Collision energy, 71–73
Collision frequency, 74
Collision-induced dissociation, 16, 42, 87
  in mass spectrometry, 87–88

Comparative reactivity method, 144, 163, 175, 176, 203
Comparator, 105
Concentration
 constituents of air, 4
 VOC determination, 14, 16, 21–22, 111–112
Confidence limits, 124
Corona discharge, 26
Creosote bush, 152
Cresols, 136, 161, 165
Crete, 142, 168
Critical orifice, 65
Crohn's disease, 278
Cross section, *see* Collision cross section
Crotonaldehyde, 202
α-Cubebene, 170
Cumulus clouds, 142, 149
Currants, 253
Custard, 241
Cuvette, 177, 185, 187, 191, 198
Cycloalkyl nitrates, 174
Cyclobutanecarbaldehyde, 175
Cyclohexanone, 19
Cyclohex-2-en-1-ol, 325
Cycloketones, 174
Cyclopropane, 29, 39
Cyclopropanecarbaldehyde, 174
Cyclopropene, 298
Cyclotetramethylenetetranitramine, *see* HMX
Cyclotrimethylenetrinitramine, *see* RDX
p-Cymene
 environmental sciences, 172
 food sciences, 245
 health sciences, 281

Dalton unit (Da), 77–78
Dark counts, 104
DART-MS, 172
Data acquisition system, 49
Data mining, 227, 253
Date rape drugs, 325
Dead space (breath analysis), 271
c-Decalactone, 242
Decanal, 301
Deepwater Horizon oil spill, 132
Density functional theory, 37, 314
Diabetes, 275
Diacetyl, 242, 244, 253
Diallyl disulfide, 227
Diallyl sulfide, 227
Diallyl trisulfide, 227
Diesel engine emissions, 164
Diethyl ether, 29, 41
Diethyl sulfide, 299
Differential mobility analyzers, 178
Differential optical absorption spectroscopy (DOAS), 5, 161
Diffusion tubes, 119
Digital signal processing, 103
Dihydro-3,5-dimethyl, 245
N,N-Dimethylacetamide, 286
Dimethylamine, 29, 41
 food sciences, 250
Dimethyl disulfide, 29, 41
 food sciences, 245
 health sciences, 293, 299
Dimethyl ether, 29
Dimethylfuran, 272
3,4-Dimethyl-3-hexen-2-one, 245
Dimethyl octane, 300
Dimethylpyrazole, 272
N,N-Dimethylpyridineamine, 272
Dimethyl sulfide, 29, 41
 environmental sciences, 134, 143, 152, 154–157, 165, 200, 201
 food sciences, 227, 245, 250, 255
 health sciences, 278, 296, 299
Dimethyl sulfone, 245
Dimethyl sulfoxide
 food sciences, 245
 health sciences, 272
Dimethylthiomethane, 245
Dimethyl thiosulfinate, 237
Dimethyl trisulfide, 293
2,3-Dimethyl trisulfide, 300
Dinitrobenzene (DNB), 312, 313
2,4-Dinitrophenylhydrazine (DNPH) measurements, 144

Dipropyl disulfide, 237
Dipropyl thiosulfinate, 237
Direct aqueous injection method, 321, 335
Discriminator, 105
Diurnal variations
 environmental sciences, 132, 141, 143, 144, 146, 148, 151–154, 156, 159, 161, 162, 191
 health sciences, 273
Drift time, *see* Reaction time
Drift tube, 9, 10, 15, 16, 20, 49, 50, 61, 62, 63
 practical aspects, 64–69
Drift velocity, 49, 69, 70–74
Drug monitoring, 281–284
Duty cycle, 10, 89
Dynamic range, 79–81

Eddy covariance, 147–150, 153, 161, 195
 disjunct, 147, 148, 150, 152, 160, 161
 relaxed, 154
 virtual disjunct, 147–149, 151, 160
Effective ion temperature, 31, 75, 76
Einzel lens, *see* Ion lenses
Electric dipole moment, 34
Electric field (definition), 51
Electric sector instruments, 77
Electron ionization (EI), 25
Electron volt (eV)
 conversion to other units, 27
 definition, 52
Elephant grass, 197
Emphysema, 277
E/N, see Reduced electric field
Enclosure methods, 147
End-expired breath, 284
Energy units, 27
E-noses, 222
Enthalpy, 28, 30, 31, 76
Entropy, 30, 76
Epilepsy, 282
Equilibrium constant, 28, 75
Escherichia coli, 256, 291
Estragole, 252
Ethane, 29, 139
Ethanethiol, 29, 41
 health sciences, 278
Ethanol, 16, 29, 39, 41
 environmental sciences, 134, 147, 154, 166, 172, 181, 189, 191, 200
 food sciences, 227, 234, 235, 236, 239, 243, 245, 255
 health sciences, 270, 293, 294, 296, 297, 298
Ethene (ethylene), 29
 environmental sciences, 136, 138
 food sciences, 251, 252
 health sciences, 301
Ethyl acetate, 29
 environmental sciences, 135, 162
 food sciences, 244, 253
Ethylamine, 29
Ethylbenzene
 environmental sciences, 135, 159, 160
 food sciences, 244
Ethyl butanoate, 223, 242, 243
Ethyl butyrate, 232
2-Ethyl-3,5-dimethylpyrazine, 43
Ethylene, *see* Ethene
Ethyl formate, 29, 41
 food sciences, 229
Ethyl hexanoate, 242, 253
Ethyl 3-methylbutanoate, 242
3-Ethyl-4-methyl hexane, 300
3-Ethyl-4-methyl-3-penten-2-one, 245
4-Ethylphenol, 166
Ethyl propionate, 232
Ethyl tiglate, 300
Ethyl vinyl ketone, 135
Eucalyptol
 environmental sciences, 172, 189
 health sciences, 283
Eucalyptus, 172, 197
European PHOtoREactor (EUPHORE) atmospheric simulation chamber, 177
Evergreen shrub, 152
Explosives, 311
 environmental aspects, 320
 trace detection, 320

Faraday plate detector, 9, 98
Farming, 164
  cattle, 165
  pigs, 165
  sheep, 165
Fast olefin sensor, 161
Field fringing effects, 53–55
Fingerprinting (mass spectral), 226
  apples, 249
  broccoli, 251
  cumin cheese, 249
  hard cheese, 253
  health sciences, 269, 289
  olive oil, 249
  strawberries, 236
  wine, 233
Fireworks, 166
Flavour perception, 227
Flavour release, 227, 240–245
Floral scent, 187, 188
Flowing afterglow, 6, 11–14, 42, 138
Flow rate, 68
Flow tube, 11, 16, 22
Fluorobenzene, 29
Fluoromethane, 29
Food
  classification, 245, 249
  control, 245
  freshness, 250
  geographical origin, 221, 226, 245–249
  quality, 245, 249
  ripening, 250
    apples, 251
    cheese, 253
    olive oil, 252
    tomatoes, 252
  texture, 240
Forests
  beech, 148
  boreal, 142, 144
  coniferous, 140, 142, 144–146, 177, 182
  deciduous forests, 146, 148, 149, 170
  holm oak, 189
  mosaic, 142
  Norway spruce, 147, 148, 155
  pine
    ponderosa, 140, 145–148
    Scots, 148
  sub-alpine, 148, 149
  temperate, 148, 149
  tropical, 132, 142–143, 148, 149, 150, 169, 171, 175, 176, 184
Formaldehyde, 29, 41
  environmental sciences, 134, 136, 137, 144, 159–161, 163, 165, 171, 174, 175, 178, 181, 191, 196, 202
  health sciences, 272, 277
Formic acid, 29, 41
  environmental sciences, 134, 171, 178
  food sciences, 226
  health sciences, 298
Fourier transform infrared (FTIR) spectroscopy, 133, 137, 169, 171, 173, 174
Fruit
  apples, 238, 245
  bananas, 243
  strawberries, 239
Fruit juices, 233
  orange, 233
  apple, 233
Funghi, 296
Furan
  environmental sciences, 134, 171–173, 182
  food sciences, 226, 229
  health sciences, 298
2(3H)-Furanone, 245
Furfural
  environmental sciences, 135, 173
  food sciences, 226, 229, 255

GABRIEL campaign, 143
Garlic, 227, 228
Gas chromatography (GC), 5, 6–9
  columns, 6, 7
  electron capture detector, 8, 171
  flame ionization detector (FID), 7, 8, 133, 143, 145, 147, 148, 150, 153, 159, 160–163, 171, 178, 190

GC-MS, 5, 8
environmental sciences, 133, 145, 146, 150, 152, 153, 162, 170–173, 176, 185, 187–192, 194, 197–201
food sciences, 223, 235, 236
health sciences, 269, 277, 293
pre-concentration, 7
PTR-MS (GC-PTR-MS), 133, 152, 196, 224, 226, 231
SPME, 145, 246, 271, 285, 296
thermal conductivity detector, 8
Gas number density (calculation), 67
Gas phase basicity, 28–30
table of illustrative values, 29–30
Gaussian distribution, 124
Gelatin, 241, 242
Geographical location, *see* Food
Geranyl acetone, 301
Gestational diabetes mellitus, 283
Gibbs energy change, 27–30, 75
Gliocladium roseum, 296
Glow discharge, 57, 62
Glucosinolate, 200, 201
Glycolaldehyde, 134, 171
Golden pothos, 202
Grassland, 142, 147, 148, 151–154, 165
Green leaf volatiles, 187, 192, 194, 200
Guard rings, 9, 64

Haemodialysis, 274
Ham, 247–248
Hantzsch monitor, 137
Hard sphere limit, 32
Helicobacter pylori, 287, 291
Helium, 3
Henry's law, 230, 331–333
determination, 331
Heptadienol, 281
Heptanal
environmental sciences, 136
health sciences, 271
Heptane, 39
2-Heptanone, 255
3-Heptanone, 282
Herb extracts, 255
Hess's law, 28
Hexanal
environmental sciences, 135, 144, 146, 151, 177, 183, 187, 188, 191, 193–197, 202
food sciences, 241, 242, 244, 249, 252
health science, 271
Hexanal, 135, 144, 146, 151, 177, 183, 187, 188, 191, 193–197, 202
Hexanoic acid, 253
Hexenal
environmental sciences, 135, 146, 153, 154, 183–187, 192–198
food sciences, 252
(Z)-3-Hexenal, 153, 183, 185, 193, 194, 196
Hexenol, 135, 153, 183, 187, 188, 190, 191, 193–197, 242, 244
Hexenyl acetate, 135, 136, 153, 187, 188, 191, 193, 194, 196, 197
HMX explosive, 312, 314
Hollow cathode discharge, 16, 57–59
Homeland security, 311
H_3O^+ production, 59–60
Humidity effects, 120–122, 137, 138, 140
HUMPPA–COPEC campaign, 144
α-Humulene, 170, 178, 188
Hydrated hydronium ions, 10, 16, 18, 20, 26, 42, 120
Hydride ion (H^-) abstraction/transfer, 26, 138, 139
Hydrogen, 4
Hydrogen cyanide
environmental sciences, 134, 136–138, 159, 169
health sciences, 272
Hydrogen sulfide
environmental sciences, 136, 165, 166
food sciences, 237
health sciences, 298
Hydroxyacetone, 135, 149, 151, 169
Hydroxyl radical (OH), 132, 143, 145, 150, 156, 163, 173–176, 179, 181, 203
3-Hydroxy-propanal, 175

Hyperemesis gravidarum, 283
Hyperglycemia, 275

Illicit drugs, 311, 322, 327
INDOEX cruise, 156, 169
Indole, 300
Infant formula, 235
Infra-red gas analyzer (IRGA), 147, 196
Innsbruck, *see* Cities
International Association for Breath Research (IABR), 270
Ion acceleration, 51
Ion cluster distribution, 73–76
Ion detectors, 49, 97
  channel electron multiplier (channeltron), 100
  discrete dynode detector, 98–100
  microchannel plate (MCP) detector, 101–103
Ionization energy (potential), 25, 44, 139–141
Ion lenses, 54–56
  aperture lenses, 55, 56
  cylindrical lenses, 55
  Einzel lens, 56
Ion mobility, 9, 10, 69
  ion mobility coefficient, 69
  reduced ion mobility, 69
Ion mobility spectrometry (IMS), 5, 6, 9–11, 320, 326
Ion optics, 50–57
Ion pulse counting, *see* Pulse counting
Ion sources, 17, 49, 57
  dc plane electrode discharge, 61–62
  hollow cathode discharge, 57–59
  radioactive, 62–63
Ion steering, 53–54
Ion trajectories
  quadrupole, 83
  simulations (SIMION®), 56
Ion trap mass spectrometers, 17, 49, 77, 85–90
  basic principles, 85–87
  environmental sciences (PTR-IT-MS), 162, 196
  linear ion trap, 90
Irradiance, 177, 186, 190
Isoamyl acetate
  food sciences, 232
  health sciences, 300
Isobaric compounds, 132, 222
Isocyanates, 326
Isoflurane, 302
Isolongifolene, 170
Isomeric compounds, 132, 140, 170, 222, 223
Isoprene, 3
  environmental sciences, 134, 140–143, 146–152, 154–156, 158–160, 162, 171, 172, 174, 176–178, 181–188, 190, 192, 194, 195, 197–199
  health sciences, 272–275, 277, 285, 286, 299
Isopropyl methylamine, 325
Isotope labelling, 184, 185
Italian ryegrass, 154

Junipene, 188

Ketones
  environmental sciences, 133, 139, 140, 144, 173, 174, 189, 202, 203
  food sciences, 229, 247, 248, 255
  health sciences, 275, 280, 281, 286, 299, 301
Kidney beans, 236
Kidney function, 280
Klebsiella, 291
Kr^+, 64, 140, 141

Lactobacilli, 250
Lactococcus lactis, 296
Larvae, 198, 200, 201
Lemon tree, 150
Ligand switching reactions, 42, 234, 316
Light–dark transitions, 185–187
Limit of detection, 122–125
Limonene
  environmental sciences, 159, 172, 188
  food sciences, 232, 245
  health science, 281

Linalool, 172, 176, 178, 198, 199
Linear ion trap, 90
Linolenic acid, 226
Lipid-lowering therapy, 273
Lipid peroxidation, 301
Liquid analysis, 331, 333
Liquid chromatography, 26
Liver cirrhosis, 227, 281
Liver disease, 280–281
Long range attractive force, 32

Macchia, 148, 152
Magnetic sector instruments, 76, 77
Maillard reaction, 224
MANCHOT campaign, 156
Mandarin tree, 150
Mass flow controller, 65
Mass resolution, 78
Meat spoilage, 250, 291
Mediterranean INtensive Oxidant Study (MINOS), 142, 168
MEGAN (emissions model), 149
MEGAPOLI winter campaign, 163
Membrane inlet system, 333
Memory effects, 65, 68
Menthol, 244
Menthone, 242, 244
Mesocosm, 155
Methacrolein, 19, 134, 139, 140, 142, 143, 145, 150–152, 155, 156, 174, 178, 182, 202
Methacrylic acid, 135, 143
Methane, 3, 4, 29, 141
Methanethiol, 29
  food sciences, 237, 245, 250
  health sciences, 293, 298
Methanol, 3, 16, 19, 29, 39, 114
  environmental sciences, 131, 134, 142–145, 147–157, 159–162, 164–169, 171–173, 177, 181, 183, 190, 194–198
  food sciences, 227, 229, 245, 250, 252, 257
  health sciences, 270, 272, 277, 281, 286, 293, 298
Methoxymethyl-benzene, 272
Methyl acetate, 29
  environmental sciences, 135, 169
  food sciences, 224, 229, 244
  health sciences, 299
  homeland security, 325
Methyl amine, 29
Methyl benzoate, 325
Methylbutanal
  food sciences, 229
  health science, 278
  2-methylbutanal, 229, 241
  3-methylbutanal, 202, 227, 241
Methyl butanoate, 253
3-Methylbutanoic acid (isovaleric acid), 253
2-Methyl-1-butanol, 299
3-Methyl-1-butanol, 299
2-Methyl-3-butene-2-ol, 134, 135, 140, 146
Methyl chavicol, 136, 145, 151, 178
Methylcyclohexane, 39
Methyl ether ketone
  environmental sciences, 134, 144, 145, 149, 155, 169, 202
  food sciences, 229
Methyl ethyl ether, 29
Methyl formate, 29, 41, 134
5-Methylfural, 229
Methyl furan
  environmental sciences, 135, 151, 152, 174
  food sciences, 226, 229
  3-methyl furan, 152
6-methyl-5-hepten-2-one, 301
Methyl hydroperoxide, 137
3-Methyl-1*H*-indole, 166
Methyl 2-methyl-butanoate, 222, 223
Methylotrophic bacteria, 190
3-Methyl-3-peten-2-one, 135
4-Methylphenol, 166
N-Methyl piperidine, 325
Methyl propanal
  environmental sciences, 134
  food sciences, 229
  2-methylpropanal, 241

2-Methylpropionic acid, 236
2-Methylpyrazine, 224
Methyl salicylate, 177, 191, 192
Methyl tert-butyl ether, 134
(Methylthio)dimethyl sulfoxide, 245
Methylthiol, 41
Methyl vinyl ketone (MVK), 134, 139, 140, 142–145, 149–152, 155, 156, 174, 178, 182
Mexico City Metropolitan Area (MCMA) field campaigns, 160–162
Microbial cultures, 290–297
Microbial growth, 256
Microbial spoilage, 250
MILAGRO campaign, 162, 169
Milk, 240
  lactic acid fermentation, 253
  storage, 251
  tainted, 255
MINOS campaign, 142, 168
Mixing ratio (definition), 22
Molar volume, 67
Monoterpenes, 3
  environmental sciences, 135, 136, 141, 144–150, 152, 154, 162, 163, 170, 174, 176–178, 181, 188–192, 194, 197–199
  health sciences, 281, 293
Motor vehicle exhaust, 157, 160–164
Mouth model, 241
Mowing (of grass), 193, 195, 196
Multi-axis differential optical absorption spectroscopy, 137
Multivariate analysis of variance (MANOVA), 235, 283
Multivariate statistics, 226, 227, 235, 246, 249, 251–253, 302
Muscodor albus, 296
Mycobacterium avium, 291, 293
Mycobacterium kansassii, 291, 293
Mycobacterium smegmatis, 293
Mycobacterium tuberculosis, 291
Mycorrhizal fungus, 201
β-Myrcene, 172

M/z (definition), 17–18, 77–78
  monoisotopic, 18
  nominal, 18, 132, 133, 138, 140, 150, 153, 170, 182, 188, 222, 252, 253, 280, 312, 315, 318

Naphthalene, 29, 136, 161, 162
Narcotics
  cocaine, 322–325
  codeine, 322–325
  ecstasy (*N*-methyl-3,4-methylenedioxyamphetamine), 322–325
  heroin, 322–325
  morphine, 322–325
NEAQS-ITCT mission, 168
Neoclovene, 170
Neon, 4
NH_3^+, 139
NH_4^+, 26, 43, 45, 139, 230
Nitrogen (N_2), 3, 4, 7, 15, 16, 29
Nitroperoxycarbonyl cycloalkanes, 174
NO^+
  contaminant ion, 18, 20
  reagent ion, 44, 63, 64
    environmental sciences, 139, 140, 146
    food sciences, 248
    homeland security, 319
Non-alcoholic steatohepatitis (NASH), 257
Nonanal
  environmental sciences, 136
  health sciences, 297
Nonane, 300
2-Nonanone, 232
Nonene, 300
Non-methane hydrocarbon (NMHC), 149, 163, 171
Nopinone, 136, 174, 178
Normal distribution, 124
Normalized counts per second (ncps), 113
Nose space analysis, 235
NO_x, 132, 141, 158, 168, 174, 177–179
Number density (of a gas), 67

O_2^+
contaminant ion, 18, 20, 138, 141, 167, 251
reagent ion, 44, 45, 63, 64
environmental sciences, 132, 138–140, 167
food science, 248
homeland security, 319
Occupational health exposure, 267
Oceans/seas, 142, 143, 154–157
Cape Grim, 155
Indian, 156
Indian Ocean Experiment (INDOEX), 156
Norwegian fjord, 156
Tropical Alantic, 156
β-Ocimene, 198, 203
Octanal, 136
2,4-Octanedione, 245
1-Octanol, 300
Octanone, 300
1-Octen-3-ol, 297
7-Octen-4-ol, 244
OH, *see* Hydroxyl radical
Olfactory epithelium, 221
Olive oil, 247
Onions, 237
Oral glucose tests, 283, 284
Orange trees, 150
Orthonasal pathways, 221
Oxygen (O_2), 3, 4, 7, 15, 16, 18, 29, 34
Ozone, 29, 132, 141, 145, 155, 157, 159, 160, 165, 168, 173, 181, 182, 190, 191, 193, 197, 250, 251

Palm oil, 244
Parent ion (definition), 18
Partial least squares analysis, 227, 237, 246, 247, 249, 251, 253
Pasteurization, 233
Pattern recognition, *see* Fingerprinting
Paul Scherrer Institute (PSI) environmental chamber, 177
Peace lily, 202
Pectin, 227, 241
Penguin emissions, 155
Pentaerythritol tetranitrate, *see* PETN
Pentanal, 135, 196
2,3-Pentanedione, 229, 255
1-Pentanol, 244
Pentan-2-one, 29
Pentan-3-one, 29
1-Penten-3-ol, 134, 146, 195, 197
Pentenone, 135, 146, 195, 281
Peppers, 236, 241
Permutation tubes, 117
Peroxyacetyl nitrate (PAN), 135, 138
Peroxy isobutyryl nitrate, 136
Peroxymethacrylic nitric anhydride, 136
Peroxypropionyl nitrate, 135
PETN, 312, 314, 316, 317
Pharmokinetics, 281–284
Phenol, 29
environmental sciences, 135, 144, 161, 166, 171, 173
health sciences, 286
homeland security, 325
Phenyl acetate, 325
2-Phenyl ethanol, 244, 293, 300
Photochemical smog, 4
Photooxidation, 133, 149, 177–179
Photosynthesis, 182, 184, 185, 199
α-Pinene, 132, 144, 155, 172, 174, 178–181, 188
β-Pinene, 19, 132, 155, 159, 172, 174, 178
Pinonaldehyde, 136, 144, 174, 178, 179
cis-Pinonic acid particles, 179
Plantations, 142, 150–151, 160
citrus, 150
Eucalyptus, 197
loblolly pine (*Pinus taeda*) plantation, 145
oil palm, 151
poplar, 197
PM_{10} particulates, 158
Poisson statistics, 100, 122, 124
Polarizability, 33
Pollination, 187
Polychlorinated biphenyls, 336

Pork, 250
Post-harvest storage, 251
Positive matrix factorization, 163
Poultry, 250
Ppbv/ppmv/pptv, *see* Mixing ratio for definition
Precision, 122–125
Principal component analysis (PCA), 227
  cheese, 236
  coffee, 231
  ham, 247, 248
  infant formula, 235
  olive oil, 252
  onions, 237
  roe (sea urchin), 247
  tomatoes, 252
  truffles, 246
  Wagyu beef, 244
  whey, 237
  wine, 233, 234
Process monitoring, 253–256
Propadiene, 298
Propanal, 29, 41
  food science, 229, 236, 237
  health science, 301
Propane, 29
Propanenitrile, 134
Propanethiol, 134
  food science, 237
  health sciences, 293
Propanoic acid, 29, 166
Propan-1-ol, 29
  health sciences, 298
Propan-2-ol, 29, 39, 41
  health science, 281, 298
i-Propanol, 277
2-Propanone, *see* Acetone
Propene, 29, 39, 134, 158
Propionaldehyde, 202
Propofol (2,6-diisopropyl phenol), 281, 282
Propyl alcohol, 41
Propylbenzene, 159
Propylene, 398
Propyl propanoate, 223
Propyne, 39, 298
Proton affinity, 15, 26, 28
  table of illustrative values, 29
  water, 29, 30, 136
  water clusters, 42
Proton transfer
  definition, 26, 28
  dehydration, 39, 41
  kinetics, 31–38
  product channels for different functional groups, 39, 40
  thermodynamics, 27–31
Pseudomonas aeruginosa, 250, 290, 291
PTR-IT-MS, 88–90
PTR-TOF-MS, 96–97
Pulse counting, 104–106
Pyrazine
  environmental sciences, 172
  food sciences, 229
Pyridine, 224, 229
Pyrrole
  environmental sciences, 134, 175
  food sciences, 224, 225

Quadrupole ion trap, 77, 80, 85–90
Quadrupole mass spectrometer (linear quadrupole), 12, 14, 17, 49, 77, 81–85

Radioactive ion source, *see* Ion sources
Raspberries, 224
Rate equation, 21
RDX explosive, 312, 313, 315
Reaction rate coefficient, 13, 21, 32, 33
  theoretical capture (collisional) rate, 32–38
    average dipole orientation (ADO), 35
    Langevin, 33–34
    locked dipole approximation, 34
    trajectory calculation, 36
Reaction time, 13, 21, 69, 71
Reagent ion switching
  homeland security, 319
Recombination energy, 139–141

Reduced electric field (*E/N*), 41, 67, 73–76
  homeland security, 315–318
  unit, *see* Townsend
Reduced mobility, 69, 70, 71, 73, 74
Relaxed eddy accumulation, 147
Resolving power (of mass spectrometer), 78–79
Response time, 68
Retronasal pathways, 221
Reverse reaction, 137, 138
Rhizobacterium
  Serratia odorifera, 293
Rhizosphere organisms, 189
Rice fields, 196
Ripening, 221, 253
Roe (sea urchin), 247
Roots, 188–189, 200
Ryegrass, 154

Sahel desert, 142
Salmonella enterica, 291
Savannahs, 142, 151, 152
Scots pine, 148
Seas, *see* Oceans
Secondary organic aerosols, 131, 133, 145, 155, 177, 178, 179, 182, 197
Selected ion flow drift tube (SIFDT), 15, 270
Selected ion flow tube (SIFT), 5, 6, 14–15, 22, 26, 32, 38, 42, 43, 44
  environmental sciences, 138
  food sciences, 222
  health science, 270
Senescence, 146, 196, 252
Sensitivity (definition), 115
Sensory panel, 226
  cheese, 236, 253
  coffee, 231
  fruit juices, 233
  milk, 240
  olive oil, 249
  wine, 234
  vodka, 235
Sesquiterpenes, 136, 141, 145, 150, 154, 170, 178, 190, 197
Sevoflurane, 302
Shelf life, 226, 251
Shipborne measurements, 133, 156
Shrubland, 146, 152, 202
Silcosteel®, 118
Simulation chamber studies, 132, 176–181
Skin emissions, 267, 301
Smoking, *see* Breath analysis
Sodorifen, 293
Soft ionization, 15, 38
Soil, 188, 189
Solid phase microextraction (SPME), 246
Source drift region, 60, 62
Soya beans, 236
Space charge effect, 87
Spider mites, 197, 201
Spontaneous reaction, 27
SPORT campaign, 163
Spruce, 144, 147, 148, 176, 177
Standard deviation, 124
Staphylococcus aureus, 291
Starch, 241, 242, 255
Static gas calibration, 116
Stomatal conductance, 192, 199, 202
STP (standard temperature and pressure) flow rate, 59, 68
Strawberry cultivar, 236
Streptococcus milleri, 291
Stress
  biotic, 185, 189
  chemical, 190, 191
  drought, 190, 192
  flooding, 190–192
  freezing, 146
  harvesting, 153, 154, 190, 193, 195–197
  irradiance, 182, 190
  leaf-drying, 146, 196
  oxidative, 182
  senescence, 196
  temperature, 182, 190
  wounding, 185
    herbivore, 190, 193, 197, 198, 202
    mechanical, 190, 193, 194. *See also* Harvesting
    weather damage, 190, 194
Stripping cell, 332

Styrene
  environmental sciences, 136, 161, 162
  health sciences, 286
Sulfur dioxide, 141
Surinam rainforest, 132, 142, 143, 175
Sycamore leaves, 185, 186
Switchgrass, 196

Tea, 231
Tedlar bags, *see* Breath sampling
Teflon® bags, *see* Breath sampling
Terpenoid, 136, 188
γ-Terpinene, 172
Terpinolene, 172
Tetrahydrofuran, 172
Texas air quality study, 158
Thermodynamics of proton transfer, 27–28
Thioacetic acid, 299
Threat agents, 311
Three-body association, *see* Adduct formation
Time-of-flight mass spectrometer, 17, 18, 20, 19, 49, 77, 80, 90–97
  advantages and limitations, 95–96
  basic principles, 90–92
  mass calibration, 94
  reflectron, 93–97
  spatial focussing, 92–93
Toluene, 29, 43, 120
  environmental sciences, 135, 150, 151, 157–164, 168, 171, 172, 177
  health sciences, 272, 286, 299
o-Toluidine, 277
Townsend unit (Td), 68
Toxic industrial chemicals, 321
Traffic emissions, 142, 157, 163
Transition state, 31
Transmission (of mass spectrometer), 79–81
Triacetone triperoxide (TATP), 312
Trichloroethanol, 325
Triisopropylbenzene, 170
Trimethylamine, 165
Trimethylbenzene, 159, 177, 179–181
Trinitrobenzene (TNB), 312, 313, 317–319
Trinitrotoluene (TNT), 73, 312, 313, 315, 317–319
Tris(methylthio)methane, 245
Tropical forest and fire emissions experiment (TROFFEE), 169, 171
Truffles, 245

Ulcerative colitis, 278
Unsaturated hydrocarbons, 136
Uptake by plants, 202
Urine, 267, 284, 301

n-Valeraldehyde, 202
Valproic acid treatment, 282
Variability-lifetime methods, 143
Venous blood, 284, 285
Venturi inlet, 14, 65, 66
2-Vinylfuran, 135
Vodka, 234
Volatile organic compounds (definition), 3
Volume mixing ratio (VMR), 112, 335
Voodoo correlations, 269

Wannier equation, 71–73
West Africa, 142
Whey, 237, 243
White clover, 153, 193
Whitethorn, 145
Wine, 233
Woodland, 142, 146, 151, 152, 167
Wounding, *see* Stress

Xylene, 29, 157–160, 162, 172

Yogurt, 242, 255